Spectral Computations for Bounded Operators

APPLIED MATHEMATICS
Editor: R.J. Knops

This series presents texts and monographs at graduate and research levels covering a wide variety of topics of current research interest in modern and traditional applied mathematics, numerical analysis, and computation.

1 Introduction to the Thermodynamics of Solids *J.L. Ericksen* (1991)

2 Order Stars *A. Iserles and S.P. Nørsett* (1991)

3 Material Inhomogeneities in Elasticity *G. Maugin* (1993)

4 Bivectors and Waves in Mechanics and Optics
Ph. Boulanger and M. Hayes (1993)

5 Mathematical Modelling of Inelastic Deformation
J.F. Besseling and E van der Geissen (1993)

6 Vortex Structures in a Stratified Fluid: Order from Chaos
Sergey I. Voropayev and Yakov D. Afanasyev (1994)

7 Numerical Hamiltonian Problems
J.M. Sanz-Serna and M.P. Calvo (1994)

8 Variational Theories for Liquid Crystals *E.G. Virga* (1994)

9 Asymptotic Treatment of Differential Equations *A. Georgescu* (1995)

10 Plasma Physics Theory *A. Sitenko and V. Malnev* (1995)

11 Wavelets and Multiscale Signal Processing
A. Cohen and R.D. Ryan (1995)

12 Numerical Solution of Convection-Diffusion Problems
K.W. Morton (1996)

13 Weak and Measure-valued Solutions to Evolutionary PDEs
J. Málek, J. Nečas, M. Rokyta and M. Růžička (1996)

14 Nonlinear Ill-Posed Problems
A.N. Tikhonov, A.S. Leonov and A.G. Yagola (1998)

15 Mathematical Models in Boundary Layer Theory
O.A. Oleinik and V.M. Samokhin (1999)

16 Robust Computational Techniques for Boundary Layers
P.A. Farrell, A.F. Hegarty, J.J.H. Miller,
E. O'Riordan and G. I. Shishkin (2000)

17 Continuous Stochastic Calculus with Applications to Finance
M. Meyer (2001)

18 Spectral Computations for Bounded Operators
Mario Ahues, Alain Largillier and Balmohan V. Limaye (2001)

(Full details concerning this series, and more information on titles in preparation are available from the publisher.)

Spectral Computations for Bounded Operators

MARIO AHUES
ALAIN LARGILLIER
BALMOHAN V. LIMAYE

CRC Press
Taylor & Francis Group
Boca Raton London New York

CRC Press is an imprint of the
Taylor & Francis Group, an **informa** business
A CHAPMAN & HALL BOOK

Published 2001 by Chapman & Hall /CRC Press
Taylor & Francis Group
6000 Broken Sound Parkway NW, Suite 300
Boca Raton, FL 33487-2742

First issued in paperback 2019

ISBN-13: 978-0-367-45535-4 (pbk)
ISBN-13: 978-1-58488-196-4 (hbk)

Visit the Taylor & Francis Web site at
http://www.taylorandfrancis.com

and the CRC Press Web site at
http://www.crcpress.com

Library of Congress Cataloging-in-Publication Data

Ahués, Mario.
Spectral computations for bounded operators / Mario Ahues, Alain Largillier, Balmohan V. Limaye.
p. cm.— (Applied mathematics ; 18)
Includes bibliographical references and index.
ISBN 1-58488-196-8 (alk. paper)
1. Spectral theory (Mathematics) 2. Operator theory. I. Largillier, Alain. II. Limaye, Balmohan Vishnu. III. Title. IV. Series.

QA320 .A37 2001
515′.7222—dc21 00-069348

Library of Congress Card Number 00-069348

Contents

Preface

This work addresses eigenvalue problems for operators on infinite dimensional spaces. Exact eigenvalues, eigenvectors, and generalized eigenvectors of operators with infinite dimensional ranges can rarely be found. It is thus imperative to approximate such operators by finite rank operators and solve the original eigenvalue problem approximately. In doing so, not just the eigenvalues but the spectral values of an infinite dimensional operator need to be considered.

Starting with a sufficiently general theoretical framework, such as the one provided by a Banach space or by a Hilbert space, concrete approximation methods are given. They lead to finite dimensional problems which can be implemented on a computer.

The first chapter contains the classical spectral theory in the setting of a Banach space with an emphasis on spectral sets of finite type, that is, sets for which spectral projections are of finite rank. The product space structure is used to develop the notions of a Gram matrix and of a block reduced resolvent operator. The Jordan Canonical Form of a matrix is avoided for computational reasons. Also, the use of transfinite induction, in the form of the Hahn-Banach Theorem, is minimized.

The second chapter shows how the approximation of a bounded operator T by a sequence (T_n) of bounded operators is useful to approximate the spectral values of T by the spectral values of T_n. The classical results concerning norm convergence and collectively compact convergence are unified with the help of a new type of convergence. It is general enough to encompass a wide variety of approximation methods and, at the same time, simple enough to be verified in practice. Under this ν-convergence, properties similar to the upper semicontinuity and the lower semicontinuity of the spectrum are shown to be available. This chapter contains error estimates for clusters of approximate eigenvalues and for approximate bases for the associated spectral subspaces.

The third chapter outlines two ways of improving the accuracy of spectral approximations: *iterative refinement* and *acceleration.* Both yield sharper error estimates for the approximations. The special case of a simple eigenvalue is considered first, before a full treatment is given for the general case of a cluster of a finite number of multiple eigenvalues.

The fourth chapter gives a number of concrete methods for constructing finite rank approximations of an operator. These satisfy the conditions required for spectral approximation and also for its iterative refinement and acceleration. Methods based on projections are presented first. They include the discretizations named after Galerkin and Sloan as well as the Finite Element Method. Special methods for integral operators involving degenerate kernels and numerical integration are considered next. They include the discretizations named after Nyström and Fredholm. A singularity subtraction technique for weakly singular integral operators is also presented. *A posteriori* error estimates are discussed along the lines of a result of Brakhage.

The fifth chapter deals with the canonical matrix formulation of the eigenvalue problem for a finite rank operator and brings the reader to calculations that can be performed on a computer. Matrix formulations of the iterative refinement technique and of the acceleration procedure are also given. Special features of this chapter include discussions of uniformly well-conditioned bases of approximate spectral subspaces, the matrix formulation of the singularity subtraction technique, and the acceleration procedure. The last section of this chapter contains numerical examples involving several infinite dimensional model problems.

The theory developed in the preceding chapters is illustrated by considering specific finite rank approximations, then formulating and solving matrix eigenvalue problems and also employing iterative refinement as well as acceleration. For this purpose, it is necessary to use intensive large-scale computation either on a personal computer or on a mainframe computer. The numerical experiments in this section make use of MATLAB.

The sixth chapter deals with eigenvalue problems for matrices. Instead of referring the reader to 'black boxes' like the standard routines given in numerical libraries such as LAPACK, an in-depth analysis of the QR Methods is presented. Also, the forward error analysis (involving condition numbers) as well as the backward error analysis (related to stability considerations) are discussed in detail. A stopping criterion for the QR Method which bounds the total relative error is proposed. Error analysis of the solution of a Sylvester equation concludes this chapter.

Starting with operators from infinite dimensional spaces to infinite dimensional spaces in the first three chapters, finite dimensional situations are progressively introduced. The fourth chapter deals with operators having infinite dimensional domains but finite dimensional ranges, while the last two chapters essentially deal with finite dimensional operators.

Definitions and major results are illustrated by elementary examples. A large number of exercises are given at the end of each chapter except for the sixth. Hints are provided whenever appropriate.

A list of symbols used in the book follows this preface. The first part contains symbols that are used freely but not defined in the text. The second part contains symbols that are introduced in the text. They are listed in the order of their appearance. Some of these symbols (like $\langle x\,,\,f\rangle$) are explained in the list itself, while only the names of the others (like $\langle \underline{x}\,,\,\underline{f}\,\rangle$) are given in the list ; the reader may consult the Index at the end of the book to find their meanings. The Index gives only the number of the page on which a term appears for the first time. We remark that matrices are denoted by letters in the sanserif font throughout the book.

Acknowledgements are due to the Service des Relations Internationales of the Université Jean Monnet de Saint-Étienne, the Curriculum Development Programme of the Indian Institute of Technology Bombay, the Research in Pairs Programme of the Mathematisches Forschungsinstitut Oberwolfach, and the Association Équipe d'Analyse Numérique de Saint-Étienne. We thank Rafikul Alam, N. Gnaneshwar, Nick Higham, Rekha Kulkarni, Françoise Tisseur and Olivier Titaud for reading parts of the manuscript and making useful suggestions, C.L. Anthony for word processing parts of the manuscript, and Nirmala Limaye for support and encouragement.

M. Ahues, A. Largillier, B.V. Limaye

Notation

$\mathbb{C}$	the field of complex numbers,
$\Re(z)$	the real part of $z \in \mathbb{C}$,
$\Im(z)$	the imaginary part of $z \in \mathbb{C}$,
$\overline{z}$	the conjugate of $z \in \mathbb{C}$,
i	the square root of -1 with imaginary part equal to 1,
$\mathbb{R}$	the field of real numbers,
$[a, b]$	the set of $t \in \mathbb{R}$ such that $a \leq t \leq b$,
$]a, b[$	the set of $t \in \mathbb{R}$ such that $a < t < b$,
$\mathbb{Z}$	the set of all integers,
$[r]$	the integral part of $r \in \mathbb{R}$,
$[\![m, n]\!]$	the set of all $k \in \mathbb{Z}$ such that $m \leq k \leq n$,
$\delta_{i,j}$	the Kronecker symbol, equals 1 if $i = j$ and equals 0 if $i \neq j$,
$\mathbb{C}^{n \times m}$	the space of all complex matrices with n rows and m columns, also called the space of $n \times m$ complex matrices,
$\mathsf{u} = [\mathsf{u}(1), \ldots, \mathsf{u}(n)]$	an element of $\mathbb{C}^{1 \times n}$,
$[\mathsf{u}(1), \ldots, \mathsf{u}(n)]^\top$	an element of $\mathbb{C}^{n \times 1}$,
$[a_{i,j}]$	a matrix having $a_{i,j}$ in the ith row and the jth column,
$\mathsf{A}(i, j)$	the entry in the ith row and the jth column of a matrix A,
$\mathsf{A}(\cdot, j)$	the jth column of a matrix A,
$\mathsf{A}(i, \cdot)$	the ith row of a matrix A,
A^{-1}	the inverse of a nonsingular matrix A,
I	the identity matrix,
I_n	the identity matrix of order n,
O	the zero matrix in $\mathbb{C}^{n \times m}$,

$\mathrm{tr}(\mathsf{A})$	the trace of a square matrix A,
$\det(\mathsf{A})$	the determinant of a square matrix A,
$\mathrm{diag}\,[\lambda_1, \ldots, \lambda_n]$	the $n \times n$ diagonal matrix having λ_j as the jth diagonal entry,
$\mathrm{dist}(E, F)$	the distance between subsets E and F in a metric space,
$\mathrm{dist}(x, E)$	$\mathrm{dist}(\{x\}, E)$,
$\mathrm{span}\{x_1, \ldots, x_n\}$	the subspace spanned by elements $x_1, \ldots, x_n$,
$\dim X$	the dimension of a linear space X,
$C^0(E)$	the space of all complex-valued bounded continuous functions defined on a metric space E with the norm $\|x\|_\infty := \sup\{\|x(t)\| \ : t \in E\}$,
$C^p(\mathcal{I})$	the space of all complex-valued functions defined on an interval $\mathcal{I}$ whose first p derivatives are continuous on $\mathcal{I}$,
$C^\infty(\mathcal{I})$	the space of all complex-valued functions defined on an interval $\mathcal{I}$ which are infinitely differentiable on $\mathcal{I}$,
x'	the first derivative of a differentiable function x,
(s_n)	a sequence whose nth term is s_n,
ℓ^p	the space of all complex sequences $x := (x(1), x(2), \ldots)$ which satisfy $\|x\|_p := \Big(\sum_{j=1}^{\infty} \|x(j)\|^p < \infty\Big)^{1/p}$, where $1 \le p < \infty$,
$\ell^2(\mathbb{Z})$	the space of all complex sequences $x := (\ldots, x(-2), x(-1), x(0), x(1), x(2), \ldots)$ which satisfy $\|x\|_2 := \Big(\sum_{j=-\infty}^{\infty} \|x(j)\|^2\Big)^{1/2} < \infty$,
e_k	the kth standard vector in ℓ^p or in $\ell^p(\mathbb{Z})$: $e_k(j) = 0$ for $j \ne k$ and $e_k(k) = 1$,
$L^p(\mathcal{I})$	the space of all equivalence classes of Lebesgue measurable functions x on an interval $\mathcal{I}$ which satisfy $\|x\|_p := \Big(\int_{\mathcal{I}} \|x(t)\|^p \, dt < \infty\Big)^{1/p}$, where $1 \le p < \infty$,
$L^\infty(\mathcal{I})$	the space of all equivalence classes of essentially bounded Lebesgue measurable functions on an interval $\mathcal{I}$ with the norm $\|x\|_\infty := \inf\{\beta > 0 : \|x\| \le \beta$ on $\mathcal{I}$ except for a subset of Lebesgue measure zero$\}$,
I	the identity operator,

O the zero operator,
$\mathcal{N}(T)$ the null space (or the kernel) of a linear operator T,
$\mathcal{R}(T)$ the range (or the image space) of a linear operator T,
$\text{rank}(T)$ the rank of a linear operator T, that is, $\dim \mathcal{R}(T)$,
$T_{|Y}$ the restriction of an operator T from X to X
to a subspace Y of X,
$T_{|Y,Z}$ the restriction of an operator T from X to X
to a subspace Y of X such that $\mathcal{R}(T)$ is contained
in a subspace Z of X.

Notation introduced in the text (in order of appearance):

$\langle \mathsf{x}, \mathsf{y} \rangle$ $\mathsf{y}^*\mathsf{x}$,
A^* the conjugate-transpose of $\mathsf{A} \in \mathbb{C}^{m\times n}$,
X a Banach space over $\mathbb{C}$,
$\text{BL}(X)$ the set of all bounded linear operators on X,
$\|T\|$ the subordinated operator norm of $T \in \text{BL}(X)$,
$\text{re}(T)$ the resolvent set of $T \in \text{BL}(X)$,
$R(T,z)$ the resolvent operator of $T \in \text{BL}(X)$ at $z \in \text{re}(T)$,
$\text{sp}(T)$ the spectrum of $T \in \text{BL}(X)$,
$\rho(T)$ the spectral radius of $T \in \text{BL}(X)$,
$\text{int}(\Gamma)$ the interior of a Jordan curve Γ,
$\text{ext}(\Gamma)$ the exterior of a Jordan curve Γ,
$X = Y \oplus Z$ the decomposition of X into its closed
subspaces Y and Z,
$\text{int}(\text{C})$ the interior of a Cauchy contour C,
$\text{ext}(\text{C})$ the exterior of a Cauchy contour C,
Λ a spectral set,
$\mathcal{C}(T,\Lambda)$ the set of all Cauchy contours which
separate Λ from $\text{sp}(T) \setminus \Lambda$,
$P(T,\Lambda)$ the spectral projection associated with
$T \in \text{BL}(X)$ and a spectral set Λ for T,
$M(T,\Lambda)$ the spectral subspace associated with
$T \in \text{BL}(X)$ and a spectral set Λ for T,
$S(T,\lambda)$ the reduced resolvent operator of
$T \in \text{BL}(X)$ at an isolated point λ of $\text{sp}(T)$,
X^* the adjoint space of X,
$\langle x, f \rangle$ $\overline{f(x)}$, where $x \in X$ and $f \in X^*$,
$\text{BL}(X,Y)$ the set of all bounded linear maps
from a Banach space X to a Banach space Y,

K^*	the adjoint operator of $K \in \mathrm{BL}(X, Y)$,
$E^\perp$	the annihilator of $E \subset X$,
$\underline{X}$	$X^{1\times m} = \{[x_1, \ldots, x_m] : x_j \in X, 1 \le j \le m\}$,
$\underline{x}$	an element of $\underline{X}$,
$\lVert \underline{x} \rVert_1$	$\sum_{j=1}^{m} \lVert x_j \rVert$,
$\lVert \underline{x} \rVert_F$	$\left(\sum_{j=1}^{m} \lVert x_j \rVert^2 \right)^{1/2}$,
$\lVert \underline{x} \rVert_\infty$	$\max_{j=1,\ldots,m} \lVert x_j \rVert$,
$\underline{T}$	the natural extension of $T \in \mathrm{BL}(X)$ to $\underline{X}$,
$\underline{x}\,\mathsf{Z}$	$\left[\sum_{i=1}^{m} z_{i,1} x_i, \ldots, \sum_{j=1}^{m} z_{i,n} x_i \right]$, where $\mathsf{Z} := [z_{i,j}] \in \mathbb{C}^{m\times n}$,
$(\underline{x}\,, \underline{f}\,)$	the Gram matrix associated with $\underline{x} \in X^{1\times m}$ and $\underline{f} \in (X^*)^{1\times n}$, the Gram product of $\underline{x} \in X^{1\times m}$ with $\underline{f} \in (X^*)^{1\times n}$,
$\lVert \mathsf{Z} \rVert_1$	the 1-norm of $\mathsf{Z} \in \mathbb{C}^{m\times n}$,
$\lVert \mathsf{Z} \rVert_2$	the 2-norm of $\mathsf{Z} \in \mathbb{C}^{m\times n}$,
$\lVert \mathsf{Z} \rVert_\infty$	the ∞-norm of $\mathsf{Z} \in \mathbb{C}^{m\times n}$,
$\lVert \mathsf{Z} \rVert_F$	the F-norm or the Frobenius norm of $\mathsf{Z} \in \mathbb{C}^{m\times n}$,
$i(\underline{f}\,)(\underline{x}\,)$	$\sum_{j=1}^{m} f_j(x_j)$, where $\underline{x} := [x_1, \ldots, x_m] \in X^{1\times m}$ and $\underline{f} := [f_1, \ldots, f_m] \in (X^*)^{1\times m}$,
$R(\underline{T}\,, \mathsf{Z})$	the block resolvent operator of $\underline{T} \in \mathrm{BL}(\underline{X}\,)$ at $\mathsf{Z} \in \mathbb{C}^{m\times m}$,
$\mathcal{S}_{\underline{\varphi}}$	the block reduced resolvent operator associated with $T \in \mathrm{BL}(X)$, a spectral set Λ of finite type for T and an ordered basis $\underline{\varphi}$ for $M(T, \Lambda)$,
$\mathcal{Z}_{\underline{\varphi}}\,\underline{x}$	$\underline{x}\,(\underline{T}\,\underline{\varphi}\,, \underline{\varphi}^*\,)$, where $\underline{\varphi}$ is an ordered basis for $M(T, \Lambda)$ and $\underline{\varphi}^*$ is the adjoint basis for $M(T^*, \overline{\Lambda})$,
$T_n \xrightarrow{\mathrm{p}} T$	the pointwise convergence of T_n to T,
$T_n \xrightarrow{\mathrm{n}} T$	the norm convergence of T_n to T,
$T_n \xrightarrow{\mathrm{cc}} T$	the collectively compact convergence of T_n to T,
$T_n \xrightarrow{\nu} T$	the ν-convergence of T_n to T,
$\ell(\mathrm{C})$	the length of a Cauchy contour C,
$\delta(\mathrm{C})$	$\min\{\lvert z \rvert : z \in \mathrm{C}\}$, where C is a Cauchy contour,
λ_n	an eigenvalue of T_n,
φ_n	an eigenvector of T_n,

$\lambda_n^{(k)}$	kth eigenvalue iterate,
$\varphi_n^{(k)}$	kth eigenvector iterate,
$\boldsymbol{X}^{[q]}$, $\boldsymbol{X}$	$X^{q\times 1} = \{[x_1,\ldots,x_q]^\top : x_i \in X, i = 1,\ldots q\}$,
$\boldsymbol{T}^{[q]}$, $\boldsymbol{T}$	the acceleratexd version of $T \in \mathrm{BL}(X)$ defined on $\boldsymbol{X}^{[q]}$,
$\boldsymbol{T}_n^{[q]}$, $\boldsymbol{T}_n$	the accelerated version of $T_n \in \mathrm{BL}(X)$ defined on $\boldsymbol{X}^{[q]}$,
$\lambda_n^{[q]}$	an eigenvalue of $\boldsymbol{T}_n$,
$\boldsymbol{\varphi}_n$	an eigenvector of $\boldsymbol{T}_n$,
$\varphi_n^{[q]}$	the first of the q components of $\boldsymbol{\varphi}_n$,
$\Lambda_n^{[q]}$	a spectral set of finite type for $\boldsymbol{T}_n$,
$\underline{\boldsymbol{\varphi}_n}$	an ordered basis for $M(\boldsymbol{T}_n, \Lambda_n^{[q]})$,
$\underline{\boldsymbol{X}}^{[q]}$, $\underline{\boldsymbol{X}}$	$X^{q\times m} = \{[\boldsymbol{x}_1,\ldots,\boldsymbol{x}_m] : \boldsymbol{x}_j \in \boldsymbol{X}^{[q]}, 1 \le j \le m\}$ $= \{[\,\underline{x}_1,\ldots,\underline{x}_q]^\top : \underline{x}_i \in X^{1\times m}, 1 \le i \le q\}$,
$\underline{\boldsymbol{x}}$	an element of $\underline{\boldsymbol{X}}$,
$\underline{\boldsymbol{T}}$	the accelerated version of $\underline{T} \in \mathrm{BL}(\underline{X})$ defined on $\underline{\boldsymbol{X}}$, also, the natural extension of $\boldsymbol{T} \in \mathrm{BL}(\boldsymbol{X})$ to $\underline{\boldsymbol{X}}$,
$\underline{\boldsymbol{T}_n}$	the accelerated version of $\underline{T_n} \in \mathrm{BL}(\underline{X})$ defined on $\underline{\boldsymbol{X}}$, also, the natural extension of $\boldsymbol{T}_n \in \mathrm{BL}(\boldsymbol{X})$ to $\underline{\boldsymbol{X}}$,
T_n^P	a projection approximation of $T \in \mathrm{BL}(X)$,
T_n^S	a Sloan approximation of $T \in \mathrm{BL}(X)$,
T_n^G	a Galerkin approximation of $T \in \mathrm{BL}(X)$,
T_n^D	a degenerate kernel approximation of an integral operator T,
T_n^N	a Nyström approximation of an integral operator T,
T_n^F	a Fredholm approximation of an integral operator T,
$\omega(x,\cdot)$	the modulus of continuity of $x \in C^0([a,b])$,
T_n^K	a Kantorowich-Krylov approximation of an integral operator T with a weakly singular kernel,
$\kappa_2(\mathsf{A})$, $\mathrm{cond}(\mathsf{A})$	condition number of $\mathsf{A} \in \mathbb{C}^{n\times n}$,
$\mathbf{c}\,(f(x))$	computed value of $f(x)$,
fl_c	floating point map in chopping arithmetic,
fl_r	floating point map in rounding arithmetic,
$\mathbf{u}_c$	unit roundoff in chopping arithmetic,
$\mathbf{u}_r$	unit roundoff in rounding arithmetic,
$\gamma_{\mathbf{u}}(k)$	$\dfrac{k\mathbf{u}}{1-k\mathbf{u}}$, where $k \in [0, 1/\mathbf{u}[$.

Chapter 1

Spectral Decomposition

This chapter gives a treatment of the spectral theory for bounded operators, the main result being the Spectral Decomposition Theorem. Spectral sets of finite type are discussed in detail. While the development of the adjoint space is along the classical lines, the product space structure is used to introduce the Gram product of a finite set of elements in the space with a finite set of elements in the adjoint space. The product space structure also allows an integral representation of the block reduced resolvent.

1.1 General Notions

We are interested in finding invariant subspaces of a linear operator: Given a linear space X over the complex field $\mathbb{C}$ and a linear operator $T : X \to X$, we want to find subspaces M of X such that $T(M) \subset M$. The operator T can be a matrix transformation, a linear integral operator, or a linear differential operator.

Let us begin by looking for one-dimensional invariant subspaces. If M is spanned by a nonzero vector x, that is, $M = \text{span}\{x\}$, then $T(M) \subset M$ if and only if there exists a scalar $\lambda \in \mathbb{C}$ such that $Tx = \lambda x$. The complex number λ is called an **eigenvalue** of T. Let I denote the **identity operator** on X. Then $\lambda \in \mathbb{C}$ is an eigenvalue of T if and only if $T - \lambda I$ is not injective. If y is another nonzero vector spanning M, then $y = cx$ for some $c \in \mathbb{C}$. Hence $Ty = T(cx) = cTx = c(\lambda x) = \lambda(cx) = \lambda y$. Thus we see that λ depends on T and M but not on the particular choice of a vector spanning M. Any nonzero vector $\varphi \in X$ satisfying $T\varphi = \lambda\varphi$

is called an **eigenvector** of T corresponding to, or associated with, the eigenvalue λ. If $E(\lambda) := \mathcal{N}(T - \lambda I)$, then $T(E(\lambda)) \subset E(\lambda)$. Moreover, the restriction of T to $E(\lambda)$, namely, $T_{|E(\lambda),E(\lambda)} : E(\lambda) \to E(\lambda)$, is a multiple of the identity operator by the scalar λ, that is, for all $x \in E(\lambda)$, $T_{|E(\lambda),E(\lambda)}x = \lambda x$. The dimension of $E(\lambda)$ is called the **geometric multiplicity** of the eigenvalue λ, and $E(\lambda)$ is called the **eigenspace** of T corresponding to λ.

We remark that the invariance of a one-dimensional subspace M under T can be written in the form of an equation, namely, $T\varphi = \lambda\varphi$, where $\text{span}\{\varphi\} = M$. Let us try to do the same thing in the case of an m dimensional invariant subspace M. If $\underline{\varphi} := [\varphi_1, \ldots, \varphi_m]$ forms an ordered basis for M, then the invariance of M under T is equivalent to the existence of m^2 complex numbers $\theta_{i,j}$, $i, j = 1, \ldots, m$, such that for each $j = 1, \ldots, m$,

$$T\varphi_j = \theta_{1,j}\varphi_1 + \cdots + \theta_{m,j}\varphi_m.$$

These m equations can be written briefly if we agree to extend the notation of matrix multiplication as follows: For $x_1, \ldots, x_m \in X$, let

$$\underline{T}\,[x_1, \ldots, x_m] := [Tx_1, \ldots, Tx_m],$$

and

$$[x_1, \ldots, x_m] \begin{bmatrix} \theta_{1,1} & \cdots & \theta_{1,m} \\ \theta_{2,1} & \cdots & \theta_{2,m} \\ \vdots & \vdots & \vdots \\ \theta_{m,1} & \cdots & \theta_{m,m} \end{bmatrix} := \Big[\sum_{i=1}^{m} \theta_{i,1}x_i, \ldots, \sum_{i=1}^{m} \theta_{i,m}x_i\Big].$$

Then the m equations defining the invariance of M under T can be written as

$$\underline{T}\,\underline{\varphi} = \underline{\varphi}\Theta,$$

where $\Theta := [\theta_{i,j}] \in \mathbb{C}^{m\times m}$. The particular case $\Theta := [\lambda] \in \mathbb{C}^{1\times 1}$ gives the **eigenequation** $T\varphi = \varphi\lambda$, traditionally written as $T\varphi = \lambda\varphi$.

Remark 1.1 The characteristic polynomial of a matrix:
Let X be a finite dimensional linear space, $n := \dim X$, $\mathcal{B}$ be an ordered basis for X and let A be a linear operator on X represented in the basis $\mathcal{B}$ by an $n\times n$ matrix A. Each $\psi \in X$ is represented by a column matrix whose entries are the coordinates of ψ with respect to the ordered basis $\mathcal{B}$. This column matrix will be called the **coordinate matrix** of ψ with

respect to $\mathcal{B}$. If x is the coordinate matrix of an eigenvector φ of A corresponding to an eigenvalue λ of A with respect to $\mathcal{B}$, then $A\varphi = \lambda\varphi$ is equivalent to $\mathsf{Ax} = \lambda\mathsf{x}$. Since φ is nonzero, x is nonzero. Hence $\det(\mathsf{A} - \lambda\,\mathsf{I}) = 0$, where I denotes the $n\times n$ identity matrix. Conversely, if $\det(\mathsf{A} - \lambda\,\mathsf{I}) = 0$, then $\mathsf{A} - \lambda\,\mathsf{I}$ is singular and there exists a nonzero column x such that $\mathsf{Ax} = \lambda\mathsf{x}$. We conclude that $\lambda \in \mathbb{C}$ is an eigenvalue of A if and only if λ satisfies the characteristic equation $\det(\mathsf{A} - \lambda\,\mathsf{I}) = 0$, that is, λ is a root of the characteristic polynomial of A:

$$\mathrm{p}(z) := \det(\mathsf{A} - z\,\mathsf{I}).$$

Thus every linear operator A on $\mathbb{C}^n$ has at least one eigenvalue, and it has at most n distinct eigenvalues. Since for each nonsingular matrix $\mathsf{V} \in \mathbb{C}^{n\times n}$ we have

$$\begin{aligned}\det(\mathsf{V}^{-1}\mathsf{AV} - \lambda\,\mathsf{I}) &= \det(\mathsf{V}^{-1}(\mathsf{A} - \lambda\,\mathsf{I})\mathsf{V}) \\ &= (\det \mathsf{V})^{-1} \det(\mathsf{A} - \lambda\,\mathsf{I}) \det \mathsf{V} \\ &= \det(\mathsf{A} - \lambda\,\mathsf{I}),\end{aligned}$$

we may choose any matrix representation of A to compute the eigenvalues of A.

We agree to say that the roots of the characteristic polynomial of A are the eigenvalues of A. Also, a nonzero vector x such that $\mathsf{Ax} = \lambda\mathsf{x}$ is called an eigenvector of A and the space $\{\mathsf{x} \in \mathbb{C}^{n\times 1} : \mathsf{Ax} = \lambda\mathsf{x}\}$ is the eigenspace of A corresponding to its eigenvalue λ.

We conclude that similar matrices have the same eigenvalues. It also follows that the eigenvalues of an upper triangular matrix are its diagonal entries and that the same holds for a lower triangular matrix.

Note that if $\mathsf{B} := \mathsf{V}^{-1}\mathsf{AV}$ and x is an eigenvector of A corresponding to an eigenvalue λ of A, then $\mathsf{V}^{-1}\mathsf{x}$ is an eigenvector of B corresponding to the eigenvalue λ of B. ∎

As remarked above, no computation is needed to obtain the eigenvalues of an upper triangular matrix. The following theorem shows that any linear map from a finite dimensional linear space into itself may be represented by an upper triangular matrix with respect to a suitable basis.

A Hermitian positive definite sesquilinear form $\langle\cdot\,,\,\cdot\rangle$ on a linear space X is called an inner product on X. We say that elements x, y in X are orthogonal if $\langle x\,,\,y\rangle = 0$. A subset E of X is said to be orthonormal if any two distinct elements of E are orthogonal, and $\|x\| := \langle x\,,\,x\rangle^{1/2}$ equals 1 for every $x \in E$.

Theorem 1.2 Schur's Theorem:
Let X be an n dimensional complex linear space and $\langle \cdot , \cdot \rangle$ be an inner product on X. If $A : X \to X$ is a linear map, then there exists an orthonormal basis for X with respect to which A is represented by an upper triangular matrix.

Proof
The proof is by induction on the dimension of X. If $\dim X = 1$, then any linear map $A : X \to X$ has a $1{\times}1$ (upper triangular) matrix representation with respect to any basis of X. Assume that the result is true for all $(n-1)$-dimensional spaces. Let $\dim X = n$ and $A : X \to X$ be a linear map. Let λ be any of the eigenvalues of A and u_1 a corresponding eigenvector such that $\langle u_1 , u_1 \rangle = 1$. Choose $n-1$ elements v_j, $j = 2, \ldots, n$, in X such that $[u_1, v_2, \ldots, v_n]$ forms an ordered basis for X satisfying $\langle u_1 , v_j \rangle = 0$ for $j = 2, \ldots, n$. Since A is linear, there exist n linear functionals $c_j : X \to \mathbb{C}$, $j = 1, \ldots, n$, such that

$$Ax = c_1(x)\, u_1 + c_2(x)\, v_2 + \cdots + c_n(x)\, v_n \quad \text{for } x \in X.$$

Let $Y := \operatorname{span}\{v_2, \ldots, v_n\}$. For $y \in Y$ define

$$By := c_2(y)\, v_2 + \cdots + c_n(y)\, v_n.$$

Then $B : Y \to Y$ is a linear map to which the induction hypothesis applies. Hence there exists an orthonormal basis $[u_2, \ldots, u_n]$ for Y with respect to which B is represented by an upper triangular matrix R. It follows that $[u_1, u_2, \ldots, u_n]$ forms an orthonormal basis for X with respect to which A is represented by an upper triangular matrix. The first diagonal entry of this matrix is λ since $Au_1 = \lambda u_1$, and its other diagonal entries are the diagonal entries of R. ∎

Remark 1.3 Schur form, square root and polar factorization of a matrix:
It should be noted that the preceding theorem has only a theoretical interest, since the proof of the existence of an orthonormal basis giving an upper triangular matrix representation is not constructive. In fact, Chapter 6 will be devoted to the numerical approximation of such an upper triangular representation. Nevertheless, the proof of this theorem points out an important fact, namely, the eigenvalues of a linear map A can be made to appear in a prescribed order along the diagonal of its upper triangular matrix representation.

Let n and m be positive integers such that $m \le n$. If $\mathsf{A} \in \mathbb{C}^{n\times m}$, then $\mathsf{A}^* \in \mathbb{C}^{m\times n}$, where $\mathsf{A}^*(i,j) := \overline{\mathsf{A}(j,i)}$. The matrix A^* is called the **adjoint matrix** or the **conjugate-transpose matrix** of A. Let I_m denote the $m\times m$ identity matrix. We say that A is a **unitary matrix** if $\mathsf{A}^*\mathsf{A} = \mathsf{I}_m$, that is, the columns of A form an orthonormal set in $\mathbb{C}^{n\times 1}$ with respect to the canonical inner product given by

$$\langle \mathsf{x}, \mathsf{y}\rangle := \mathsf{y}^* \mathsf{x} \quad \text{for } \mathsf{x}, \mathsf{y} \in \mathbb{C}^{n\times 1}.$$

Then Schur's Theorem can be stated as follows:

Given any matrix $\mathsf{A} \in \mathbb{C}^{n\times n}$, *there exists a unitary matrix* $\mathsf{Q} \in \mathbb{C}^{n\times n}$ *such that* $\mathsf{Q}^*\mathsf{A}\mathsf{Q}$ *is an upper triangular matrix.*

The upper triangular matrix $\mathsf{Q}^*\mathsf{A}\mathsf{Q}$ is called a **Schur form** of A. Simple examples, like the following one, show that Schur form of a matrix is not uniquely determined: Let

$$\mathsf{A} := \begin{bmatrix} 1 & 1 \\ -1 & 3 \end{bmatrix}, \quad \mathsf{Q} := \frac{1}{\sqrt{2}}\begin{bmatrix} 1 & 1 \\ 1 & -1 \end{bmatrix}, \quad \widetilde{\mathsf{Q}} := \frac{1}{\sqrt{2}}\begin{bmatrix} 1 & -1 \\ 1 & 1 \end{bmatrix}.$$

Then both Q and $\widetilde{\mathsf{Q}}$ are unitary matrices and both

$$\mathsf{Q}^*\mathsf{A}\mathsf{Q} = 2\begin{bmatrix} 1 & -1 \\ 0 & 1 \end{bmatrix} \quad \text{and} \quad \widetilde{\mathsf{Q}}^*\mathsf{A}\widetilde{\mathsf{Q}} = 2\begin{bmatrix} 1 & 1 \\ 0 & 1 \end{bmatrix}$$

are Schur forms of A.

Let $\mathsf{B} \in \mathbb{C}^{n\times n}$ be a **normal matrix**, that is, $\mathsf{B}^*\mathsf{B} = \mathsf{B}\mathsf{B}^*$, and let $\mathsf{T} := \mathsf{Q}^*\mathsf{B}\mathsf{Q}$ be a Schur form of B, where Q is a unitary matrix. Then $\mathsf{T}^*\mathsf{T} = \mathsf{T}\mathsf{T}^*$ and hence T is a diagonal matrix. If, in fact, B is a **Hermitian matrix**, that is, $\mathsf{B}^* = \mathsf{B}$, then T is also Hermitian, that is, T is a diagonal matrix with real entries. If, moreover, B is **positive definite**, that is, all the eigenvalues of B are positive numbers, then so is T. Define $\mathsf{T}^{1/2} := \operatorname{diag}\left(\mathsf{T}(1,1)^{1/2}, \ldots, \mathsf{T}(n,n)^{1/2}\right)$ and $\mathsf{P} := \mathsf{Q}\mathsf{T}^{1/2}\mathsf{Q}^*$. Then P is a positive definite Hermitian matrix, it commutes with B and $\mathsf{P}^2 = \mathsf{B}$. We now show that such a matrix P is unique. Let $\widetilde{\mathsf{P}}$ be another positive definite Hermitian matrix such that $\widetilde{\mathsf{P}}^2 = \mathsf{B}$. Let q be any column of Q and t the corresponding diagonal entry of T. Then $\widetilde{\mathsf{P}}^2\mathsf{q} = t\mathsf{q}$, that is, $(\widetilde{\mathsf{P}} + \sqrt{t}\,\mathsf{I}_n)(\widetilde{\mathsf{P}} - \sqrt{t}\,\mathsf{I}_n)\mathsf{q} = 0$. Since the eigenvalues of $\widetilde{\mathsf{P}}$ are positive numbers, the matrix $\widetilde{\mathsf{P}} + \sqrt{t}\,\mathsf{I}_n$ is nonsingular. Hence $(\widetilde{\mathsf{P}} - \sqrt{t}\,\mathsf{I}_n)\mathsf{q} = 0$. Thus $\widetilde{\mathsf{P}}\mathsf{q} = \mathsf{P}\mathsf{q}$ for each column q of Q, so that $\widetilde{\mathsf{P}}\mathsf{Q} = \mathsf{P}\mathsf{Q}$, or $\widetilde{\mathsf{P}} = \mathsf{P}$. Thus if B is a positive definite Hermitian matrix, then there is a unique positive definite Hermitian matrix whose square is equal to B. It will be denoted by $\mathsf{B}^{1/2}$ and called the **square root** of B. We remark that $\mathsf{B}^{1/2}$ commutes with B.

If $\mathsf{A} \in \mathbb{C}^{n \times n}$ is a nonsingular matrix, then $\mathsf{A}^*\mathsf{A}$ is Hermitian positive definite, so that $\mathsf{P} := (\mathsf{A}^*\mathsf{A})^{1/2}$ exists, is nonsingular, and $\mathsf{Q} := \mathsf{A}\mathsf{P}^{-1}$ is unitary. This gives $\mathsf{A} = \mathsf{Q}\mathsf{P}$, which is known as the Polar Factorization of A. This factorization can be viewed as the matrix version of the polar representation $z = e^{\mathrm{i}\,\theta} r$ of $z \in \mathbb{C}$. ▮

Let us give some examples of eigenvalues of operators defined on infinite dimensional spaces.

Example 1.4 In this example, every complex number is an eigenvalue: Let $X := C^\infty(\mathbb{R})$, the linear space of complex-valued functions defined on $\mathbb{R}$ whose derivatives of all orders exist. We define $T : X \to X$ by $Tx := x'$ for $x \in X$. Then every complex number λ is an eigenvalue of T, and if $\varphi(t) := \exp(\lambda t)$ for $t \in \mathbb{R}$, then φ is an eigenvector of T corresponding to the eigenvalue λ. ▮

Example 1.5 In this example, no complex number is an eigenvalue: Let X be the linear space of all functions $x \in C^\infty(\mathbb{R})$ such that $x(t) = 0$ if $|t| \geq 1$. Other than the null function, this linear space contains, for instance,

$$x(t) := \begin{cases} \exp\left[1/(t^2-1)\right] & \text{if } |t| < 1, \\ 0 & \text{otherwise.} \end{cases}$$

Define $T : X \to X$ by $Tx := x'$ for $x \in X$. Were $\lambda \in \mathbb{C}$ an eigenvalue of T and $\varphi \in X$ a corresponding eigenvector, then

$$\varphi' = \lambda\varphi, \qquad \text{so that} \qquad \varphi(t) = c\exp(\lambda t), \quad t \in \mathbb{R},$$

for some complex constant $c \neq 0$. But the condition '$\varphi(t) = 0$ if $|t| \geq 1$' implies that $c = 0$. Hence T has no eigenvalues at all. ▮

Example 1.6 In this example, no nonzero real number is an eigenvalue: Let X be the linear space of all bounded functions $x \in C^\infty(\mathbb{R})$ such that for each positive integer j, the jth derivative of x, denoted by $x^{(j)}$, is also bounded. Let T be defined by $Tx := x'$ for $x \in X$. Since $\varphi(t) := \exp(\lambda t)$ defines a bounded function if and only if $\Re(\lambda) = 0$, we conclude that the set of eigenvalues of T is $\{\lambda \in \mathbb{C} : \Re(\lambda) = 0\}$. ▮

It is evident that if M_1 and M_2 are invariant subspaces of T, then so is $M_1 + M_2 := \{x_1 + x_2 : x_1 \in M_1, x_2 \in M_2\}$. In particular, if λ_1

and λ_2 are distinct eigenvalues of T, then $M := E(\lambda_1) + E(\lambda_2)$ is an invariant subspace of T. However, the restriction $T_{|M,M} : M \to M$ is not a scalar multiple of the identity operator on M. In fact, $T_{|M,M}$ will have λ_1 and λ_2 as eigenvalues (and only those).

Given an eigenvalue λ of T, it is clear that $E(\lambda)$ is the largest subspace M of X such that $T_{|M,M}$ is λ times the identity operator on M. Let us ask the following question:

Is $E(\lambda)$ the largest subspace M of X such that M is invariant under T and $T_{|M,M}$ has λ as the only eigenvalue?

The answer is in the negative, as the following example shows.

Example 1.7 In this example, $\mathcal{N}\left[(T - \lambda I)^2\right]$ is an invariant subspace which is larger than the eigenspace $\mathcal{N}(T - \lambda I)$:
Let X be a linear space of dimension 5. Let $\mathcal{B} := [\varphi_1, \varphi_2, \psi_1, \psi_2, \xi]$ form an ordered basis for X, and let $T : X \to X$ be the linear operator defined by the following equations in which λ, μ and ν are three different complex numbers:

$$T\varphi_1 = \lambda\varphi_1,\ T\varphi_2 = \varphi_1 + \lambda\varphi_2,\ T\psi_1 = \mu\psi_1,\ T\psi_2 = \mu\psi_2 \text{ and } T\xi = \nu\xi.$$

The eigenvalues of T are λ, μ and ν. In fact, φ_1 is an eigenvector of T corresponding to the eigenvalue λ, φ_2 is not an eigenvector of T, ψ_1 and ψ_2 are two linearly independent eigenvectors of T corresponding to the eigenvalue μ, and ξ is an eigenvector of T corresponding to the eigenvalue ν. The eigenspaces of T are

$$E(\lambda) = \operatorname{span}\{\varphi_1\}, \quad E(\mu) = \operatorname{span}\{\psi_1, \psi_2\} \quad \text{and} \quad E(\nu) = \operatorname{span}\{\xi\}.$$

The subspace $M(\lambda) := \operatorname{span}\{\varphi_1, \varphi_2\}$ is invariant under T. The restriction $T_{|M(\lambda),M(\lambda)}$ is represented in the ordered basis $[\varphi_1, \varphi_2]$ of $M(\lambda)$ by the matrix

$$\mathsf{A}_1 := \begin{bmatrix} \lambda & 1 \\ 0 & \lambda \end{bmatrix},$$

which has λ as the only eigenvalue but $\mathsf{A}_1 \neq \lambda\,\mathsf{I}_2$, where I_2 denotes the 2×2 identity matrix. A change of basis will not alter this situation, that is to say, for any nonsingular 2×2 matrix V, $\mathsf{V}^{-1}\mathsf{A}_1\mathsf{V}$ is different from $\lambda\,\mathsf{I}_2$. We remark that $\varphi_1 = (T - \lambda I)\varphi_2$ and since φ_1 is an eigenvector of T associated with λ, $(T - \lambda I)^2\varphi_2 = 0$. Hence

$$E(\lambda) = \mathcal{N}(T - \lambda I) \subsetneq \mathcal{N}\left[(T - \lambda I)^2\right] = M(\lambda).$$

On the other hand,

$$E(\mu) = \mathcal{N}(T - \mu I) = \mathcal{N}\left[(T - \mu I)^2\right].$$

We remark that

$$\underline{T}\,[\varphi_1, \varphi_2] = [\varphi_1, \varphi_2]\begin{bmatrix} \lambda & 1 \\ 0 & \lambda \end{bmatrix} \quad \text{and} \quad \underline{T}\,[\psi_1, \psi_2] = [\psi_1, \psi_2]\begin{bmatrix} \mu & 0 \\ 0 & \mu \end{bmatrix}.$$

We observe that μ is such that the restriction of T to $E(\mu)$ can be represented by $\mu\, \mathsf{I}_2$, which is not the case for λ. On the other hand, $E(\nu) = \mathcal{N}(T - \nu I) = \mathcal{N}\left[(T - \nu I)^2\right]$ is one-dimensional. ∎

Our goal is to compute several eigenvalues of T. Also, for a given eigenvalue λ of T, we want to find the largest invariant subspace M such that the restriction $T_{|M,M}$ has λ as the only eigenvalue.

More generally, we are interested in computing a set Λ of eigenvalues of T as well as the largest invariant subspace M such that the restriction $T_{|M,M}$ has Λ as the set of its eigenvalues.

The reason for considering this more general situation is that, in most of the applications, the eigenvalues and the corresponding maximal invariant subspaces of the operator T are not computable in exact arithmetic. Hence techniques of Numerical Analysis are needed to approximate the desired eigenvalues and the corresponding maximal invariant subspaces. In this context, the operator T will be approximated by an operator $\widetilde{T}$, which is often of finite rank. The eigenvalues of $\widetilde{T}$ may be computed by reducing the eigenproblem for $\widetilde{T}$ to a matrix eigenvalue problem and then by using standard routines. Then a single multiple eigenvalue of T is approximated by a cluster of eigenvalues of $\widetilde{T}$, and the corresponding maximal invariant subspace of T is approximated by the direct sum of all the maximal invariant subspaces of $\widetilde{T}$ corresponding to the individual eigenvalues in that cluster. The word 'cluster' is used here in a very primitive sense. It only signifies a subset of the eigenvalues of $\widetilde{T}$ which may be identified by the user by noting that its elements are sufficiently close to one another and the subset itself is sufficiently far away from the rest of the eigenvalues of $\widetilde{T}$.

If we were interested only in the finite dimensional case, so that T admits a matrix representation, then a somewhat brief theory of spectral analysis would be sufficient. Since our goal includes the approximation of eigenvalues and of maximal invariant subspaces of integral and differential operators whose domains are infinite dimensional linear spaces, we need more sophisticated notions.

Our abstract setting is a complete normed space X over the complex field $\mathbb{C}$, that is, a complex Banach space. It is true that some of the results that we shall state and/or prove do not need the completeness of the norm on X. However, all our applications will take place in complete normed spaces.

A linear map from a finite dimensional linear space to a linear space of the same dimension is injective if and only if it is surjective. This equivalence does not hold in infinite dimensional linear spaces. This means that to consider the bijectivity of $T - \lambda I$, $\lambda \in \mathbb{C}$, the notion of an eigenvalue is not enough.

Example 1.8 An injective or a surjective operator may not be bijective: Let $X := \ell^2$ and (e_k) be the standard basis for X. For $x := \sum_{k=1}^{\infty} x(k)e_k$, define

$$T_1 x := \sum_{k=1}^{\infty} x(k)e_{k+1} \quad \text{and} \quad T_2 x := \sum_{k=1}^{\infty} x(k+1)e_k.$$

It is clear that T_1 is injective but not surjective, while T_2 is surjective but not injective. ∎

New definitions are thus needed, and things become more interesting. In the rest of the book, X will denote a complex Banach space and $\mathrm{BL}(X)$ the Banach algebra of all bounded linear operators from X into itself. The operator that sends each $x \in X$ to 0 will be denoted by O and, as we have mentioned before, the identity operator will be denoted by I. Let $\|\cdot\|$ denote the norm on X. The so-called subordinated operator norm is defined as follows:

$$\|T\| := \sup\{\|Tx\| \,:\, x \in X,\, \|x\| \le 1\} \quad \text{for } T \in \mathrm{BL}(X).$$

The following definitions concern an operator $T \in \mathrm{BL}(X)$:

The resolvent set of T is defined by

$$\mathrm{re}(T) := \{z \in \mathbb{C} \,:\, T - zI \text{ is bijective}\}.$$

For $z \in \mathrm{re}(T)$,

$$R(T, z) := (T - zI)^{-1}$$

is called the resolvent operator of T at z. We shall write $R(z)$ for $R(T, z)$ when there is no ambiguity.

The spectrum of T is the set

$$\mathrm{sp}(T) := \mathbb{C} \setminus \mathrm{re}(T).$$

An element of $\mathrm{sp}(T)$ will be called a spectral value of T.

If X is finite dimensional, then $\mathrm{sp}(T)$ consists of the eigenvalues of T.

The spectral radius of T is defined by

$$\rho(T) := \sup\{|\lambda| : \lambda \in \mathrm{sp}(T)\}.$$

We shall prove in Proposition 1.15(a) that if $X \neq \{0\}$, then $\mathrm{sp}(T) \neq \emptyset$ and $\rho(T)$ is finite.

Remark 1.9
Let $T \in \mathrm{BL}(X)$. If $z \in \mathbb{C}$ is given, then for every $y \in X$, a solution $x \in X$ of the linear equation

$$(T - zI)x = y$$

will be uniquely determined by y if and only if $z \in \mathrm{re}(T)$ and in that case, x is given by

$$x = R(T, z)y.$$

This explains the name given to the operator $R(T, z)$. If $\lambda \in \mathrm{sp}(T)$ and $y \in X$, then the existence and the uniqueness of solutions of the equation

$$(T - \lambda I)x = y$$

can be considered with the help of the 'reduced resolvent operator' of T associated with λ. See Remark 1.25. ∎

Proposition 1.10 Properties of the resolvent operator:
Let $T \in \mathrm{BL}(X)$.

(a) *If $z \in \mathrm{re}(T)$, then $R(z) \in \mathrm{BL}(X)$.*

(b) *For $z \in \mathrm{re}(T)$ and $z_0 \in \mathbb{C}$,*

$$z_0 \in \mathrm{re}(R(z)) \textit{ if and only if } z_0 = 0 \textit{ or } z + \frac{1}{z_0} \in \mathrm{re}(T).$$

(c) *For* $z \in \text{re}(T)$,

$$\rho(R(z)) = \frac{1}{\text{dist}(z, \text{sp}(T))}.$$

(d) First Resolvent Identity: *For* $z_1, z_2 \in \text{re}(T)$,

$$R(z_1) - R(z_2) = (z_1 - z_2)R(z_1)R(z_2).$$

(e) *For* $z \in \text{re}(T)$,

$$TR(z) = I + zR(z).$$

(f) *For* $z_1, z_2 \in \text{re}(T)$, *the operators* T, $R(z_1)$ *and* $R(z_2)$ *commute with each other.*

Proof

(a) This is a consequence of the Open Mapping Theorem. (See Theorem 11.1 of [55] or Theorem 4.12-2 of [48].)

(b) Let $z \in \text{re}(T)$. Since $R(z)$ is invertible, $z_0 = 0$ belongs to $\text{re}(R(z))$. If $z_0 \neq 0$, then

$$R(z) - z_0 I = R(z)\,[I - z_0(T - zI)] = -z_0 R(z)\left[T - \left(z + \frac{1}{z_0}\right) I\right],$$

so that

$$z_0 \in \text{re}(R(z)) \text{ if and only if } z + \frac{1}{z_0} \in \text{re}(T).$$

(c) By (b) above, for $z \in \text{re}(T)$,

$$\lambda \in \text{sp}(T) \;\text{ if and only if }\; \lambda \neq z \text{ and } \frac{1}{\lambda - z} \in \text{sp}(R(z)).$$

Hence

$$\rho(R(z)) = \sup\left\{\frac{1}{|\lambda - z|} : \lambda \in \text{sp}(T)\right\} = \frac{1}{\text{dist}(z, \text{sp}(T))}.$$

(d) If z_1 and z_2 belong to $\text{re}(T)$, then

$$\begin{aligned} R(z_1) - R(z_2) &= R(z_1)[I - (T - z_1 I)R(z_2)] \\ &= R(z_1)[T - z_2 I - (T - z_1 I)]R(z_2) \\ &= (z_1 - z_2)R(z_1)R(z_2). \end{aligned}$$

(e) For $z \in \text{re}(T)$,

$$I = (T - zI)R(z) = TR(z) - zR(z).$$

(f) Let z_1 and z_2 belong to $\text{re}(T)$. Since $T(T - z_1 I) = (T - z_1 I)T$, multiplication on the left and on the right by $R(z_1)$ gives $R(z_1)T = TR(z_1)$. The commutativity of $R(z_1)$ and $R(z_2)$ follows from (d). ∎

Functions of a complex variable with values in a complex Banach space will play an important role in what follows. Let Y be a complex Banach space and Ω an open subset of $\mathbb{C}$. We say that $F : \Omega \to Y$ is **analytic** in Ω if for every $z_0 \in \Omega$, the limit

$$F'(z_0) := \lim_{z \to z_0} \frac{F(z) - F(z_0)}{z - z_0}$$

exists. In that case, the function F' is called the first **derivative** of F. It can be shown that derivatives of higher order, defined recursively, also exist. The theory of contour integration can be developed in this more general context with only minor changes in the case of complex-valued functions of a complex variable.

A curve Γ in $\mathbb{C}$ is defined to be a continuous function from a compact interval $[a, b]$ in $\mathbb{R}$ to $\mathbb{C}$. It is said to be closed if $\Gamma(b) = \Gamma(a)$. A closed curve Γ is called a **simple curve** if for all t, $\widetilde{t} \in [a, b]$, $t < \widetilde{t}$ and $\Gamma(t) = \Gamma(\widetilde{t})$ imply that $t = a$ and $\widetilde{t} = b$. In this case we shall identify Γ with its image $\Gamma([a, b])$.

A curve Γ is said to be **rectifiable** if the supremum of the set

$$\left\{ \sum_{i=1}^{n} |\Gamma(t_{n,i}) - \Gamma(t_{n,i-1})| : a = t_{n,0} < \cdots < t_{n,n} = b,\ n = 1, 2, \ldots \right\}$$

is finite. For example, a piecewise continuously differentiable curve is rectifiable.

By a **Jordan curve** we mean a simple closed rectifiable curve.

The Jordan Curve Theorem states that if Γ is a simple closed curve, then $\mathbb{C} \setminus \Gamma([a, b])$ has two connected components, one bounded and the other unbounded. The **interior of a Jordan curve** Γ is the bounded component of $\mathbb{C} \setminus \Gamma([a, b])$ and it is denoted by $\text{int}(\Gamma)$. The **exterior of a Jordan curve** Γ is the unbounded component of $\mathbb{C} \setminus \Gamma([a, b])$ and it is denoted by $\text{ext}(\Gamma)$.

A Jordan curve Γ is said to be **positively oriented** if $\text{int}(\Gamma)$ lies to the left as the curve is traced out.

A **domain** in $\mathbb{C}$ is an open connected subset of $\mathbb{C}$. A domain Ω is said to be **simply connected** if the complement of Ω in $\mathbb{C}\cup\{\infty\}$ is a connected set.

Let $\Gamma : [a,b] \to \mathbb{C}$ be a rectifiable curve in $\mathbb{C}$. If $F : \Gamma([a,b]) \to Y$ is continuous, then we define

$$\int_\Gamma F(z)\,dz := \int_a^b F(\Gamma(t))\,d\Gamma(t),$$

where the integral on the right is the limit of Riemann-Stieltjes sums of the form $\sum_{i=1}^{n} F(\Gamma(s_{n,i}))\,[\Gamma(t_{n.i}) - \Gamma(t_{n,i-1})]$, where $a = t_{n,0} < \cdots < t_{n,n} = b$ and $s_{n,i} \in [t_{n,i-1}, t_{n,i}]$, as the mesh of the subdivision of $[a,b]$ tends to zero.

Remark 1.11
The most fundamental result about analytic functions is as follows.

Cauchy's Theorem: Let Ω be a domain in $\mathbb{C}$. If Ω is simply connected and $F : \Omega \to Y$ is analytic in Ω, then

$$\int_\Gamma F(z)\,dz = 0$$

for every Jordan curve Γ in Ω.

It can be proved that for any positively oriented Jordan curve Γ in Ω and every $z_0 \in \operatorname{int}(\Gamma)$,

$$F^{(k)}(z_0) = \frac{k!}{2\pi \mathrm{i}} \int_\Gamma \frac{F(z)}{(z-z_0)^{k+1}}\,dz \quad \text{for each nonnegative integer } k.$$

These are known as Cauchy's Integral Formulæ.

Let $\Gamma, \Gamma_1, \ldots, \Gamma_r$ be positively oriented Jordan curves in Ω such that $\Gamma_1, \ldots, \Gamma_r$ lie in $\operatorname{int}(\Gamma)$ and for $i \neq j$, each Γ_i lies in $\operatorname{ext}(\Gamma_j)$. If

$$\widetilde{\Omega} := \operatorname{int}(\Gamma) \cap \operatorname{ext}(\Gamma_1) \cap \cdots \cap \operatorname{ext}(\Gamma_r) \subset \Omega$$

and $F : \Omega \to Y$ is continuous in Ω and analytic in $\widetilde{\Omega}$, then

$$\int_\Gamma F(z)\,dz = \sum_{j=1}^{r} \int_{\Gamma_j} F(z)\,dz.$$

This result is known as Cauchy's Theorem for Multiply Connected Domains. ∎

We shall now consider power series with coefficients in a complex Banach space Y. As in the special case $Y := \mathbb{C}$, the classical results about the convergence of power series remain valid in general.

Consider $z_0 \in \mathbb{C}$ and $y_0, y_1 \ldots$ in Y. Let

$$r := \left(\limsup_{k\to\infty} \|y_k\|^{1/k}\right)^{-1}.$$

If $z \in \mathbb{C}$ satisfies $|z - z_0| < r$, then there exists $y(z) \in Y$ such that

$$\lim_{n\to\infty} \left\| \sum_{k=0}^{n} y_k(z-z_0)^k - y(z) \right\| = 0.$$

Also, if $z \in \mathbb{C}$ satisfies $|z - z_0| > r$, then the series $\sum_{k=0}^{\infty} y_k(z-z_0)^k$ is not convergent. Hence r is called the **radius of convergence** of the **power series** $\sum_{k=0}^{\infty} y_k(z-z_0)^k$.

Similarly, let $t := \limsup_{k\to\infty} \|y_k\|^{1/k}$. If $z \in \mathbb{C}$ satisfies $|z| > t$, then there exists $y(z) \in Y$ such that

$$\lim_{n\to\infty} \left\| \sum_{k=0}^{n} y_k z^{-k-1} - y(z) \right\| = 0,$$

and if $|z| < t$, then the series $\sum_{k=0}^{\infty} y_k z^{-k-1}$ is not convergent.

Remark 1.12

Let Ω be a domain in $\mathbb{C}$ and $F : \Omega \to Y$ be analytic.

Taylor's Theorem: Let $z_0 \in \Omega$ and $r > 0$ be such that the open ball with center z_0 and radius r is contained in Ω. Then

$$F(z) = \sum_{k=0}^{\infty} a_k(z-z_0)^k \quad \text{when } |z-z_0| < r,$$

where

$$a_k := \frac{F^{(k)}(z_0)}{k!} \in Y \quad \text{for } k = 0, 1, \ldots$$

Laurent's Theorem: If $z_0 \in \Omega$, $0 < r_1 < r_2$ are such that $\{z \in \mathbb{C} : r_1 < |z - z_0| < r_2\} \subset \Omega$ and Γ is a positively oriented Jordan curve such that

$$\{z \in \mathbb{C} : |z-z_0| = r_1\} \subset \operatorname{int}(\Gamma), \quad \{z \in \mathbb{C} : |z-z_0| = r_2\} \subset \operatorname{ext}(\Gamma),$$

then

$$F(z) = \sum_{k=-\infty}^{\infty} c_k (z - z_0)^k \quad \text{when } r_1 < |z - z_0| < r_2,$$

where

$$c_k := \frac{1}{2\pi \mathrm{i}} \int_\Gamma \frac{F(\tilde{z})}{(\tilde{z} - z_0)^{k+1}} \, d\tilde{z} \quad \text{for } k = 0, \pm 1, \pm 2, \ldots$$

Liouville's Theorem: If $F : \mathbb{C} \to Y$ is bounded and analytic, then F is a constant function. ∎

The results stated in Remarks 1.11 and 1.12 can be proved in exactly the same manner as the corresponding results for the complex-valued functions of a complex variable are proved.

Consider now the case $Y := \mathrm{BL}(X)$. Let Γ be a rectifiable curve in $\mathbb{C}$ and $F : \Gamma([a, b]) \to \mathrm{BL}(X)$ be continuous. Let $x \in X$ and $F_x : \Gamma([a, b]) \to X$ be defined by

$$F_x(z) := F(z)x, \quad z \in \Gamma([a, b]).$$

Then F_x is continuous, and by the definition of the integral $\int_\Gamma F(z)\, dz$, we obtain

$$\Big(\int_\Gamma F(z)\, dz\Big)x = \int_\Gamma F_x(z)\, dz = \int_\Gamma F(z)x\, dz.$$

In particular, if $T : X \to X$, then for every $x \in X$, we have

$$\Big(\int_\Gamma F(z)\, dz\Big)Tx = \int_\Gamma F(z)Tx\, dz.$$

Further, if $T \in \mathrm{BL}(X)$, then by the continuity and the linearity of T, we have

$$T\Big(\int_\Gamma F(z)\, dz\Big) = \int_\Gamma TF(z)\, dz.$$

We now use the preceding concepts to study the resolvent set and the resolvent operator of $T \in \mathrm{BL}(X)$.

Theorem 1.13
Let $T \in \mathrm{BL}(X)$.

(a) *The set* $\mathrm{re}(T)$ *is open in* $\mathbb{C}$.

(b) *The function* $R(\cdot) : \mathrm{re}(T) \to \mathrm{BL}(X)$ *is analytic in* $\mathrm{re}(T)$.

(c) First Neumann Expansion: *For* $z_0 \in \mathrm{re}(T)$ *and* $z \in \mathbb{C}$ *such that* $|z - z_0| < \mathrm{dist}(z_0, \mathrm{sp}(T))$, *we have* $z \in \mathrm{re}(T)$ *and* $R(z)$ *has the* Taylor Expansion

$$R(z) = \sum_{k=0}^{\infty} R(z_0)^{k+1}(z - z_0)^k.$$

(d) *For* $z \in \mathbb{C}$ *such that* $|z| > \rho(T)$, *we have* $z \in \mathrm{re}(T)$ *and* $R(z)$ *has the* Laurent Expansion

$$R(z) = -\sum_{k=0}^{\infty} T^k z^{-k-1}.$$

If in fact $|z| > \|T\|$, *then*

$$\|R(z)\| \leq \frac{1}{|z| - \|T\|}.$$

Proof

(a) Let $z_0 \in \mathrm{re}(T)$. The radius of convergence of a power series in $\mathrm{BL}(X)$ with kth coefficient $R(z_0)^{k+1}$ is

$$r(z_0) := \frac{1}{\limsup\limits_{k\to\infty} \|R(z_0)^{k+1}\|^{1/k}} = \frac{1}{\limsup\limits_{k\to\infty} \|R(z_0)^{k}\|^{1/k}} \geq \frac{1}{\|R(z_0)\|} > 0.$$

Hence the series $\sum\limits_{k=0}^{\infty} R(z_0)^{k+1}(z - z_0)^k$ converges in $\mathrm{BL}(X)$ for all $z \in \mathbb{C}$ satisfying $|z - z_0| < r(z_0)$. Setting

$$A_n(z) := \sum_{k=0}^{n} R(z_0)^{k+1}(z - z_0)^k,$$

we have

$$\begin{aligned} A_n(z)(T - zI) = (T - zI)A_n(z) &= [(T - z_0 I) - (z - z_0)I]A_n(z) \\ &= I - R(z_0)^{n+1}(z - z_0)^{n+1}, \end{aligned}$$

which tends to I as n tends to infinity, provided $|z-z_0| < r(z_0)$, and then the sum of the series is the operator $R(z)$. Thus for each $z_0 \in \mathrm{re}(T)$, the set $\mathrm{re}(T)$ contains the open ball with center z_0 and radius $r(z_0)$. Hence $\mathrm{re}(T)$ is an open set.

(b) Let $z_0 \in \mathrm{re}(T)$. As in part (a) above,

$$R(z) = R(z_0) + [R(z_0)]^2(z - z_0) + [R(z_0)]^3(z - z_0)^2 + \dots$$

if $|z - z_0| < r(z_0)$. Hence if $0 < |z - z_0| < 1/\|R(z_0)\|$, then

$$\begin{aligned}\left\|\frac{R(z) - R(z_0)}{z - z_0} - [R(z_0)]^2\right\| &= \left\|[R(z_0)]^2 \sum_{k=1}^{\infty}[R(z_0)]^k(z - z_0)^k\right\| \\ &\leq \frac{\|R(z_0)\|^3|z - z_0|}{1 - \|R(z_0)\|\,|z - z_0|}.\end{aligned}$$

Thus

$$\lim_{z \to z_0} \frac{R(z) - R(z_0)}{z - z_0} = R(z_0)^2.$$

This shows that the function $R(\cdot)$ is analytic in $\mathrm{re}(T)$ and its derivative is $R(\cdot)^2$.

(c) Let $z_0 \in \mathrm{re}(T)$. Since $\mathrm{re}(T)$ is open, $\mathrm{sp}(T)$ is closed and we have $\mathrm{dist}(z_0, \mathrm{sp}(T)) > 0$. Now the open ball centered at z_0 and with radius $\mathrm{dist}(z_0, \mathrm{sp}(T)))$ is contained in $\mathrm{re}(T)$ and the function $R(\cdot)$ is analytic on it. Hence by Taylor's Theorem, the expansion

$$R(z) = \sum_{k=0}^{\infty} R(z_0)^{k+1}(z - z_0)^k,$$

obtained in (a) above is valid for all z such that $|z-z_0| < \mathrm{dist}(z_0, \mathrm{sp}(T))$.

(d) Let $t := \limsup_{k\to\infty} \|T^k\|^{1/k}$. Then $t \leq \|T\|$. For $z \in \mathbb{C}$ such that $|z| > t$, it follows that the power series $\sum_{k=0}^{\infty} T^k z^{-k-1}$ converges in $\mathrm{BL}(X)$. Setting

$$B_n(z) := -\sum_{k=0}^{n} T^k z^{-k-1},$$

we have

$$B_n(z)(T - zI) = (T - zI)B_n(z) = I - \left(\frac{T}{z}\right)^{n+1},$$

which tends to I as $n \to \infty$ provided $|z| > t$, and then the sum of the series is the operator $R(z)$. If $|z| > \rho(T)$, then $z \notin \text{sp}(T)$. Thus $\{z \in \mathbb{C} : |z| > \rho(T)\}$ is contained in $\text{re}(T)$ and the function $R(\cdot)$ is analytic in this set. Hence the Laurent Expansion

$$R(z) = -\sum_{k=0}^{\infty} T^k z^{-k-1}$$

is valid for all z such that $|z| > \rho(T)$.

For $|z| > ||T||$, we have

$$||R(z)|| = \left\| -\sum_{k=0}^{\infty} T^k z^{-k-1} \right\| \leq \frac{1}{|z|} \sum_{k=0}^{\infty} \left(\frac{||T||}{|z|} \right)^k \leq \frac{1}{|z| - ||T||},$$

as desired. ∎

Remark 1.14

By Theorem 1.13(b), $z \in \text{re}(T) \mapsto ||R(z)|| \in \mathbb{R}$ is a continuous function which takes only positive values. Hence it attains a maximum value and a positive minimum value on each compact subset of $\mathbb{C}$ contained in $\text{re}(T)$. ∎

If $X \neq \{0\}$, then $\text{sp}(T)$ is nonempty and the spectral radius of T is related to the powers of T in a peculiar manner.

Proposition 1.15

Let $X \neq \{0\}$ and $T \in \text{BL}(X)$.

(a) Gelfand-Mazur Theorem: *The subset* $\text{sp}(T)$ *of* $\mathbb{C}$ *is nonempty and compact.*

(b) Spectral radius formula:

$$\rho(T) = \lim_{k \to \infty} ||T^k||^{1/k} = \inf_{k=1,2,\ldots} ||T^k||^{1/k}.$$

Proof

(a) By Theorem 1.13(d), $\text{sp}(T)$ is a closed subset of $\mathbb{C}$ and $|\lambda| \leq ||T||$ for all $\lambda \in \text{sp}(T)$. Hence $\text{sp}(T)$ is compact. If it were empty, then $R(\cdot)$ would be analytic in the whole complex plane; and, by Theorem

1.13(d), it follows that $\|R(z)\| \to 0$ as $|z|$ tends to infinity. Hence $R(\cdot)$ is a bounded entire function. By Liouville's theorem, it must be a constant function; and in fact this constant must be equal to 0, but this is impossible because $X \neq \{0\}$.

(b) The proof of Theorem 1.13(d) shows that

$$\rho(T) \leq \limsup_{k \to \infty} \|T^k\|^{1/k} = t.$$

In fact, $\rho(T) = t$ since the Laurent Expansion must converge for $|z| > \rho(T)$ and it must diverge for $|z| < t$. The identity

$$T^k - z^k I = (T - zI) \sum_{j=0}^{k-1} T^{k-1-j} z^j,$$

which is valid for each positive integer k, shows that $z \in \mathrm{re}(T)$ whenever z^k belongs to $\mathrm{re}(T^k)$. Hence $\lambda^k \in \mathrm{sp}(T^k)$ whenever $\lambda \in \mathrm{sp}(T)$, so that $|\lambda| \leq \|T^k\|^{1/k}$. Thus

$$\rho(T) \leq \inf_{k=1,2,\ldots} \|T^k\|^{1/k} \leq \liminf_{k \to \infty} \|T^k\|^{1/k} \leq \limsup_{k \to \infty} \|T^k\|^{1/k} = \rho(T),$$

which completes the proof. ∎

Example 1.16 In this example, some spectral values are not eigenvalues: Let $X := \ell^2$. Let (e_k) be the standard basis for X. Consider the left shift operator on X:

$$Tx := \sum_{k=1}^{\infty} x(k+1) e_k, \quad x := \sum_{k=1}^{\infty} x(k) e_k \in \ell^2.$$

Let us first compute the eigenvalues of T. The equation $Tx = \lambda x$ leads to $x(k+1) = \lambda x(k)$ for each positive integer k. Fixing $x(1) := 1$, we get

$$x = \sum_{k=1}^{\infty} \lambda^{k-1} e_k.$$

Now $x \in X$ if and only if $|\lambda| < 1$. Hence any complex number λ such that $|\lambda| < 1$ is an eigenvalue of T. Since $\mathrm{sp}(T)$ is closed, this implies that $\mathrm{sp}(T)$ contains $\{\lambda \in \mathbb{C} : |\lambda| \leq 1\}$. It can be easily seen that for any integer $k \geq 0$, $||T^k|| = 1$, so that $\rho(T) = \lim_{k \to \infty} ||T^k||^{1/k} = 1$ and hence

$$\mathrm{sp}(T) = \{\lambda \in \mathbb{C} : |\lambda| \leq 1\}.$$

However, no point λ on the boundary of the spectrum is an eigenvalue, because then $\sum_{k=1}^{\infty} \lambda^{k-1} e_k \notin \ell^2$. ∎

Example 1.17 An example in which no spectral value is an eigenvalue: Any eigenvalue of T belongs to $\mathrm{sp}(T)$, since $T - \lambda I$ is not one-to-one when λ is an eigenvalue of T. When X is finite dimensional, $\mathrm{sp}(T)$ is the set of all eigenvalues of T. However, when X is infinite dimensional, $\mathrm{sp}(T)$ may contain scalars which are not eigenvalues of T. Consider the space $X := C^0([0,1])$. Let $x_0 \in X$ be given. We define an operator T on X by

$$(Tx)(t) := x_0(t)x(t), \quad t \in [0,1].$$

It is easily seen that $||T|| = ||x_0||_\infty$ and hence $T \in \mathrm{BL}(X)$.

First we prove that $\mathrm{sp}(T) = x_0([0,1])$. Let z be a complex number not in $x_0([0,1])$. Then for any $y \in X$, the equation $(T - zI)x = y$ has a unique solution $x \in X$ defined by $x(t) = y(t)/(x_0(t) - z)$ for all $t \in [0,1]$. Hence $z \in \mathrm{re}(T)$. This shows that $\mathrm{sp}(T) \subset x_0([0,1])$.

Now let $\lambda \in x_0([0,1])$. The operator $T - \lambda I$ is not surjective, because the function $y \in X$ defined by $y(t) := 1$ for $t \in [0,1]$ is not in $\mathcal{R}(T - \lambda I)$. In fact, if $t_0 \in [0,1]$ is such that $x_0(t_0) = \lambda$, and x were a function in X such that $(T - \lambda I)x = y$, then $0 = (T - \lambda I)x(t_0) = 1$.

We now show that $\lambda \in x_0([0,1])$ is an eigenvalue of T if and only if x_0 takes the value λ in some nonempty open interval.

Let $\lambda \in \mathrm{sp}(T)$. Suppose that there exist a and b such that $0 \leq a < b \leq 1$ and $x_0(t) = \lambda$ for all $t \in]a,b[$. Then any nonzero continuous function vanishing on $[0,1] \setminus [a,b]$ is an eigenvector of T corresponding to λ; for example, the function $\varphi : [0,1] \to \mathbb{C}$ defined by

$$\varphi(t) := \begin{cases} (t-a)(b-t) & \text{if } t \in [a,b], \\ 0 & \text{if } t \in [0,1] \setminus [a,b]. \end{cases}$$

On the other hand, let $\lambda \in \mathrm{sp}(T)$ be an eigenvalue of T and φ a corresponding eigenvector. Then $\varphi \neq 0$ but $(x_0 - \lambda)\varphi = 0$. Let $s \in [0,1]$ be such that $\varphi(s) \neq 0$. Since φ is a continuous function, there exist a and b such that $0 \leq a < b \leq 1$, $s \in [a,b]$ and $\varphi(t) \neq 0$ for all $t \in [a,b]$. This implies that $x_0(t) = \lambda$ for all $t \in]a,b[$.

In the particular case $x_0(t) := t$ for $t \in [0, 1]$, we have $\mathrm{sp}(T) = [0, 1]$ and no spectral value of T is an eigenvalue because $T\varphi = \lambda\varphi$ implies that $\varphi(t) = 0$ for all $t \in [0, 1]$. ∎

1.2 Decompositions

Let Y and Z be closed subspaces of X. We say that (Y, Z) is a **decomposition** of X, or **decomposes** X if

$$X = Y \oplus Z,$$

that is, each $x \in X$ has a unique decomposition

$$x = x_{_Y} + x_{_Z} \text{ with } x_{_Y} \in Y \text{ and } x_{_Z} \in Z.$$

The operator $P : X \to X$ defined by $x \mapsto Px := x_{_Y}$ is linear and satisfies $P^2 = P$, that is, P is a **projection**. Since $\mathcal{R}(P) = Y$ and $\mathcal{N}(P) = Z$, we say that P **projects onto** Y **along** Z. Because Y and Z are closed, the Closed Graph Theorem (see Theorem 10.2 of [55] or Theorem 4.13-2 of [48]) implies that P is continuous. Thus $P \in \mathrm{BL}(X)$.

Let (Y, Z) decompose X. Consider $T \in \mathrm{BL}(X)$. We say that (Y, Z) is a decomposition of T (or **decomposes** T) if both Y and Z are invariant under T:

$$T(Y) \subset Y \quad \text{and} \quad T(Z) \subset Z.$$

If P projects onto Y along Z, then it is easy to see that (Y, Z) decomposes T if and only if P commutes with T.

Proposition 1.18
If (Y, Z) decomposes T, then

$$\mathrm{sp}(T) = \mathrm{sp}(T_{|Y,Y}) \cup \mathrm{sp}(T_{|Z,Z}).$$

Proof
Let $z \in \mathrm{re}(T)$. First notice that if $x \in Y$, then $y := Tx - zx \in Y$. Also, if $y \in Y$, $Tx - zx = y$ for $x \in X$ and $x = x_{_Y} + x_{_Z}$ for $x_{_Y} \in Y$ and $x_{_Z} \in Z$, then

$$(Tx_{_Y} - zx_{_Y}) + (Tx_{_Z} - zx_{_Z}) = y,$$

so that $Tx_Z - zx_Z = 0$, that is, $x_Z = 0$ and $x = x_Y \in Y$. Thus $T - zI$ is a bijection from Y onto itself. The same argument applies for Z. This shows that $\mathrm{re}(T) \subset \mathrm{re}(T_{|Y,Y}) \cap \mathrm{re}(T_{|Z,Z})$. Since Y and Z are invariant under $R(T, z)$, we have

$$R(T, z)_{|Y,Y} = R(T_{|Y,Y}, z) \quad \text{and} \quad R(T, z)_{|Z,Z} = R(T_{|Z,Z}, z).$$

Conversely, let P be the projection onto Y along Z. For $z \in \mathrm{re}(T_{|Y,Y}) \cap \mathrm{re}(T_{|Z,Z})$ and $x \in X$, it can be seen that

$$(T - zI)[R(T_{|Y,Y}, z)P + R(T_{|Z,Z}, z)(I - P)]x = x$$

and

$$[R(T_{|Y,Y}, z)P + R(T_{|Z,Z}, z)(I - P)](T - zI)x = x,$$

so that $z \in \mathrm{re}(T)$ and

$$R(T, z) = R(T_{|Y,Y}, z)P + R(T_{|Z,Z}, z)(I - P).$$

Thus

$$\mathrm{re}(T) = \mathrm{re}(T_{|Y,Y}) \cap \mathrm{re}(T_{|Z,Z}).$$

Hence we obtain the desired result. ▮

In general, $\mathrm{sp}(T_{|Y,Y})$ and $\mathrm{sp}(T_{|Z,Z})$ may not be disjoint as the following example shows.

Example 1.19 A simple example of decomposition:
Let X be a two-dimensional complex space and an operator T on X be represented by the matrix

$$\mathsf{A} := \begin{bmatrix} \lambda & 0 \\ 0 & \mu \end{bmatrix}$$

with respect to an ordered basis $[v_1, v_2]$. Let $Y := \mathrm{span}\{v_1\}$ and $Z := \mathrm{span}\{v_2\}$. The pair (Y, Z) decomposes T. Further, $T_{|Y,Y}$ and $T_{|Z,Z}$ have disjoint spectra if and only if $\lambda \neq \mu$. ▮

Our aim now is to find a decomposition (Y, Z) of X such that the restrictions $T_{|Y,Y} : Y \to Y$ and $T_{|Z,Z} : Z \to Z$ have disjoint spectra:

$$\mathrm{sp}(T_{|Y,Y}) \cap \mathrm{sp}(T_{|Z,Z}) = \emptyset.$$

Further, if $\Lambda \subset \operatorname{sp}(T)$ and $\operatorname{sp}(T) \setminus \Lambda$ are closed in $\mathbb{C}$, then we would like to find a decomposition (Y, Z) of X such that $\operatorname{sp}(T_{|Y,Y}) = \Lambda$ and $\operatorname{sp}(T_{|Z,Z}) = \operatorname{sp}(T) \setminus \Lambda$. With this in mind, we introduce the following concept.

A subset Λ of $\operatorname{sp}(T)$ such that Λ as well as $\operatorname{sp}(T) \setminus \Lambda$ are closed in $\mathbb{C}$ is called a **spectral set** for T.

Since $\operatorname{sp}(T)$ is a closed set in $\mathbb{C}$, it follows that a subset Λ of $\operatorname{sp}(T)$ is a spectral set for T if and only if Λ is closed as well as open in $\operatorname{sp}(T)$.

Note that a singleton set $\{\lambda\}$ is a spectral set for T if and only if λ is an isolated point of $\operatorname{sp}(T)$.

If Λ is a spectral set for T, and Γ is a Jordan curve in $\operatorname{re}(T)$ satisfying $\operatorname{sp}(T) \cap \operatorname{int}(\Gamma) = \Lambda$, then we say that the Jordan curve Γ separates Λ from $\operatorname{sp}(T) \setminus \Lambda$.

We remark that for some spectral sets Λ for T, there may be no Jordan curve which separates Λ from $\operatorname{sp}(T) \setminus \Lambda$.

Example 1.20 In this example, a single Jordan curve is not enough to separate a spectral set Λ for T from $\operatorname{sp}(T) \setminus \Lambda$:
Let $X := \ell^2$ and (e_k) be the standard basis for X. Let $\{\theta_1, \theta_2, \ldots\} := [0, 1] \cap \mathbb{Q}$. We define $T \in \operatorname{BL}(X)$ by

$$Tx := 2x(2)e_2 + \sum_{k=1}^{\infty} e^{2\pi \mathrm{i}\, \theta_k} x(k+2) e_{k+2} \quad \text{for } x := \sum_{k=1}^{\infty} x(k) e_k \in X.$$

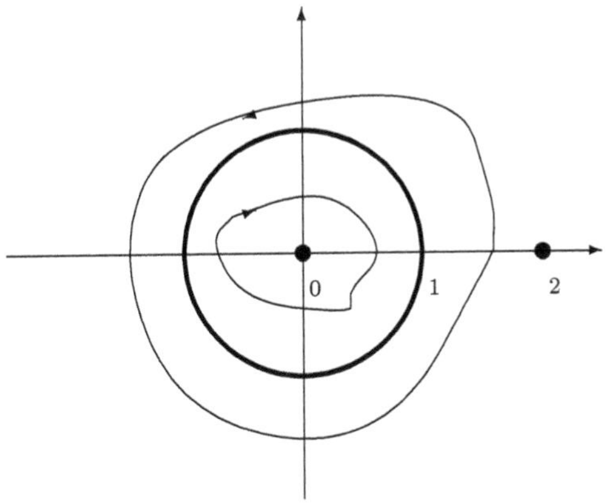

Clearly, 0, 2 and $e^{2\pi \mathrm{i}\, \theta_j}$ for each positive integer j are eigenvalues of T. In fact, they are all the eigenvalues of T. Since $\{\theta_1, \theta_2, \ldots\}$ is dense in $[0, 1]$, we obtain $\mathrm{sp}(T) = \{0, 2\} \cup \{\lambda \in \mathbb{C} : |\lambda| = 1\}$. If $\Lambda := \{\lambda \in \mathbb{C} : |\lambda| = 1\}$, then at least two disjoint Jordan curves are needed to separate Λ from $\mathrm{sp}(T) \setminus \Lambda$. The same holds if $\Lambda := \{0, 2\}$. ∎

The situation illustrated in Example 1.20 motivates the notion of a **Cauchy contour**. We first introduce the notion of a Cauchy domain.

An **elementary Cauchy domain** is a bounded open connected subset of $\mathbb{C}$ whose boundary is the union of a finite number of nonintersecting Jordan curves. A finite union of elementary Cauchy domains having disjoint closures is called a **Cauchy domain**.

Theorem 1.21
If E is a compact subset of $\mathbb{C}$ contained in an open subset Ω, then there exists a Cauchy domain D such that E is a subset of D and the closure of D is a subset of Ω.

Proof
Let $F := \mathbb{C} \setminus \Omega$ and $\delta := \mathrm{dist}(E, F)$. Then $\delta > 0$. For all integers i, j, let

$$z_{i,j} := \frac{(i + \mathrm{i}\, j)\delta}{3}$$

and

$$B_{i,j} := \{z \in \mathbb{C} : |z - z_{i,j}| < \frac{\delta\sqrt{3}}{6}\}.$$

Then $\mathbb{C} \subset \bigcup_{i,j} B_{i,j}$. Since E is compact, it can be covered with a finite number of such open discs having nonempty intersection with E, say $E \subset D := \bigcup_{k=1}^{n} B_{i_k, j_k}$. Let z be in the closure of D. Find $w \in D$ such that $|z-w| < \delta/3$. Then there exists $k \in \{1, \ldots, n\}$ such that $w \in B_{i_k, j_k}$, and $|z - z_{i_k,j_k}| \le |z - w| + |w - z_{i_k,j_k}| < \delta + 2\sqrt{3}\delta/6 = \delta$. This implies that $z \notin F$ and hence $z \in \Omega$. The boundary of D is contained in the union of the boundaries of the discs B_{i_k, j_k}, $k = 1, \ldots, n$. Moreover, the part of the boundary of D inside each square of the grid defined by the centers z_{i_k, j_k}, $k = 1, \ldots, n$, is composed of an arc of a circle, or of two arcs of circles having a common endpoint, or of a disjoint pair of two such arcs. Hence the boundary of each connected component of D is a union of a finite number of Jordan curves.

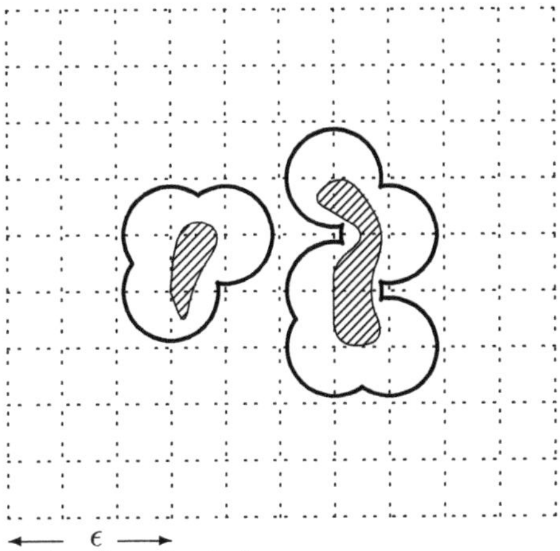

E is the shaded set,
Ω is the entire open rectangle,
D is the union of the open disks and
$E \subset D \subset$ closure of $D \subset \Omega$.

Thus each connected component of D is an elementary Cauchy domain. As a consequence, D is a Cauchy domain. ∎

Let D be a Cauchy domain. If each Jordan curve involved in the boundary of D is oriented in such a way that points in D lie to the left as the curve is traced out, then the oriented boundary C of D is called a Cauchy contour. The **interior of a Cauchy contour** C determined by a Cauchy domain D is defined to be $\text{int}(\mathrm{C}) := D$, and the **exterior of a Cauchy contour** C determined by a Cauchy domain D is defined to be $\text{ext}(\mathrm{C}) := \mathbb{C} \setminus (D \cup \mathrm{C})$. We remark that this notation is consistent with our earlier notation $\text{int}(\Gamma)$ and $\text{ext}(\Gamma)$ for a Jordan curve Γ.

Let C be a Cauchy contour. If E and $\widetilde{E}$ are subsets of $\mathbb{C}$ such that $E \subset \text{int}(\mathrm{C})$ and $\widetilde{E} \subset \text{ext}(\mathrm{C})$, then we say that C separates E from $\widetilde{E}$.

Corollary 1.22
Let E be a compact subset of $\mathbb{C}$ and $\widetilde{E}$ be a closed subset of $\mathbb{C}$. If $E \cap \widetilde{E} = \emptyset$, then there is a Cauchy contour C separating E from $\widetilde{E}$. Further, there is a Cauchy contour separating $E \cup \mathrm{C}$ from $\widetilde{E}$, and there is a Cauchy contour separating E from $\widetilde{E} \cup \mathrm{C}$.

Proof
Letting $\Omega := \mathbb{C} \setminus \widetilde{E}$ in Theorem 1.21, we obtain a Cauchy domain D

such that E is a subset of D and the closure of D does not intersect $\widetilde{E}$. Now let C be the oriented boundary of D. Then $E \subset \text{int}(\text{C})$ and $\widetilde{E} \subset \mathbb{C} \setminus (D \cup \text{C}) = \text{ext}(\text{C})$, that is, the Cauchy contour C separates E from $\widetilde{E}$.

Next, since $E \cup \text{C}$ is compact and $(E \cup \text{C}) \cap \widetilde{E} = \emptyset$, we may replace E by $E \cup \text{C}$ to obtain a Cauchy contour separating $E \cup \text{C}$ from $\widetilde{E}$. Similarly, since $\widetilde{E} \cup \text{C}$ is closed and $E \cap (\widetilde{E} \cup \text{C}) = \emptyset$, we may replace $\widetilde{E}$ by $\widetilde{E} \cup \text{C}$ to obtain a Cauchy contour separating E from $\widetilde{E} \cup \text{C}$. ∎

By Corollary 1.22, for every spectral set Λ for T, there is a Cauchy contour C that separates Λ from $\text{sp}(T) \setminus \Lambda$. If Λ is a spectral set for T, the set of all Cauchy contours separating Λ from $\text{sp}(T) \setminus \Lambda$ will be denoted by $\mathcal{C}(T, \Lambda)$.

For a spectral set Λ for T and $\text{C} \in \mathcal{C}(T, \Lambda)$, define

$$P(T, \Lambda) := -\frac{1}{2\pi \text{i}} \int_{\text{C}} R(T, z)\, dz,$$

which is the contour integral along C of the $\text{BL}(X)$-valued function $z \mapsto R(T, z)$ of the complex variable z. Since any $\widetilde{\text{C}} \in \mathcal{C}(T, \Lambda)$ can be continuously deformed in $\text{re}(T)$ to C, the bounded operator $P(T, \Lambda)$ does not depend on the choice of $\text{C} \in \mathcal{C}(T, \Lambda)$.

This operator allows us to obtain a decomposition of T for which the restricted operators have disjoint spectra. For this purpose a number of properties of $P(T, \Lambda)$ need to be proved first.

Proposition 1.23

Let Λ be a spectral set for T and $P := P(T, \Lambda)$. Then P is a bounded projection and it commutes with T. Moreover, if $\widetilde{P} \in \text{BL}(X)$ is any projection such that $\widetilde{P}$ commutes with T and $\mathcal{R}(\widetilde{P}) = \mathcal{R}(P)$, then $\widetilde{P} = P$.

Proof

Let $\text{C} \in \mathcal{C}(T, \Lambda)$. By Corollary 1.22, there exists a Cauchy contour $\widetilde{\text{C}}$ that separates $\Lambda \cup \text{C}$ from $\text{sp}(T) \setminus \Lambda$. Then $\widetilde{\text{C}} \in \mathcal{C}(T, \Lambda)$ and hence

$$P^2 = \Big(-\frac{1}{2\pi \text{i}}\Big)^2 \int_{\text{C}} \Big[\int_{\widetilde{\text{C}}} R(z) R(\widetilde{z})\, d\widetilde{z}\Big]\, dz.$$

By the First Resolvent Identity (Proposition 1.10(d)), we have

$$R(z) - R(\widetilde{z}) = (z - \widetilde{z}) R(z) R(\widetilde{z}) \quad \text{for } z \in \text{C},\ \widetilde{z} \in \widetilde{\text{C}}.$$

Hence

$$P^2 = \Big(-\frac{1}{2\pi i}\Big)^2 \int_{C}\Big[\int_{\widetilde{C}} [R(z) - R(\widetilde{z})]\frac{d\widetilde{z}}{z-\widetilde{z}}\Big]\,dz$$

$$= \Big(-\frac{1}{2\pi i}\Big)^2 \int_{C}\Big[R(z)\int_{\widetilde{C}} \frac{d\widetilde{z}}{z-\widetilde{z}}\Big]\,dz$$

$$-\Big(-\frac{1}{2\pi i}\Big)^2 \int_{\widetilde{C}}\Big[R(\widetilde{z})\int_{C} \frac{dz}{z-\widetilde{z}}\Big]\,d\widetilde{z}$$

by interchanging the order of integrations. But C lies in $\mathrm{int}(\widetilde{C})$, so that for each $z \in C$ and each $\widetilde{z} \in \widetilde{C}$, we have by Cauchy's theorem,

$$\int_{\widetilde{C}} \frac{d\widetilde{z}}{z-\widetilde{z}} = -2\pi i \quad \text{and} \quad \int_{C} \frac{dz}{z-\widetilde{z}} = 0.$$

This proves that $P^2 = P$. Also, $P \in \mathrm{BL}(X)$ because it is the contour integral of a $\mathrm{BL}(X)$-valued function. The operators P and T commute because T commutes with $R(z)$ for each z on C.

Let now $\widetilde{P} \in \mathrm{BL}(X)$ be such that $\widetilde{P}^2 = \widetilde{P}$, $\widetilde{P}T = T\widetilde{P}$ and $\mathcal{R}(\widetilde{P}) = \mathcal{R}(P)$. Then $\widetilde{P}R(z) = R(z)\widetilde{P}$ for every z on $C \in \mathcal{C}(T,\Lambda)$, so that $\widetilde{P}P = P\widetilde{P}$. Let $x \in X$. Since $\widetilde{P}x \in \mathcal{R}(P)$ and $Px \in \mathcal{R}(\widetilde{P})$, we have

$$\widetilde{P}x = P\widetilde{P}x = \widetilde{P}Px = Px.$$

Thus $\widetilde{P} = P$. ∎

Let Λ be a spectral set for T and $P := P(T,\Lambda)$. The fact that P and T commute implies that $M := \mathcal{R}(P)$ and $N := \mathcal{N}(P) = \mathcal{R}(I-P)$ are invariant under T, and this pair of invariant subspaces decomposes T. Moreover, we shall show that they satisfy $\mathrm{sp}(T_{|M,M}) \cap \mathrm{sp}(T_{|N,N}) = \emptyset$. For this purpose, we need an additional concept:

Let $C \in \mathcal{C}(T,\Lambda)$ and for $\lambda_0 \in \mathrm{sp}(T)$ define

$$S(T,\Lambda,\lambda_0) := -\frac{1}{2\pi i}\int_{C} R(T,z)\frac{dz}{\lambda_0 - z}.$$

Clearly, $S(T,\Lambda,\lambda_0) \in \mathrm{BL}(X)$ and it does not depend on $C \in \mathcal{C}(T,\Lambda)$ in the same way as $P(T,\Lambda)$.

Proposition 1.24

Let Λ be a spectral set for T, $C \in \mathcal{C}(T,\Lambda)$, $P := P(T,\Lambda)$, $M := \mathcal{R}(P)$, $N := \mathcal{N}(P)$, $\lambda_0 \in \mathrm{sp}(T)$ and $S := S(T,\Lambda,\lambda_0)$. Then

(a) *S commutes with T and with P.*

(b) *If $\lambda_0 \in \Lambda$, then*

$$S(T - \lambda_0 I) = I - P \quad \textit{and} \quad SP = O,$$

so that $\lambda_0 \in \mathrm{re}(T_{|N,N})$,

$$R\left(T_{|N,N}, \lambda_0\right) = S_{|N,N} \textit{ and } S_{|M,M} = O_{|M,M}.$$

(c) *If $\lambda_0 \in \mathrm{sp}(T) \setminus \Lambda$, then*

$$S(T - \lambda_0 I) = -P \quad \textit{and} \quad SP = S,$$

so that $\lambda_0 \in \mathrm{re}(T_{|M,M})$,

$$R\left(T_{|M,M}, \lambda_0\right) = -S_{|M,M} \textit{ and } S_{|N,N} = O_{|N,N}.$$

Proof

(a) Since T is continuous and commutes with $R(z)$ for each $z \in \mathrm{C}$, T and S commute. Similarly, since P is continuous and commutes with $R(z)$ for $z \in \mathrm{C}$, P and S commute.

(b) Let $\lambda_0 \in \Lambda$ and $\mathrm{C} \in \mathcal{C}(T, \Lambda)$. Then

$$\begin{aligned} S(T - \lambda_0 I) &= -\frac{1}{2\pi \mathrm{i}} \int_{\mathrm{C}} R(z)(T - \lambda_0 I) \frac{dz}{\lambda_0 - z} \\ &= -\frac{1}{2\pi \mathrm{i}} \int_{\mathrm{C}} R(z)(T - zI + zI - \lambda_0 I) \frac{dz}{\lambda_0 - z} \\ &= -\frac{1}{2\pi \mathrm{i}} \int_{\mathrm{C}} \frac{dz}{\lambda_0 - z} I + \frac{1}{2\pi \mathrm{i}} \int_{\mathrm{C}} R(z)\, dz = I - P, \end{aligned}$$

since $\lambda_0 \in \mathrm{int}(\mathrm{C})$ and $\displaystyle\int_{\mathrm{C}} \frac{dz}{\lambda_0 - z} = -2\pi \mathrm{i}$.

Next, by Corollary 1.22, there exists a Cauchy contour $\widetilde{\mathrm{C}}$ separating $\Lambda \cup \mathrm{C}$ from $\mathrm{sp}(T) \setminus \Lambda$. Using C in the integral defining S and $\widetilde{\mathrm{C}}$ in the integral defining P, and taking into account the First Resolvent Identity, we obtain

$$\begin{aligned} SP &= -\frac{1}{2\pi \mathrm{i}} \int_{\mathrm{C}} R(z) P \frac{dz}{\lambda_0 - z} \\ &= -\frac{1}{2\pi \mathrm{i}} \int_{\mathrm{C}} R(z) \Big[-\frac{1}{2\pi \mathrm{i}} \int_{\widetilde{\mathrm{C}}} R(\widetilde{z})\, d\widetilde{z} \Big] \frac{dz}{\lambda_0 - z} \end{aligned}$$

$$= \Big(-\frac{1}{2\pi \mathrm{i}}\Big)^2 \int_{\mathrm{C}} \Big[\int_{\widetilde{\mathrm{C}}} [R(z) - R(\widetilde{z})] \frac{d\widetilde{z}}{(\lambda_0 - z)(z - \widetilde{z})} \Big] \, dz$$

$$= \Big(-\frac{1}{2\pi \mathrm{i}}\Big)^2 \int_{\mathrm{C}} \Big[\frac{R(z)}{\lambda_0 - z} \int_{\widetilde{\mathrm{C}}} \frac{d\widetilde{z}}{z - \widetilde{z}} \Big] \, dz$$

$$- \Big(-\frac{1}{2\pi \mathrm{i}}\Big)^2 \int_{\widetilde{\mathrm{C}}} \Big[R(\widetilde{z}) \int_{\mathrm{C}} \frac{dz}{(\lambda_0 - z)(z - \widetilde{z})} \Big] \, d\widetilde{z} = O,$$

because $\mathrm{C} \subset \mathrm{int}(\widetilde{\mathrm{C}})$, so $\displaystyle\int_{\widetilde{\mathrm{C}}} \frac{d\widetilde{z}}{z - \widetilde{z}} = -2\pi\mathrm{i}$ for every $z \in \mathrm{C}$, $\displaystyle\int_{\mathrm{C}} \frac{dz}{z - \widetilde{z}} = 0$ for every $\widetilde{z} \in \widetilde{\mathrm{C}}$ and

$$\int_{\mathrm{C}} \frac{dz}{(\lambda_0 - z)(z - \widetilde{z})} = \frac{1}{\lambda_0 - \widetilde{z}} \int_{\mathrm{C}} \frac{dz}{z - \widetilde{z}} + \frac{1}{\lambda_0 - \widetilde{z}} \int_{\mathrm{C}} \frac{dz}{\lambda_0 - z} = -\frac{2\pi\mathrm{i}}{\lambda_0 - \widetilde{z}}$$

for every $\widetilde{z} \in \widetilde{\mathrm{C}}$.

Now, S maps M into M, and it maps N into N, since S and P commute. Hence

$$S_{|M,M} = O_{|M,M} \text{ and } S_{|N,N}(T - \lambda_0 I)_{|N,N} = I_{|N,N}.$$

In particular $\lambda_0 \in \mathrm{re}(T_{|N,N})$ and $R\left(T_{|N,N}, \lambda_0\right) = S_{|N,N}$.

(c) Let $\lambda_0 \in \mathrm{sp}(T) \setminus \Lambda$. As in part (b),

$$S(T - \lambda_0 I) = -\frac{1}{2\pi\mathrm{i}} \int_{\mathrm{C}} \frac{dz}{\lambda_0 - z} I + \frac{1}{2\pi\mathrm{i}} \int_{\mathrm{C}} R(z) \, dz.$$

But now $\lambda_0 \in \mathrm{ext}(\mathrm{C})$, so $\displaystyle\int_{\mathrm{C}} \frac{dz}{\lambda_0 - z} = 0$ and hence $S(T - \lambda_0 I) = -P$.

Next, by Corollary 1.22, there exists a Cauchy contour $\widetilde{\mathrm{C}}$ separating $\Lambda \cup \mathrm{C}$ from $\mathrm{sp}(T) \setminus \Lambda$. Using C in the integral defining S and $\widetilde{\mathrm{C}}$ in the integral defining P, and taking into account the First Resolvent Identity, we obtain, as in part (b),

$$SP = \Big(-\frac{1}{2\pi \mathrm{i}}\Big)^2 \int_{\mathrm{C}} \Big[\frac{R(z)}{\lambda_0 - z} \int_{\widetilde{\mathrm{C}}} \frac{d\widetilde{z}}{z - \widetilde{z}} \Big] \, dz$$

$$- \Big(-\frac{1}{2\pi \mathrm{i}}\Big)^2 \int_{\widetilde{\mathrm{C}}} \Big[R(\widetilde{z}) \int_{\mathrm{C}} \frac{dz}{(\lambda_0 - z)(z - \widetilde{z})} \Big] \, d\widetilde{z}.$$

Since $\lambda_0 \in \mathrm{sp}(T) \setminus \Lambda \subset \mathrm{ext}(\mathrm{C})$, $\displaystyle\int_{\mathrm{C}} \frac{dz}{(\lambda_0 - z)(z - \widetilde{z})} = 0$ and $\displaystyle\int_{\widetilde{\mathrm{C}}} \frac{d\widetilde{z}}{z - \widetilde{z}} = -2\pi\mathrm{i}$. Hence $SP = S$.

Again,

$$S_{|M,M}(T - \lambda_0 I)_{|M,M} = -I_{|M,M} \text{ and } S_{|N,N} = O_{|N,N}.$$

In particular $\lambda_0 \in \text{re}(T_{|M,M})$ and $R\left(T_{|M,M}, \lambda_0\right) = -S_{|M,M}$. ∎

Remark 1.25

Let T, Λ, λ_0, P and S be as in Proposition 1.24, and assume that $\lambda_0 \in \Lambda$. Part (b) of Proposition 1.24 suggests a method of finding Sy for a given $y \in X$. In fact, $Sy = x$ if and only if x satisfies the equations $(T - \lambda_0 I)x = y - Py$ and $Px = 0$. This method will be useful later in computing successive iterates of certain refinement schemes. ∎

Theorem 1.26 Spectral Decomposition Theorem:

Let $T \in \text{BL}(X)$, Λ be a spectral set for T, $P := P(T, \Lambda)$, $M := \mathcal{R}(P)$ and $N := \mathcal{N}(P)$. Then T is decomposed by (M, N),

$$\text{sp}(T_{|M,M}) = \Lambda \quad \textit{and} \quad \text{sp}(T_{|N,N}) = \text{sp}(T) \setminus \Lambda.$$

In particular, $\text{sp}(T_{|M,M}) \cap \text{sp}(T_{|N,N}) = \emptyset$.

Proof

Since T and P commute, the pair (M, N) decomposes T. By Proposition 1.18, we obtain

$$\text{sp}(T) = \text{sp}(T_{|M,M}) \cup \text{sp}(T_{|N,N}).$$

By Proposition 1.24, we have

$$\Lambda \subset \text{re}(T_{|N,N}) \text{ and } \text{sp}(T) \setminus \Lambda \subset \text{re}(T_{|M,M}).$$

Considering the complements of the preceding sets in $\mathbb{C}$ and noting that $\text{sp}(T_{|N,N})$ as well as $\text{sp}(T_{|M,M})$ are subsets of $\text{sp}(T)$, we obtain

$$\text{sp}(T_{|N,N}) \subset \text{sp}(T) \setminus \Lambda \quad \text{and} \quad \text{sp}(T_{|M,M}) \subset \Lambda.$$

Again considering the complements of the preceding sets in $\text{sp}(T)$, we see that $\Lambda \subset \text{sp}(T) \setminus \text{sp}(T_{|N,N})$ and $\text{sp}(T) \setminus \Lambda \subset \text{sp}(T) \setminus \text{sp}(T_{|M,M})$. Thus $\Lambda \subset \text{sp}(T_{|M,M})$ and $\text{sp}(T) \setminus \Lambda \subset \text{sp}(T_{|N,N})$. ∎

Corollary 1.27

Let $T \in \text{BL}(X)$, Λ be a spectral set for T and $P := P(T, \Lambda)$. Then

$$P = O \quad \textit{if and only if} \quad \Lambda = \emptyset,$$
$$P = I \quad \textit{if and only if} \quad \Lambda = \text{sp}(T).$$

Proof
Let $\mathrm{C} \in \mathcal{C}(T, \Lambda)$. If $\Lambda = \emptyset$, then the function $z \mapsto R(z)$ is analytic in $\mathrm{int}(\mathrm{C})$, so $P = O$ by part (a) of Cauchy's Theorem. Conversely, if $P = O$, then $M := \mathcal{R}(P) = \{0\}$ and so $\Lambda = \mathrm{sp}(T_{|M,M}) = \emptyset$. Next, if $\Lambda = \mathrm{sp}(T)$ and we let $N := \mathcal{N}(P)$, then $\mathrm{sp}(T_{|N,N}) = \mathrm{sp}(T) \setminus \Lambda = \emptyset$, so $N = \{0\}$ by the Gelfand-Mazur Theorem (Proposition 1.15(a)) and hence $P = I$. Conversely, if $P = I$, then $N = \{0\}$ and $\mathrm{sp}(T) \setminus \Lambda = \mathrm{sp}(T_{|N,N}) = \emptyset$, that is, $\Lambda = \mathrm{sp}(T)$. ▮

The projection $P := P(T, \Lambda)$ is called the **spectral projection** associated with T and Λ, and the range of P is known as the **spectral subspace** associated with T and Λ. This subspace will be denoted by $M(T, \Lambda)$, or simply by M if no confusion is likely to arise. We now prove that it is the largest closed subspace of X which is invariant under T such that the spectrum of the restricted operator is contained in Λ.

Proposition 1.28
Let Λ be a spectral set for T and $P := P(T, \Lambda)$. If Y is a closed subspace of X such that $T(Y) \subset Y$ and $\mathrm{sp}(T_{|Y,Y}) \subset \Lambda$, then $Y \subset \mathcal{R}(P)$.

Proof
Let $\mathrm{C} \in \mathcal{C}(T, \Lambda)$. Since $\mathrm{sp}(T_{|Y,Y}) \subset \Lambda \subset \mathrm{int}(\mathrm{C})$, we have $\mathrm{C} \subset \mathrm{re}(T_{|Y,Y})$. As in the proof of Proposition 1.18, $R(T_{|Y,Y}, z) = R(T, z)_{|Y,Y}$ for $z \in \mathrm{C}$. As a consequence, by Corollary 1.27,

$$\begin{aligned} P_{|Y,Y} &= \left(-\frac{1}{2\pi\mathrm{i}} \int_{\mathrm{C}} R(T, z)\, dz\right)_{|Y,Y} = -\frac{1}{2\pi\mathrm{i}} \int_{\mathrm{C}} R(T, z)_{|Y,Y}\, dz \\ &= -\frac{1}{2\pi\mathrm{i}} \int_{\mathrm{C}} R(T_{|Y,Y}, z)\, dz = I_{|Y,Y}. \end{aligned}$$

Hence $y = Py$ for all $y \in Y$. This shows that $Y \subset \mathcal{R}(P)$. ▮

Let Λ be a spectral set for $T \in \mathrm{BL}(X)$. The definition of the spectral projection associated with T and Λ, namely

$$P(T, \Lambda) = -\frac{1}{2\pi\mathrm{i}} \int_{\mathrm{C}} R(T, z)\, dz,$$

where $\mathrm{C} \in \mathcal{C}(T, \Lambda)$, is admittedly mysterious! The following result provides a motivation for this definition. If a pair (Y, Z) of closed subspaces of X decomposes T and $\text{sp}(T_{|Y,Y}) \cap \text{sp}(T_{|Z,Z}) = \emptyset$, then the projection P onto Y along Z must equal $P(T, \Lambda)$, where $\Lambda := \text{sp}(T_{|Y,Y})$. (See Exercise 1.6.)

Finally, we note that if λ is an isolated point of $\text{sp}(T)$, then $\{\lambda\}$ is a spectral set for T, and the operator

$$S(T, \lambda) := S(T, \{\lambda\}, \lambda) = -\frac{1}{2\pi \mathrm{i}} \int_{\mathrm{C}} R(T, z) \frac{dz}{\lambda - z}$$

does not depend on $\mathrm{C} \in \mathcal{C}(T, \{\lambda\})$. $S(T, \lambda)$ is known as the **reduced resolvent operator** of T associated with λ. The reason for this name is that if $N := \mathcal{N}(P(T, \{\lambda\}))$, then

$$S(T, \lambda)_{|N,N} = R(T_{|N,N}, \lambda) = (T_{|N,N} - \lambda I_{|N,N})^{-1},$$

as can be seen by letting $\lambda_0 = \lambda$ in Proposition 1.24(b).

We conclude this section with a result which will be of use on several occasions.

Proposition 1.29
Let X and Y be Banach spaces over $\mathbb{C}$, $T \in \text{BL}(X)$ and $U \in \text{BL}(Y)$. Consider $E \subset \mathbb{C}$ such that $\text{sp}(T) \setminus E = \text{sp}(U) \setminus E$, and let $\Lambda \subset \mathbb{C}$ satisfy $\Lambda \cap E = \emptyset$. Then Λ is a spectral set for T if and only if Λ is a spectral set for U. In this case, there is some $\mathrm{C} \in \mathcal{C}(T, \Lambda)$ such that $E \subset \text{ext}(\mathrm{C})$, and every such C belongs to $\mathcal{C}(U, \Lambda)$ as well.

Proof
Let Λ be a spectral set for T. Then Λ is a closed subset of $\text{sp}(T)$, and $\text{sp}(T) \setminus \Lambda$ is also a closed set. Since $\Lambda \cap E = \emptyset$ and $\text{sp}(T) \setminus E \subset \text{sp}(U) \setminus E$, it follows that Λ is a closed subset of $\text{sp}(U)$. To show that $\text{sp}(U) \setminus \Lambda$ is also a closed set, let (μ_n) be a sequence in $\text{sp}(U) \setminus \Lambda$ such that $\mu_n \to \mu$ in $\mathbb{C}$. As $\mu_n \in \text{sp}(U)$ for each n, $\mu_n \to \mu$ and $\text{sp}(U)$ is a closed set, we see that $\mu \in \text{sp}(U)$. We prove that $\mu \notin \Lambda$. If $\mu \in E$, then clearly $\mu \notin \Lambda$ since $\Lambda \cap E = \emptyset$. Next, assume that $\mu \notin E$. Since E is a closed set, there is some $\epsilon > 0$ such that $\{z \in \mathbb{C} : |z - \mu| < \epsilon\} \cap E = \emptyset$. Also, as $\mu_n \to \mu$, there is a positive integer n_0 such that $|\mu_n - \mu| < \epsilon$ and hence $\mu_n \notin E$ for all $n > n_0$. Thus $\mu_n \in \text{sp}(U) \setminus E = \text{sp}(T) \setminus E$ and so $\mu_n \in \text{sp}(T) \setminus \Lambda$ for all $n > n_0$. Since $\text{sp}(T) \setminus \Lambda$ is a closed set and $\mu_n \to \mu$, it follows that $\mu \in \text{sp}(T) \setminus \Lambda$. In particular, $\mu \notin \Lambda$ as desired. This shows that Λ is a spectral set for U. The converse follows by interchanging the roles of T and U.

Let now Λ be a spectral set for T (and hence for U). Letting $E_1 := \Lambda$ and $E_2 := E \cup (\mathrm{sp}(T) \setminus \Lambda)$ in Corollary 1.22, we find a Cauchy contour C which separates Λ from $E \cup (\mathrm{sp}(T) \setminus \Lambda)$. As $\Lambda \subset \mathrm{int}(\mathrm{C})$ and $E \cup (\mathrm{sp}(T) \setminus \Lambda) \subset \mathrm{ext}(\mathrm{C})$, we see that $\mathrm{C} \in \mathcal{C}(T, \Lambda)$ and $E \subset \mathrm{ext}(\mathrm{C})$. Finally, since $E_2 = E \cup (\mathrm{sp}(U) \setminus \Lambda)$, we obtain $\mathrm{C} \in \mathcal{C}(U, \Lambda)$ as well. ∎

1.3 Spectral Sets of Finite Type

Suppose that λ is an isolated point of $\mathrm{sp}(T)$. Clearly, $\Lambda := \{\lambda\}$ is a spectral set for T. Let $P := P(T, \{\lambda\})$.

Three questions arise naturally:

1. Must λ be an eigenvalue of T?
2. Can we relate $M := \mathcal{R}(P)$ to the null space of $T - \lambda I$ or to the null space of some power of $T - \lambda I$?
3. Is the invariant subspace M finite dimensional?

Matrix arguments allow us to give an affirmative answer to all of these questions if X is finite dimensional. In the general case, the answers are not always affirmative.

Let $T \in \mathrm{BL}(X)$. If there exists a positive integer n such that $T^n = O$, then T is called a **nilpotent** operator of degree n. If $\rho(T) = 0$, then T is called a **quasinilpotent** operator. By the Spectral Radius Formula (Proposition 1.15(b)), each nilpotent operator is quasinilpotent. The following example shows that the converse does not hold.

Example 1.30 In this example, $\mathrm{sp}(T) = \{0\}$ but 0 is not an eigenvalue of T:
Let $X := C^0([0,1])$ and $T : X \to X$ be defined by

$$(Tx)(s) := \int_0^s x(t)\,dt, \quad x \in X, \quad s \in [0,1].$$

Clearly, T is a linear operator. Let $x \in X$. Then $|(Tx)(s)| \leq s\|x\|_\infty$ for $s \in [0,1]$. Let us suppose that for a positive integer n, $|(T^n x)(s)| \leq$

$\dfrac{\|x\|_\infty}{n!} s^n$ for $s \in [0,1]$. Then

$$|(T^{n+1}x)(s)| \le \frac{\|x\|_\infty}{n!} \int_0^s t^n \, dt = \frac{\|x\|_\infty}{(n+1)!} s^{n+1} \qquad \text{for } s \in [0,1].$$

This proves that for all positive integers n and all $x \in X$, $\|T^n x\|_\infty \le \dfrac{\|x\|_\infty}{n!}$ and hence $\|T^n\| \le \dfrac{1}{n!}$. So $\rho(T) = \lim_{n\to\infty} \|T^n\|^{1/n} \le \lim_{n\to\infty} \dfrac{1}{\sqrt[n]{n!}} = 0$. This implies that $\mathrm{sp}(T) = \{0\}$. Although 0 is the only spectral value of T, it is not an eigenvalue since for $x \in X$, $(Tx)'(s) = x(s)$ for $s \in [0,1]$. Thus $Tx = 0$ implies that $x = 0$. This shows that T is quasinilpotent but not nilpotent. By Corollary 1.27, $P := P(T, \{0\}) = I$ and hence $M := \mathcal{R}(P) = C^0([0,1])$. Note that M is infinite dimensional and $\mathcal{N}(T^n) = \{0\}$ for $n = 1, 2, \ldots$ ∎

Let λ be an isolated point of $\mathrm{sp}(T)$. The dimension of the spectral subspace $M(T, \{\lambda\})$ associated with T and λ is called the **algebraic multiplicity** of λ, and we will denote it by $m(T, \lambda)$ or more simply by m when the context is clear. If $m(T, \lambda)$ is finite, then λ is called a **spectral value of finite type**. We have taken this terminology from [38].

Proposition 1.31
Let λ be an isolated point of $\mathrm{sp}(T)$ and denote by P the corresponding spectral projection. Let $M := \mathcal{R}(P)$ and $N := \mathcal{N}(P)$. Then

(a) $\bigcup_{j=1}^{\infty} \mathcal{N}\left[(T-\lambda I)^j\right] \subset M$ *and* $N \subset \bigcap_{j=1}^{\infty} \mathcal{R}\left[(T-\lambda I)^j\right]$,

(b) *if λ is a spectral value of finite type with algebraic multiplicity m, then*

$$M = \mathcal{N}\left[(T-\lambda I)^m\right], \quad N = \mathcal{R}\left[(T-\lambda I)^m\right]$$

and

$$X = \mathcal{N}\left[(T-\lambda I)^m\right] \oplus \mathcal{R}\left[(T-\lambda I)^m\right].$$

Further, λ is an eigenvalue of T.

Proof

(a) Let $x \in \mathcal{N}\left[(T-\lambda I)^j\right]$ for some $j \ge 1$. Since T and P commute, we have $(T-\lambda I)^j(x - Px) = 0$. Now $(I-P)x \in N$ and $(T-\lambda I)^j_{|N,N} =$

$\left[(T-\lambda I)_{|N,N}\right]^j$ is invertible by Proposition 1.24(b). Hence $(I-P)x = 0$, that is, $x \in M$. Thus $\bigcup_{j=1}^{\infty} \mathcal{N}\left[(T-\lambda I)^j\right] \subset M$.

Next, let $x \in N$ and $j \geq 1$. Again by the invertibility of $\left[(T - \lambda I)_{|N,N}\right]^j$, there is some $u \in N$ such that

$$x = \left[(T-\lambda I)_{|N,N}\right]^j u = (T-\lambda I)^j u.$$

Hence $x \in \mathcal{R}\left[(T-\lambda I)^j\right]$. Thus $N \subset \bigcap_{j=1}^{\infty} \mathcal{N}\left[(T-\lambda I)^j\right]$.

(b) Let the algebraic multiplicity of λ be $m < \infty$. By Theorem 1.2, the operator $T_{|M,M} : M \to M$ can be represented with respect to an ordered basis for M by an upper triangular $m \times m$ matrix whose only eigenvalue is λ. Hence $\left[(T-\lambda I)_{|M,M}\right]^m = O_{|M,M}$. So $M \subset \mathcal{N}\left[\left((T-\lambda I)_{|M,M}\right)^m\right] \subset \mathcal{N}\left[(T-\lambda I)^m\right]$.

Next, let $x \in \mathcal{R}\left[(T-\lambda I)^m\right]$. Then $x = \left[(T-\lambda I)\right]^m u$ for some $u \in X$. Since P commutes with T and $Pu \in M$, we have

$$Px = P(T-\lambda I)^m u = (T-\lambda I)^m Pu = 0,$$

that is, $x \in \mathcal{N}(P) = N$. Thus $\mathcal{R}\left[(T-\lambda I)^m\right] \subset N$. Now part (a) shows that

$$M = \mathcal{N}\left[(T-\lambda I)^m\right] \quad \text{and} \quad N = \mathcal{R}\left[(T-\lambda I)^m\right].$$

Hence $X = M \oplus N = \mathcal{N}\left[(T-\lambda I)^m\right] \oplus \mathcal{R}\left[(T-\lambda I)^m\right]$.

Finally, since $\lambda \in \mathrm{sp}(T_{|M,M})$ and M is finite dimensional, we see that λ is an eigenvalue of $T_{|M,M}$ and hence of T. ∎

Let $\lambda \in \mathbb{C}$ and $\varphi \in X$ be such that

$$(T-\lambda I)^{k-1}\varphi \neq 0 \quad \text{and} \quad (T-\lambda I)^k \varphi = 0$$

for some positive integer k. Then $\psi := (T-\lambda I)^{k-1}\varphi$ satisfies

$$\psi \neq 0 \quad \text{and} \quad T\psi = \lambda\psi,$$

so that λ is an eigenvalue of T and ψ a corresponding eigenvector. In this case, we say that φ is a **generalized eigenvector** of T corresponding to the eigenvalue λ. The integer k is called the **grade of the generalized eigenvector** φ. Eigenvectors themselves are generalized eigenvectors of grade 1.

The preceding proposition shows that if λ is a spectral value of T of finite type, then λ is an eigenvalue of T and the corresponding spectral subspace $M(T, \{\lambda\})$ consists of all generalized eigenvectors of T corresponding to λ together with the null vector of X.

Also, since $\mathcal{N}(T - \lambda I) \subset M$, the geometric multiplicity g of λ is less than or equal to its algebraic multiplicity m. Further, we have

$$\{0\} \subsetneq \mathcal{N}(T - \lambda I) \subset \mathcal{N}\left[(T - \lambda I)^2\right] \subset \cdots$$

$$\cdots \subset \mathcal{N}\left[(T - \lambda I)^m\right] = \mathcal{N}\left[(T - \lambda I)^{m+1}\right].$$

If we let ℓ be the smallest positive integer such that

$$\mathcal{N}\left[(T - \lambda I)^\ell\right] = \mathcal{N}\left[(T - \lambda I)^{\ell+1}\right],$$

then ℓ is called the **ascent** of the eigenvalue λ. Clearly, $\ell \leq m$. Note that the highest grade of a generalized eigenvector of T corresponding to λ is ℓ. If $m = 1$, λ is called a **simple eigenvalue.** If $\ell = 1$, then $g = m$ and λ is called a **semisimple** eigenvalue. If $\ell > 1$ then $g < m$ and λ is called a **defective** eigenvalue.

For relationships between the integers m, g and ℓ, see Exercise 1.16.

Since we shall be interested in clusters of spectral values of finite type, we introduce the following concept. If Λ is a nonempty spectral set for T and the corresponding spectral subspace is finite dimensional, then Λ is called a **spectral set of finite type.**

Theorem 1.32

Let Λ be a nonempty spectral set for T and denote by P the corresponding spectral projection. Then Λ is a spectral set of finite type for T if and only if Λ consists of a finite number of spectral values of T, each of which is of finite type. If $\Lambda = \{\lambda_1, \ldots, \lambda_r\}$ and P_j denotes the spectral projection corresponding to T and λ_j for $j = 1, \ldots, r$, then

$$P = P_1 + \cdots + P_r, \quad P_i P_j = O \quad \textit{if} \quad i \neq j$$

and $\text{rank}(P) = m_1 + \cdots + m_r$, *where m_j is the algebraic multiplicity of λ_j, $j = 1, \ldots, r$.*

Proof

Let $M := \mathcal{R}(P)$. By the Spectral Decomposition Theorem,

$$\text{sp}(T_{|M,M}) = \Lambda.$$

Assume first that $\text{rank}(P) = \dim M := m < \infty$. Now the spectrum of $T_{|M,M}$ consists of a finite number of eigenvalues. Thus $\Lambda = \{\lambda_1, \ldots, \lambda_r\}$ for some positive integer r. Let $\mathrm{C} \in \mathcal{C}(T, \Lambda)$ and for $j = 1, \ldots, r$, let Γ_j denote a Jordan curve such that $\text{sp}(T) \cap \text{int}(\Gamma_j) = \{\lambda_j\}$ and $\text{int}(\Gamma_j) \cup \Gamma_j \subset \text{ext}(\Gamma_i)$ for every $i = 1, \ldots, r$, $i \neq j$. Then by Cauchy's theorem,

$$P := -\frac{1}{2\pi \mathrm{i}} \int_{\mathrm{C}} R(z)\, dz = -\frac{1}{2\pi \mathrm{i}} \sum_{j=1}^{r} \int_{\Gamma_j} R(z)\, dz = \sum_{j=1}^{r} P_j.$$

Also, if $i \neq j$, then by the First Resolvent Identity, we have

$$\begin{aligned} P_i P_j &= -\frac{1}{2\pi \mathrm{i}} \int_{\Gamma_i} R(z) P_j dz \\ &= -\frac{1}{2\pi \mathrm{i}} \int_{\Gamma_i} R(z) \Big[-\frac{1}{2\pi \mathrm{i}} \int_{\Gamma_j} R(w) dw \Big] dz \\ &= \Big(-\frac{1}{2\pi \mathrm{i}} \Big)^2 \Big\{ \int_{\Gamma_i} \Big[\int_{\Gamma_j} [R(z) - R(w)] \frac{dw}{z - w} \Big] dz \Big\} \\ &= \Big(-\frac{1}{2\pi \mathrm{i}} \Big)^2 \Big\{ \int_{\Gamma_i} R(z) \Big[\int_{\Gamma_j} \frac{dw}{z - w} \Big] dz \\ &\quad - \int_{\Gamma_j} R(w) \Big[\int_{\Gamma_i} \frac{dz}{z - w} \Big] dw \Big\}. \end{aligned}$$

But since Γ_i lies in $\text{ext}(\Gamma_j)$ and Γ_j lies in $\text{ext}(\Gamma_i)$, we have

$$\int_{\Gamma_j} \frac{dw}{z - w} = 0 \text{ for all } z \in \Gamma_i \quad \text{and} \quad \int_{\Gamma_i} \frac{dz}{z - w} = 0 \text{ for all } w \in \Gamma_j.$$

Thus $P_i P_j = O$ if $i \neq j$, and

$$\mathcal{R}(P) = \mathcal{R}(P_1) \oplus \cdots \oplus \mathcal{R}(P_r).$$

Hence $\text{rank}(P) = \sum_{j=1}^{r} \text{rank}(P_j)$. In particular, $\text{rank}(P_j)$ is finite for each $j = 1, \ldots, r$, showing that each λ_j is a spectral value of T of finite type, and if $m_j := \text{rank}(P_j)$ is its algebraic multiplicity, then $\text{rank}(P) = m_1 + \cdots + m_r$.

Conversely, if $\Lambda = \{\lambda_1, \ldots, \lambda_r\}$, where each λ_j is a spectral value of T of finite type, then the argument given above shows that $\text{rank}(P) = \sum_{j=1}^{r} \text{rank}(P_j)$, which is finite since each $\text{rank}(P_j)$ is finite. ∎

If X is finite dimensional and $T \in \mathrm{BL}(X)$, then every subset of $\mathrm{sp}(T)$ is a spectral set of finite type for T. Let an $n\times n$ matrix A represent T with respect to some basis $\mathcal{B}$ for X. Thus for $z \in \mathrm{re}(T)$, the matrix $(\mathsf{A} - z\mathsf{I})^{-1}$ represents $R(T, z)$ with respect to $\mathcal{B}$. Also, if $\Lambda \subset \mathrm{sp}(T)$ and $\mathrm{C} \in \mathcal{C}(T, \Lambda)$, then the matrix $-\dfrac{1}{2\pi \mathrm{i}}\displaystyle\int_{\mathrm{C}} (\mathsf{A} - z\mathsf{I})^{-1}\, dz$ represents $P(T, \Lambda)$ with respect to $\mathcal{B}$. We therefore, agree to say that the set of coordinate matrices of elements in $P(T, \Lambda)$ with respect to $\mathcal{B}$ constitutes the spectral subspace associated with the matrix A and the spectral set Λ. (Compare our comments about the eigenvalues, eigenvectors and eigenspaces of a matrix A in Remark 1.1.)

Remark 1.33 An upper triangular block diagonal form of a square matrix:
Let X be an n dimensional linear space and $T : X \to X$ be a linear operator. Let $\mathsf{A} \in \mathbb{C}^{n\times n}$ be the matrix representing T with respect to an ordered basis $\mathcal{B}$ of X. We keep the notation of Theorem 1.32. Let $\Lambda := \mathrm{sp}(\mathsf{A})$. By Corollary 1.27, $P(T, \Lambda) = I$. For each eigenvalue λ_j of A, let $P_j := P(T, \{\lambda_j\})$. Then

$$P_1 + \cdots + P_r = I \quad \text{and} \quad \mathcal{R}(P_1) \oplus \cdots \oplus \mathcal{R}(P_r) = X.$$

Let $\mathsf{U}_j \in \mathbb{C}^{n\times m_j}$ be a matrix whose columns contain the coordinates relative to $\mathcal{B}$ of an ordered basis for $\mathcal{R}(P_j)$, $j = 1, \ldots, r$. Since the spectral subspace $\mathcal{R}(P_j)$ is invariant under T and $\mathrm{sp}(T_{|\mathcal{R}(P_j),\mathcal{R}(P_j)}) = \{\lambda_j\}$, there exists $\mathsf{A}_j \in \mathbb{C}^{m_j\times m_j}$ such that

$$\mathsf{A}\mathsf{U}_j = \mathsf{U}_j\mathsf{A}_j, \quad \mathrm{sp}(\mathsf{A}_j) = \{\lambda_j\}, \quad j = 1, \ldots, r.$$

In fact, A_j is the matrix which represents $T_{|\mathcal{R}(P_j),\mathcal{R}(P_j)} : \mathcal{R}(P_j) \to \mathcal{R}(P_j)$ with respect to the ordered basis of $\mathcal{R}(P_j)$, whose coordinates relative to $\mathcal{B}$ are given by the columns of U_j.

By Schur's Theorem (1.2), there exists a unitary matrix $\mathsf{Q}_j \in \mathbb{C}^{m_j\times m_j}$ such that $\mathsf{T}_j := \mathsf{Q}_j^*\mathsf{A}_j\mathsf{Q}_j$ is an upper triangular matrix. Then

$$\mathsf{T}_j = \lambda_j\, \mathsf{I}_{m_j} + \mathsf{N}_j, \quad j = 1, \ldots, r,$$

where $\mathsf{N}_j \in \mathbb{C}^{m_j\times m_j}$ is strictly upper triangular. In fact, N_j is a nilpotent matrix of degree ℓ_j, the ascent of the eigenvalue λ_j of A.

Let $\mathsf{U} := [\mathsf{U}_1, \ldots, \mathsf{U}_r] \in \mathbb{C}^{n\times n}$ and $\mathsf{Q} := \mathrm{diag}\,[\mathsf{Q}_1, \ldots, \mathsf{Q}_r] \in \mathbb{C}^{n\times n}$. Then U is a nonsingular matrix, Q is a unitary matrix and

$$\mathsf{T} := (\mathsf{U}\mathsf{Q})^{-1}\, \mathsf{A}\, (\mathsf{U}\mathsf{Q}) = \mathrm{diag}\,[\mathsf{T}_1, \ldots, \mathsf{T}_r],$$

since

$$\begin{aligned}\mathsf{A}(\mathsf{UQ}) &= [\mathsf{AU}_1, \ldots, \mathsf{AU}_r]\,\mathrm{diag}\,[\mathsf{Q}_1, \ldots, \mathsf{Q}_r] \\ &= [\mathsf{U}_1\mathsf{A}_1, \ldots, \mathsf{U}_r\mathsf{A}_r]\,\mathrm{diag}\,[\mathsf{Q}_1, \ldots, \mathsf{Q}_r] \\ &= [\mathsf{U}_1\mathsf{A}_1\mathsf{Q}_1, \ldots, \mathsf{U}_r\mathsf{A}_r\mathsf{Q}_r] \\ &= [\mathsf{U}_1\mathsf{Q}_1\mathsf{T}_1, \ldots, \mathsf{U}_r\mathsf{Q}_r\mathsf{T}_r] \\ &= [\mathsf{U}_1\mathsf{Q}_1, \ldots, \mathsf{U}_r\mathsf{Q}_r]\,\mathrm{diag}\,[\mathsf{T}_1, \ldots, \mathsf{T}_r] \\ &= \mathsf{UQ}\,\mathrm{diag}\,[\mathsf{T}_1, \ldots, \mathsf{T}_r] \\ &= \mathsf{UQT}.\end{aligned}$$

Thus T is an upper triangular block diagonal matrix which is similar to A, and for $j = 1, \ldots, n$, all the diagonal entries of the j th $m_j \times m_j$ diagonal block are equal to λ_j. It follows that the characteristic polynomial of A is

$$\mathrm{p}(z) = \prod_{j=1}^{r} (\lambda_j - z)^{m_j}.$$

Hence the algebraic multiplicity m_j of λ_j as an eigenvalue of T is equal to the multiplicity of λ_j as a root of the characteristic polynomial of A. As a consequence, we deduce the **Cayley-Hamilton Theorem**:

$$\begin{aligned}\mathrm{p}(\mathsf{A}) &:= (\lambda_1\,\mathsf{I}_n - \mathsf{A})^{m_1} \cdots (\lambda_r\,\mathsf{I}_n - \mathsf{A})^{m_r} \\ &= \mathsf{UQ}(\lambda_1\,\mathsf{I}_n - \mathsf{T})^{m_1} \cdots (\lambda_r\,\mathsf{I}_n - \mathsf{T})^{m_r}(\mathsf{UQ})^{-1} = \mathsf{O}.\end{aligned}$$

Let the characteristic polynomial of A be written as

$$\mathrm{p}(z) = a_0 + \sum_{i=1}^{n} a_i z^i.$$

If A is nonsingular, then $0 \neq \det \mathsf{A} = a_0$ and hence

$$\Big[-\frac{1}{a_0}\Big(a_1\mathsf{I} + \sum_{i=1}^{n-1} a_{i+1}\mathsf{A}^i\Big)\Big]\mathsf{A} = \mathsf{I}_n = \mathsf{A}\Big[-\frac{1}{a_0}\Big(a_1\mathsf{I} + \sum_{i=1}^{n-1} a_{i+1}\mathsf{A}^i\Big)\Big].$$

This shows that

$$\mathsf{A}^{-1} = -\frac{1}{a_0}\Big(a_1\mathsf{I} + \sum_{i=1}^{n-1} a_{i+1}\mathsf{A}^i\Big),$$

which is a polynomial in A of degree less than n.

There is another upper triangular block diagonal form similar to A called the Jordan canonical form of A. This form will not be used in this book. The main reason for this is that, in general, the Jordan canonical form J(A) of a matrix A is not a continuous function of A. (See Section 3.1 in [43] and Exercise 1.17.) ∎

Remark 1.34 Compact operators:
Important examples of spectral values of finite type are obtained by considering a class of operators which we now describe.

Let $T : X \to X$ be a linear operator. Then T is called a compact operator if the set $E := \{Tx \in X : x \in X, \|x\| \leq 1\}$ is relatively compact, that is, its closure is a compact subset of X. Hence, if T is compact, then $T \in \mathrm{BL}(X)$. The set of all compact operators is a closed subset of $\mathrm{BL}(X)$ and every bounded finite rank operator is compact. (See 17.2(c) and 17.1(c) of [55] or Theorems 8.1-5 and 8.1-5 of [48].) If $T \in \mathrm{BL}(X)$ is compact and $A \in \mathrm{BL}(X)$, then AT and TA are compact. Also, for each nonzero $z \in \mathbb{C}$, $\mathcal{R}(T - zI)$ is a closed subspace of X.

A projection $P \in \mathrm{BL}(X)$ is compact if and only if $\mathrm{rank}(P)$ is finite, since the restriction of P to $\mathcal{R}(P)$ is the identity operator on $\mathcal{R}(P)$ and the closed unit ball of a normed space is compact if and only if the normed space is finite dimensional. (See 5.5 of [55] or Lemma 8.1-2 of [48].)

Let $T \in \mathrm{BL}(X)$ be a compact operator. Then $\mathrm{sp}(T)$ is a countable (that is, a finite or a denumerable) set having 0 as the only possible limit point. In particular, each nonzero spectral value λ of T is an isolated point of $\mathrm{sp}(T)$. (See 18.5(a) of [55] or Section 8.3 of [48].)

Let Λ be a spectral set for T not containing zero. Since $\{0\} \cup (\mathrm{sp}(T) \setminus \Lambda)$ and Λ are disjoint compact sets, there is, by Corollary 1.22, a Cauchy contour C separating Λ from $\{0\} \cup \mathrm{sp}(T) \setminus \Lambda$. Clearly $\mathrm{C} \in \mathcal{C}(T, \Lambda)$ and $0 \in \mathrm{ext}(\mathrm{C})$. Since $R(T, z) = \dfrac{1}{z}\Big[T\,R(T, z) - I\Big]$ for all nonzero $z \in \mathrm{re}(T)$, we have

$$P(T, \Lambda) = -\frac{1}{2\pi \mathrm{i}} \int_{\mathrm{C}} \Big[T\,R(T, z) - I\Big] \frac{dz}{z} = -\frac{1}{2\pi \mathrm{i}}\, T \int_{\mathrm{C}} \frac{R(T, z)}{z}\, dz,$$

because $\displaystyle\int_{\mathrm{C}} \frac{dz}{z} = 0$. This proves that $P(T, \Lambda)$ is a compact projection and hence its rank is finite. Thus every spectral set for T not containing 0 is of finite type. In particular, each nonzero spectral value of T is of finite type and, by Proposition 1.31, it is an eigenvalue of T.

On the other hand, if X is infinite dimensional and Λ is a spectral set for T containing 0, then $\text{rank}(P) = \infty$. (See Exercise 1.15.) ∎

1.4 Adjoint and Product Spaces

1.4.1 Adjoint Space

Let X be a complex Banach space. We say that $f : X \to \mathbb{C}$ is a **conjugate-linear functional** if for all $x, y \in X$ and $c \in \mathbb{C}$,

$$f(cx + y) = \overline{c}f(x) + f(y).$$

The set of all conjugate-linear continuous functionals on X is called the **adjoint space** of X, and it will be denoted by X^*. It is easy to show that X^* is a complex linear space. We shall use the notation

$$\langle x\,, f\rangle := \overline{f(x)}, \quad x \in X,\ f \in X^*.$$

Since the continuity of $f \in X^*$ on X is equivalent to its continuity at 0, and since this is equivalent to the boundedness of f on the closed unit ball of X, we define the following norm on X^*:

$$\|f\| := \sup\{|\langle x\,, f\rangle| : \|x\| \le 1\}.$$

This definition leads to the inequality

$$|\langle x\,, f\rangle| \le \|x\|\,\|f\|, \quad x \in X,\ f \in X^*.$$

Thus the function $\langle\cdot\,, \cdot\rangle : X \times X^* \to \mathbb{C}$ is linear with respect to its first argument, conjugate-linear with respect to its second argument, and continuous on $X \times X^*$. Since $\mathbb{C}$ is complete, X^* is a complex Banach space.

Let Y be a Banach space and $K \in \text{BL}(X, Y)$, the space of all bounded linear maps from X to Y. We define the **adjoint operator** of K to be the operator $K^* : Y^* \to X^*$ given by

$$K^*f := fK \quad \text{for all } f \in Y^*.$$

Then

$$\langle x\,, K^*f\rangle = \langle Kx\,, f\rangle \quad \text{for all } x \in X \text{ and } f \in Y^*.$$

Given a nonempty subset E of X, we define its annihilator to be the following subset of X^*:

$$E^\perp := \{f \in X^* : \langle x\,,\, f\rangle = 0 \text{ for all } x \in E\}.$$

Remark 1.35
The following properties of the adjoint operator can be verified easily:

(i) If $K \in \mathrm{BL}(X, Y)$, then $K^* \in \mathrm{BL}(Y^*, X^*)$. In fact, it can be easily seen that $\|K^*\| \le \|K\|$. The inequality $\|K\| \le \|K^*\|$ can be proved by using the Hahn-Banach Extension Theorem. Hence $\|K^*\| = \|K\|$.

(ii) I^* is the identity operator on X^*.

(iii) If K_1, $K_2 \in \mathrm{BL}(X, Y)$, then for all $c \in \mathbb{C}$, $(cK_1 + K_2)^* = \overline{c}K_1^* + K_2^*$.

(iv) If $K \in \mathrm{BL}(X, Y)$ and $L \in \mathrm{BL}(Y, X)$, then $(KL)^* = L^*K^*$.

(v) If $E \subset X$, then $E^\perp$ is a closed subspace of X^*.

(vi) If $K \in \mathrm{BL}(X, Y)$, then $\mathcal{N}(K^*) = \mathcal{R}(K)^\perp$ and $\mathcal{R}(K^*) \subset \mathcal{N}(K)^\perp$.

(vii) $K^* \in \mathrm{BL}(X^*)$ is compact whenever $K \in \mathrm{BL}(X)$ is compact. The converse also holds.

The proofs of the first five assertions can be found in Section 13 of [55] or Section 4.5 of [48], and of assertion (vii) in 17.3 of [55] or Theorem 8.2-5 of [48]. ∎

The following results concern spectral values of T and of T^*:

Proposition 1.36
Let $T \in \mathrm{BL}(X)$.

(a) *A complex number z belongs to $\mathrm{re}(T)$ if and only if its conjugate $\overline{z}$ belongs to $\mathrm{re}(T^*)$, and then $R(T^*, \overline{z}) = R(T, z)^*$.*

(b) *A subset Λ of $\mathbb{C}$ is a spectral set for T if and only if $\overline{\Lambda} = \{\overline{\lambda} : \lambda \in \Lambda\}$ is a spectral set for T^*, and then $P(T^*, \overline{\Lambda}) = P(T, \Lambda)^*$.*

Proof

(a) Let $z \in \mathrm{re}(T)$. Taking adjoints of both sides of

$$(T - zI)(T - zI)^{-1} = I = (T - zI)^{-1}(T - zI),$$

we see that $\overline{z} \in \mathrm{re}(T^*)$ and $R(T^*, \overline{z}) = R(T, z)^*$.

Conversely, let $\overline{z} \in \mathrm{re}(T^*)$. Since the operator $T^* - \overline{z}I^*$ is injective,

$$\mathcal{R}(T - zI)^{\perp} = \mathcal{N}(T^* - \overline{z}I^*) = \{0\},$$

that is, $\mathcal{R}(T - zI)$ is dense in X. Consider now $x \in X$. By the Hahn-Banach Extension Theorem, there is some $f \in X^*$ such that $\langle x\,,\, f\rangle = \|x\|$ and $\|f\| = 1$. Then

$$\begin{aligned}\|x\| = \langle x\,,\, f\rangle &= \langle x\,,\, (T^* - \overline{z}I^*)(T^* - \overline{z}I^*)^{-1} f\rangle \\ &= \langle (T - zI)x\,,\, (T^* - \overline{z}I^*)^{-1} f\rangle \\ &\le \|(T - zI)x\|\, \|(T^* - \overline{z}I^*)^{-1}\|.\end{aligned}$$

This shows that the operator $T - zI$ is injective. Moreover, since X is complete, it can be seen that $\mathcal{R}(T - zI)$ is closed. As $\mathcal{R}(T - zI)$ is dense in X as well, the operator $T - zI$ is surjective. Thus $z \in \mathrm{re}(T)$.

(b) We have $\mathrm{sp}(T^*) = \{\overline{\lambda} : \lambda \in \mathrm{sp}(T)\}$ by (a) above. It follows that a subset Λ of $\mathbb{C}$ is a spectral set for T if and only if $\overline{\Lambda}$ is a spectral set for T^*.

For each Jordan curve $\Gamma : [a, b] \to \mathbb{C}$, define the **conjugate Jordan curve** $\Gamma^* : [a, b] \to \mathbb{C}$ by

$$\Gamma^*(t) := \overline{\Gamma(a + b - t)}, \quad t \in [a, b].$$

Then Γ^* is a Jordan curve in $\mathrm{re}(T^*)$. Note that if Γ is positively oriented, then so is Γ^*. Now if Γ lies in $\mathrm{re}(T)$, then Γ^* lies in $\mathrm{re}(T^*)$ and

$$\begin{aligned}\frac{1}{2\pi \mathrm{i}} \int_{\Gamma^*} R(T^*, z)dz &= \frac{1}{2\pi \mathrm{i}} \int_a^b R\left(T^*, \overline{\Gamma(a + b - t)}\right) d\,\overline{\Gamma(a + b - t)} \\ &= -\frac{1}{2\pi \mathrm{i}} \int_a^b R\left(T^*, \overline{\Gamma(s)}\right) d\,\overline{\Gamma(s)} \\ &= -\frac{1}{2\pi \mathrm{i}} \int_a^b R(T, \Gamma(s))^* d\,\overline{\Gamma(s)}\end{aligned}$$

$$= \left(\frac{1}{2\pi \mathrm{i}} \int_a^b R(T, \Gamma(s)) d\Gamma(s) \right)^*$$

$$= \left(\frac{1}{2\pi \mathrm{i}} \int_\Gamma R(T, z) dz \right)^*,$$

where we have used the change of variable $s := a + b - t$, $t \in [a, b]$.

Let $\mathrm{C} \in \mathcal{C}(T, \Lambda)$ and denote by C^* the Cauchy contour comprising the conjugate curves of the curves involved in C. Then $\overline{\Lambda} \subset \mathrm{int}(\mathrm{C}^*)$ and $\mathrm{sp}(T^*) \setminus \overline{\Lambda} \subset \mathrm{ext}(\mathrm{C}^*)$, so that $\mathrm{C}^* \in \mathcal{C}(T^*, \overline{\Lambda})$.
It follows that

$$\begin{aligned} P(T^*, \overline{\Lambda}) &= -\frac{1}{2\pi \mathrm{i}} \int_{\mathrm{C}^*} R(T^*, z) dz \\ &= \left(-\frac{1}{2\pi \mathrm{i}} \int_{\mathrm{C}} R(T, z) dz \right)^* \\ &= P(T, \Lambda)^*, \end{aligned}$$

as desired. ∎

Special properties concerning the adjoint space and the adjoint operator hold when the Banach space X has the following additional structure.

A Banach space X is called a **Hilbert space** if the norm on X is induced by an inner product, that is, by a Hermitian positive definite sesquilinear form $\langle \cdot \, , \cdot \rangle$, as follows: $\|x\| := \langle x \, , x \rangle^{1/2}$ for $x \in X$.

As we have mentioned earlier, elements x, $y \in X$ are said to be orthogonal if $\langle x \, , y \rangle = 0$. The subspace $E^\perp := \{x \in X \, : \, \langle x \, , y \rangle = 0$ for all $y \in E\}$ is known as the **orthogonal complement** of the subset $E \subset X$.

Remark 1.37
The following results hold in a Hilbert space X:

(i) **Polarization Identity:** For each linear operator $A : X \to X$ and all x, $y \in X$,

$$\begin{aligned} 4\langle Ax \, , y \rangle &= \langle A(x+y) \, , x+y \rangle - \langle A(x-y) \, , x-y \rangle \\ &\quad + \mathrm{i} \, \langle A(x+\mathrm{i}\, y) \, , x+\mathrm{i}\, y \rangle - \mathrm{i} \, \langle A(x-\mathrm{i}\, y) \, , x-\mathrm{i}\, y \rangle. \end{aligned}$$

(ii) **Schwarz Inequality:** For all x, $y \in X$,

$$|\langle x \, , y \rangle| \le \|x\| \, \|y\|.$$

As a result, for each fixed $y \in X$, the conjugate-linear map $f : X \to \mathbb{C}$ given by $f(x) := \langle y\,,\, x\rangle$ for $x \in X$, is continuous and in fact $\|f\| = \|y\|$.

(iii) Riesz Representation Theorem: For each $f \in X^*$, there exists a unique $y \in X$ such that $f(x) = \langle y\,,\, x\rangle$ for all $x \in X$.

We see that the adjoint space X^* is linearly isometric with X itself. We shall identify X^* with X and write $X^* = X$. Thus the notation $\langle \cdot\,,\, \cdot\rangle$ introduced for Banach spaces is consistent with the notation of an inner product on a Hilbert space. Also the annihilator is identified with the orthogonal complement.

(iv) If Y is a closed subspace of X, then $X = Y \oplus Y^\perp$ and $(Y^\perp)^\perp = Y$.

(v) The following three conditions are equivalent to each other if π is a bounded projection defined on X:

(a) $\pi^* = \pi$, (b) $\|\pi\| \le 1$, (c) $\mathcal{N}(\pi) = \mathcal{R}(\pi)^\perp$.

If one of these conditions (and hence each of them) is satisfied, then π is called an orthogonal projection. ∎

Proposition 1.38
Let X be a Hilbert space and $T \in \mathrm{BL}(X)$. Then

$$\|T\| = \sup\{|\langle Tx\,,\, y\rangle| \,:\, x,\, y \in X,\, \|x\| \le 1,\, \|y\| \le 1\} = \|T^*T\|^{1/2}.$$

Proof
If $\|x\| \le 1$ and $\|y\| \le 1$, then clearly $|\langle Tx\,,\, y\rangle| \le \|T\|$. Now, if $T \ne O$, there exists x_0 such that $\|x_0\| = 1$ and $Tx_0 \ne 0$. Hence for $y_0 := Tx_0/\|Tx_0\|$, we have $|\langle Tx_0\,,\, y_0\rangle| = \|Tx_0\|$. The first equality is proved. The second is proved as follows: $\|T^*T\| \le \|T^*\|\,\|T\| = \|T\|^2 = \sup_{\|x\|\le 1} \|Tx\|^2 = \sup_{\|x\|\le 1} \langle Tx\,,\, Tx\rangle = \sup_{\|x\|\le 1} \langle T^*Tx\,,\, x\rangle \le \|T^*T\|$. ∎

Proposition 1.39 Gram-Schmidt Process:
Let X be a Hilbert space. If $\{x_1, \ldots, x_n\}$ is a linearly independent set, then the following procedure leads to an orthonormal basis *for the subspace* $\mathrm{span}\{x_1, \ldots, x_n\}$*, that is, to an ordered basis $[e_1, \ldots, e_n]$ of elements of X which have norm 1 and which are orthogonal to each other:*

Since $x_1 \ne 0$, define $e_1 := x_1/\|x_1\|$.

Suppose that for some integer k satisfying $2 \le k < n$, $e_1, \ldots, e_{k-1}$ have been constructed.
Define $\widetilde{e}_k := x_k - \sum_{i=1}^{k-1} \langle x_k\,,\, e_i\rangle e_i$. *Then* $\widetilde{e}_k \neq 0$.
Define $e_k := \widetilde{e}_k/\|\widetilde{e}_k\|$.

Proof
Since $x_1 \neq 0$, $e_1 := x_1/\|x_1\| \neq 0$ and $\mathrm{span}\{e_1\} = \mathrm{span}\{x_1\}$. Suppose that for some integer k satisfying $2 \le k < n$, we have $\mathrm{span}\{e_1, \ldots, e_j\} = \mathrm{span}\{x_1, \ldots, x_j\}$ for $j = 1, \ldots, k-1$. If $\widetilde{e}_k := x_k - \sum_{i=1}^{k-1} \langle x_k\,,\, e_i\rangle e_i$, then $\widetilde{e}_k \neq 0$, since otherwise $x_k \in \mathrm{span}\{x_1, \ldots, x_{k-1}\}$, which is impossible. Let $e_k := \widetilde{e}_k/\|\widetilde{e}_k\|$. Clearly, $\{e_1, \ldots, e_k\}$ is an orthonormal set and $\mathrm{span}\{e_1, \ldots, e_k\} = \mathrm{span}\{x_1, \ldots, x_k\}$. ∎

Remark 1.40 Fourier series in separable Hilbert spaces:
Suppose that the Hilbert space X is a separable space, that is, X has a countable dense subset. Then X is equal to the closure of the span of some countable linearly independent set $\{f_1, f_2, \ldots\}$ and the Gram-Schmidt Process yields a countable orthonormal basis for X. Then each $x \in X$ has a Fourier Expansion

$$x = \sum_{j=1}^{\infty} \langle x\,,\, e_j\rangle\, e_j = \lim_{n\to\infty} \pi_n x,$$

where $\pi_n : X \to X$, defined by $\pi_n x := \sum_{j=1}^{n} \langle x\,,\, e_j\rangle e_j$ for $x \in X$, is an orthogonal bounded finite rank projection onto $\mathrm{span}\{e_1, \ldots, e_n\}$. ∎

Let $T \in \mathrm{BL}(X)$. We say that T is a normal operator if $T^*T = TT^*$, that T is a selfadjoint operator if $T^* = T$, and that T is a unitary operator if $T^*T = I = TT^*$.

Proposition 1.41
Let X be a Hilbert space and $T \in \mathrm{BL}(X)$ be normal.

(a) *For each positive integer k, we have*

$$\mathcal{N}(T^k) = \mathcal{N}(T) \quad \textit{and} \quad \|T^{2^k}\| = \|T\|^{2^k}.$$

Also, $\rho(T) = ||T||$.

(b) *If* λ *is an isolated point of* $\mathrm{sp}(T)$, *then* λ *is an eigenvalue of* T; *and the corresponding eigenspace coincides with the spectral subspace associated with* T *and* λ.

(c) Krylov-Weinstein Inequality:
Let $z \in \mathbb{C}$. *Then there exists* $\lambda \in \mathrm{sp}(T)$ *such that for all nonzero* $x \in X$,

$$|z - \lambda| \leq \frac{||Tx - zx||}{||x||}.$$

Proof

(a) For $y \in X$, we have

$$||Ty||^2 = \langle Ty\,,\, Ty\rangle = \langle T^*Ty\,,\, y\rangle = \langle TT^*y\,,\, y\rangle = \langle T^*y\,,\, T^*y\rangle = ||T^*y||^2.$$

Let $x \in X$. Considering $y = Tx$, we obtain

$$||Tx||^2 = \langle T^*Tx\,,\, x\rangle \leq ||T^*(Tx)||\; ||x|| = ||T(Tx)||\; ||x|| = ||T^2x||\; ||x||.$$

Hence $\mathcal{N}(T^2) \subset \mathcal{N}(T)$. Since $\mathcal{N}(T) \subset \mathcal{N}(T^2)$ always, we have $\mathcal{N}(T^2) = \mathcal{N}(T)$. By induction, it follows that $\mathcal{N}(T^k) = \mathcal{N}(T)$ for $k = 1, 2, \ldots$

Also, the preceding inequality shows that $||T||^2 \leq ||T^2||$. Since $||T^2|| \leq ||T||^2$ always, we obtain $||T^2|| = ||T||^2$. Noting that the operator T^2 is normal, it follows by induction that $||T^{2^k}|| = ||T||^{2^k}$ for $k = 1, 2, \ldots$

By the Spectral Radius formula (Proposition 1.15(b)),

$$\rho(T) = \lim_{n\to\infty} ||T^n||^{1/n} = \lim_{k\to\infty} ||T^{2^k}||^{1/2^k} = ||T||.$$

(b) Let λ be an isolated point of $\mathrm{sp}(T)$, $P := P(T, \{\lambda\})$ and $x \in \mathcal{R}(P)$. Consider $0 < \epsilon < \mathrm{dist}(\lambda, \mathrm{sp}(T) \setminus \{\lambda\})$, and let Γ denote the positively oriented circle $\{z \in \mathbb{C} : |z - \lambda| = \epsilon\}$. Then $\Gamma \in \mathcal{C}(T, \{\lambda\})$, and

$$P = -\frac{1}{2\pi \mathrm{i}} \int_\Gamma R(z)dz, \quad TP = -\frac{1}{2\pi \mathrm{i}} \int_\Gamma TR(z)dz.$$

For $z \in \mathrm{re}(T)$, we have by Proposition 1.10(e),

$$TR(z) = I + zR(z) = I + (z - \lambda)R(z) + \lambda R(z),$$

so that $(T - \lambda I)R(z) = I + (z - \lambda)R(z)$. Hence

$$\begin{aligned} TP - \lambda P &= -\frac{1}{2\pi \mathrm{i}} \int_\Gamma (T - \lambda I)R(z)dz \\ &= -\frac{1}{2\pi \mathrm{i}} \int_\Gamma [I + (z - \lambda)R(z)]dz \\ &= -\frac{1}{2\pi \mathrm{i}} \int_\Gamma (z - \lambda)R(z)dz. \end{aligned}$$

Since T is normal, $R(z)$ is normal for each $z \in \mathrm{re}(T)$. By (b) above and Proposition 1.10(c), we obtain

$$\|R(z)\| = \rho(R(z)) = \frac{1}{\mathrm{dist}(z, \mathrm{sp}(T))}, \quad z \in \mathrm{re}(T).$$

Now assume that $\epsilon \leq \mathrm{dist}(\lambda, \mathrm{sp}(T) \setminus \{\lambda\})/2$. Then $\mathrm{dist}(z, \mathrm{sp}(T)) = \epsilon$ for all z on Γ. As a result,

$$\|TP - \lambda P\| \leq \frac{\ell(\Gamma)}{2\pi} \max_{|z-\lambda|=\epsilon} \{|z - \lambda| \, \|R(z)\|\} = \frac{2\pi\epsilon}{2\pi} \frac{\epsilon}{\epsilon} = \epsilon.$$

Letting $\epsilon \to 0$, we obtain $TP = \lambda P$. It follows that $\mathcal{R}(P) \subset \mathcal{N}(T - \lambda I)$. Also, $\mathcal{N}(T - \lambda I) \subset \mathcal{R}(P)$ by Proposition 1.31. Thus $\mathcal{R}(P) = \mathcal{N}(T - \lambda I)$. Also, $\mathcal{N}(T - \lambda I) \neq \{0\}$, since $P \neq 0$ by Corollary 1.27. Thus λ is an eigenvalue of T.

(c) If $z \in \mathrm{sp}(T)$, then let $\lambda := z$. If $z \in \mathrm{re}(T)$ and $x \in X$, we have

$$\|x\| = \|R(z)(Tx - zx)\| \leq \frac{\|Tx - zx\|}{\mathrm{dist}(z, \mathrm{sp}(T))}.$$

Since $\mathrm{sp}(T)$ is a compact set, there is some $\lambda \in \mathrm{sp}(T)$ such that

$$\mathrm{dist}(z, \mathrm{sp}(T)) = |z - \lambda|,$$

so that

$$|z - \lambda| \leq \frac{\|Tx - zx\|}{\|x\|},$$

for all nonzero $x \in X$. ∎

1.4.2 Product Space

We consider now a finite Cartesian product of the Banach space X. Let m be a positive integer and

$$X^{1\times m} := \{[x_1, \ldots, x_m] : x_j \in X \text{ for } j = 1, \ldots, m\}.$$

For $\underline{x} := [x_1, \ldots, x_m] \in X^{1\times m}$, we define

$$\|\underline{x}\|_1 := \sum_{j=1}^{m} \|x_j\|, \ \|\underline{x}\|_F := \Big(\sum_{j=1}^{m} \|x_j\|^2\Big)^{1/2}, \ \|\underline{x}\|_\infty := \max_{j=1,\ldots,m} \|x_j\|.$$

Then $\|\cdot\|_1$, $\|\cdot\|_F$ and $\|\cdot\|_\infty$ are equivalent norms on $X^{1\times m}$, and $X^{1\times m}$ is a Banach space.

Remark 1.42
Let $X := \mathbb{C}^{n\times 1}$ with the 2-norm. We identify the product space $X^{1\times m}$ with $\mathbb{C}^{n\times m}$. For $\mathsf{A} := [\mathsf{a}_1, \ldots, \mathsf{a}_m] = [a_{i,j}] \in \mathbb{C}^{n\times m}$, we obtain

$$\|\mathsf{A}\|_F := \sqrt{\sum_{j=1}^{m} \|\mathsf{a}_j\|_2^2} = \sqrt{\sum_{j=1}^{m}\sum_{i=1}^{n} |a_{i,j}|^2},$$

which is called the **Frobenius norm** or the F-**norm** of the matrix A. That is why we have chosen the notation $\|\cdot\|_F$ rather than $\|\cdot\|_2$. (See also Exercise 1.20.) ∎

Let n be a positive integer. In the Banach algebra $\mathrm{BL}(X^{1\times m}, X^{1\times n})$ of bounded linear operators from $X^{1\times m}$ into $X^{1\times n}$, we consider the corresponding subordinated norm denoted simply by $\|\cdot\|$.

Let $T \in \mathrm{BL}(X)$. Then T induces an operator $\underline{T} \in \mathrm{BL}(X^{1\times m})$, called the **natural extension** of T to $X^{1\times m}$, as follows:

$$\underline{T}\,[x_1, \ldots, x_m] := [Tx_1, \ldots, Tx_m] \quad \text{for } [x_1, \ldots, x_m] \in X^{1\times m}.$$

Clearly, $\underline{I}$ is the identity operator on $X^{1\times m}$. Also, $\|\underline{T}\| = \|T\|$.

We recall that for $\underline{x} := [x_1, \ldots, x_m] \in X^{1\times m}$ and $\mathsf{Z} = [z_{i,j}] \in \mathbb{C}^{m\times n}$,

$$\underline{x}\,\mathsf{Z} := \Big[\sum_{i=1}^{m} z_{i,1}x_i, \ldots, \sum_{i=1}^{m} z_{i,n}x_i\Big].$$

It easy to see that

$$\underline{T}\,(\underline{x}\,\mathsf{Z}) = (\underline{T}\,\underline{x})\mathsf{Z} \quad \text{for all } \underline{x} \in X^{1\times m} \text{ and } \mathsf{Z} \in \mathbb{C}^{m\times m}.$$

For this reason, it is unambiguous to write $\underline{T}\,\underline{x}\,\mathsf{Z}$.

Let $\underline{x} := [x_1, \dots, x_m] \in X^{1\times m}$ and $\underline{f} := [f_1, \dots, f_n] \in (X^*)^{1\times n}$. Then the Gram matrix G associated with $\underline{x}$ and $\underline{f}$, or the Gram product of $\underline{x}$ with $\underline{f}$, is defined by

$$\mathsf{G}(i,j) := \langle x_j \,, f_i \rangle.$$

It will be denoted by $(\underline{x}\,, \underline{f}\,)$.

Proposition 1.43
If $\underline{x} := [x_1, \dots, x_m] \in X^{1\times m}$ and $\underline{f} := [f_1, \dots, f_m] \in (X^*)^{1\times m}$ are such that the $m\times m$ matrix $(\underline{x}\,, \underline{f}\,)$ is nonsingular, then the set $\{x_1, \dots, x_m\}$ is linearly independent in X and the set $\{f_1, \dots, f_m\}$ is linearly independent in X^*.

Proof
If $f_i \in \text{span}\{f_j \,:\, j = 1, \dots, m,\, j \neq i\}$, then the ith row of $(\underline{x}\,, \underline{f}\,)$ is a linear combination of the others rows. If $x_j \in \text{span}\{x_i \,:\, i = 1, \dots, m,\, i \neq j\}$, then the jth column of $(\underline{x}\,, \underline{f}\,)$ is a linear combination of the others columns. In any of these events, $(\underline{x}\,, \underline{f}\,)$ would be a singular matrix. ∎

Let $\underline{x} \in X^{1\times m}$ and $\underline{x}^* \in (X^*)^{1\times m}$. We say that $\underline{x}^*$ is adjoint to $\underline{x}$ if

$$(\underline{x}\,, \underline{x}^*) = \mathsf{I}_m,$$

the $m\times m$ identity matrix.

Remark 1.44
Let k, ℓ, m and n be positive integers. We observe that for all $\underline{x} \in X^{1\times m}$, $\underline{f} \in (X^*)^{1\times k}$, $\mathsf{A} \in \mathbb{C}^{m\times n}$ and $\mathsf{B} \in \mathbb{C}^{k\times \ell}$, we have

$$(\underline{x}\mathsf{A}, \underline{f}\mathsf{B}) = \mathsf{B}^*(\underline{x}\,, \underline{f}\,)\mathsf{A}.$$

This shows that if $\underline{x} \in X^{1\times m}$ and $\underline{f} \in (X^*)^{1\times m}$ are such that $\mathsf{V} := (\underline{x}\,, \underline{f}\,)$ is a nonsingular matrix, then $\underline{y}^* := \underline{f}$ is adjoint to $\underline{y} := \underline{x}\mathsf{V}^{-1}$ and $\underline{x}^* := \underline{f}(\mathsf{V}^*)^{-1}$ is adjoint to $\underline{x}$. ∎

Example 1.45 The Gram-Schmidt process revisited:
Let $\{f_1, \dots, f_n\}$ be a linearly independent set in a Hilbert space X, and

let $\underline{e} := [e_1, \ldots, e_n]$ be the orthonormal set obtained by applying the Gram-Schmidt Process to the ordered set $\underline{f} := [f_1, \ldots, f_n]$. It follows that $(\underline{e}, \underline{e}) = \mathsf{I}_n$ and that $\underline{f} = \underline{e}\,\mathsf{R}$ for some upper triangular matrix R with positive diagonal entries. Moreover,

$$(\underline{f}, \underline{e}) = (\underline{e}\,\mathsf{R}, \underline{e}) = (\underline{e}, \underline{e})\mathsf{R} = \mathsf{R},$$

which gives R in terms of $\underline{e}$ and $\underline{f}$. ∎

For $\mathsf{Z} = [z_{i,j}] \in \mathbb{C}^{m \times n}$, consider the following matrix norms:

$$\|\mathsf{Z}\|_1 := \max_{j=1,\ldots,n} \sum_{i=1}^{m} |z_{i,j}|, \quad \text{which is called the 1-norm,}$$

$$\|\mathsf{Z}\|_2 := \rho(\mathsf{Z}^*\mathsf{Z})^{1/2}, \quad \text{which is called the 2-norm,}$$

$$\|\mathsf{Z}\|_\infty := \max_{i=1,\ldots,m} \sum_{j=1}^{n} |z_{i,j}|, \quad \text{which is called the } \infty\text{-norm.}$$

These matrix norms are subordinated operator norms. (See Exercise 1.21.) We shall also consider the F-norm of a matrix, defined in Remark 1.42,

$$\|\mathsf{Z}\|_{\mathrm{F}} := \mathrm{tr}(\mathsf{Z}^*\mathsf{Z})^{1/2} = \Big(\sum_{i=1}^{m}\sum_{j=1}^{n} |z_{i,j}|^2\Big)^{1/2}.$$

This norm is not a subordinated operator norm since $\|\mathsf{I}_n\|_{\mathrm{F}} = \sqrt{n} \neq 1$, unless $n = 1$.

Proposition 1.46
For all $\underline{x} \in X^{1\times m}$ and all $\underline{f} \in (X^*)^{1\times n}$,

$$\|(\underline{x}, \underline{f})\|_1 \le \|\underline{x}\|_\infty \|\underline{f}\|_1,$$
$$\|(\underline{x}, \underline{f})\|_\infty \le \|\underline{x}\|_1 \|\underline{f}\|_\infty,$$
$$\|(\underline{x}, \underline{f})\|_{\mathrm{F}} \le \|\underline{x}\|_{\mathrm{F}} \|\underline{f}\|_{\mathrm{F}}.$$

Proof
For $\underline{x} := [x_1, \ldots, x_m] \in X^{1\times m}$ and $\underline{f} = [f_1, \ldots, f_n] \in (X^*)^{1\times n}$,

$$\begin{aligned}\|(\underline{x}, \underline{f})\|_1 &= \max_{j=1,\ldots,m} \sum_{i=1}^{n} |\langle x_j\,, f_i\rangle| \le \max_{j=1,\ldots,m} \sum_{i=1}^{n} \|x_j\|\,\|f_i\| \\ &= \|\underline{x}\|_\infty \|\underline{f}\|_1,\end{aligned}$$

$$\|\,(\underline{x}\,,\,\underline{f}\,)\|_{\infty} = \max_{i=1,\ldots,n} \sum_{j=1}^{m} |\langle x_j\,,\, f_i\rangle| \le \max_{i=1,\ldots,n} \sum_{j=1}^{m} \|x_j\|\,\|f_i\|$$
$$= \|\,\underline{x}\,\|_{1}\,\|\,\underline{f}\,\|_{\infty},$$

and

$$\|\,(\underline{x}\,,\,\underline{f}\,)\|_{\mathrm{F}}^2 = \sum_{i=1}^{n}\sum_{j=1}^{m} |\langle x_j\,,\, f_i\rangle|^2 \le \sum_{i=1}^{n}\sum_{j=1}^{m} \|x_j\|^2\|f_i\|^2$$
$$\le \Big(\sum_{j=1}^{m} \|x_j\|^2\Big)\Big(\sum_{i=1}^{n} \|f_i\|^2\Big) = \|\,\underline{x}\,\|_{\mathrm{F}}^2\,\|\,\underline{f}\,\|_{\mathrm{F}}^2.$$

This completes the proof. ∎

We shall show that the adjoint space $(X^{1\times m})^*$ of $X^{1\times m}$ can be identified with the product space $(X^*)^{1\times m}$ by means of a linear isometry.

For $\underline{f} := [f_1, \ldots, f_m] \in (X^*)^{1\times m}$, let $i(\underline{f}\,) \in (X^{1\times m})^*$ be defined by

$$i(\underline{f}\,)(\underline{x}\,) := \sum_{j=1}^{m} f_j(x_j) \quad \text{for all } \underline{x} := [x_1, \ldots, x_m] \in X^{1\times m}.$$

Proposition 1.47
The map $i : (X^*)^{1\times m} \to (X^{1\times m})^*$ is linear, bijective and

$$\langle\, \underline{x}\,,\, i(\underline{f}\,)\rangle = \mathrm{tr}\,(\,\underline{x}\,,\,\underline{f}\,) \quad \textit{for all } \underline{x} \in X^{1\times m},\ \underline{f} \in (X^*)^{1\times m}.$$

Also,

$$\|i(\underline{f}\,)\| = \begin{cases} \|\,\underline{f}\,\|_{\infty} & \textit{if } X^{1\times m} \textit{ is given the 1-norm,}\\ \|\,\underline{f}\,\|_{1} & \textit{if } X^{1\times m} \textit{ is given the } \infty\textit{-norm,}\\ \|\,\underline{f}\,\|_{\mathrm{F}} & \textit{if } X^{1\times m} \textit{ is given the } \mathrm{F}\textit{-norm.}\end{cases}$$

Proof
It is clear that the map i is linear, bijective and that $\langle\, \underline{x}\,,\, i(\underline{f}\,)\rangle = \sum_{j=1}^{m} \overline{f_j(x_j)} = \mathrm{tr}\,(\,\underline{x}\,,\,\underline{f}\,)$ for all $\underline{x} \in X^{1\times m}$ and $\underline{f} \in (X^*)^{1\times m}$.

Let $X^{1\times m}$ be given the 1-norm. Then for $\underline{f} \in (X^*)^{1\times m}$,

$$\begin{aligned}\|i(\underline{f})\| &:= \sup\{|\langle \underline{x}\,,\, i(\underline{f})\rangle| \,:\, \|\underline{x}\|_1 \le 1\}\\ &= \sup\left\{\Big|\sum_{i=1}^m \langle x_i\,,\, f_i\rangle\Big| \,:\, \sum_{i=1}^m \|x_i\| \le 1\right\}\\ &\le \sup\left\{\|\underline{f}\|_\infty \sum_{i=1}^m \|x_i\| \,:\, \sum_{i=1}^m \|x_i\| \le 1\right\} = \|\underline{f}\|_\infty.\end{aligned}$$

For $\underline{f} \in (X^*)^{1\times m}$, let j be such that $\|\underline{f}\|_\infty = \|f_j\|$. For $x \in X$ satisfying $\|x\| \le 1$, we have $|\langle x\,,\, f_j\rangle| = |i(\underline{f})[0,\ldots,0,x,0,\ldots,0]|$, where x is the jth entry. Taking supremum over all $x \in X$ satisfying $\|x\| \le 1$, we obtain

$$\|\underline{f}\|_\infty = \|f_j\| \le \|i(\underline{f})\| := \sup\{|\langle \underline{x}\,,\, i(\underline{f})\rangle| \,:\, \|\underline{x}\|_1 \le 1\}.$$

Thus $\|i(\underline{f})\| = \|\underline{f}\|_\infty$.

Let $X^{1\times m}$ be given the ∞-norm. Then for $\underline{f} \in (X^*)^{1\times m}$,

$$\begin{aligned}\|i(\underline{f})\| &:= \sup\{|\langle \underline{x}\,,\, i(\underline{f})\rangle| \,:\, \|\underline{x}\|_\infty \le 1\}\\ &= \sup\left\{\Big|\sum_{i=1}^m \langle x_i\,,\, f_i\rangle\Big| \,:\, \max\{\|x_i\| \,:\, i=1,\ldots,m\} \le 1\right\}\\ &\le \sup\left\{\|\underline{x}\|_\infty \sum_{i=1}^m \|f_i\| \,:\, \max\{\|x_i\| \,:\, i=1,\ldots,m\} \le 1\right\}\\ &= \sum_{i=1}^m \|f_i\| = \|\underline{f}\|_1.\end{aligned}$$

Now, given $\epsilon > 0$, let $x_i \in X$ be such that $\|x_i\| \le 1$, $\langle x_i\,,\, f_i\rangle \ge 0$ and $\|f_i\| < \langle x_i\,,\, f_i\rangle + \epsilon$ for $i = 1,\ldots,m$. Define $\underline{x} := [x_1,\ldots,x_m]$. Then

$$|\langle \underline{x}\,,\, i(\underline{f})\rangle| = \sum_{i=1}^m \langle x_i\,,\, f_i\rangle > \sum_{i=1}^m \|f_i\| - m\epsilon.$$

Hence $\|i(\underline{f})\| \ge \sum_{i=1}^m \|f_i\| = \|\underline{f}\|_1$. Thus $\|i(\underline{f})\| = \|\underline{f}\|_1$.

Let $X^{1\times m}$ be given the F-norm. Then for $\underline{f} \in (X^*)^{1\times m}$,

$$\|i(\underline{f})\|^2 := \sup\{|\langle \underline{x}\,,\, i(\underline{f})\rangle|^2 \,:\, \|\underline{x}\|_F \le 1\}$$

$$\leq \sup\Big\{\Big(\sum_{i=1}^{m}|\langle x_i\,,\,f_i\rangle|\Big)^2 \,:\, \sum_{i=1}^{m}\|x_i\|^2 \leq 1\Big\}$$

$$\leq \sup\Big\{\sum_{i=1}^{m}\|x_i\|^2\sum_{i=1}^{m}\|f_i\|^2 \,:\, \sum_{i=1}^{m}\|x_i\|^2 \leq 1\Big\} \;=\; \|\,\underline{f}\,\|_{\mathrm{F}}^2.$$

Now, given $\epsilon > 0$, let $x_i \in X$ be such that $\|x_i\| \leq 1$ and

$$\|f_i\|^2 < |\langle x_i\,,\,f_i\rangle|^2 + \epsilon \quad \text{for } i = 1,\ldots,m.$$

Then

$$\|\,\underline{f}\,\|_{\mathrm{F}}^2 = \sum_{i=1}^{m}\|f_i\|^2 \leq \sum_{i=1}^{m}|\langle x_i\,,\,f_i\rangle|^2 + m\epsilon.$$

This shows that

$$\|\,\underline{f}\,\|_{\mathrm{F}}^2 \leq \sup\Big\{\sum_{i=1}^{m}|\langle x_i\,,\,f_i\rangle|^2 \,:\, \|x_i\| \leq 1,\, i = 1,\ldots,m\Big\}.$$

Consider $x_i \in X$ such that $\|x_i\| \leq 1$ for $i = 1,\ldots,m$ and

$$\delta := \Big(\sum_{i=1}^{m}|\langle x_i\,,\,f_i\rangle|^2\Big)^{1/2} > 0.$$

Define $y_i := \dfrac{\overline{\langle x_i\,,\,f_i\rangle}}{\delta}\,x_i$, $i = 1,\ldots,m$. Then

$$\sum_{i=1}^{m}\|y_i\|^2 = \frac{1}{\delta^2}\sum_{i=1}^{m}|\langle x_i\,,\,f_i\rangle|^2\|x_i\|^2 \leq \frac{1}{\delta^2}\sum_{i=1}^{m}|\langle x_i\,,\,f_i\rangle|^2 = 1.$$

Also,

$$\sum_{i=1}^{m}\langle y_i\,,\,f_i\rangle = \frac{1}{\delta}\sum_{i=1}^{m}|\langle x_i\,,\,f_i\rangle|^2 = \delta$$

and so

$$\Big|\sum_{i=1}^{m}\langle y_i\,,\,f_i\rangle\Big|^2 = \delta^2 = \sum_{i=1}^{m}|\langle x_i\,,\,f_i\rangle|^2.$$

Hence

$$\|\,\underline{f}\,\|_{\mathrm{F}}^2 \leq \sup\Big\{\sum_{i=1}^{m}|\langle x_i\,,\,f_i\rangle|^2 \,:\, \|x_i\| \leq 1,\, i = 1,\ldots,m\Big\}$$

$$\leq \sup\Big\{\Big|\sum_{i=1}^{m}\langle y_i\,,\,f_i\rangle\Big|^2 \,:\, \sum_{i=1}^{m}\|y_i\|^2 \leq 1\Big\} \;=\; \|i(\,\underline{f}\,)\|^2.$$

Thus $\|i(\underline{f})\| = \|\underline{f}\|_F$. ∎

Proposition 1.48

Let $T \in \mathrm{BL}(X)$. The adjoint of the extension $\underline{T}$ and the extension of the adjoint T^* are related by

$$(\underline{T})^* i(\underline{f}) = i[\underline{(T^*)}\,\underline{f}] \quad \textit{for } \underline{f} \in (X^*)^{1\times m}.$$

Proof

Let $\underline{x} \in X^{1\times m}$ and $\underline{f} \in (X^*)^{1\times m}$. Then by Proposition 1.47,

$$\begin{aligned}\langle \underline{x}\,,\,(\underline{T})^* i(\underline{f})\rangle &= \langle \underline{T}\,\underline{x}\,,\,i(\underline{f})\rangle = \mathrm{tr}\,(\underline{T}\,\underline{x}, \underline{f}) \\ &= \mathrm{tr}\,(\underline{x}, \underline{(T^*)}\,\underline{f}) = \langle \underline{x}\,,\,i[\underline{(T^*)}\,\underline{f}]\rangle,\end{aligned}$$

as we wanted to prove. ∎

Proposition 1.49

Let $\mathsf{Z} \in \mathbb{C}^{m\times m}$ and $\mathcal{Z} : X^{1\times m} \to X^{1\times m}$ be defined by $\mathcal{Z}\,\underline{x} := \underline{x}\,\mathsf{Z}$. Suppose $X^{1\times m}$ is given the p-norm, $p = 1, \infty$ or F. Denote by $\|\cdot\|$ the corresponding subordinated operator norm on $\mathrm{BL}(X^{1\times m})$, as usual. Then $\mathcal{Z} \in \mathrm{BL}(X^{1\times m})$ and

$$\|\mathcal{Z}\| \le \begin{cases} \|\mathsf{Z}\|_\infty & \textit{if } X^{1\times m} \textit{ is given the 1-norm,} \\ \|\mathsf{Z}\|_1 & \textit{if } X^{1\times m} \textit{ is given the } \infty\textit{-norm,} \\ \|\mathsf{Z}\|_F & \textit{if } X^{1\times m} \textit{ is given the F-norm.} \end{cases}$$

Moreover, the adjoint $\mathcal{Z}^*$ of $\mathcal{Z}$ satisfies

$$\mathcal{Z}^* i(\underline{f}) = i(\underline{f}\,\mathsf{Z}^*) \quad \textit{for } \underline{f} \in (X^*)^{1\times m}.$$

Proof

Let $z_{i,j}$ denote the (i,j)th entry of Z and $\underline{x} := [x_1, \ldots, x_m] \in X^{1\times m}$. Then

$$\|\mathcal{Z}\,\underline{x}\|_1 = \sum_{j=1}^m \Big\|\sum_{i=1}^m z_{i,j}x_i\Big\| \le \sum_{i=1}^m \|x_i\| \sum_{j=1}^m |z_{i,j}| \le \|\mathsf{Z}\|_\infty \|\underline{x}\|_1,$$

$$\|\mathcal{Z}\,\underline{x}\|_\infty = \max\Big\{\Big\|\sum_{i=1}^m z_{i,j}x_i\Big\| \,:\, j = 1, \ldots, m\Big\}$$

$$\leq \max\Big\{\sum_{i=1}^{m}|z_{i,j}|\,\|x_i\| \;:\; j=1,\ldots,m\Big\}$$

$$\leq \|\underline{x}\|_\infty \max\Big\{\sum_{i=1}^{m}|z_{i,j}| \;:\; j=1,\ldots,m\Big\} = \|\mathsf{Z}\|_1\|\underline{x}\|_\infty,$$

and

$$\|\mathcal{Z}\,\underline{x}\|_{\mathrm{F}}^2 = \sum_{j=1}^{m}\Big\|\sum_{i=1}^{m} z_{i,j}x_i\Big\|^2 \leq \sum_{j=1}^{m}\Big(\sum_{i=1}^{m}|z_{i,j}|^2\Big)\sum_{j=1}^{m}\|x_i\|^2 \leq \|\mathsf{Z}\|_{\mathrm{F}}^2\|\underline{x}\|_{\mathrm{F}}^2.$$

Hence the bounds for $\|\mathcal{Z}\|$ follow.

Let $\underline{x} \in X^{1\times m}$ and $\underline{f} \in (X^*)^{1\times m}$. Then by Proposition 1.47,

$$\begin{aligned}\langle \mathcal{Z}\,\underline{x}\,,\, i(\underline{f})\rangle &= \mathrm{tr}\,(\mathcal{Z}\,\underline{x}, \underline{f}\,) = \mathrm{tr}\,(\,\underline{x}\,\mathsf{Z}, \underline{f}\,)\\ &= \mathrm{tr}\,(\,\underline{x}, \underline{f}\,)\mathsf{Z} = \mathrm{tr}\,\mathsf{Z}(\,\underline{x}, \underline{f}\,)\\ &= \mathrm{tr}\,(\,\underline{x}, \underline{f}\,\mathsf{Z}^*) = \langle \underline{x}\,,\, i(\underline{f}\,\mathsf{Z}^*)\rangle,\end{aligned}$$

as we wanted to prove. ∎

We shall identify $(X^{1\times m})^*$ with $(X^*)^{1\times m}$ without writing explicitly the linear isometry $i : (X^*)^{1\times m} \to (X^{1\times m})^*$ and, as a consequence, the adjoint of the extension of T will be identified with the extension of the adjoint of T. It will be denoted by $\underline{T}^*$. Then for $\underline{x} \in X^{1\times m}$ and $\underline{f} \in (X^*)^{1\times m}$, we have

$$(\,\underline{T}\,\underline{x}, \underline{f}\,) = (\,\underline{x}, \underline{T}^*\underline{f}\,).$$

The notions of the resolvent operator and the reduced resolvent operator can be extended to the product space in the following way.

Instead of considering the equation $Tx - zx = y$, where $z \in \mathbb{C}$ and $y \in X$, we consider the Sylvester equation

$$\underline{T}\,\underline{x} - \underline{x}\,\mathsf{Z} = \underline{y}$$

where $\mathsf{Z} \in \mathbb{C}^{m\times m}$ and $\underline{y} \in X^{1\times m}$.

Proposition 1.50

Let $T \in \mathrm{BL}(X)$, $\mathsf{Z} \in \mathbb{C}^{m\times m}$ and $\mathcal{S} : X^{1\times m} \to X^{1\times m}$ be defined by

$$\mathcal{S}(\,\underline{x}\,) := \underline{T}\,\underline{x} - \underline{x}\,\mathsf{Z} \quad \textit{for } \underline{x} \in X^{1\times m}.$$

Then $\mathrm{sp}(\mathcal{S}) = \{\mu - \lambda : \mu \in \mathrm{sp}(T),\ \lambda \in \mathrm{sp}(\mathsf{Z})\}$.

As a consequence, for every $\underline{y} \in X^{1\times m}$ *the equation*

$$\underline{T}\,\underline{x} - \underline{x}\,\mathsf{Z} = \underline{y}$$

has a unique solution $\underline{x} \in X^{1\times m}$ *if and only if*

$$\mathrm{sp}(T) \cap \mathrm{sp}(\mathsf{Z}) = \emptyset.$$

Proof

By Remark 1.3, there exists a unitary matrix $\mathsf{Q} \in \mathbb{C}^{m\times m}$ such that $\mathsf{R} := \mathsf{Q}^*\mathsf{Z}\mathsf{Q}$ is an upper triangular matrix. For $\underline{x}$, $\underline{y} \in X^{1\times m}$, let

$$\underline{u} = [u_1, \ldots, u_m] := \underline{x}\,\mathsf{Q} \quad \text{and} \quad \underline{v} = [v_1, \ldots, v_m] := \underline{y}\,\mathsf{Q}.$$

Let $\sigma \in \mathbb{C}$. Then $(\mathcal{S} - \sigma\,\underline{I})\,\underline{x} = \underline{y}$ if and only if $(\underline{T} - \sigma\,\underline{I})\,\underline{u} - \underline{u}\,\mathsf{R} = \underline{v}$. Also, $(\underline{T} - \sigma\,\underline{I})\,\underline{u} - \underline{u}\,\mathsf{R} = \underline{v}$ if and only if the following m equations in X are satisfied:

$(E1)$ $\quad (T - (\sigma + \mathsf{R}(1,1))I)u_1 = v_1$

(Ej) $\quad (T - (\sigma + \mathsf{R}(j,j))I)u_j = v_j + \sum_{i=1}^{j-1} \mathsf{R}(i,j)u_i \quad$ for $j = 2, \ldots, m$.

Suppose that there exists $\lambda \in \mathrm{sp}(\mathsf{Z})$ such that $\mu := \sigma + \lambda \in \mathrm{sp}(T)$. By Remark 1.3, Q can be so chosen that $\mathsf{R}(1,1) = \lambda$. Then equation $(E1)$ has either a nonzero solution in X when $v_1 = 0$, or it has no solution in X for some v_1. Hence $(\mathcal{S} - \sigma\,\underline{I})\,\underline{x} = \underline{y}$ has either a nonzero solution in $X^{1\times m}$ when $\underline{y} = \underline{0}$, or it has no solution in $X^{1\times m}$ for some $\underline{y}$. Thus $\mu - \lambda = \sigma \in \mathrm{sp}(\mathcal{S})$.

Conversely, suppose that $\sigma \in \mathrm{sp}(\mathcal{S})$. Then there exists an integer j such that $1 \le j \le m$ and $\sigma + \mathsf{R}(j,j) \in \mathrm{sp}(T)$ since, otherwise, equation $(E1)$ has a unique solution for each $v_1 \in X$ and equation (Ej) has a unique solution for each $v_j \in X$ for $j = 2, \ldots, m$, that is, $(\mathcal{S} - \sigma\,\underline{I})\,\underline{x} = \underline{y}$ has a unique solution for each $\underline{y} \in X^{1\times m}$. Let $\lambda := R(j,j)$ and $\mu = \sigma + \lambda$. Then $\sigma = \mu - \lambda$, where $\lambda \in \mathrm{sp}(\mathsf{Z})$ and $\mu \in \mathrm{sp}(T)$. This proves that $\mathrm{sp}(\mathcal{S}) = \{\mu - \lambda : \mu \in \mathrm{sp}(T),\ \lambda \in \mathrm{sp}(\mathsf{Z})\}$.

As a consequence, $0 \in \mathrm{re}(\mathcal{S})$ if and only if $\mathrm{sp}(T) \cap \mathrm{sp}(\mathsf{Z}) = \emptyset$, proving the second assertion of the proposition. ∎

Let $\mathsf{Z} \in \mathbb{C}^{m\times m}$ and $\mathcal{Z}$ be the operator defined in Proposition 1.49: $\mathcal{Z}\,\underline{x} := \underline{x}\,\mathsf{Z}$ for $\underline{x} \in X^{1\times m}$. Proposition 1.50 says that the operator $\underline{T} - \mathcal{Z}$ is bijective if and only if $\mathrm{sp}(\mathsf{Z}) \cap \mathrm{sp}(T) = \emptyset$. In that case, it has a bounded inverse in $\mathrm{BL}(X^{1\times m})$, which we denote by $R(\underline{T}, \mathsf{Z})$ and call it the **block resolvent operator** of $\underline{T}$ at Z.

Although general finite rank operators will be studied in Chapter 4, the following result concerning bounded finite rank projections is needed for representing the spectral projection associated with a spectral set of finite type.

Lemma 1.51
Let $P \in \mathrm{BL}(X)$ be a projection with finite rank equal to m and let $\underline{\psi} \in X^{1\times m}$ form an ordered basis for $\mathcal{R}(P)$. Then there exists a unique ordered basis $\underline{\psi}^* \in (X^*)^{1\times m}$ for $\mathcal{R}(P^*)$, which is adjoint to $\underline{\psi}$. Also,

$$\begin{aligned} Px &= \underline{\psi}\,(x, \underline{\psi}^*) \quad \text{for all } x \in X, \\ \underline{P}\,\underline{x} &= \underline{\psi}\,(\underline{x}, \underline{\psi}^*) \quad \text{for all } \underline{x} \in X^{1\times m}, \\ P^* f &= \underline{\psi}^*\,(\underline{\psi}, f)^* \quad \text{for all } f \in X^*, \\ \underline{P}^*\,\underline{f} &= \underline{\psi}^*\,(\underline{\psi}, \underline{f})^* \quad \text{for all } \underline{f} \in (X^*)^{1\times m}. \end{aligned}$$

In particular, P^* is a projection and $\mathrm{rank}(P^*) = \mathrm{rank}(P)$.

Proof
Let $\underline{\psi} := [\psi_1, \ldots, \psi_m]$. Given $x \in X$, there exist m scalars $c_i(x)$, $i = 1, \ldots, m$, uniquely defined by x, such that

$$Px = \sum_{i=1}^{m} c_i(x)\psi_i.$$

Since P is linear, given $x, y \in X$ and $c \in \mathbb{C}$, we have $P(cx + y) = cPx + Py$, and so

$$\sum_{i=1}^{m} [c_i(cx + y) - cc_i(x) - c_i(y)]\psi_i = 0.$$

As the set $\{\psi_1, \ldots, \psi_m\}$ is linearly independent, we conclude that the functionals c_i, $i = 1, \ldots, m$, are linear. Given $a_1\psi_1 + \cdots + a_m\psi_m \in \mathcal{R}(P)$, we define

$$\|a_1\psi_1 + \cdots + a_m\psi_m\|_\infty := \max\{|a_i| \, : \, i = 1, \ldots, m\}.$$

Then $\|\cdot\|_\infty$ is a norm on $\mathcal{R}(P)$ and since all norms on $\mathcal{R}(P)$ must be equivalent, $\|\cdot\|_\infty$ is equivalent to the norm induced on $\mathcal{R}(P)$ by the norm of X. Hence there exists a constant α such that for all $x \in X$,

$$\|Px\|_\infty = \max\{|c_i(x)| : i = 1, \ldots, m\} \leq \alpha\|Px\| \leq \alpha\|P\| \, \|x\|.$$

This proves that each linear functional c_i is continuous. Define $\psi_i^* \in X^*$ by $\langle x \,,\, \psi_i^* \rangle := c_i(x)$ for $x \in X$, $i = 1, \ldots, m$, and $\underline{\psi}^* := [\psi_1^*, \ldots, \psi_m^*]$. Let $x \in X$. Then

$$Px = \sum_{i=1}^{m} \langle x \,,\, \psi_i^* \rangle \psi_i = \underline{\psi} \, (x, \underline{\psi}^*).$$

Thus the formulæ for Px and $\underline{P} \; \underline{x}$ follow.

Let $f \in X^*$. Then for $x \in X$,

$$\begin{aligned}(P^* f)x := f(Px) &= f\Big(\sum_{i=1}^{m} c_i(x)\psi_i\Big) = \sum_{i=1}^{m} \overline{\langle x \,,\, \psi_i^* \rangle} f(\psi_i) \\ &= \sum_{i=1}^{m} \overline{\langle \psi_i \,,\, f \rangle} \psi_i^*(x).\end{aligned}$$

and hence $P^* f = \sum_{i=1}^{m} \overline{\langle \psi_i \,,\, f \rangle} \psi_i^*$.

Now the formulæ for $P^* f$ and $\underline{P}^* \, \underline{f}$ follow. This also shows that $\mathcal{R}(P^*) \subset \operatorname{span}\{\psi_1^*, \ldots, \psi_m^*\}$.

Moreover, since P is a projection, $P\psi_i = \psi_i$ for $i = 1, \ldots, m$. Thus $(\underline{\psi} \,,\, \underline{\psi}^*) = \mathsf{I}_m$, that is, $\underline{\psi}^*$ is adjoint to $\underline{\psi}$. By Proposition 1.43, this equality implies the linear independence of the set $\{\psi_1^*, \ldots, \psi_m^*\}$ in X^*. But since

$$P^* \psi_j^* = \sum_{i=1}^{m} \overline{\langle \psi_i \,,\, \psi_j^* \rangle} \psi_i^* = \psi_j^*,$$

we have $\psi_j^* \in \mathcal{R}(P^*)$ for $j = 1, \ldots, m$. This proves that $\underline{\psi}^*$ forms an ordered basis for $\mathcal{R}(P^*)$.

To prove the uniqueness of $\underline{\psi}^*$, let $\underline{\widetilde{\psi}}^*$ form an ordered basis for $\mathcal{R}(P^*)$ which is adjoint to $\underline{\psi}$. Then there exists a matrix $\mathsf{V} \in \mathbb{C}^{m \times m}$ such that $\underline{\widetilde{\psi}}^* = \underline{\psi}^* \mathsf{V}$ and

$$\mathsf{I}_m = (\underline{\psi} \,,\, \underline{\widetilde{\psi}}^*) = (\underline{\psi} \,,\, \underline{\psi}^* \mathsf{V}) = \mathsf{V}^* (\underline{\psi} \,,\, \underline{\psi}^*) = \mathsf{V}^*.$$

Hence $\mathsf{V} = \mathsf{I}_m$, that is, $\underline{\widetilde{\psi}}^* = \underline{\psi}^*$. ∎

We note that a much shorter proof of the preceding result can be given if we use the Hahn Banach Extension Theorem: Let $f_1, \ldots, f_m$ in X^* be such that $\langle \psi_j \,, f_i \rangle = \delta_{i,j}$, $i, j = 1, \ldots, m$. Let us define $\psi_j^* := P^* f_j$, $j = 1, \ldots, m$. Then $\underline{\psi}^* := [\psi_1^*, \ldots, \psi_m^*]$ forms an ordered basis for $\mathcal{R}(P^*)$, which is adjoint to $\underline{\psi}$.

Theorem 1.52
Let $T \in \mathrm{BL}(X)$, Λ be a spectral set of finite type for T, $P := P(T, \Lambda)$, $m := \mathrm{rank}(P)$ and $\underline{\varphi}$ be an ordered basis for $M := \mathcal{R}(P)$. Then there is a unique ordered basis $\underline{\varphi}^*$ for $\mathcal{R}(P^*)$ which is adjoint to $\underline{\varphi}$. Let $\Theta := (\underline{T}\,\underline{\varphi}, \underline{\varphi}^*)$. Then $\underline{T}\,\underline{\varphi} = \underline{\varphi}\Theta$, $\underline{T}^*\underline{\varphi}^* = \underline{\varphi}^*\Theta^*$ and $\mathrm{sp}(\Theta) = \Lambda$. Also,

$$\underline{P}\,\underline{x} = \underline{\varphi}\,(\underline{x}, \underline{\varphi}^*) \quad \textit{and} \quad \underline{P}\,\underline{T}\,\underline{x} = \underline{\varphi}\Theta(\underline{x}, \underline{\varphi}^*) \quad \textit{for } \underline{x} \in X^{1\times m}.$$

Proof
The existence and uniqueness of $\underline{\varphi}^*$ follows from Lemma 1.51, as also the relation $\underline{P}\,\underline{x} = \underline{\varphi}\,(\underline{x}, \underline{\varphi}^*)$ for all $\underline{x} \in X^{1\times m}$. Since the subspace M is invariant under T, there exists a matrix $\mathsf{Z} \in \mathbb{C}^{m\times m}$ such that $\underline{T}\,\underline{\varphi} = \underline{\varphi}\mathsf{Z}$. As $(\underline{\varphi}, \underline{\varphi}^*) = \mathsf{I}_m$, we obtain

$$\Theta = (\underline{T}\,\underline{\varphi}, \underline{\varphi}^*) = (\underline{\varphi}\mathsf{Z}, \underline{\varphi}^*) = \mathsf{I}_m\mathsf{Z} = \mathsf{Z}.$$

Thus $\underline{T}\,\underline{\varphi} = \underline{\varphi}\Theta$.

Now the matrix Θ represents the operator $T_{|M,M}$ with respect to the ordered basis $\underline{\varphi}$ for M. Hence $\mathrm{sp}(\Theta) = \mathrm{sp}(T_{|M,M})$. But $\mathrm{sp}(T_{|M,M}) = \Lambda$ by the Spectral Decomposition Theorem. Thus $\mathrm{sp}(\Theta) = \Lambda$.

Again, since $\mathcal{R}(P^*)$ is invariant under T^*, there is a matrix $\mathsf{W} \in \mathbb{C}^{m\times m}$ such that $\underline{T}^*\underline{\varphi}^* = \underline{\varphi}^*\mathsf{W}$. But as $(\underline{\varphi}, \underline{\varphi}^*) = \mathsf{I}_m$,

$$\mathsf{W}^* = (\underline{\varphi}, \underline{\varphi}^*\mathsf{W}) = (\underline{\varphi}, \underline{T}^*\underline{\varphi}^*) = (\underline{T}\,\underline{\varphi}, \underline{\varphi}^*) = \Theta.$$

Thus $\underline{T}^*\underline{\varphi}^* = \underline{\varphi}^*\Theta^*$. Finally, for all $\underline{x} \in X^{1\times m}$, we have

$$\begin{aligned} \underline{P}\,\underline{T}\,\underline{x} &= \underline{\varphi}\,(\underline{T}\,\underline{x}, \underline{\varphi}^*) = \underline{\varphi}\,(\underline{x}, \underline{T}^*\underline{\varphi}^*) \\ &= \underline{\varphi}\,(\underline{x}, \underline{\varphi}^*\Theta^*) = \underline{\varphi}\Theta(\underline{x}, \underline{\varphi}^*), \end{aligned}$$

as desired. ∎

Remark 1.53 Dependence of Θ on $\underline{\varphi}$:
The matrix Θ in Theorem 1.52 does not depend on $\underline{\varphi}^* \in (X^*)^{1\times m}$ but

only on T and $\underline{\varphi}$. Indeed, if $\underline{f} \in (X^*)^{1\times m}$ is adjoint to $\underline{\varphi}$, then also

$$\langle \underline{T}\,\underline{\varphi}, \underline{f} \rangle = \langle \underline{\varphi}\Theta, \underline{f} \rangle = \langle \underline{\varphi}, \underline{f} \rangle \Theta = \mathsf{I}_m\Theta = \Theta.$$

If $\underline{\psi}$ forms another ordered basis for M, then there exists a nonsingular matrix $\mathsf{V} \in \mathbb{C}^{m\times m}$ such that $\underline{\psi} = \underline{\varphi}\mathsf{V}$, that is, the columns of V contain the coordinates of the elements in $\underline{\psi}$ relative to the ordered basis $\underline{\varphi}$. Since the corresponding ordered basis $\underline{\psi}^*$ for $\mathcal{R}(P^*)$ adjoint to $\underline{\psi}$ must satisfy $\underline{\psi}^* = \underline{\varphi}^*\mathsf{B}$ for some matrix $\mathsf{B} \in \mathbb{C}^{m\times m}$, we have

$$\mathsf{I}_m = \langle \underline{\psi}, \underline{\psi}^* \rangle = \langle \underline{\varphi}\mathsf{V}, \underline{\varphi}^*\mathsf{B} \rangle = \mathsf{B}^* \langle \underline{\varphi}, \underline{\varphi}^* \rangle \mathsf{V} = \mathsf{B}^*\mathsf{V}$$

and hence $\mathsf{B} = \left(\mathsf{V}^{-1}\right)^*$. Thus the matrix representing $T_{|M,M}$ with respect to the ordered basis $\underline{\psi}$ is

$$\langle \underline{T}\,\underline{\psi}, \underline{\psi}^* \rangle = \langle \underline{T}\,\underline{\varphi}\mathsf{V}, \underline{\varphi}^* \left(\mathsf{V}^{-1}\right)^* \rangle = \mathsf{V}^{-1}\Theta\mathsf{V},$$

as we learn in our first course in linear algebra! ∎

Let T, Λ, P and Θ be as in Theorem 1.52 As $\mathrm{sp}(\Theta) = \Lambda \subset \mathrm{sp}(T)$, we have $\mathrm{re}(T) \subset \mathrm{re}(\Theta)$. Since $\underline{T}\,\underline{\varphi} = \underline{\varphi}\,\Theta$, we observe that for all $z \in \mathrm{re}(T)$,

$$\underline{R(T,z)}\,\underline{\varphi} = \underline{\varphi}\,R(\Theta,z)$$

and hence

$$R(\Theta,z) = \langle \underline{R(T,z)}\,\underline{\varphi}, \underline{\varphi}^* \rangle.$$

We define the block reduced resolvent operator $\mathcal{S}_{\underline{\varphi}} : X^{1\times m} \to X^{1\times m}$ associated to T, Λ and $\underline{\varphi}$ as follows. Let $\mathrm{C} \in \mathcal{C}(T,\Lambda)$ and

$$\mathcal{S}_{\underline{\varphi}}\,\underline{x} := -\frac{1}{2\pi\mathrm{i}}\int_{\mathrm{C}} \underline{R(T,z)}\,\underline{x}\,R(\Theta,z)\,dz \quad \text{for } \underline{x} \in X^{1\times m}.$$

Note that the operator $\mathcal{S}_{\underline{\varphi}}$ does not depend on $\mathrm{C} \in \mathcal{C}(T,\Lambda)$.

Remark 1.54 The semisimple case:
If λ is a semisimple eigenvalue of T, then $\mathcal{R}(P) = \mathcal{N}(T-\lambda I)$, and any ordered basis $\underline{\varphi}$ of $\mathcal{R}(P)$ contains only eigenvectors of T corresponding to λ. Hence $\Theta = \lambda\,\mathsf{I}_m$, $R(\Theta,z) = \dfrac{1}{\lambda - z}\,\mathsf{I}_m$ for all $z \neq \lambda$ and

$$\mathcal{S}_{\underline{\varphi}}\,\underline{x} := -\frac{1}{2\pi\mathrm{i}}\int_{\mathrm{C}} \underline{R(T,z)}\,\underline{x}\,\frac{dz}{\lambda - z} = \underline{S(T,\lambda)}\,\underline{x} \quad \text{for } \underline{x} \in X^{1\times m}.$$

Thus the operator $\underline{\mathcal{S}_\varphi}$ is a generalization of the operator $\underline{S(T,\lambda)}$ to the case of a cluster of possibly defective eigenvalues of T. ∎

Remark 1.55 Dependence of $\underline{\mathcal{S}_\varphi}$ on $\underline{\varphi}$:
By Remark 1.53, if we change the basis $\underline{\varphi}$ for $M(T,\Lambda)$ to a basis $\underline{\psi} := \underline{\varphi}\mathsf{V}$, where V is a nonsingular matrix, then the matrix $R(\Theta,z)$ in the definition of $\underline{\mathcal{S}_\varphi}$ must be replaced by the matrix $\mathsf{V}^{-1}R(\Theta,z)\mathsf{V}$. Hence

$$\underline{\mathcal{S}_\psi}\,\underline{x} = -\frac{1}{2\pi \mathrm{i}}\int_{\mathrm{C}} \underline{R(T,z)}\,\underline{x}\,\mathsf{V}^{-1}R(\Theta,z)\mathsf{V}\,dz = \left(\underline{\mathcal{S}_\varphi}\,(\underline{x}\,\mathsf{V}^{-1})\right)\mathsf{V}$$

for all $\underline{x} \in X^{1\times m}$. ∎

Let $\underline{\varphi}$ form an ordered basis for $\mathcal{R}(P)$, $\underline{\varphi^*}$ form the corresponding adjoint ordered basis for $\mathcal{R}(P^*)$, and let $\underline{\mathcal{Z}_\varphi} : X^{1\times m} \to X^{1\times m}$ be defined by

$$\underline{\mathcal{Z}_\varphi}\,\underline{x} := \underline{x}\,(\underline{T}\,\underline{\varphi}\,,\underline{\varphi^*}\,) \quad \text{for } \underline{x} \in X^{1\times m}.$$

Proposition 1.56
The following properties hold:

(a) *The operators* $\underline{T}$, $\underline{P}$, $\underline{\mathcal{S}_\varphi}$ *and* $\underline{\mathcal{Z}_\varphi}$ *commute.*

(b) $(\underline{T} - \underline{\mathcal{Z}_\varphi})\underline{\mathcal{S}_\varphi} = \underline{I} - \underline{P}$.

(c) $\underline{\mathcal{S}_\varphi}\,\underline{P} = \underline{O}$.

Proof
Let $\Theta := (\underline{T}\,\underline{\varphi}\,,\underline{\varphi^*}\,)$.

(a) Since $(\underline{T}\,\underline{x}\,)\Theta = \underline{T}\,(\underline{x}\,\Theta)$ for $\underline{x} \in X^{1\times m}$, $\underline{\mathcal{Z}_\varphi}$ and $\underline{T}$ commute. The same argument holds for $\underline{P}$ and $\underline{\mathcal{Z}_\varphi}$. Since T and P commute, so do $\underline{T}$ and $\underline{P}$. Since T is continuous and commutes with $R(T,z)$ for $z \in \mathrm{C}$, $\underline{T}$ and $\underline{\mathcal{S}_\varphi}$ commute. The same argument holds for $\underline{P}$ and $\underline{\mathcal{S}_\varphi}$. Since Θ and $R(\Theta,z)$ commute for $z \in \mathrm{C}$, so do $\underline{\mathcal{Z}_\varphi}$ and $\underline{\mathcal{S}_\varphi}$.

(b) Let $\mathrm{C} \in \mathcal{C}(T,\Lambda)$. Then

$$(\underline{T} - \underline{\mathcal{Z}_\varphi})\underline{\mathcal{S}_\varphi}\,\underline{x} = -\frac{1}{2\pi \mathrm{i}}\int_{\mathrm{C}} \underline{R(T,z)}\,(\underline{T}\,\underline{x} - \underline{x}\,\Theta)R(\Theta,z)\,dz$$

$$
\begin{aligned}
&= -\frac{1}{2\pi \mathrm{i}}\int_{\mathrm{C}} \underline{R(T,z)}\,(\,\underline{T}\;\underline{x} - z\,\underline{x} + z\,\underline{x} - \underline{x}\,\Theta)R(\Theta,z)\,dz \\
&= \underline{x}\Big[-\frac{1}{2\pi \mathrm{i}}\int_{\mathrm{C}} R(\Theta,z)\,dz\Big] + \Big[\frac{1}{2\pi \mathrm{i}}\int_{\mathrm{C}} \underline{R(T,z)}\,dz\Big]\,\underline{x} \\
&= (\,\underline{I} - \underline{P}\,)\,\underline{x}\,,
\end{aligned}
$$

since $\mathrm{sp}(\Theta) \subset \mathrm{int}(\mathrm{C})$ and hence $\displaystyle\int_{\mathrm{C}} R(\Theta,z)\,dz = -2\pi\mathrm{i}\;\mathsf{I}_m$.

(c) Let $\mathrm{C} \in \mathcal{C}(T,\Lambda)$. By Corollary 1.22, there exists $\widetilde{\mathrm{C}}$ separating $\Lambda \cup \mathrm{C}$ from $\mathrm{sp}(T)\setminus\Lambda$. Using C in the integral defining $\underline{\mathcal{S}_\varphi}$ and $\widetilde{\mathrm{C}}$ in the integral defining $\underline{P}$, and taking into account the First Resolvent Identity, we obtain for $\underline{x} \in X^{1\times m}$,

$$
\begin{aligned}
\underline{\mathcal{S}_\varphi}\;\underline{P}\;\underline{x} &= -\frac{1}{2\pi\mathrm{i}}\int_{\mathrm{C}} \underline{R(T,z)}\,(\,\underline{P}\;\underline{x}\,)R(\Theta,z)\,dz \\
&= -\frac{1}{2\pi\mathrm{i}}\int_{\mathrm{C}} \underline{R(T,z)}\Big[-\frac{1}{2\pi\mathrm{i}}\int_{\widetilde{\mathrm{C}}} \underline{R(T,\widetilde{z})}\;\underline{x}\;d\widetilde{z}\Big]R(\Theta,z)\,dz \\
&= \Big(-\frac{1}{2\pi\mathrm{i}}\Big)^2\int_{\mathrm{C}}\Big[\int_{\widetilde{\mathrm{C}}}[\,\underline{R(T,z)} - \underline{R(T,\widetilde{z})}\,]\,\underline{x}\,\frac{R(\Theta,z)}{z-\widetilde{z}}\,d\widetilde{z}\Big]\,dz \\
&= \Big(-\frac{1}{2\pi\mathrm{i}}\Big)^2\int_{\mathrm{C}}\Big[\underline{R(T,z)}\;\underline{x}\,R(\Theta,z)\int_{\widetilde{\mathrm{C}}}\frac{d\widetilde{z}}{z-\widetilde{z}}\Big]\,dz \\
&\quad -\Big(-\frac{1}{2\pi\mathrm{i}}\Big)^2\int_{\widetilde{\mathrm{C}}}\Big[\underline{R(T,\widetilde{z})}\;\underline{x}\int_{\mathrm{C}}\frac{R(\Theta,z)}{z-\widetilde{z}}\,dz\Big]\,d\widetilde{z}.
\end{aligned}
$$

But $\mathrm{C} \subset \mathrm{int}(\widetilde{\mathrm{C}})$, so $\displaystyle\int_{\widetilde{\mathrm{C}}}\frac{d\widetilde{z}}{z-\widetilde{z}} = -2\pi\mathrm{i}$ for every $z \in \mathrm{C}$ and $\displaystyle\int_{\mathrm{C}}\frac{dz}{z-\widetilde{z}} = 0$ for every $\widetilde{z} \in \widetilde{\mathrm{C}}$. Also, by the First Resolvent Identity, we have for $\widetilde{z} \in \widetilde{\mathrm{C}}$,

$$
\int_{\mathrm{C}}\frac{R(\Theta,z)}{z-\widetilde{z}}\,dz = R(\Theta,\widetilde{z})\int_{\mathrm{C}}\frac{dz}{z-\widetilde{z}} + R(\Theta,\widetilde{z})\int_{\mathrm{C}} R(\Theta,z)\,dz = -2\pi\mathrm{i}\,R(\Theta,\widetilde{z}),
$$

since $\mathrm{sp}(\Theta) \subset \mathrm{int}(\mathrm{C})$. Thus for $\underline{x} \in X^{1\times m}$,

$$
\underline{\mathcal{S}_\varphi}\;\underline{P}\;\underline{x} = -\frac{1}{2\pi\mathrm{i}}\int_{\mathrm{C}} \underline{R(T,z)}\;\underline{x}\,R(\Theta,z)\,dz + \frac{1}{2\pi\mathrm{i}}\int_{\widetilde{\mathrm{C}}} \underline{R(T,\widetilde{z})}\;\underline{x}\,R(\Theta,\widetilde{z})\,d\widetilde{z} = \underline{0}
$$

(Compare the proof of part (b) of Proposition 1.24.) ∎

The operator $\underline{\mathcal{S}_\varphi}$ can be described with the help of the operator $\underline{\mathcal{G}_\varphi}$: $X^{1\times m} \to X^{1\times m}$ defined by

$$
\underline{\mathcal{G}_\varphi} := (\,\underline{I} - \underline{P}\,)\,\underline{T} - \underline{\mathcal{Z}_\varphi}\,.
$$

Proposition 1.57
Let Λ be a spectral set of finite type for T, $P := P(T, \Lambda)$ and $N := \mathcal{N}(P)$. If $\underline{\varphi}$ forms an ordered basis for $\mathcal{R}(P)$, then $(\mathcal{G}_{\underline{\varphi}})_{|N^{1\times m}, N^{1\times m}}$ is bijective and

$$\mathcal{S}_{\underline{\varphi}} = \left((\mathcal{G}_{\underline{\varphi}})_{|N^{1\times m}, N^{1\times m}}\right)^{-1}(\underline{I} - \underline{P}).$$

Proof
By Proposition 1.56, $\mathcal{S}_{\underline{\varphi}}\,\underline{x} \in N^{1\times m}$ for all $\underline{x} \in N^{1\times m}$ and

$$\mathcal{G}_{\underline{\varphi}}\mathcal{S}_{\underline{\varphi}}\,\underline{x} = (\underline{T} - \mathcal{Z}_{\underline{\varphi}})\mathcal{S}_{\underline{\varphi}}\,\underline{x} = (\underline{I} - \underline{P})\,\underline{x}.$$

Let $\underline{\varphi}^*$ form a basis for $\mathcal{R}(P^*)$ which is adjoint to $\underline{\varphi}$ and consider $\Theta := (\underline{T}\,\underline{\varphi}, \underline{\varphi}^*)$. Replacing $X^{1\times m}$ by $N^{1\times m} = \mathcal{R}(\underline{I} - \underline{P}) = [\mathcal{R}(I - P)]^{1\times m}$ in Proposition 1.50 and noting that

$$\mathrm{sp}(\Theta) \cap \mathrm{sp}(T_{|N,N}) = \mathrm{sp}(T_{|M,M}) \cap \mathrm{sp}(T_{|N,N}) = \emptyset,$$

we see that the operator $(\mathcal{G}_{\underline{\varphi}})_{|N^{1\times m}, N^{1\times m}}$ is invertible in $\mathrm{BL}(N^{1\times m})$. Hence the desired result follows. ∎

Remark 1.58
The preceding result suggests a method of finding $\mathcal{S}_{\underline{\varphi}}\,\underline{y}$ for a given $\underline{y} \in X^{1\times m}$. In fact, since $\mathcal{S}_{\underline{\varphi}}\,\underline{y} = \mathcal{S}_{\underline{\varphi}}(\underline{y} - \underline{P}\,\underline{y})$ and $\underline{y} - \underline{P}\,\underline{y} \in \mathcal{N}(\underline{P}) = \mathcal{N}(P)^{1\times m}$, we observe that $\mathcal{S}_{\underline{\varphi}}\,\underline{y}$ is the unique element $\underline{x} \in X$ which satisfies $\underline{T}\,\underline{x} - \underline{x}\,(\underline{T}\,\underline{\varphi}, \underline{\varphi}^*) = \underline{y} - \underline{P}\,\underline{y}$ and $\underline{P}\,\underline{x} = \underline{0}$. (Compare Remark 1.25.) Assume now that $0 \notin \Lambda$. Since

$$\mathrm{sp}((I-P)T) = \mathrm{sp}((I-P)T(I-P)) = \mathrm{sp}(T_{|N,N}) \cup \{0\} = (\mathrm{sp}(T)\setminus\Lambda) \cup \{0\}$$

and $\mathrm{sp}(\Theta) = \Lambda$, we see that $\mathrm{sp}((I-P)T) \cap \mathrm{sp}(\Theta) = \emptyset$. Then, by Proposition 1.50, the operator $\mathcal{G}_{\underline{\varphi}}$ is invertible in $\mathrm{BL}(X^{1\times m})$ and hence $\mathcal{S}_{\underline{\varphi}} = \mathcal{G}_{\underline{\varphi}}^{-1}(\underline{I} - \underline{P})$. ∎

1.5 Exercises

Unless otherwise stated, X denotes a Banach space over $\mathbb{C}$, $X \neq \{0\}$ and $T \in \mathrm{BL}(X)$. Prove the following assertions.

1.1 Let Y be a closed subspace of X which is invariant under T. Then $\rho(T_{|Y,Y}) \le \rho(T)$. In general, $\mathrm{sp}(T_{|Y,Y})$ may not contain $\mathrm{sp}(T)$, and $\mathrm{sp}(T)$ may not contain $\mathrm{sp}(T_{|Y,Y})$.

1.2 Let $K \in \mathrm{BL}(X,Y)$ and $L \in \mathrm{BL}(Y,X)$. Then $\mathrm{re}(LK) \setminus \{0\} = \mathrm{re}(KL) \setminus \{0\}$ and for z in this set,

$$R(KL,z) = \frac{1}{z}[K\,R(LK,z)L - I_Y],$$
$$R(LK,z) = \frac{1}{z}[L\,R(KL,z)K - I_X].$$

1.3 Let C be a Cauchy contour contained in $\mathrm{re}(T)$. Then

(a) If $\mathrm{sp}(T) \subset \mathrm{int}(\mathrm{C})$, then $\displaystyle\int_{\mathrm{C}} R(z)\,dz = -2\pi\mathrm{i}\,I$.

(b) If $\mathrm{sp}(T) \subset \mathrm{ext}(\mathrm{C})$, then $R(z_0) = \displaystyle\frac{1}{2\pi\mathrm{i}}\int_{\mathrm{C}} \frac{R(z)}{z-z_0}\,dz$ for every $z_0 \in \mathrm{int}(\mathrm{C})$.

1.4 Let λ be an isolated point of $\mathrm{sp}(T)$ and define $M := \mathcal{R}(P(T,\{\lambda\}))$. Then $M = \{x \in X : \|(T-\lambda I)^n x\|^{1/n} \to 0 \text{ as } n \to \infty\}$. $\Big($Hint: For $x \in X$ such that $\|(T-\lambda I)^n x\|^{1/n} \to 0$, and $z \in \mathrm{re}(T)$, $R(z)x = -\displaystyle\sum_{k=0}^{\infty} \frac{(T-\lambda I)^k x}{(z-\lambda)^{k+1}}$.$\Big)$

1.5 Let Λ be a spectral set for T, $\lambda \in \Lambda$ and j be any positive integer. Then $\mathcal{N}[(T-\lambda I)^j] \subset M(T,\Lambda)$. (Hint: $Y := \mathcal{N}[(T-\lambda I)^j]$ is a closed subspace of X which is invariant under T and $\mathrm{sp}(T_{|Y,Y}) = \{\lambda\} \subset \Lambda$. Use Proposition 1.28.)

1.6 Let (Y,Z) decompose $T \in \mathrm{BL}(X)$, P denote the projection on Y along Z, and C be a Cauchy contour in $\mathrm{re}(T)$ such that $\mathrm{sp}(T_{|Y,Y}) \subset \mathrm{int}(\mathrm{C})$ and $\mathrm{sp}(T_{|Z,Z}) \subset \mathrm{ext}(\mathrm{C})$. Then $P = -\displaystyle\frac{1}{2\pi\mathrm{i}}\int_{\mathrm{C}} R(T,z)\,dz$, and that for every $z_0 \in \mathrm{int}(\mathrm{C})$, $R(T_{|Z,Z},z_0) = \displaystyle\frac{1}{2\pi\mathrm{i}}\int_{\mathrm{C}} \frac{R(T_{|Z,Z},z)}{z-z_0}\,dz$. (These results provide a motivation for the formulæ of the spectral projection and the reduced resolvent operator associated with T.)

1.7 Let U be in $\mathrm{BL}(X)$. As for matrices, T is said to be similar to U, if there exists an invertible operator $V \in \mathrm{BL}(X)$ such that $U = V^{-1}TV$. Let Λ be a spectral set for T and T be similar to U. Then Λ is a spectral set for U and $\mathrm{rank}\,P(U,\Lambda) = \mathrm{rank}\,P(T,\Lambda)$.

1.8 Let Λ be a spectral set for T, $\mathrm{C} \in \mathcal{C}(T, \Lambda)$, $P := P(T, \Lambda)$ and k be a nonnegative integer. Then

$$T^k P = -\frac{1}{2\pi \mathrm{i}} \int_{\mathrm{C}} z^k R(z)\, dz \quad \text{and} \quad P = -\frac{T^k}{2\pi \mathrm{i}} \int_{\mathrm{C}} \frac{R(z)}{z^k}\, dz.$$

If T^k is a compact operator, then the spectral set Λ is of finite type.

1.9 Let T be compact, $U \in \mathrm{BL}(X)$, $\lambda \in \mathrm{sp}(U)$. Then $\lambda \in \mathrm{sp}(T + U)$, if λ is not an eigenvalue of U. (Hint: If $z \in \mathrm{re}(T + U)$, then $U - zI = (T + U - zI)[I - (T + U - zI)^{-1}T]$.)

1.10 Let $\lambda \in \mathbb{C}$.

(a) λ is an isolated point of $\mathrm{sp}(T)$ if and only if there are closed subspaces Y and Z of X such that (Y, Z) decomposes T, $\mathrm{sp}(T_{|Y,Y}) = \{\lambda\}$ and $\lambda \notin \mathrm{sp}(T_{|Z,Z})$. In this case, $S(T, \{\lambda\}) = \lim_{z \to \lambda} R(T, z)(I - P(T, \{\lambda\}))$.

(b) λ is a spectral value of finite type for T if and only if there are closed subspaces Y and Z of X such that (Y, Z) decomposes T, $\dim Y$ is finite, $\mathrm{sp}(T_{|Y,Y}) = \{\lambda\}$ and $\lambda \notin \mathrm{sp}(T_{|Z,Z})$.

1.11 Not every eigenvalue of a bounded operator is a spectral value of finite type.

1.12 Let Λ be a spectral set for T and $\widetilde{\Lambda} := \mathrm{sp}(T) \setminus \Lambda$. Then $P(T, \Lambda) + P(T, \widetilde{\Lambda}) = I$ and $P(T, \Lambda)P(T, \widetilde{\Lambda}) = O$.

1.13 Let Λ_1, Λ_2 be spectral subsets for T. Then

$$\Lambda_1 \subset \Lambda_2 \quad \text{if and only if} \quad M(T, \Lambda_1) \subset M(T, \Lambda_2).$$

(Hint: For the 'only if' assertion, $E := \Lambda_1 \setminus \Lambda_2$ is a spectral set for T, $\mathcal{R}(P(T, E)) \subset M(T, \Lambda_1)$ and $\mathcal{R}(P(T, E)) \cap M(T, \Lambda_2) = \{0\}$.)

1.14 Let λ be an isolated point of $\mathrm{sp}(T)$ and $z \in \mathbb{C}$ be such that $0 < |z - \lambda| < \mathrm{dist}(\lambda, \mathrm{sp}(T) \setminus \{\lambda\})$. Then $R(z) = \sum\limits_{k=0}^{\infty} S^k (z - \lambda)^k - \dfrac{P}{z - \lambda} - \sum\limits_{k=2}^{\infty} \dfrac{D^{k-1}}{(z - \lambda)^k}$, where $P := P(T, \{\lambda\})$, $S := S(T, \{\lambda\})$ and $D := (T - \lambda I)P$. Hence λ is a pole of $R(\cdot)$ if and only if D is nilpotent. In that case, the order ℓ of the pole is the smallest positive integer such that $D^\ell = O$. It is also the smallest positive integer ℓ such that $\mathcal{N}[(T - \lambda I)^\ell] = \mathcal{N}[(T - \lambda I)^{\ell+1}]$. Further, $\mathcal{R}(P(T, \{\lambda\})) = \mathcal{N}[(T - \lambda I)^\ell]$ and $\mathcal{N}(P(T, \{\lambda\})) = \mathcal{R}[(T - \lambda I)^\ell]$ and λ is, in fact, an eigenvalue of T with ascent ℓ.

1.15 Let T be a compact operator and Λ a spectral set for T containing 0. Then if $\dim X$ is infinite, $\operatorname{rank} P(T, \Lambda)$ is infinite.

1.16 Let g, ℓ and m be the geometric multiplicity, the ascent and the algebraic multiplicity of an isolated spectral value λ of T. Then $g + \ell - 1 \leq gl \leq m$. Conversely, if g, ℓ and m are positive integers such that $g + \ell - 1 \leq g\ell \leq m$, then there is a matrix A and an eigenvalue λ of A such that g, ℓ and m are its geometric multiplicity, ascent, and algebraic multiplicity, respectively. (See [59].)

1.17 The Jordan canonical form may be discontinuous: Let $\mathsf{A} := \begin{bmatrix} 1 & 1 \\ 0 & 1 \end{bmatrix}$ and $\mathsf{A}_n := \begin{bmatrix} 1 & 1 \\ 1/n & 1 \end{bmatrix}$, $n = 1, 2, \ldots$ Let J and J_n, $n = 1, 2, \ldots$ be their Jordan canonical forms, respectively. If $\|\cdot\|$ be any norm in $\mathbb{C}^{2\times 2}$, then $\lim\limits_{n\to\infty} \|\mathsf{A}_n - \mathsf{A}\| = 0$ but $\lim\limits_{n\to\infty} \|\mathsf{J}_n - \mathsf{J}\| \neq 0$.

1.18 Let Λ be a spectral set for T. Then

$$S(T^*, \overline{\Lambda}, \overline{\lambda}_0) = S(T, \Lambda, \lambda_0)^* \quad \text{for each } \lambda \in \operatorname{sp}(T).$$

1.19 Let X be Hilbert space and $\langle\cdot\,,\,\cdot\rangle$ its inner product. If T is self-adjoint, then $\operatorname{sp}(T)$ is a subset of $\mathbb{R}$.

1.20 Let X be a Hilbert space and $\langle\cdot\,,\,\cdot\rangle$ its inner product. Consider the product space $X^{1\times m}$ and define, for $\underline{x} \in X^{1\times m}$, $\|\underline{x}\|_2 := \rho(\left(\underline{x}\,,\,\underline{x}\right))^{1/2}$. Then $\|\cdot\|_2$ is a norm on $X^{1\times m}$, which is equivalent to the norm $\|\cdot\|_p$ for $p = 1, \mathrm{F}, \infty$. (Hint: $\|\underline{x}\|_2 = \max\{\|\underline{x}\,\mathsf{u}\| \,:\, \mathsf{u} \in \mathbb{C}^{m\times 1}, \|\mathsf{u}\|_2 = 1\}$.)

1.21 Let $\mathsf{Z} := [z_{i,j}] \in \mathbb{C}^{m\times n}$. Then
(a) $\|\mathsf{Z}\|_1 = \max\{\|\mathsf{Z}\mathsf{x}\|_1 \,:\, \mathsf{x} \in \mathbb{C}^{n\times 1}, \|\mathsf{x}\|_1 \leq 1\}$.
(b) $\|\mathsf{Z}\|_2 = \max\{\|\mathsf{Z}\mathsf{x}\|_2 \,:\, \mathsf{x} \in \mathbb{C}^{n\times 1}, \|\mathsf{x}\|_2 \leq 1\}$.
(c) $\|\mathsf{Z}\|_\infty = \max\{\|\mathsf{Z}\mathsf{x}\|_\infty \,:\, \mathsf{x} \in \mathbb{C}^{n\times 1}, \|\mathsf{x}\|_\infty \leq 1\}$.

1.22 Let $\mathsf{Z} := [z_{i,j}] \in \mathbb{C}^{m\times n}$ and define the matrix $|\mathsf{Z}| \in \mathbb{C}^{m\times n}$ by $|\mathsf{Z}| := [\,|z_{i,j}|\,]$. Then $\|\,|\cdot|\,\|_p$, $p = 1, 2, \infty, \mathrm{F}$, defines a norm on $\mathbb{C}^{m\times n}$, and for all $\mathsf{Z} := [z_{i,j}] \in \mathbb{C}^{m\times n}$, we have
(a) $\|\mathsf{Z}\|_1 = \|\,|\mathsf{Z}|\,\|_1 = \|\mathsf{Z}^*\|_\infty$.
(b) $\|\mathsf{Z}\|_\infty = \|\,|\mathsf{Z}|\,\|_\infty = \|\mathsf{Z}^*\|_1$.
(c) $\|\,|\mathsf{Z}|\,\|_2 \leq \|\,|\mathsf{Z}|\,\|_F = \|\mathsf{Z}\|_F$.
(d) $\dfrac{1}{\sqrt{m}}\|\mathsf{Z}\|_2 \leq \|\,|\mathsf{Z}|\,\|_2 \leq \sqrt{m}\,\|\mathsf{Z}\|_2$.

1.23 Let $X := \mathbb{C}^{n\times 1}$ with a norm $\|\cdot\|$ such that for each standard basis vector e_j, $\|\mathsf{e}_j\| = 1$, $j = 1,\ldots,n$. Let $\mathsf{Z} \in \mathbb{C}^{m\times m}$ and define $\mathcal{Z} : \mathbb{C}^{n\times m} \to \mathbb{C}^{n\times m}$ by $\mathcal{Z}\,\underline{x} := \underline{x}\,\mathsf{Z}$ for $\underline{x} := [x_1,\ldots,x_m] \in \mathbb{C}^{n\times m}$. Then

(a) $\|\mathcal{Z}\| = \|\mathsf{Z}\|_\infty$ if $\mathbb{C}^{n\times m}$ is given the 1-norm, $\|\,\underline{x}\,\|_1 := \sum_{j=1}^{m} \|x_j\|$.

(b) $\|\mathcal{Z}\| = \|\mathsf{Z}\|_1$ if $\mathbb{C}^{n\times m}$ is given the ∞-norm, $\|\,\underline{x}\,\|_\infty := \max_{j=1,\ldots,m} \|x_j\|$.

1.24 Let m, n and p be integers.

(a) If $\mathsf{A} \in \mathbb{C}^{n\times m}$ and $\mathsf{B} \in \mathbb{C}^{m\times p}$, then $\|\mathsf{AB}\|_{\mathrm{F}} \le \|\mathsf{A}\|_{\mathrm{F}}\|\mathsf{B}\|_2$ and $\|\mathsf{AB}\|_{\mathrm{F}} \le \|\mathsf{A}\|_2\|\mathsf{B}\|_{\mathrm{F}}\}$.

(b) Let m and n be such that $m \le n$. Let $\mathsf{V} \in \mathbb{C}^{n\times m}$ and $[\mathsf{e}_1,\ldots,\mathsf{e}_m]$ be the standard basis for $\mathbb{C}^{m\times 1}$. Then V is a unitary matrix if and only if $\|\mathsf{V}\|_2 = \|\mathsf{V}\mathsf{e}_j\|_2 = 1$ for all $j = 1,\ldots,m$.

(c) A matrix $\mathsf{A} \in \mathbb{C}^{n\times n}$ is normal if and only if there exists a nonsingular matrix V such that $\mathsf{V}^{-1}\mathsf{AV}$ is a diagonal matrix and $\kappa_2(\mathsf{V}) := \|\mathsf{V}\|_2\,\|\mathsf{V}^{-1}\|_2 = 1$.

1.25 Let Λ be a spectral set for T. Consider a positive integer p, $\underline{x} := [x_1,\ldots,x_p] \in X^{1\times p}$ and $\mathsf{Z} \in \mathbb{C}^{p\times p}$ such that $\underline{T}\,\underline{x} = \underline{x}\,\mathsf{Z}$. Let $Y := \mathrm{span}\{x_1,\ldots,x_p\}$. Then

(a) Y is a closed invariant subspace for T.

(b) $\mathrm{sp}\left(T_{|Y,Y}\right) \subset \Lambda$ if and only if $Y \subset M(T,\Lambda)$.

(c) If $\dim Y = p$, then $\mathrm{sp}\left(T_{|Y,Y}\right) = \mathrm{sp}(\mathsf{Z})$.

Chapter 2

Spectral Approximation

Our main interest lies in determining clusters of spectral values of finite type and the associated spectral subspaces of a given bounded linear operator T. Since exact computations are almost always impossible, we attempt to obtain numerical approximations. Usually these approximations are the exact results of spectral computations on a bounded linear operator $\widetilde{T}$ which is close to the given operator T. Often the operator $\widetilde{T}$ is chosen to be a member of a sequence (T_n) of bounded linear operators which converges to T pointwise, or with respect to the operator norm, or in a 'collectively compact' manner. Other than these three modes of convergence, we study a new mode of convergence which is strong enough to yield the desired spectral results and general enough to be applicable in a number of situations. Under this mode of convergence, properties similar to the upper semicontinuity and the lower semicontinuity of the spectrum are proved. The chapter concludes with error bounds first for the approximation of a simple eigenvalue and the corresponding eigenvector, and then for the approximation of a cluster of multiple eigenvalues and a basis for the associated spectral subspace.

2.1 Convergence of Operators

In this chapter, T and T_n denote bounded linear operators on a complex Banach space X, that is, $T, T_n \in \mathrm{BL}(X)$. Unless otherwise mentioned, the convergence is as $n \to \infty$.

If (T_n) converges to T in some sense, the following questions arise naturally:

1. If $\lambda_n \in \mathrm{sp}(T_n)$ and $\lambda_n \to \lambda$, does $\lambda \in \mathrm{sp}(T)$?
2. If $\lambda \in \mathrm{sp}(T)$, does there exist $\lambda_n \in \mathrm{sp}(T_n)$ for each large n such that $\lambda_n \to \lambda$?

We shall say that under a given mode of convergence, denoted by $\to$,

> **Property U** holds if, whenever $T_n \to T$, λ_n belongs to $\mathrm{sp}(T_n)$ and (λ_n) converges to λ, we have $\lambda \in \mathrm{sp}(T)$;
>
> **Property L** holds if, whenever $T_n \to T$ and $\lambda \in \mathrm{sp}(T)$, there exists some λ_n belonging to $\mathrm{sp}(T_n)$ for each large enough n such that (λ_n) converges to λ.

Let us consider the following property:

'Whenever $T_n \to T$, $\sup\{\mathrm{dist}(\mu, \mathrm{sp}(T)) : \mu \in \mathrm{sp}(T_n)\} \to 0$.'

It is known as the **upper semicontinuity of the spectrum** under the given mode of convergence, and Property U is a consequence of it.

Property L is known as the **lower semicontinuity of the spectrum** under the given mode of convergence. It can also be stated as follows:

'Whenever $T_n \to T$ and $\lambda \in \mathrm{sp}(T)$, $\mathrm{dist}(\lambda, \mathrm{sp}(T_n)) \to 0$.

The upper semicontinuity of the spectrum and the lower semicontinuity of the spectrum, taken together, give the **continuity of the spectrum** in the sense of Kuratowski.

If $T_n \to T$, λ is a spectral value of T of finite type, λ_n is a spectral value of T_n of finite type, and (λ_n) converges to λ in $\mathbb{C}$, we would like to know how the algebraic multiplicities of the λ_n's are related to the algebraic multiplicity of λ and whether a basis for a spectral subspace of T corresponding to λ can be approximated by bases for the corresponding spectral subspaces of T_n.

Let us consider three well-known modes of convergence.

The **pointwise convergence**, denoted by $T_n \xrightarrow{\mathrm{p}} T$:

$$\|T_n x - Tx\| \to 0 \text{ for every } x \in X.$$

The **norm convergence**, denoted by $T_n \xrightarrow{\mathrm{n}} T$:

$$\|T_n - T\| \to 0.$$

The collectively compact convergence, denoted by $T_n \xrightarrow{\text{cc}} T$:

$T_n \xrightarrow{\text{p}} T$, and for some positive integer n_0,

$$\bigcup_{n \geq n_0} \{(T_n - T)x \,:\, x \in X,\, \|x\| \leq 1\}$$

is a relatively compact subset of X. If T is compact, then the latter condition is equivalent to the condition that for some positive integer n_0, the set

$$\bigcup_{n \geq n_0} \{T_n x \,:\, x \in X,\, \|x\| \leq 1\}$$

is a relatively compact subset of X.

While pointwise convergence and norm convergence are classical concepts, the concept of a 'collectively compact' convergence was developed by Atkinson and by Anselone. (See [21] and [17].)

If $T_n \xrightarrow{\text{n}} T$ or $T_n \xrightarrow{\text{cc}} T$, then clearly $T_n \xrightarrow{\text{p}} T$. But the converse is not true.

Example 2.1 In this example, $T_n \xrightarrow{\text{p}} I$ but $T_n \overset{\text{n}}{\not\to} I$ and $T_n \overset{\text{cc}}{\not\to} I$: Consider $X := \ell^p$, $1 \leq p < \infty$. For $n = 1, 2, \ldots$ and $x := \sum_{k=1}^{\infty} x(k)e_k$ in X, let

$$T_n x := \sum_{k=1}^{n} x(k)e_k.$$

Then each T_n is a bounded finite rank operator on X and $T_n \xrightarrow{\text{p}} I$. But $T_n \overset{\text{n}}{\not\to} I$ since $\|T_n - I\| = 1$ for each n, and $T_n \overset{\text{cc}}{\not\to} I$, since given any positive integer n_0,

$$e_k \in \{(I - T_n)x \,:\, x \in X,\, \|x\| \leq 1\} \quad \text{for } k = n_0 + 1, n_0 + 2, \ldots$$

but the sequence (e_k) has no convergent subsequence. ∎

If X is an infinite dimensional Banach space, then neither norm convergence nor collectively compact convergence is stronger than the other. For example, if $T := I$ and $T_n := c_n I$, where $c_n \to 1$ in $\mathbb{C}$, then $T_n \xrightarrow{\text{n}} T$ but $T_n \overset{\text{cc}}{\not\to} T$, unless $c_n := 1$ for all large n. On the other hand, by a result of Josefson and Nissenzweig ([36], Chapter XII), there is a sequence (f_n)

in X^* such that $\|f_n\| = 1$ for each n and $\langle x\,,\, f_n\rangle \to 0$ for each $x \in X$. For a fixed nonzero $x_0 \in X$, let $T_n x := \langle x\,,\, f_n\rangle x_0$, $x \in X$, and $T := O$. Then $T_n \stackrel{\text{cc}}{\to} T$, but $T_n \stackrel{\text{n}}{\not\to} T$.

Under additional hypotheses, norm convergence may imply collectively compact convergence, or collectively compact convergence may imply norm convergence. (See Exercise 2.1.)

We shall see that under pointwise convergence neither Property U nor Property L holds. On the other hand, we shall prove that under norm convergence Property U holds, and Property L holds 'at each isolated point of the spectrum'. However, there are operators of practical interest and there are natural approximations of such an operator T which are not norm convergent. An example of this kind is provided by the Nyström approximation of a Fredholm integral operator on $C^0([a,b])$ with a continuous kernel. (cf. Proposition 4.6.) Many of these situations are covered by collectively compact convergence. However, compactness is an essential feature of this mode of convergence. This will not be the case of the *new* mode of convergence which we define as follows:

The ν-convergence, denoted by $T_n \stackrel{\nu}{\to} T$:

$$(\|T_n\|) \text{ is bounded}, \quad \|(T_n - T)T\| \to 0 \quad \text{and} \quad \|(T_n - T)T_n\| \to 0.$$

These conditions on the norms of the operators T_n, $(T_n - T)T$ and $(T_n - T)T_n$ have evolved from [1] and [61].

Simple examples show that none of these conditions is implied by the remaining two. For example, let $X := \mathbb{C}^{2\times 1}$ with any norm, $\mathsf{A} := \begin{bmatrix} 0 & 1 \\ 0 & 0 \end{bmatrix}$ and $\mathsf{A}_n := \begin{bmatrix} 0 & n \\ 0 & 0 \end{bmatrix}$ for $n = 1, 2, \ldots$ Then $\|(\mathsf{A}_n - \mathsf{A})\mathsf{A}\| = 0 = \|(\mathsf{A}_n - \mathsf{A})\mathsf{A}_n\|$, but $(\|\mathsf{A}_n\|)$ is unbounded. For other examples, see Exercise 2.18.

We shall show that under ν-convergence, Property U holds (Corollary 2.7) and Property L holds 'at each nonzero isolated point of the spectrum' (Corollary 2.13 and the remark thereafter). In Chapter 3 we shall give several practically useful examples of ν-convergence.

Lemma 2.2

(a) *If $T_n \stackrel{\text{n}}{\to} T$, then $T_n \stackrel{\nu}{\to} T$. Conversely, if $0 \notin \operatorname{sp}(T)$ and $T_n \stackrel{\nu}{\to} T$, then $T_n \stackrel{\text{n}}{\to} T$.*

(b) *Let* $T_n \xrightarrow{\nu} T$ *and* $U_n \xrightarrow{\text{n}} U$. *Then* $T_n + U_n \xrightarrow{\nu} T + U$ *if and only if* $(T_n - T)U \xrightarrow{\text{n}} O$. *In particular,* (i) *if* $T_n \xrightarrow{\nu} T$ *and* $U_n \xrightarrow{\text{n}} O$, *then* $T_n + U_n \xrightarrow{\nu} T$, *and* (ii) *if* $T_n \xrightarrow{\nu} O$, $U_n \xrightarrow{\text{n}} U$ *and* $T_nU \xrightarrow{\text{n}} O$, *then* $T_n + U_n \xrightarrow{\nu} U$.

(c) *If* $T_n \xrightarrow{\text{cc}} T$ *and* T *is a compact operator, then* $T_n \xrightarrow{\nu} T$.

(d) *Let* $T_n, U_n \in \text{BL}(X)$, $T_n \xrightarrow{\text{cc}} T$, T *be a compact operator,* $U_n \xrightarrow{\text{n}} O$ *and* $\widehat{T}_n := T_n + U_n$. *Then* $\widehat{T}_n \xrightarrow{\nu} T$. *In addition, if* $T_n \overset{\text{n}}{\not\to} T$, *then* $\widehat{T}_n \overset{\text{n}}{\not\to} T$; *and if* $U_n \overset{\text{cc}}{\not\to} O$, *then* $\widehat{T}_n \overset{\text{cc}}{\not\to} T$.

Proof

(a) Let $T_n \xrightarrow{\text{n}} T$. Since $\|T_n\| \leq \|T_n - T\| + \|T\|$, $\|(T_n - T)T\| \leq \|T_n - T\| \, \|T\|$ and $\|(T_n - T)T_n\| \leq \|T_n - T\| \, \|T_n\|$, we see that $T_n \xrightarrow{\nu} T$.

Conversely, let $0 \notin \text{sp}(T)$ and $T_n \xrightarrow{\nu} T$. Then T is invertible and $\|T_n - T\| = \|(T_n - T)TT^{-1}\| \leq \|(T_n - T)T\| \, \|T^{-1}\|$, so that $T_n \xrightarrow{\text{n}} T$.

(b) Since $\|T_n + U_n\| \leq \|T_n\| + \|U_n\|$, we see that the sequence $(\|T_n + U_n\|)$ is bounded. Assume that $(T_n - T)U \xrightarrow{\text{n}} O$. As

$$\begin{aligned}\|(T_n + U_n - T - U)(T + U)\| &\leq \|(T_n - T)T\| \\ &+\|(T_n - T)U\| + \|U_n - U\| \, \|T + U\|,\end{aligned}$$

$$\begin{aligned}\|(T_n + U_n - T - U)(T_n + U_n)\| &\leq \|(T_n - T)T_n\| \\ &+\|(T_n - T)U_n\| + \|U_n - U\|(\|T_n\| + \|U_n\|),\end{aligned}$$

where $\|(T_n - T)U_n\| \leq \|T_n - T\| \, \|U_n - U\| + \|(T_n - T)U\|$, we see that $T_n + U_n \xrightarrow{\nu} T + U$.

Conversely, assume that $T_n + U_n \xrightarrow{\nu} T + U$. Since

$$\begin{aligned}(T_n - T)U = {} & (T_n + U_n - T - U)(T + U) \\ & -(T_n - T)T - (U_n - U)(T + U),\end{aligned}$$

we obtain $(T_n - T)U \xrightarrow{\text{n}} O$.

The particular cases (i) and (ii) follow easily.

(c) Let $T_n \xrightarrow{\text{cc}} T$. By the Uniform Boundedness Principle (Theorem 9.1 of [55] or Theorem 4.7-3 of [48]), the sequence $(\|T_n\|)$ is bounded and the pointwise convergence of (T_n) to T is uniform on the relatively compact sets $\{Tx : x \in X, \|x\| \leq 1\}$ and $\bigcup_{n \geq n_0} \{T_nx : x \in X, \|x\| \leq 1\}$.

Hence $\|(T_n - T)T\| \to 0$ and $\|(T_n - T)T_n\| \to 0$. Thus $T_n \xrightarrow{\nu} T$.

(d) By (c) above, $T_n \xrightarrow{\nu} T$ and hence by (b)(i) above, $\widehat{T}_n \xrightarrow{\nu} T$. If $T_n \overset{\text{n}}{\not\to} T$, then $\widehat{T}_n \overset{\text{n}}{\not\to} T$ since $\|T_n - T\| = \|\widehat{T}_n - U_n - T\| \leq \|\widehat{T}_n - T\| + \|U_n\|$. Finally, we show that if $\widehat{T}_n \xrightarrow{\text{cc}} T$, then $U_n \xrightarrow{\text{cc}} O$. Clearly, $U_n \xrightarrow{\text{p}} O$. Also, for any positive integer n_0, the set $F := \bigcup_{n \geq n_0} \{U_n x : x \in X, \|x\| \leq 1\}$ is a subset of $\widehat{E} - E$, where $\widehat{E}$ is the closure of the set $\bigcup_{n \geq n_0} \{\widehat{T}_n x : x \in X, \|x\| \leq 1\}$ and E is the closure of the set $\bigcup_{n \geq n_0} \{T_n x : x \in X, \|x\| \leq 1\}$. Since $T_n \xrightarrow{\text{cc}} T$ and T is a compact operator, the set E is compact for some n_0. If $\widehat{T}_n \xrightarrow{\text{cc}} T$, the set $\widehat{E}$ would also be compact for some n_0, implying the compactness of the set $\widehat{E} - E$ and, in turn, the relative compactness of the set F for some n_0. ∎

We make some remarks on the preceding lemma.

Part (a) shows that norm convergence implies ν-convergence. Part (c) shows that collectively compact convergence to a compact operator implies ν-convergence. On the other hand, if we let $X := \ell^2$, $T := O$, $T_n x := x(n+1) e_n$ for $n = 1, 2, \ldots$ and $x := \sum_{k=1}^{\infty} x(k) e_k \in \ell^2$, then $T_n \xrightarrow{\text{p}} T$, $T_n \xrightarrow{\nu} T$, but $T_n \overset{\text{n}}{\not\to} T$, $T_n \overset{\text{cc}}{\not\to} T$. In this example, each T_n is of finite rank.

Part (a) says that $T_n \xrightarrow{\nu} T$ is equivalent to $T_n \xrightarrow{\text{n}} T$ if $0 \notin \operatorname{sp}(T)$. However, we shall be often interested in situations where $0 \in \operatorname{sp}(T)$. For example, if X is infinite dimensional and T is compact, then $0 \in \operatorname{sp}(T)$.

Part (b) implies that ν-convergence is stable under norm perturbations.

Part (d) shows how ν-convergence can be available even in the absence of norm convergence and collectively compact convergence. For instance, let T be a Fredholm integral operator on $C^0([a, b])$ with a continuous kernel, T_n a Nyström approximation of T and $U_n := c_n I$, where $c_n \to 0$ in $\mathbb{C}$. Then the conditions of part (d) are satisfied. In Subsection 4.2.3, we shall consider Fredholm integral operators with 'weakly singular' kernels. A modification of the Nyström approximation of such an operator will then yield a useful example in which $\widehat{T}_n \xrightarrow{\nu} T$, but $\widehat{T}_n \overset{\text{n}}{\not\to} T$ and $\widehat{T}_n \overset{\text{cc}}{\not\to} T$.

We note that if $T_n \xrightarrow{\text{n}} T$, or $T_n \xrightarrow{\text{cc}} T$ and each T_n is a compact operator, then T is necessarily compact. On the other hand, when $T_n \xrightarrow{\nu} T$ and each T_n is a compact operator (or even a bounded finite rank operator), the operator T need not be compact. We give a simple example to illustrate this feature.

Example 2.3 In this example, $T_n \xrightarrow{\nu} T$, $\mathrm{rank}(T_n) < \infty$, but T is not compact:

Consider $1 \leq p < \infty$. For $x := \sum_{k=1}^{\infty} x(k)e_k \in \ell^p$, let

$$Tx := \sum_{k=1}^{\infty} x(2k)e_{2k-1}$$

and for each positive integer n,

$$T_n x := \sum_{k=1}^{n} x(2k)e_{2k-1}.$$

Clearly, $T, T_n \in \mathrm{BL}(\ell^p)$, $\|T_n\| = 1$ and $\mathrm{rank}(T_n) = n$. Since for all $x \in \ell^p$,

$$(T_n - T)x = -\sum_{k=n+1}^{\infty} x(2k)e_{2k-1},$$

we see that $(T_n - T)T = O = (T_n - T)T_n$. Hence $T_n \xrightarrow{\nu} T$. But T is not a compact operator. This follows since for the bounded sequence (e_{2k}) in ℓ^p, the sequence $(Te_{2k}) = (e_{2k-1})$ does not have a convergent subsequence. As a result, we have $T_n \overset{\mathrm{n}}{\not\to} T$ and $T_n \overset{\mathrm{cc}}{\not\to} T$. ∎

There are several other notions of convergence of a sequence of operators which yield spectral results: compact convergence and regular convergence ([19]), stable and strongly stable convergence ([25]), resolvent operator convergence ([54]), spectral convergence ([4]), convergence of (T_n) to T in the sense that the spectral radius $\rho(T_n - T)$ of $T_n - T$ tends to zero and $\|(T_n - T)T_n\|$ tends to zero ([12]), asymptotically compact convergence ([20]) and also discrete versions of some of these ([68], [40], [69]), to name a few. Our reason for considering the three conditions which comprise ν-convergence is twofold. On the one hand, these conditions are general enough to encompass a wide variety of approximation methods; and on the other hand, they are simple enough to be verified in practice, as we shall see in Chapter 3. We point out that the operations of addition and composition in $\mathrm{BL}(X)$ are not compatible with ν-convergence. (See Exercise 2.3.)

Also, ν-convergence is a 'pseudo-convergence' in the sense that it is possible to have $T_n \xrightarrow{\nu} T$ and $T_n \xrightarrow{\nu} U$, where $U \neq T$. (See Exercise 2.4.) However, if $T_n \xrightarrow{\nu} T$ and $T_n \xrightarrow{\nu} U$, then $\mathrm{sp}(U) = \mathrm{sp}(T)$ and $Ux = Tx$ whenever x belongs to a spectral subspace of U corresponding to a spectral set not containing 0. (See Exercise 2.12.)

2.2 Property U

If $T_n \xrightarrow{\mathrm{p}} T$, then we may have $\lambda_n \in \mathrm{sp}(T_n)$ with $\lambda_n \to \lambda$, but $\lambda \notin \mathrm{sp}(T)$.

Example 2.4 Property U does not hold under pointwise convergence: Consider $X := \ell^2$. For $x := \sum_{k=1}^{\infty} x(k)e_k \in X$, let $Tx := x(1)e_1$ and for each integer $n \geq 2$,

$$T_n x := x(1)e_1 - x(n)e_n.$$

Since $\|T_n x - Tx\|_2 = |x(n)| \to 0$ for every $x \in X$, we see that $T_n \xrightarrow{\mathrm{p}} T$. Now

$$\mathrm{sp}(T) = \{0, 1\} \text{ and } \mathrm{sp}(T_n) = \{-1, 0, 1\}.$$

Since $\lambda_n := -1 \in \mathrm{sp}(T_n)$ for each n, but $-1 \notin \mathrm{sp}(T)$, we see that Property U does not hold. Also, if for $x \in X$, we let

$$T_n x := x(1)e_1 + x(n)e_n,$$

then $T_n \xrightarrow{\mathrm{p}} T$, 1 is an eigenvalue of T_n of algebraic multiplicity 2, but 1 is an eigenvalue of T of algebraic multiplicity 1.

On the other hand, if we let for $x \in X$,

$$\widetilde{T}x := x(1)e_1 + x(2)e_2 \quad \text{and} \quad \widetilde{T}_n x := x(1)e_1 + \frac{n-1}{n}\, x(2)e_2,$$

then again $\|\widetilde{T}_n x - \widetilde{T}x\|_2 = |x(2)|/n \to 0$ for every $x \in X$, so that $\widetilde{T}_n \xrightarrow{\mathrm{p}} \widetilde{T}$, 1 is an eigenvalue of each $\widetilde{T}_n$ of algebraic multiplicity 1, while 1 is eigenvalue of $\widetilde{T}$ of algebraic multiplicity 2.

We conclude that the algebraic multiplicities are not preserved under pointwise convergence. ∎

We now wish to show that Property U holds under ν-convergence. For this purpose, we prove some preliminary results.

For T and $\widetilde{T}$ in $\mathrm{BL}(X)$, we shall denote $R(T,\cdot)$ and $R(\widetilde{T},\cdot)$ by $R(\cdot)$ and $\widetilde{R}(\cdot)$, respectively.

Proposition 2.5
Let T and $\widetilde{T}$ be in $\mathrm{BL}(X)$.

(a) Second Resolvent Identity: *Let $z \in \mathrm{re}(T) \cap \mathrm{re}(\widetilde{T})$. Then*

$$\widetilde{R}(z) - R(z) = \widetilde{R}(z)(T - \widetilde{T})R(z) = R(z)(T - \widetilde{T})\widetilde{R}(z).$$

(b) Second Neumann Expansion: *Let $z \in \mathrm{re}(T)$ be such that $\rho((T - \widetilde{T})R(z)) < 1$. Then $z \in \mathrm{re}(\widetilde{T})$ and*

$$\widetilde{R}(z) = R(z)\sum_{k=0}^{\infty}[(T - \widetilde{T})R(z)]^k.$$

If in fact $\|(T - \widetilde{T})R(z)\| < 1$, then

$$\|\widetilde{R}(z)\| \le \frac{\|R(z)\|}{1 - \|(T - \widetilde{T})R(z)\|},$$

$$\|\widetilde{R}(z) - R(z)\| \le \frac{\|R(z)\|\,\|(T - \widetilde{T})R(z)\|}{1 - \|(T - \widetilde{T})R(z)\|}.$$

Also, if $\|[(T - \widetilde{T})R(z)]^2\| < 1$, then

$$\|\widetilde{R}(z)\| \le \frac{\|R(z)\|\,(1 + \|(T - \widetilde{T})R(z)\|)}{1 - \|[(T - \widetilde{T})R(z)]^2\|},$$

$$\|\widetilde{R}(z) - R(z)\| \le \frac{\|R(z)\|\,\|(T - \widetilde{T})R(z)\|\,(1 + \|(T - \widetilde{T})R(z)\|)}{1 - \|[(T - \widetilde{T})R(z)]^2\|}.$$

Proof

(a) For $z \in \mathrm{re}(T) \cap \mathrm{re}(\widetilde{T})$, we have

$$\widetilde{R}(z)(T - \widetilde{T})R(z) = \widetilde{R}(z)[(T - zI) - (\widetilde{T} - zI)]R(z) = \widetilde{R}(z) - R(z).$$

Interchanging T and $\widetilde{T}$, we obtain the other equality.

(b) For $z \in \mathrm{re}(T)$, consider the identity

$$\widetilde{T} - zI = T - zI - (T - \widetilde{T}) = [I - (T - \widetilde{T})R(z)](T - zI).$$

Since $\rho((T - \widetilde{T})R(z)) < 1$, the operator $I - (T - \widetilde{T})R(z)$ is invertible. The identity stated above shows that $z \in \mathrm{re}(\widetilde{T})$ and by Theorem 1.13(d),

$$\begin{aligned}\widetilde{R}(z) &= (T - zI)^{-1}[I - (T - \widetilde{T})R(z)]^{-1}\\ &= R(z)\sum_{k=0}^{\infty}[(T - \widetilde{T})R(z)]^k.\end{aligned}$$

Let $\|(T-\widetilde{T})R(z)\| < 1$. Then $\rho((T-\widetilde{T})R(z)) \le \|(T-\widetilde{T})R(z)\| < 1$ and we have

$$\|\widetilde{R}(z)\| \le \|R(z)\|\sum_{k=0}^{\infty}\|(T - \widetilde{T})R(z)\|^k = \frac{\|R(z)\|}{1 - \|(T - \widetilde{T})R(z)\|}.$$

Let now $\|[(T - \widetilde{T})R(z)]^2\| < 1$. Then by Proposition 1.15(b),

$$\rho((T - \widetilde{T})R(z)) \le \|[(T - \widetilde{T})R(z)]^2\|^{1/2} < 1,$$

so that $z \in \mathrm{re}(\widetilde{T})$ and we have

$$\begin{aligned}\widetilde{R}(z) &= R(z)\Big(\sum_{j=0}^{\infty}[(T - \widetilde{T})R(z)]^{2j} + \sum_{j=0}^{\infty}[(T - \widetilde{T})R(z)]^{2j+1}\Big)\\ &= R(z)[I + (T - \widetilde{T})R(z)]\sum_{j=0}^{\infty}\left[\left((T - \widetilde{T})R(z)\right)^2\right]^j.\end{aligned}$$

Hence

$$\|\widetilde{R}(z)\| \le \frac{\|R(z)\|\,(1 + \|(T - \widetilde{T})R(z)\|)}{1 - \|[(T - \widetilde{T})R(z)]^2\|}.$$

Also, by (a) above,

$$\|\widetilde{R}(z) - R(z)\| \le \|\widetilde{R}(z)\|\,\|(T - \widetilde{T})R(z)\|.$$

Thus the desired bounds for $\|\widetilde{R}(z) - R(z)\|$ follow. ∎

For T_n in $\mathrm{BL}(X)$, we denote $R(T_n, \cdot)$ by $R_n(\cdot)$.

Theorem 2.6
Let $T \in \mathrm{BL}(X)$ and E be a nonempty closed subset of $\mathrm{re}(T)$. Then

$$\alpha_1(E) := \sup\{\|R(z)\| \,:\, z \in E\} < \infty.$$

If $T_n \xrightarrow{\nu} T$, then there is a positive integer n_0 such that $E \subset \mathrm{re}(T_n)$ for all $n \geq n_0$ and

$$\alpha_2(E) := \sup\{\|R_n(z)\| \, : \, z \in E, \, n \geq n_0\} < \infty.$$

Proof

If $|z| > \|T\|$, then by Theorem 1.13(d),

$$\|R(z)\| \leq \frac{1}{|z| - \|T\|}.$$

Hence $\|R(z)\| \to 0$ as $|z| \to \infty$ and there is some $\alpha > 0$ such that $\|R(z)\| \leq 1$ for all $z \in \mathbb{C}$ with $|z| > \alpha$. Now $E_0 := \{z \in E \, : \, |z| \leq \alpha\}$ is a compact subset of $\mathrm{re}(T)$ and the function $z \longmapsto \|R(z)\|$ is continuous on E_0 by Remark 1.14. There is, therefore, some $\beta > 0$ such that $\|R(z)\| \leq \beta$ for all $z \in E_0$. Thus

$$\alpha_1(E) := \sup\{\|R(z)\| \, : \, z \in E\} \leq \max\{1\,,\,\beta\} < \infty.$$

Let $T_n \xrightarrow{\nu} T$.

Case (i): $0 \in E$.

Since $E \subset \mathrm{re}(T)$, we see that $0 \notin \mathrm{sp}(T)$. It follows from Lemma 2.2(a) that $T_n \xrightarrow{\mathrm{n}} T$. Find n_0 such that $\|T_n - T\| < 1/(2\alpha_1(E))$ for all $n \geq n_0$. Then for $z \in E$ and $n \geq n_0$,

$$\|(T_n - T)R(z)\| \leq \|T_n - T\|\alpha_1(E) \leq \frac{1}{2}$$

and by Proposition 2.5(b), $z \in \mathrm{re}(T_n)$ with

$$\|R_n(z)\| \leq \frac{\|R(z)\|}{1 - \|(T_n - T)R(z)\|} \leq 2\alpha_1(E).$$

This shows that

$$\alpha_2(E) := \sup\{\|R_n(z)\| \, : \, z \in E, \, n \geq n_0\} \leq 2\alpha_1(E) < \infty.$$

Case (ii): $0 \notin E$.

As E is a closed subset of $\mathbb{C}$, there is some $\delta > 0$ such that $|z| \geq \delta$ for all $z \in E$. For $z \in E$, we have by Proposition 1.10(e)

$$\begin{aligned}
&[(T - T_n)R(z)]^2 = (T - T_n)\frac{TR(z) - I}{z}(T - T_n)R(z) \\
&= \frac{1}{z}\left[(T - T_n)TR(z)(T - T_n) - (T - T_n)T + (T - T_n)T_n\right]R(z).
\end{aligned}$$

Since the sequence $(\|T_n\|)$ is bounded, there is some $t \geq 0$ such that $\|T - T_n\| \leq t$ for all n. As $\|(T - T_n)T\| \to 0$ and $\|(T - T_n)T_n\| \to 0$, find n_0 such that for $n \geq n_0$,

$$\big[(\alpha_1(E)t + 1)\|(T - T_n)T\| + \|(T - T_n)T_n\|\big]\alpha_1(E) \leq \frac{\delta}{2}.$$

Then for all $z \in E$ and $n \geq n_0$,

$$\|[(T - T_n)R(z)]^2\| \leq \frac{1}{2},$$

and by Proposition 2.5(b), $z \in \mathrm{re}(T_n)$ with

$$\|R_n(z)\| \leq \frac{\|R(z)\|\,(1 + \|(T - T_n)R(z)\|)}{1 - \|((T - T_n)R(z))^2\|} \leq 2\alpha_1(E)(1 + \alpha_1(E)t).$$

The theorem is proved. ∎

Corollary 2.7
Property U holds under ν-convergence, that is, if $T_n \xrightarrow{\nu} T$, $\lambda_n \in \mathrm{sp}(T_n)$ and $\lambda_n \to \lambda$, then $\lambda \in \mathrm{sp}(T)$.

Proof
Suppose for a moment that $\lambda \in \mathrm{re}(T)$. Since the set $\mathrm{re}(T)$ is open in $\mathbb{C}$ by Theorem 1.13(a), there is some $r > 0$ such that $E := \{z \in \mathbb{C} : |z - \lambda| \leq r\} \subset \mathrm{re}(T)$. By Theorem 2.6, $E \subset \mathrm{re}(T_n)$ for all large n. Since $\lambda_n \to \lambda$, we see that $\lambda_n \in E \subset \mathrm{re}(T_n)$ for all large n, which is contradictory to the hypothesis that $\lambda_n \in \mathrm{sp}(T_n)$ for each n. Hence λ must belong to $\mathrm{sp}(T)$. ∎

Under ν-convergence one can in fact prove the upper semicontinuity of each spectral set for T. (See Exercise 2.6.)

We note that Corollary 2.7 does not hold if, from the definition of the ν-convergence, one of the two conditions $\|(T_n - T)T\| \to 0$, $\|(T_n - T)T_n\| \to 0$ is omitted or if these two conditions are replaced by the condition $\|(T_n - T)^2\| \to 0$. (See Exercise 2.17 and parts (a), (b) and (c) of Exercise 2.18.)

2.3 Property L

Some spectral values of T may not be approximable by spectral values of T_n, even when $T_n \xrightarrow{\text{n}} T$, as the following example shows.

Example 2.8 Property L does not hold under norm convergence: Let $X := \ell^2(\mathbb{Z})$. For $x \in X$, let

$$(Tx)(k) := \begin{cases} x(k+1) & \text{if } k \neq -1, \\ 0 & \text{if } k = -1, \end{cases}$$

and for each positive integer n,

$$(T_n x)(k) := \begin{cases} x(k+1) & \text{if } k \neq -1, \\ \dfrac{x(0)}{n} & \text{if } k = -1. \end{cases}$$

Since $\|T_n x - Tx\|_2 = |x(0)|/n$ for all $x \in X$, we see that $\|T_n - T\| = 1/n \to 0$.

We now show that $\text{sp}(T) = \{\lambda \in \mathbb{C} : |\lambda| \leq 1\}$. If $|\lambda| < 1$, consider $x_\lambda \in X$ defined by $x_\lambda(k) := 0$ for all $k \leq -1$, $x_\lambda(0) := 1$ and $x_\lambda(k) := \lambda^k$ for all $k \geq 1$, and note that $Tx_\lambda = \lambda x_\lambda$ with $x_\lambda \neq 0$. Thus every λ satisfying $|\lambda| < 1$ is an eigenvalue and hence a spectral value of T. Since $\text{sp}(T)$ is closed, it follows that $\{\lambda \in \mathbb{C} : |\lambda| \leq 1\} \subset \text{sp}(T)$. But since $\rho(T) \leq \|T\| = 1$, we see that $\text{sp}(T)$ is contained in $\{\lambda \in \mathbb{C} : |\lambda| \leq 1\}$, and we are through.

On the other hand, we claim that $\text{sp}(T_n) = \{\lambda \in \mathbb{C} : |\lambda| = 1\}$ for each positive integer n. If $|\lambda| = 1$, consider $y \in X$ defined by $y(k) := 0$ for all $k \neq -1$ and $y(-1) := 1$, and note that there is no $x \in X$ with $T_n x - \lambda x = y$. Thus every λ satisfying $|\lambda| = 1$ is a spectral value of T. Also, since $\rho(T_n) \leq \|T_n\| = 1$,

$$\{\lambda \in \mathbb{C} : |\lambda| = 1\} \subset \text{sp}(T_n) \subset \{\lambda \in \mathbb{C} : |\lambda| \leq 1\}.$$

It can be seen that T_n is bijective and for $x \in X$,

$$(T_n^{-1} x)(k) = \begin{cases} x(k-1) & \text{if } k \neq 0, \\ n\, x(-1) & \text{if } k = 0, \end{cases}$$

so that $\|T_n^{-1}\| = n$. Similarly, for $j = 2, 3, \ldots$ and $x \in X$,

$$(T_n^{-j}x)(k) = \begin{cases} x(k-j) & \text{if } k \neq 0, 1, \ldots, j-1, \\ n\, x(k-j) & \text{if } k = 0, 1, \ldots, j-1, \end{cases}$$

so that $\|T_n^{-j}\| = n$ for each positive integer j. Hence

$$\rho(T_n^{-1}) = \lim_{j\to\infty} \|T_n^{-j}\|^{1/j} = \lim_{j\to\infty} n^{1/j} = 1.$$

Thus $\mathrm{sp}(T_n^{-1}) \subset \{\lambda \in \mathbb{C} : |\lambda| \leq 1\}$, $\mathrm{sp}(T_n) \subset \{\lambda \in \mathbb{C} : |\lambda| \geq 1\}$, and we are through.

It is now clear that if $|\lambda| < 1$, then $\lambda \in \mathrm{sp}(T)$, but there is no λ_n in $\mathrm{sp}(T_n)$ such that $\lambda_n \to \lambda$. ∎

The preceding example shows that the spectrum of an operator can suddenly shrink if the operator is subjected to an arbitrarily small norm perturbation. For examples of very drastic shrinkage, see Exercises 2.9 and 2.10.

We shall prove that (i) if λ is an isolated point of $\mathrm{sp}(T)$ and $T_n \xrightarrow{\text{n}} T$, or (ii) if λ is a nonzero isolated point of $\mathrm{sp}(T)$ and $T_n \xrightarrow{\nu} T$, then there is some $\lambda_n \in \mathrm{sp}(T_n)$ for each large n such that $\lambda_n \to \lambda$ (Corollary 2.13). Further, we shall prove that if λ is an eigenvalue having finite algebraic multiplicity m, then for each large n, there are a finite number of eigenvalues of T_n near λ and the sum of their algebraic multiplicities is m, and that the associated spectral subspaces corresponding to T and T_n are close to each other. In addition, we shall give error estimates for these spectral approximations. With this in mind, we consider a general set-up.

Let $T \in \mathrm{BL}(X)$ and Λ be a spectral set for T, that is, Λ is a subset of $\mathrm{sp}(T)$ such that the sets Λ and $\mathrm{sp}(T) \setminus \Lambda$ are also closed in $\mathbb{C}$. Then there is a Cauchy contour $\mathrm{C} \in \mathcal{C}(T, \Lambda)$, that is, C satisfies $\Lambda \subset \mathrm{int}(\mathrm{C})$ and $(\mathrm{sp}(T) \setminus \Lambda) \subset \mathrm{ext}(\mathrm{C})$.

Recall that the spectral projection associated with T and Λ is given by

$$P := P(T, \Lambda) := -\frac{1}{2\pi \mathrm{i}} \int_{\mathrm{C}} R(z)\, dz$$

and that it does not depend on $\mathrm{C} \in \mathcal{C}(T, \Lambda)$.

Let $\ell(\mathrm{C})$ denote the length of a Cauchy contour C and

$$\delta(\mathrm{C}) := \min\{|z| : z \in \mathrm{C}\}.$$

Proposition 2.9
Let Λ be a spectral set for T, $\mathrm{C} \in \mathcal{C}(T,\Lambda)$ and $T_n \xrightarrow{\nu} T$.

(a) *There is a positive integer n_0 such that for each $n \geq n_0$, C lies in $\mathrm{re}(T_n)$. Further, if*

$$\Lambda_n := \mathrm{sp}(T_n) \cap \mathrm{int}(\mathrm{C}) \quad \textit{and} \quad P_n := -\frac{1}{2\pi \mathrm{i}} \int_{\mathrm{C}} R_n(z)\, dz,$$

then Λ_n and P_n do not depend on $\mathrm{C} \in \mathcal{C}(T,\Lambda)$ for each large n.

(b) *In Theorem 2.6, let $E := \mathrm{C}$. Then for each $n \geq n_0$, the following estimates hold:*

$$\|P\| \leq \frac{\ell(\mathrm{C})}{2\pi}\alpha_1(\mathrm{C}), \quad \|P_n\| \leq \frac{\ell(\mathrm{C})}{2\pi}\alpha_2(\mathrm{C}),$$

and

$$\|P_n - P\| \leq \frac{\ell(\mathrm{C})}{2\pi}\alpha_1(\mathrm{C})\alpha_2(\mathrm{C})\|T_n - T\|,$$
$$\|(P_n - P)P\| \leq \frac{\ell(\mathrm{C})}{2\pi}\alpha_1(\mathrm{C})\alpha_2(\mathrm{C})\|(T_n - T)P\|,$$
$$\|(P_n - P)P_n\| \leq \frac{\ell(\mathrm{C})}{2\pi}\alpha_1(\mathrm{C})\alpha_2(\mathrm{C})\|(T_n - T)P_n\|.$$

Further, if $0 \in \mathrm{ext}(\mathrm{C})$, then

$$\|(T_n - T)P\| \leq \frac{\ell(\mathrm{C})}{2\pi}\frac{\alpha_1(\mathrm{C})}{\delta(\mathrm{C})}\|(T_n - T)T\|,$$
$$\|(T_n - T)P_n\| \leq \frac{\ell(\mathrm{C})}{2\pi}\frac{\alpha_2(\mathrm{C})}{\delta(\mathrm{C})}\|(T_n - T)T_n\|.$$

Proof

(a) Letting $E := \mathrm{C}$ in Theorem 2.6, we see that there exists some n_0 such that $\mathrm{C} \subset \mathrm{re}(T_n)$ for each $n \geq n_0$.

Consider another Cauchy contour $\widetilde{\mathrm{C}}$ in $\mathcal{C}(T,\Lambda)$. Then again there exists some $\widetilde{n}_0$ such that $\widetilde{\mathrm{C}} \subset \mathrm{re}(T_n)$ for each $n \geq \widetilde{n}_0$, and we let

$$\widetilde{\Lambda}_n := \mathrm{sp}(T_n) \cap \mathrm{int}(\widetilde{\mathrm{C}}).$$

We claim that $\Lambda_n \subset \widetilde{\Lambda}_n$ for all large n. Suppose for a moment that this is not the case. Then there exist positive integers $k(1) < k(2) < \cdots$ and complex numbers $\lambda_{k(1)}, \lambda_{k(2)}, \dots$ such that $\lambda_{k(n)} \in \Lambda_{k(n)} \setminus \text{int}(\widetilde{\mathrm{C}})$ for $n = 1, 2, \dots$. Since $\Lambda_{k(n)}$ is contained in the compact set $\text{int}(\mathrm{C}) \cup \mathrm{C}$, there are positive integers $j(1) < j(2) < \cdots$ such that the subsequence $(\lambda_{k(j(n))})$ converges to some $\lambda \in \text{int}(\mathrm{C}) \cup \mathrm{C}$ as $n \to \infty$. Since $T_{k(j(n))} \xrightarrow{\nu} T$ as $n \to \infty$, we see that $\lambda \in \text{sp}(T)$ by Property U as proved in Corollary 2.7. As C is contained in $\text{re}(T)$, we see that

$$\lambda \in \text{sp}(T) \cap \text{int}(\mathrm{C}) = \Lambda \subset \text{int}(\widetilde{\mathrm{C}}).$$

Since $\text{int}(\widetilde{\mathrm{C}})$ is an open subset of $\mathbb{C}$ containing λ and $\lambda_{k(j(n))} \to \lambda$ as $n \to \infty$, it follows that

$$\{\lambda_{k(j(1))}, \lambda_{k(j(2))}, \dots\} \cap \text{int}(\widetilde{\mathrm{C}}) \neq \emptyset.$$

This contradicts our choice of $\lambda_{k(j(1))}, \lambda_{k(j(2))}, \dots$ Hence $\Lambda_n \subset \widetilde{\Lambda}_n$ for all large n. Interchanging the roles of C and $\widetilde{\mathrm{C}}$, we obtain $\widetilde{\Lambda}_n \subset \Lambda_n$ for all large n. Thus $\widetilde{\Lambda}_n = \Lambda_n$ for all large n. It follows that Λ_n is a spectral set for T_n and $\widetilde{\mathrm{C}} \in \mathcal{C}(T_n, \Lambda_n)$ for all large n, so that

$$-\frac{1}{2\pi\mathrm{i}} \int_{\widetilde{\mathrm{C}}} R(\widetilde{z})\, d\widetilde{z} = P(T_n, \Lambda_n) = P_n.$$

(b) Let $n \geq n_0$. The estimates for $\|P\|$ and $\|P_n\|$ follow from the definitions of P and P_n. Next, by the Second Resolvent Identity (Proposition 2.5(a)),

$$\begin{aligned} P_n - P &= -\frac{1}{2\pi\mathrm{i}} \int_{\mathrm{C}} [R_n(z) - R(z)]\, dz \\ &= \frac{1}{2\pi\mathrm{i}} \int_{\mathrm{C}} R_n(z)(T_n - T)R(z)\, dz \\ &= \frac{1}{2\pi\mathrm{i}} \int_{\mathrm{C}} R(z)(T_n - T)R_n(z)\, dz. \end{aligned}$$

Hence the estimate for $\|P_n - P\|$ also follows easily. As P and $R(z)$ commute for all $z \in \mathrm{C}$,

$$(P_n - P)P = \frac{1}{2\pi\mathrm{i}} \int_{\mathrm{C}} R_n(z)(T_n - T)PR(z)\, dz$$

and as P_n and $R_n(z)$ commute for all $z \in \mathrm{C}$,

$$(P_n - P)P_n = \frac{1}{2\pi\mathrm{i}} \int_{\mathrm{C}} R(z)(T_n - T)P_nR_n(z)\, dz.$$

Hence the estimates for $\|(P_n - P)P\|$ and $\|(P_n - P)P_n\|$ follow.

Finally, assume that $0 \in \text{ext}(\mathrm{C})$. Then

$$P = -\frac{1}{2\pi\mathrm{i}}\int_{\mathrm{C}} [TR(z) - I]\frac{dz}{z} = -\frac{T}{2\pi\mathrm{i}}\int_{\mathrm{C}} R(z)\frac{dz}{z},$$
$$P_n = -\frac{1}{2\pi\mathrm{i}}\int_{\mathrm{C}} [T_nR_n(z) - I]\frac{dz}{z} = -\frac{T_n}{2\pi\mathrm{i}}\int_{\mathrm{C}} R_n(z)\frac{dz}{z},$$

since $\displaystyle\int_{\mathrm{C}} \frac{dz}{z} = 0$. Hence

$$\begin{aligned}
\|(T_n - T)P\| &= \left\|\frac{(T_n - T)T}{2\pi\mathrm{i}}\int_{\mathrm{C}} \frac{R(z)}{z}\,dz\right\| \\
&\leq \frac{\ell(\mathrm{C})}{2\pi}\frac{\alpha_1(\mathrm{C})}{\delta(\mathrm{C})}\|(T_n - T)T\|, \\
\|(T_n - T)P_n\| &= \left\|\frac{(T_n - T)T_n}{2\pi\mathrm{i}}\int_{\mathrm{C}} \frac{R_n(z)}{z}\,dz\right\| \\
&\leq \frac{\ell(\mathrm{C})}{2\pi}\frac{\alpha_2(\mathrm{C})}{\delta(\mathrm{C})}\|(T_n - T)T_n\|.
\end{aligned}$$

This completes the proof. ∎

In Chapter 3, we shall consider some situations where the rates at which $\|T_n - T\|$, $\|(T_n - T)T\|$ and $\|(T_n - T)T_n\|$ tend to 0 can be estimated. (See Theorem 4.12(a), Example 4.13 and Exercise 4.12.)

Corollary 2.10

Let Λ be a spectral set for T. With the notation of Proposition 2.9,

(a) *if $T_n \xrightarrow{\mathrm{n}} T$, then $P_n \xrightarrow{\mathrm{n}} P$, and*

(b) *if $0 \notin \Lambda$ and $T_n \xrightarrow{\nu} T$, then $\|(T_n - T)P\| \to 0$, $\|(T_n - T)P_n\| \to 0$ and $P_n \xrightarrow{\nu} P$.*

Proof

(a) Let $T_n \xrightarrow{\mathrm{n}} T$. The estimate for $\|P_n - P\|$ given in Proposition 2.9(b) shows that $P_n \xrightarrow{\mathrm{n}} P$.

(b) Let now $0 \notin \Lambda$ and $T_n \xrightarrow{\nu} T$. If $0 \notin \mathrm{sp}(T)$, then $T_n \xrightarrow{\mathrm{n}} T$ by Lemma 2.2(a) and hence $P_n \xrightarrow{\mathrm{n}} P$ by (a) above. Thus we are through.

If $0 \in \mathrm{sp}(T)$, then since $\mathrm{C} \in \mathcal{C}(T,\Lambda)$ separates the spectral set Λ from the spectral value 0 of T, we see that $0 \in \mathrm{ext}(\mathrm{C})$; and the estimates for $\|P_n\|$, $\|(T_n-T)P\|$, $\|(T_n-T)P_n\|$, $\|(P_n-P)P\|$ and $\|(P_n-P)P_n\|$ given in Proposition 2.9(b) yield the desired results. ∎

The following elementary result will be used in the proof of the next theorem.

Lemma 2.11
Let P and $\widetilde{P}$ be projections in $\mathrm{BL}(X)$ such that $\rho(P-\widetilde{P}) < 1$, $Y := \mathcal{R}(P)$, and $\widetilde{Y} := \mathcal{R}(\widetilde{P})$. Then the linear map $\widetilde{P}_{|Y,\widetilde{Y}}$ from Y to $\widetilde{Y}$ is bijective and $\mathrm{rank}(P) = \mathrm{rank}(\widetilde{P})$.

Proof
Let $y \in Y$ with $\widetilde{P}y = 0$. Since $\rho(P-\widetilde{P}) < 1$, we have $1 \in \mathrm{re}(P-\widetilde{P})$, that is, the map $I-(P-\widetilde{P})$ is invertible in $\mathrm{BL}(X)$. In particular, it is injective. But $[I-(P-\widetilde{P})]y = y - Py + \widetilde{P}y = Py - Py + 0 = 0$. Hence $y = 0$. Thus the linear map $\widetilde{P}_{|Y,\widetilde{Y}}$ is injective.

Next, consider $\widetilde{y} \in \widetilde{Y}$. Again, since $\rho(\widetilde{P}-P) < 1$, the map $I-(\widetilde{P}-P)$ is invertible in $\mathrm{BL}(X)$. In particular, it is surjective. Hence there is some $x \in X$ such that $\widetilde{y} = [I-(\widetilde{P}-P)]x$. But then $\widetilde{y} = \widetilde{P}\widetilde{y} = \widetilde{P}[I-(\widetilde{P}-P)]x = \widetilde{P}x - \widetilde{P}x + \widetilde{P}Px = \widetilde{P}_{|Y,\widetilde{Y}}Px$. Thus the map $\widetilde{P}_{|Y,\widetilde{Y}}$ is surjective. As this map is an isomorphism from Y onto $\widetilde{Y}$, $\mathrm{rank}(P) = \dim Y = \dim \widetilde{Y} = \mathrm{rank}(\widetilde{P})$. ∎

Theorem 2.12
Let Λ be a spectral set for T. Assume that

$$\text{(i) } T_n \xrightarrow{\mathrm{n}} T \quad \textit{or} \quad \text{(ii) } 0 \notin \Lambda \textit{ and } T_n \xrightarrow{\nu} T.$$

With the notation of Proposition 2.9,

- **(a)** *$\rho(P_n - P) \to 0$ and for all large n,* $\mathrm{rank}(P_n) = \mathrm{rank}(P)$, *where we identify all infinite cardinals,*
- **(b)** *$\Lambda \neq \emptyset$ if and only if $\Lambda_n \neq \emptyset$ for all large n, and in this case,*

$$\mathrm{dist}(\Lambda_n, \Lambda) \to 0,$$

so that there are $\lambda_n \in \Lambda_n$, $\lambda'_n \in \Lambda$ such that $\lambda_n - \lambda'_n \to 0$.

Proof

(a) Since

$$\rho(P_n - P) \leq \|(P_n - P)^2\|^{1/2} \leq [\|(P_n - P)P_n\| + \|(P_n - P)P\|]^{1/2},$$

it follows from Corollary 2.10 that $\rho(P_n - P) \to 0$. Hence $\rho(P_n - P) < 1$ for all large n, so that $\text{rank}(P_n) = \text{rank}(P)$ by Lemma 2.11.

(b) Note that $\Lambda \neq \emptyset$ if and only if $P \neq O$, that is, $\text{rank}(P) > 0$. A similar statement holds for Λ_n. Hence by (a) above, $\Lambda \neq \emptyset$ if and only if $\Lambda_n \neq \emptyset$ for all large n.

Let $\Lambda \neq \emptyset$. Suppose for a moment that $\text{dist}(\Lambda_n, \Lambda) \not\to 0$. Then there is some $\epsilon > 0$ such that $\text{dist}(\Lambda_{k(n)}, \Lambda) \geq \epsilon$ for some positive integers $k(1) < k(2) < \cdots$. Since $\Lambda_{k(n)} \neq \emptyset$, let $\mu_{k(n)} \in \Lambda_{k(n)}$ be such that $\text{dist}(\mu_{k(n)}, \Lambda) = \text{dist}(\Lambda_{k(n)}, \Lambda)$. As $\bigcup_{n=1}^{\infty} \Lambda_{k(n)} \subset \text{int(C)} \cup \text{C}$, and $\text{int(C)} \cup \text{C}$ is a compact subset of $\mathbb{C}$, there exist positive integers $j(1) < j(2) < \cdots$ such that $(\mu_{k(j(n))})$ converges to some $\mu \in \text{int(C)} \cup \text{C}$. By Property U proved in Corollary 2.7, $\mu \in \text{sp}(T)$. Thus $\mu \in \text{sp}(T) \cap (\text{int(C)} \cup \text{C}) = \Lambda$. But as $\text{dist}(\mu_{k(j(n))}, \Lambda) \geq \epsilon$ for all n, we have $\mu \notin \Lambda$. This contradiction shows that $\text{dist}(\Lambda_n, \Lambda) \to 0$.

Now since Λ and Λ_n are nonempty compact subsets of $\mathbb{C}$, there are $\lambda_n \in \Lambda_n$ and $\lambda'_n \in \Lambda$ such that $|\lambda_n - \lambda'_{(n)}| = \text{dist}(\Lambda_n, \Lambda)$, so that $|\lambda_n - \lambda'_n| \to 0$. ∎

Corollary 2.13

Let λ be an isolated point of $\text{sp}(T)$. *Assume that*

$$\text{(i) } T_n \xrightarrow{\text{n}} T \quad \textit{or} \quad \text{(ii) } \lambda \neq 0 \textit{ and } T_n \xrightarrow{\nu} T.$$

For each positive $\epsilon < \text{dist}(\lambda, \text{sp}(T) \setminus \{\lambda\})$, let

$$\Lambda_n := \{\lambda_n \in \text{sp}(T_n) \, : \, |\lambda_n - \lambda| < \epsilon\}.$$

Then for all large n, $\Lambda_n \neq \emptyset$ and if $\lambda_n \in \Lambda_n$, the sequence (λ_n) converges to λ.

If the algebraic multiplicity of λ is finite, then for each large n, every λ_n in Λ_n is an eigenvalue of T_n of finite algebraic multiplicity; and the sum of the algebraic multiplicities of the eigenvalues of T_n in Λ_n is equal to the algebraic multiplicity of λ.

Proof

In Proposition 2.9, let $\Lambda := \{\lambda\}$ and C be the positively oriented circle with center λ and radius ϵ. Then by Theorem 2.12(b), we obtain $\Lambda_n \neq \emptyset$ for all large n. Let $\lambda_n \in \Lambda_n$. We prove that $\lambda_n \to \lambda$. Since the sequence (λ_n) lies in the compact set $E := \{z \in \mathbb{C} : |z - \lambda| \leq \epsilon\}$, it is enough to show that every convergent subsequence of (λ_n) converges to λ itself. Let a subsequence $(\lambda_{k(n)})$ converge to $\widetilde{\lambda} \in \mathbb{C}$. By Property U proved in Corollary 2.7, we see that $\widetilde{\lambda} \in \text{sp}(T)$. But $\widetilde{\lambda} \in E$ and $\text{sp}(T) \cap E = \{\lambda\}$. Hence $\widetilde{\lambda} = \lambda$. Thus $\lambda_n \to \lambda$.

Now let the algebraic multiplicity of λ be finite. By Theorem 2.12(a), $\text{rank}\, P(T_n, \Lambda_n) = \text{rank}\, P(T, \{\lambda\})$ for all large n, so that Λ_n is a spectral set of finite type for T_n. Theorem 1.32 shows that every λ_n in Λ_n is an eigenvalue of T_n of finite algebraic multiplicity, and their sum equals the algebraic multiplicity of λ. ∎

The preceding result allows us to say the following:
Under norm convergence, Property L holds at each isolated point of the spectrum; and under ν-convergence, Property L holds at each nonzero isolated point of the spectrum.

It would be interesting to know whether under ν-convergence Property L holds at 0 whenever 0 is an isolated point of the spectrum.

Example 2.14 Under the pointwise convergence, Property L does not hold even at a nonzero isolated point of the spectrum:
Let $X := \ell^1$. For $x \in X$, let

$$(Tx)(k) := \begin{cases} x(1) & \text{if } k = 1, \\ 2x(k) & \text{if } k \neq 1 \text{ and } k \text{ is odd}, \\ 0 & \text{otherwise}. \end{cases}$$

Then $T \in \text{BL}(X)$ and $\text{sp}(T) = \{0, 1, 2\}$. For each positive integer n and $x \in X$, let

$$(T_n x)(k) := \begin{cases} x(1) + x(2n) & \text{if } k = 1, \\ 2x(k) & \text{if } k \neq 1 \text{ and } k \text{ is odd}, \\ x(k) + x(k+2) & \text{if } 2n \leq k \leq 4n-6 \text{ and } k \text{ is even}, \\ x(4n-4) - \dfrac{x(1)}{n} & \text{if } k = 4n-4, \\ 0 & \text{otherwise}. \end{cases}$$

For $x \in X$, we have

$$(T_n x - Tx)(k) = \begin{cases} x(2n) & \text{if } k = 1, \\ x(k) + x(k+2) & \text{if } 2n \le k \le 4n-6 \text{ and } k \text{ is even,} \\ x(4n-4) - \dfrac{x(1)}{n} & \text{if } k = 4n-4, \\ 0 & \text{otherwise,} \end{cases}$$

so that

$$\|(T_n - T)x\|_1 \le \frac{|x(1)|}{n} + 2\sum_{k=n}^{2n-2} |x(2k)| \to 0 \quad \text{as } n \to \infty.$$

Thus $T_n \xrightarrow{\text{p}} T$. Also, as $\|(T_n - T)Tx\|_1 = |x(1)|/n$, we have

$$\|(T_n - T)T\| \le 1/n \to 0.$$

For each integer $n > 1$ and $x \in X$, consider

$$(\pi_n x)(k) := \begin{cases} x(1) & \text{if } k = 1, \\ x(k) & \text{if } 2n \le k \le 4n-4 \text{ and } k \text{ is even,} \\ 0 & \text{otherwise.} \end{cases}$$

It can be easily seen that $T_n, \pi_n \in \mathrm{BL}(X)$ and $\pi_n^2 = \pi_n$.

Let $Y_n := \mathcal{R}(\pi_n) = \mathrm{span}\{e_1, e_{2n}, e_{2n+2}, \ldots, e_{4n-4}\}$ and $Z_n := \mathcal{N}(\pi_n)$, the closure of span $\big(\{e_i : 2 \le i \le 2n-1\} \cup \{e_{2n+2j+1} : 0 \le j \le n-2\} \cup \{e_k : k > 4n-4\}\big)$. Since the closed subspaces Y_n and Z_n are invariant under T_n, we see that $\pi_n T_n = T_n \pi_n$. Hence by Proposition 1.18,

$$\mathrm{sp}(T_n) = \mathrm{sp}(T_{n|Y_n,Y_n}) \cup \mathrm{sp}(T_{n|Z_n,Z_n}).$$

Consider the $n \times n$ matrix A_n representing the operator $T_{n|Y_n,Y_n}$ with respect to the standard basis:

$$\mathsf{A}_n = \begin{bmatrix} 1 & 1 & 0 \cdots\cdots 0 \\ 0 & 1 & 1 \quad 0 \cdots 0 \\ \vdots & \ddots & \ddots \quad \ddots \quad \ddots \quad \vdots \\ 0 \cdots & 0 & 1 \quad 1 \; 0 \\ 0 \cdots\cdots & & 0 \quad 1 \; 1 \\ -1/n & 0 \cdots & 0 \quad 0 \; 1 \end{bmatrix}.$$

Developing $\det(\mathsf{A}_n - z\mathsf{I})$ by the first column, we obtain

$$\det(\mathsf{A}_n - z\mathsf{I}) = (1-z)^n - (-1)^{n-1}\frac{1}{n}.$$

Hence $\mathrm{sp}(T_{n|Y_n,Y_n}) = \mathrm{sp}(\mathsf{A}_n) \subset \{z \in \mathbb{C} : |1 - z| = n^{-1/n}\}$. Also, since $T_n x = Tx$ for every $x \in Z_n$, we see that

$$\mathrm{sp}(T_{n|Z_n,Z_n}) = \mathrm{sp}(T_{|Z_n,Z_n}) = \{0, 2\}.$$

Thus $|1 - \lambda_n| \geq n^{-1/n} \geq e^{-1/e}$ for every $\lambda_n \in \mathrm{sp}(T_n)$, and hence the isolated spectral value 1 of T cannot be approximated by any of the spectral values of T_n. ∎

Another example of this kind is given in Exercise 2.15. We remark that the situation described in the preceding example is not possible if each T_n is a normal operator on a Hilbert space. (See Exercise 2.16.)

We also remark that Corollary 2.13 does not hold if, from the definition of ν-convergence, one of the two conditions $\|(T_n - T)T\| \to 0$, $\|(T_n - T)T_n\| \to 0$ is omitted or if these two conditions are replaced by the condition $\|(T_n - T)^2\| \to 0$. (See Example 2.14, Exercise 2.17 and parts (b), (c) of Exercise 2.18.)

We note that under the assumptions of Corollary 2.13, for any eigenvalue of T of finite algebraic multiplicity, there is an *eigenvalue* λ_n of T_n such that $\lambda_n \to \lambda$. This statement may not hold if the algebraic multiplicity of λ is infinite. For example, let $X := \ell^2$ and (e_k) be its standard orthonormal basis. For $x := \sum_{k=1}^{\infty} x(k)e_k \in X$, define $Tx := \sum_{k=2}^{\infty} x(k)e_k$ and $T_n x := \sum_{k=2}^{\infty} (x(k) + x(k-1)/n)e_k$. Then $\|T_n - T\| = 1/n \to 0$, $\mathrm{sp}(T) = \{0, 1\}$, 1 is an eigenvalue of T, but the only eigenvalue of T_n is 0 for $n = 1, 2, \ldots$

2.4 Error Estimates

Let Λ be a spectral set for T and consider $\mathrm{C} \in \mathcal{C}(T, \Lambda)$. Assume that $T_n \xrightarrow{\nu} T$. By Proposition 2.9(a), there is an integer n_0 such that $\Lambda_n := \mathrm{sp}(T_n) \cap \mathrm{int}(\mathrm{C})$ is a spectral set for T_n for each $n \geq n_0$. Theorem 2.6 shows that

$$\alpha_1(\mathrm{C}) := \sup\{\|R(z)\| : z \in \mathrm{C}\} < \infty,$$
$$\alpha_2(\mathrm{C}) := \sup\{\|R_n(z)\| : z \in \mathrm{C},\ n \geq n_0\} < \infty,$$

where, as usual, $R(z) := R(T, z)$ for $z \in \text{re}(T)$, and $R_n(z) := R(T_n, z)$ for $z \in \text{re}(T_n)$.

Let $P := P(T, \Lambda)$ and for $n \geq n_0$, $P_n := P(T_n, \Lambda_n)$ denote the spectral projections. Also, let $M := M(T, \Lambda)$ and for $n \geq n_0$, $M_n := M(T_n, \Lambda_n)$ denote the spectral subspaces, and

$$\widetilde{\alpha}_1(\text{C}) := \sup\{\|R(z)_{|M}\| : z \in \text{C}\},$$
$$\widetilde{\alpha}_2(\text{C}) := \sup\{\|R_n(z)_{|M_n}\| : z \in \text{C}, \ n \geq n_0\}.$$

Clearly, $\widetilde{\alpha}_1(\text{C}) \leq \alpha_1(\text{C})$ and $\widetilde{\alpha}_2(\text{C}) \leq \alpha_2(\text{C})$.

In case $m := \text{rank}\, P < \infty$, let $\underline{\varphi}$ form an ordered basis for M, $\underline{\varphi_n}$ form an ordered basis for M_n, and $\underline{T\,\varphi} = \underline{\varphi}\,\Theta$, $\underline{T_n\,\varphi_n} = \underline{\varphi_n}\,\Theta_n$, where $\Theta, \Theta_n \in \mathbb{C}^{m\times m}$. Since $\text{sp}(\Theta) = \text{sp}(T_{|M,M}) = \Lambda$, $\text{sp}(\Theta_n) = \text{sp}(T_{n|M_n,M_n}) = \Lambda_n$ and $\Lambda \cup \Lambda_n \subset \text{int}(\text{C})$, we have $\text{C} \subset \text{re}(\Theta) \cap \text{re}(\Theta_n)$. Define

$$\beta_1(\text{C}) := \sup\{\|R(\Theta, z)\| : z \in \text{C}\},$$
$$\beta_{2,n}(\text{C}) := \sup\{\|R(\Theta_n, z)\| : z \in \text{C}\}.$$

Clearly, $\beta_1(\text{C})$ is finite and for each large n, $\beta_{2,n}(\text{C})$ is finite.

Recall the notation $\underline{x} := [x_1, \ldots, x_m]$ for $x_1, \ldots, x_m \in X$ and

$$\|\,\underline{x}\,\|_\infty := \max\{\|x_1\|, \ldots, \|x_m\|\}$$

from Subsection 1.4.2.

We shall use the above-mentioned notation throughout this section.

The following result is crucial for the error estimates we shall give.

Proposition 2.15
Let Λ be a spectral set for T. Assume that

$$\text{(i) } T_n \xrightarrow{\text{n}} T \quad \textit{or} \quad \text{(ii) } 0 \notin \Lambda \textit{ and } T_n \xrightarrow{\nu} T.$$

Then

$$\|(T_n - T)_{|M}\| \to 0 \quad \textit{and} \quad \|(T_n - T)_{|M_n}\| \to 0.$$

Also, for all large n,

$$\|(P_n - P)_{|M}\| \leq \frac{\ell(\text{C})}{2\pi}\widetilde{\alpha}_1(\text{C})\alpha_2(\text{C})\|(T_n - T)_{|M}\|,$$
$$\|(P_n - P)_{|M_n}\| \leq \frac{\ell(\text{C})}{2\pi}\alpha_1(\text{C})\widetilde{\alpha}_2(\text{C})\|(T_n - T)_{|M_n}\|.$$

Further, if rank P *is finite, then*

$$\|(\underline{P_n} - \underline{P})\underline{\varphi}\|_\infty \leq \frac{\ell(\mathrm{C})}{2\pi}\beta_1(\mathrm{C})\alpha_2(\mathrm{C})\|(\underline{T_n} - \underline{T})\underline{\varphi}\|_\infty,$$
$$\|(\underline{P_n} - \underline{P})\underline{\varphi_n}\|_\infty \leq \frac{\ell(\mathrm{C})}{2\pi}\alpha_1(\mathrm{C})\beta_{2,n}(\mathrm{C})\|(\underline{T_n} - \underline{T})\underline{\varphi_n}\|_\infty.$$

Proof
If $T_n \xrightarrow{\mathrm{n}} T$, then $\|(T_n - T)_{|M}\| \leq \|T_n - T\| \to 0$ and $\|(T_n - T)_{|M_n}\| \leq \|T_n - T\| \to 0$. Let now $0 \notin \Lambda$ and $T_n \xrightarrow{\nu} T$. If $0 \notin \mathrm{sp}(T)$, then by Lemma 2.2(a), $T_n \xrightarrow{\mathrm{n}} T$. If $0 \in \mathrm{sp}(T)$, then $0 \in \mathrm{ext}(\mathrm{C})$, and Proposition 2.9(b) shows that

$$\|(T_n - T)_{|M}\| \leq \|(T_n - T)P\| \to 0,$$
$$\|(T_n - T)_{|M_n}\| \leq \|(T_n - T)P_n\| \to 0.$$

Now, for each large n, we have

$$(P_n - P)_{|M} = \Big(\frac{1}{2\pi\mathrm{i}}\int_{\mathrm{C}} R_n(z)(T_n - T)R(z)\,dz\Big)\Big|_M$$
$$= \frac{1}{2\pi\mathrm{i}}\int_{\mathrm{C}} R_n(z)(T_n - T)_{|M}R(z)_{|M}\,dz,$$

since $R(z)(M) \subset M$ for all $z \in \mathrm{C}$. Hence

$$\|(P_n - P)_{|M}\| \leq \frac{\ell(\mathrm{C})}{2\pi}\alpha_2(\mathrm{C})\|(T_n - T)_{|M}\|\widetilde{\alpha}_1(\mathrm{C}).$$

Similarly, as $R_n(z)(M_n) \subset M_n$ for all $z \in \mathrm{C}$, we obtain

$$(P_n - P)_{|M_n} = \frac{1}{2\pi\mathrm{i}}\int_{\mathrm{C}} R(z)(T_n - T)_{|M_n}R_n(z)_{|M_n}\,dz,$$

so that

$$\|(P_n - P)_{|M_n}\| \leq \frac{\ell(\mathrm{C})}{2\pi}\alpha_1(\mathrm{C})\|(T_n - T)_{|M_n}\|\widetilde{\alpha}_2(\mathrm{C}).$$

Next, for $z \in \mathrm{C}$, we have $\underline{R(z)}\,\underline{\varphi} = \underline{\varphi}\,R(\Theta, z)$ and $\underline{R_n(z)}\,\underline{\varphi_n} = \underline{\varphi_n}\,R(\Theta_n, z)$, so that

$$(\underline{P_n} - \underline{P})\underline{\varphi} = \frac{1}{2\pi\mathrm{i}}\int_{\mathrm{C}} \underline{R_n(z)}(\underline{T_n} - \underline{T})\underline{\varphi}\,R(\Theta, z)\,dz,$$
$$(\underline{P_n} - \underline{P})\underline{\varphi_n} = \frac{1}{2\pi\mathrm{i}}\int_{\mathrm{C}} \underline{R(z)}(\underline{T_n} - \underline{T})\underline{\varphi_n}\,R(\Theta_n, z)\,dz.$$

Hence the desired estimates for $\|(\underline{P_n} - \underline{P})\underline{\varphi}\|_\infty$ and $\|(\underline{P_n} - \underline{P})\underline{\varphi_n}\|_\infty$ follow. ∎

Remark 2.16
The preceding proposition allows us to consider the nearness of the spectral subspaces M_n and M. For this purpose we introduce the notion of **gap** between two closed subspaces Y and $\widetilde{Y}$ of a Banach space X. Let

$$\delta(Y, \widetilde{Y}) := \sup\left\{\operatorname{dist}(y, \widetilde{Y}) \,:\, y \in Y,\, \|y\| = 1\right\},$$
$$\operatorname{gap}(Y, \widetilde{Y}) := \max\left\{\delta(Y, \widetilde{Y}),\, \delta(\widetilde{Y}, Y)\right\}.$$

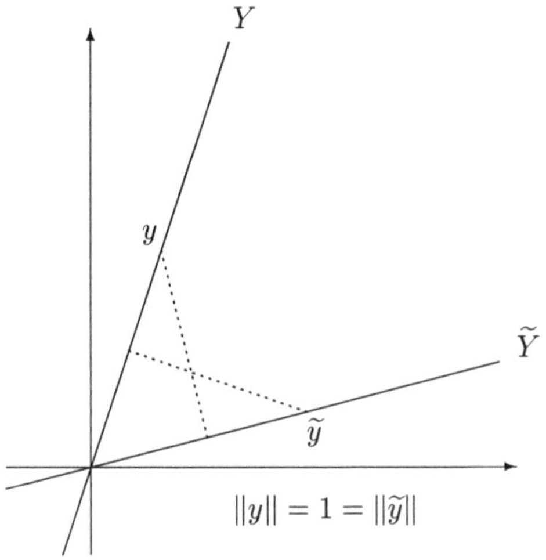

The following properties can be proved easily.

(i) $0 \le \operatorname{gap}(Y, \widetilde{Y}) \le 1$.

(ii) $\operatorname{gap}(Y, \widetilde{Y}) = 0$ if and only if $Y = \widetilde{Y}$.

(iii) $\operatorname{gap}(Y, \widetilde{Y}) = \operatorname{gap}(\widetilde{Y}, Y)$.

However, the triangle inequality does not hold, as Exercise 2.21 shows.

Let P and $\widetilde{P}$ be projections in $\operatorname{BL}(X)$ such that $\mathcal{R}(P) = Y$ and $\mathcal{R}(\widetilde{P}) = \widetilde{Y}$. Then for every $y \in Y$ with $\|y\| = 1$,

$$\operatorname{dist}(y, \widetilde{Y}) \le \|y - \widetilde{P}y\| = \|Py - \widetilde{P}y\| \le \|(P - \widetilde{P})_{|Y}\|,$$

so that $\delta(Y, \widetilde{Y}) \le \|(P - \widetilde{P})_{|Y}\|$. Hence

$$\text{gap}(Y, \widetilde{Y}) \le \max\left\{\|(P - \widetilde{P})_{|Y}\|, \|(P - \widetilde{P})_{|\widetilde{Y}}\|\right\}.$$

Under the hypothesis of Proposition 2.15, we have

$$\begin{aligned}\text{gap}(M, M_n) &\le \max\{\|(P_n - P)_{|M}\|, \|(P_n - P)_{|M_n}\|\} \\ &\le \frac{\ell(\mathrm{C})}{2\pi}\alpha_1(\mathrm{C})\alpha_2(\mathrm{C}) \max\{\|(T_n - T)_{|M}\|, \|(T_n - T)_{|M_n}\|\}\end{aligned}$$

which tends to zero as $n \to \infty$. ▮

Let now the spectral set Λ be of finite type for T. First we consider the case when Λ consists of a single simple eigenvalue of T.

Theorem 2.17
Let λ be a simple eigenvalue of T, ϵ such that $0 < \epsilon < \text{dist}(\lambda, \text{sp}(T)\setminus\{\lambda\})$ and C_ϵ the positively oriented circle with center λ and radius ϵ.
Assume that

$$\text{(i) } T_n \xrightarrow{\text{n}} T \quad or \quad \text{(ii) } \lambda \ne 0 \text{ and } T_n \xrightarrow{\nu} T.$$

(a) *There is a positive integer n_0 such that for each $n \ge n_0$, we have a unique $\lambda_n \in \text{sp}(T_n)$ satisfying $|\lambda_n - \lambda| < \epsilon$. Also, λ_n is a simple eigenvalue of T_n for each $n \ge n_0$ and $\lambda_n \to \lambda$.*

(b) *Let φ be an eigenvector of T corresponding to λ and for $n \ge n_0$, let φ_n be an eigenvector of T_n corresponding to λ_n. Then for each large n, $P\varphi_n$ is an eigenvector of T corresponding to λ, $P_n\varphi$ is an eigenvector of T_n corresponding to λ_n, and*

$$|\lambda_n - \lambda| \le 2\epsilon \min\left\{\alpha_1(\mathrm{C}_\epsilon)\frac{\|(T_n - T)\varphi_n\|}{\|\varphi_n\|}, \alpha_2(\mathrm{C}_\epsilon)\frac{\|(T_n - T)\varphi\|}{\|\varphi\|}\right\},$$

which tends to zero. Also, for all large n,

$$\|\varphi_n - P\varphi_n\| \le \frac{\epsilon\,\alpha_1(\mathrm{C}_\epsilon)}{\text{dist}(\lambda_n, \mathrm{C}_\epsilon)}\|(T_n - T)\varphi_n\|.$$

If the sequence $(\|\varphi_n\|)$ is bounded, then $\|(T_n - T)\varphi_n\| \to 0$ and $\|\varphi_n - P\varphi_n\| \to 0$.

For each large n, let c_n be the nonzero complex number such that $P_n\varphi = c_n\varphi_n$. Then

$$||c_n\varphi_n - \varphi|| \to 0.$$

Also, if $\varphi_{(n)} := \varphi/c_n$, then $\varphi_{(n)}$ is the eigenvector of T corresponding to λ such that $P_n\varphi_{(n)} = \varphi_n$. We have $||\varphi_{(n)}|| \leq 2||\varphi_n||$ for all large n and

$$\frac{||\varphi_n - \varphi_{(n)}||}{||\varphi_{(n)}||} \leq \alpha_2(\mathrm{C}_\epsilon)\,\frac{||(T_n - T)\varphi||}{||\varphi||}.$$

In fact, if φ_n^* is the eigenvector of T_n^* corresponding to its (simple) eigenvalue $\overline{\lambda}_n$ such that $\langle \varphi_n\, , \varphi_n^*\rangle = 1$, then $c_n := \langle \varphi\, , \varphi_n^*\rangle$.

Proof

(a) As the algebraic multiplicity of the spectral value λ of T is 1, the results follow from Corollary 2.13.

(b) Since for all large n, $\dim M = 1 = \dim M_n$, we see that $M = \mathcal{N}(T - \lambda I)$ and $M_n = \mathcal{N}(T_n - \lambda_n I)$ by Proposition 1.31(b).

Theorem 2.12(a) and Lemma 2.11 show that for all large n, the linear map $Q_n := P_{|M_n,M}$ is bijective. In particular, $P\varphi_n$ is a nonzero element of $\mathcal{N}(T - \lambda I)$, that is, $P\varphi_n$ is an eigenvector of T corresponding to λ. Also, by Corollary 2.10, $||(P_n - P)P_n|| \leq 1/2$ for all large n, and hence for $x \in M_n$,

$$||x|| - ||Px|| \leq ||x - Px|| = ||(P_n - P)P_n x|| \leq \frac{||x||}{2}.$$

Thus $||Q_n^{-1}|| \leq 2$ for all large n. Noting that $Q_n^{-1}P\varphi_n = \varphi_n$, $P^2 = P$ and $PT = \lambda P$, we obtain

$$\lambda_n\varphi_n - \lambda\varphi_n = Q_n^{-1}P(T_n\varphi_n - PT\varphi_n) = Q_n^{-1}P(T_n - T)\varphi_n.$$

Hence for all large n, $||\lambda_n\varphi_n - \lambda\varphi_n|| \leq 2||P||\,||(T_n - T)\varphi_n||$, so that

$$|\lambda_n - \lambda| \leq 2\epsilon\,\alpha_1(\mathrm{C}_\epsilon)\frac{||(T_n - T)\varphi_n||}{||\varphi_n||}.$$

Also, considering $\Theta_n := [\lambda_n] \in \mathbb{C}^{1\times 1}$, we see that

$$\beta_{2,n}(\mathrm{C}_\epsilon) = \sup_{z\in \mathrm{C}_\epsilon} \frac{1}{|\lambda_n - z|}$$

and hence by Proposition 2.15, we obtain

$$\begin{aligned}\|\varphi_n - P\varphi_n\| &= \|(P_n - P)\varphi_n\| \\ &\leq \frac{\ell(\mathrm{C}_\epsilon)}{2\pi}\alpha_1(\mathrm{C}_\epsilon) \sup_{z\in \mathrm{C}_\epsilon} \frac{1}{|\lambda_n - z|}\|(T_n - T)\varphi_n\| \\ &= \frac{\epsilon\, \alpha_1(\mathrm{C}_\epsilon)}{\mathrm{dist}(\lambda_n, \mathrm{C}_\epsilon)}\|(T_n - T)\varphi_n\|.\end{aligned}$$

Since $\lambda_n \to \lambda$, we have $\mathrm{dist}(\lambda_n, \mathrm{C}_\epsilon) \geq \epsilon/2$ for all large n. If $\|\varphi_n\| \leq c$, then by Proposition 2.15,

$$\|(T_n - T)\varphi_n\| \leq c\|(T_n - T)_{|M_n}\| \to 0$$

and hence $\|\varphi_n - P\varphi_n\| \to 0$.

By interchanging the roles of T and T_n, we see that for all large n, the map $P_{n|M,M_n} : M \to M_n$ is bijective, $P_n\varphi$ is an eigenvector of T_n corresponding to λ_n, $\left\|\left(P_{n|M,M_n}\right)^{-1}\right\| \leq 2$, and

$$|\lambda_n - \lambda| \leq 2\epsilon\, \alpha_2(\mathrm{C}_\epsilon)\frac{\|(T_n - T)\varphi\|}{\|\varphi\|}.$$

Since $\dim M_n = 1$, we have $P_n\varphi = c_n\varphi_n$ for some nonzero scalar c_n. Also,

$$\|c_n\varphi_n - \varphi\| = \|P_n\varphi - P\varphi\| = \|(P_n - P)P\varphi\| \leq \|(P_n - P)P\|\,\|\varphi\| \to 0$$

by Corollary 2.10. Since $\varphi_{(n)} := \varphi/c_n$, we see that $P_n\varphi_{(n)} = P_n\varphi/c_n = \varphi_n$. As $P_{n|M,M_n}$ is injective, $\varphi_{(n)}$ is the unique element of M such that $P_n\varphi_{(n)} = \varphi_n$; and we have $\|\varphi_{(n)}\| \leq 2\|\varphi_n\|$. By considering $\Theta := [\lambda] \in \mathbb{C}^{1\times 1}$, we see that

$$\beta_1(\mathrm{C}_\epsilon) = \sup_{z\in \mathrm{C}_\epsilon} \frac{1}{|\lambda - z|}$$

and hence by Proposition 2.15, we obtain

$$\begin{aligned}\|\varphi_n - \varphi_{(n)}\| &= \frac{1}{|c_n|}\|(P_n - P)\varphi\| \\ &\leq \frac{\ell(\mathrm{C}_\epsilon)}{2\pi|c_n|}\alpha_2(\mathrm{C}_\epsilon) \sup_{z\in \mathrm{C}_\epsilon} \frac{1}{|\lambda - z|}\|(T_n - T)\varphi\| \\ &= \frac{\alpha_2(\mathrm{C}_\epsilon)}{|c_n|}\|(T_n - T)\varphi\| \\ &= \alpha_2(\mathrm{C}_\epsilon)\, \frac{\|(T_n - T)\varphi\|}{\|\varphi\|}\|\varphi_{(n)}\|,\end{aligned}$$

so that

$$\frac{\|\varphi_n - \varphi_{(n)}\|}{\|\varphi_{(n)}\|} \le \alpha_2(\mathrm{C}_\epsilon)\,\frac{\|(T_n - T)\varphi\|}{\|\varphi\|}.$$

Since the subspaces M, M_n are one dimensional, and since $0 \neq \varphi \in M$, $0 \neq \varphi_n \in M_n$, we have

$$\|(T_n - T)_{|M}\| = \frac{\|(T_n - T)\varphi\|}{\|\varphi\|}, \qquad \|(T_n - T)_{|M_n}\| = \frac{\|(T_n - T)\varphi_n\|}{\|\varphi_n\|},$$

both of which tend to zero by Proposition 2.15.

Proposition 1.36(b) shows that $\overline{\lambda}_n$ is a simple eigenvalue of T_n^* and P_n^* is the associated spectral projection. By Lemma 1.51, there is an eigenvector φ_n^* of T_n^* corresponding to $\overline{\lambda}_n$ such that $\langle \varphi_n , \varphi_n^* \rangle = 1$. Then

$$c_n = \langle c_n\varphi_n , \varphi_n^* \rangle = \langle P_n\varphi , \varphi_n^* \rangle = \langle \varphi , P_n^*\varphi_n^* \rangle = \langle \varphi , \varphi_n^* \rangle.$$

The proof is now complete. ∎

We now take up the general case of a spectral set of finite type for T. As one may expect, the general case is much more complex as compared to the case of a single simple eigenvalue of T considered above; but the arguments are similar.

Theorem 2.18
Let $\Lambda := \{\lambda_1, \ldots, \lambda_r\}$, where each λ_j is a spectral value of T of finite type.

Let m_j be the algebraic multiplicity of λ_j, $m := m_1 + \cdots + m_r$. Assume that

$$\text{(i) } T_n \xrightarrow{\text{n}} T \quad \textit{or} \quad \text{(ii) } \textit{each } \lambda_j \neq 0 \textit{ and } T_n \xrightarrow{\nu} T.$$

(a) *There is a positive integer n_0 such that for all $n \ge n_0$, $\mathrm{C} \subset \mathrm{re}(T_n)$ and if $\Lambda_n := \mathrm{sp}(T_n) \cap \mathrm{int}(\mathrm{C})$, then*

$$\Lambda_n = \{\lambda_{n,1}, \ldots, \lambda_{n,r(n)}\},$$

where each $\lambda_{n,j}$ is spectral value of T_n of finite type. Further, if $m_{n,j}$ is the algebraic multiplicity of $\lambda_{n,j}$, then $m_{n,1} + \cdots + m_{n,r(n)} = m$.

(b) *For $n \ge n_0$, consider the weighted arithmetic means*

$$\widehat{\lambda} := \frac{m_1\lambda_1 + \cdots + m_r\lambda_r}{m}$$

and

$$\widehat{\lambda}_n := \frac{m_{n,1}\lambda_{n,1} + \cdots + m_{n,r(n)}\lambda_{n,r(n)}}{m}$$

of elements in Λ *and* Λ_n*, respectively. Then*

$$|\widehat{\lambda}_n - \widehat{\lambda}| \leq \frac{\ell(\mathsf{C})}{\pi} \min\left\{\alpha_1(\mathsf{C})\|(T_n - T)_{|M_n}\|, \alpha_2(\mathsf{C})\|(T_n - T)_{|M}\|\right\}$$
$$\to 0.$$

(c) *For each large* n*,* $\underline{P}\,\underline{\varphi_n}$ *forms an ordered basis for* M *and*

$$\frac{\|\underline{\varphi_n} - \underline{P}\,\underline{\varphi_n}\|_\infty}{\|\underline{\varphi_n}\|_\infty} \leq \frac{\ell(\mathsf{C})}{2\pi}\alpha_1(\mathsf{C})\widetilde{\alpha}_2(\mathsf{C})\|(T_n - T)_{|M_n}\| \to 0.$$

For each large n*,* $\underline{P_n}\,\underline{\varphi}$ *forms an ordered basis for* M_n *and there is an* $m\times m$ *nonsingular matrix* C_n *such that* $\underline{P_n}\,\underline{\varphi} = \underline{\varphi_n}\,\mathsf{C}_n$*. Then*

$$\|\underline{\varphi_n}\,\mathsf{C}_n - \underline{\varphi}\|_\infty \to 0.$$

Also, if $\underline{\varphi_{(n)}} := \underline{\varphi}\,\mathsf{C}_n^{-1}$*, then* $\underline{\varphi_{(n)}}$ *is the unique ordered basis for* M *such that* $\underline{P_n}\,\underline{\varphi_{(n)}} = \underline{\varphi_n}$*. We have* $\|\underline{\varphi_{(n)}}\|_\infty \leq 2\|\underline{\varphi_n}\|_\infty$ *for all large* n *and*

$$\frac{\|\underline{\varphi_n} - \underline{\varphi_{(n)}}\|_\infty}{\|\underline{\varphi_{(n)}}\|_\infty} \leq \frac{\ell(\mathsf{C})}{2\pi}\widetilde{\alpha}_1(\mathsf{C})\alpha_2(\mathsf{C})\|(T_n - T)_{|M}\| \to 0.$$

In fact, if $\underline{\varphi_n^*}$ *forms the ordered basis for* $\mathcal{R}(P_n^*)$ *which is adjoint to* $\underline{\varphi_n}$*, then* $\mathsf{C}_n := (\underline{\varphi}\,,\underline{\varphi_n^*})$*.*

(d) *For* $n \geq n_0$*, let* $\psi_n \in M_n$ *such that* $(\|\psi_n\|)$ *is bounded. Then the sequence* (ψ_n) *has a convergent subsequence, and every convergent subsequence of* (ψ_n) *converges to an element of* M*.*

Proof

(a) Since each λ_j is an isolated point of $\mathrm{sp}(T)$, $\Lambda := \{\lambda_1, \ldots, \lambda_r\}$ is a spectral set for T. As we have seen in Theorem 1.32, $\mathrm{rank}(P) = m$, the sum of the algebraic multiplicities of $\lambda_1, \ldots, \lambda_r$. By Theorem 2.12(a), there is a positive integer n_0 such that $\mathrm{rank}(P_n) = \mathrm{rank}(P) = m$ for all $n \geq n_0$. Hence again by Theorem 1.32, $\Lambda_n := \mathrm{sp}(T_n) \cap \mathrm{int}(\mathsf{C})$ consists of a finite number of spectral values $\lambda_{n,1}, \ldots, \lambda_{n,r(n)}$ of T_n of finite type, and the sum of their respective algebraic multiplicities $m_{n,1}, \ldots, m_{n,r(n)}$ is m.

(b) Again, by Theorem 2.12(a), $\rho(P_n - P) < 1$ for all large n, and hence by Lemma 2.11, the linear map $P_{n|M,M_n} : M \to M_n$ is bijective. Also, by Corollary 2.10, $\|(P_n - P)P\| \leq 1/2$ for all large n, and so for $x \in M$,

$$\|x\| - \|P_n x\| \leq \|x - P_n x\| = \|(P - P_n)Px\| \leq \frac{\|x\|}{2}.$$

Thus $\|(P_{n|M,M_n})^{-1}\| \leq 2$ for all large n.

Let us define $A := T_{|M,M}$ and for each large n, $A_n := T_{n|M_n,M_n}$, $A'_n := (P_{n|M,M_n})^{-1} T_n (P_{n|M,M_n})$. By the Spectral Decomposition Theorem 1.26, $\mathrm{sp}(A) = \Lambda$ and the algebraic multiplicity of $\lambda_j \in \mathrm{sp}(A)$ is m_j, $j = 1, \ldots, r$. Similarly, $\mathrm{sp}(A_n) = \Lambda_n$ and the algebraic multiplicity of $\lambda_{n,j} \in \mathrm{sp}(A_n)$ is $m_{n,j}$, $j = 1, \ldots, r(n)$. The matrices Θ and Θ_n represent the linear maps A and A_n with respect to the ordered basis $\underline{\varphi}$ for M and $\underline{\varphi_n}$ for M_n, respectively. Let the matrix Θ'_n represent the linear map A'_n with respect to the ordered basis $\underline{\varphi}$ for M. Since $A'_n = (P_{n|M,M_n})^{-1} A_n (P_{n|M,M_n})$, the matrices Θ_n and Θ'_n are similar. Hence it can be seen that $\mathrm{sp}(\Theta'_n) = \mathrm{sp}(\Theta_n)$ and that the algebraic multiplicity of $\lambda_{n,j}$ as an eigenvalue of Θ'_n is equal to the algebraic multiplicity of $\lambda_{n,j}$ as an eigenvalue of Θ_n, namely $m_{n,j}$, $j = 1, \ldots, r(n)$. Thus

$$\widehat{\lambda} := \frac{m_1\lambda_1 + \cdots + m_r\lambda_r}{m} = \frac{\mathrm{tr}\,(\Theta)}{m},$$
$$\widehat{\lambda}_n := \frac{m_{n,1}\lambda_{n,1} + \cdots + m_{n,r(n)}\lambda_{n,r(n)}}{m} = \frac{\mathrm{tr}\,(\Theta_n)}{m} = \frac{\mathrm{tr}\,(\Theta'_n)}{m}$$

for all large n. Consequently,

$$\begin{aligned}
|\widehat{\lambda}_n - \widehat{\lambda}| &= \frac{|\,\mathrm{tr}(\Theta'_n - \Theta)|}{m} \leq \rho(\Theta'_n - \Theta) = \rho(A'_n - A) \\
&\leq \|A'_n - A\| \\
&= \sup\left\{\|(P_{n|M,M_n})^{-1} P_n (T_n - T)x \, : \, x \in M, \, \|x\| \leq 1\right\} \\
&\leq \|(P_{n|M,M_n})^{-1}\| \, \|P_n\| \, \|(T_n - T)_{|M}\| \\
&\leq \frac{\ell(\mathrm{C})}{\pi} \alpha_2(\mathrm{C}) \|(T_n - T)_{|M}\| \to 0
\end{aligned}$$

by Proposition 2.15. Interchanging the roles of T and T_n, we obtain

$$|\widehat{\lambda}_n - \widehat{\lambda}| \leq \frac{\ell(\mathrm{C})}{\pi} \alpha_1(\mathrm{C}) \|(T_n - T)_{|M_n}\| \to 0.$$

Hence

$$|\widehat{\lambda}_n - \widehat{\lambda}| \leq \frac{\ell(\mathrm{C})}{\pi} \min\left\{\alpha_1(\mathrm{C}) \|(T_n - T)_{|M_n}\|, \, \alpha_2(\mathrm{C}) \|(T_n - T)_{|M}\|\right\}.$$

(c) Since the linear map $P_{|M_n,M}$ is bijective for all large n, $\underline{P}\,\underline{\varphi_n}$ forms an ordered basis for M. Also, by Proposition 2.15,

$$\begin{aligned}\|\underline{\varphi_n} - \underline{P}\,\underline{\varphi_n}\|_\infty &= \|(\underline{P_n} - \underline{P})\,\underline{\varphi_n}\|_\infty \leq \|(P_n - P)_{|M_n}\|\,\|\underline{\varphi_n}\|_\infty \\ &\leq \frac{\ell(\mathrm{C})}{2\pi}\alpha_1(\mathrm{C})\widetilde{\alpha}_2(\mathrm{C})\|(T_n - T)_{|M_n}\|\,\|\underline{\varphi_n}\|_\infty.\end{aligned}$$

Hence by Proposition 2.15,

$$\frac{\|\underline{\varphi_n} - \underline{P}\,\underline{\varphi_n}\|_\infty}{\|\underline{\varphi_n}\|_\infty} \leq \frac{\ell(\mathrm{C})}{2\pi}\alpha_1(\mathrm{C})\widetilde{\alpha}_2(\mathrm{C})\|(T_n - T)_{|M_n}\| \to 0.$$

Since the linear map $P_{n|M,M_n}$ is bijective for all large n, $\underline{P_n}\,\underline{\varphi}$ forms an ordered basis for M_n. Hence $\underline{P_n}\,\underline{\varphi} = \underline{\varphi_n}\,\mathsf{C}_n$ for some nonsingular $m\times m$ matrix C_n. Then

$$\begin{aligned}\|\underline{\varphi_n}\,\mathsf{C}_n - \underline{\varphi}\|_\infty &= \|\underline{P_n}\,\underline{\varphi} - \underline{P}\,\underline{\varphi}\|_\infty = \|(\underline{P_n} - \underline{P})\,\underline{P}\,\underline{\varphi}\|_\infty \\ &\leq \|(P_n - P)P\|\,\|\underline{\varphi}\|_\infty \to 0\end{aligned}$$

by Corollary 2.10.

Also, since $\underline{\varphi_{(n)}} := \underline{\varphi}\,\mathsf{C}_n^{-1}$, $\underline{\varphi_{(n)}}$ is the unique ordered basis for M such that $\underline{P_n}\,\underline{\varphi_{(n)}} = \underline{\varphi_n}$. We have $\|\underline{\varphi_{(n)}}\|_\infty \leq 2\|\underline{\varphi_n}\|_\infty$ and

$$\|\underline{\varphi_n} - \underline{\varphi_{(n)}}\|_\infty = \|(\underline{P_n} - \underline{P})\,\underline{\varphi_{(n)}}\|_\infty \leq \|(P_n - P)_{|M}\|\,\|\underline{\varphi_{(n)}}\|_\infty.$$

Hence by Proposition 2.15,

$$\frac{\|\underline{\varphi_n} - \underline{\varphi_{(n)}}\|_\infty}{\|\underline{\varphi_{(n)}}\|_\infty} \leq \frac{\ell(\mathrm{C})}{2\pi}\widetilde{\alpha}_1(\mathrm{C})\alpha_2(\mathrm{C})\|(T_n - T)_{|M}\| \to 0.$$

Let $\underline{\varphi_n^*}$ form the ordered basis for $\mathcal{R}(P_n^*)$ which is adjoint to $\underline{\varphi_n}$. Note that $\mathcal{R}(P_n^*)$ is the spectral subspace associated with T_n^* and its spectral set $\overline{\Lambda}_n$. Then

$$\mathsf{C}_n = (\underline{\varphi_n}\,\mathsf{C}_n, \underline{\varphi_n^*}) = (\underline{P_n}\,\underline{\varphi}, \underline{\varphi_n^*}) = (\underline{\varphi}, \underline{P_n^*}\,\underline{\varphi_n^*}) = (\underline{\varphi}, \underline{\varphi_n^*}).$$

(d) Let (ψ_n) be a bounded sequence in X with $\psi_n \in M_n$ for all large n. Then as in (c) above,

$$\|\psi_n - P\psi_n\| \leq \frac{\ell(\mathrm{C})}{2\pi}\alpha_1(\mathrm{C})\widetilde{\alpha}_2(\mathrm{C})\|(T_n - T)_{|M_n}\|\,\|\psi_n\| \to 0.$$

Consider a subsequence $(\psi_{k(n)})$ of (ψ_n). The estimate given above shows that $(\psi_{k(n)})$ converges to $\psi \in X$ if and only if $(P\psi_{k(n)})$ converges to $\psi \in X$. In that event,

$$P\psi = P\left(\lim_{n\to\infty} \psi_{k(n)}\right) = \lim_{n\to\infty} P\psi_{k(n)} = \psi,$$

so that $\psi \in M$.

Since $P \in \mathrm{BL}(X)$ and $\mathrm{rank}(P)$ is finite, P is a compact operator on X. Therefore the bounded sequence (ψ_n) does have a subsequence $(\psi_{k(n)})$ such that $(P\psi_{k(n)})$ converges in X. ∎

We make some remarks about the conclusions of Theorem 2.18.

Firstly, the weighted arithmetic mean $\widehat{\lambda}$ of a cluster Λ of spectral values of T of finite type is well approximated by the weighted arithmetic mean $\widehat{\lambda}_n$ of a cluster Λ_n of spectral values of T_n of finite type. When Λ consists of a single multiple spectral value λ of T (so that $\widehat{\lambda} = \lambda$), the arithmetic mean $\widehat{\lambda}_n$ of Λ_n provides, in general, a better approximation of λ as compared to the individual elements of Λ_n. In fact, if $\Lambda_n := \{\lambda_{n,1}, \ldots, \lambda_{n,r(n)}\}$ and ℓ is the ascent of λ, then it can be proved that for all large n and $j = 1, \ldots, r(n)$,

$$|\lambda_{n,j} - \lambda|^\ell \le c \min\left\{\|(T_n - T)_{|M_n}\|,\, \|(T_n - T)_{|M}\|\right\},$$

where c is a constant independent of n. See [62] and [30] for the details.

Secondly, let (ψ_n) be a bounded sequence in X with $\psi_n \in M_n$. Part (d) guarantees that (ψ_n) has a subsequence which converges to an element of M. If the given sequence (ψ_n) is also bounded away from zero, that is, $\|\psi_n\| \ge \delta$ for all large enough n and some constant $\delta > 0$, then the limit of this convergent subsequence will be a *nonzero* element of M, something that we are looking for! Consider the special case where $\Lambda := \{\lambda\}$. Then $\widehat{\lambda} = \lambda$ and for $\lambda_n \in \mathrm{sp}(T_n)$, we see that $\lambda_n \to \lambda$ if and only if $\lambda_n \in \Lambda_n$ for all large n by Corollary 2.13. Let $\lambda_n \in \Lambda_n$ and ψ_n be a corresponding generalized eigenvector of T_n of grade at most p, that is,

$$0 \ne \psi_n \in \mathcal{N}\left[(T_n - \lambda_n I)^p\right].$$

Assume that $0 < \delta \le \|\psi_n\| \le \gamma$ for all large n. Then there is a subsequence $(\psi_{k(n)})$ which converges to a nonzero element ψ of M. We show that ψ is in fact a generalized eigenvector of T corresponding to λ and its grade is at most p, that is, $\psi \in \mathcal{N}\left[(T - \lambda I)^p\right]$.

Since $\|(T_n - T)P\| \to 0$, $T_n x \to Tx$ for each $x \in M$ as $n \to \infty$. Now for each $j = 1, \ldots, p$,

$$\begin{aligned} T_n^j - T^j &= T_n^j - T_n^{j-1}T + T_n^{j-1}T - T_n^{j-2}T^2 + T_n^{j-2}T^2 \\ &\quad + \cdots + T_n T^{j-1} - T^j \\ &= \sum_{i=1}^{j} T_n^{j-i}(T_n - T)T^{i-1}. \end{aligned}$$

As $\psi \in M$ and M is invariant under T, it follows that $T_n^j \psi \to T^j \psi$ for each $j = 1, \ldots, p$. But

$$\|T_{k(n)}^j \psi_{k(n)} - T^j \psi\| \leq \|T_{k(n)}^j\| \, \|\psi_{k(n)} - \psi\| + \|T_{k(n)}^j \psi - T^j \psi\| \to 0,$$

so that $T_{k(n)}^j \psi_{k(n)} \to T^j \psi$ as $n \to \infty$. Also, $\lambda_{k(n)} \to \lambda$ as $n \to \infty$. Hence

$$(T - \lambda I)^p \psi = \lim_{n \to \infty} \left(T_{k(n)} - \lambda_{k(n)} I\right)^p \psi_{k(n)} = 0,$$

as desired.

It is significant to observe that, when $\psi_n \in M_n$ and $0 < \delta \leq \|\psi_n\| \leq \gamma$ for some constants δ and γ, the entire sequence (ψ_n) may not converge, although it has a convergent subsequence. In the following simple example, the sequence $(c_n \psi_n)$ does not converge in X for any scalars $c_1, c_2, \ldots$ except when $c_n \to 0$.

Example 2.19 Only a subsequence of approximate eigenvectors may converge:
Let $X := \mathbb{C}^{2\times 1}$, the linear space of all 2×1 matrices with complex entries, $\mathsf{A} := \mathsf{I}$ and for each positive integer n, $\mathsf{A}_n := \begin{bmatrix} 1 & 1/n \\ 1/n & 1 \end{bmatrix}$. Then $\lambda := 1$ is the only spectral value of A and its algebraic multiplicity is 2. Also, $\lambda_{n,1} := 1 + 1/n$ and $\lambda_{n,2} := 1 - 1/n$ are the only spectral values of A_n, each having algebraic multiplicity equal to 1. Further,

$$\psi_{n,1} := [1, 1]^\top \quad \text{and} \quad \psi_{n,2} := [1, -1]^\top$$

are eigenvectors of A_n corresponding to $\lambda_{n,1}$ and $\lambda_{n,2}$, respectively, and $M_n = \text{span}\{\psi_{n,1}, \psi_{n,2}\}$. Let $\psi_n := [1, (-1)^n]^\top$. Then $\psi_n \in M_n$ and the sequence (ψ_n) is bounded as well as bounded away from zero. The subsequence (ψ_{2n}) of (ψ_n) converges to the eigenvector $[1, 1]^\top$ of A. Also, the subsequence (ψ_{2n+1}) of (ψ_n) converges to the eigenvector $[1, -1]^\top$

of A. However, if (c_n) is any sequence in $\mathbb{C}$ which does not converge to 0, then the entire sequence $(c_n\psi_n)$ does not converge in X. ▮

Before we conclude this section, we seek estimates for $\| \underline{\varphi_n} - \underline{P}\, \underline{\varphi_n} \|_\infty$ and $\dfrac{\| \underline{\varphi_n} - \underline{\varphi_{(n)}} \|_\infty}{\| \underline{\varphi_{(n)}} \|_\infty}$ in terms of $\|(\underline{T_n} - \underline{T})\, \underline{\varphi_n}\|_\infty$ and $\|(\underline{T_n} - \underline{T})\, \underline{\varphi}\|_\infty$, respectively. We note that this was easily accomplished when $\dim M = 1$ in Theorem 2.17. For the general case, we need a result proved by Kato in [46] which says that if a sequence of the norms of $m \times m$ matrices is bounded and if all the eigenvalues of these matrices are bounded away from zero, then the sequence of the norms of the inverses of these matrices is also bounded. This result will also be useful when we consider uniformly well-conditioned bases in Chapter 3.

Lemma 2.20
Let $\mathsf{Z} \in \mathbb{C}^{m\times m}$ be a nonsingular matrix. Then

(a) $\|\mathsf{Z}^{-1}\|_2 \le \dfrac{\|\mathsf{Z}\|_2^{m-1}}{|\det \mathsf{Z}|}$.

(b) *Let $\epsilon := \min\{|\lambda| \,:\, \lambda \in \mathrm{sp}(\mathsf{Z})\}$. Then*

$$\|\mathsf{Z}^{-1}\|_1 \le \frac{m^{m/2}}{\epsilon^m}\|\mathsf{Z}\|_1^{m-1}.$$

Proof

(a) Let $\mathsf{Z} := \mathsf{QP}$, where Q is a unitary matrix and P is a Hermitian positive definite matrix. (See Remark 1.3.) Since Q and Q^{-1} are unitary, and $\mathsf{Z}^{-1} = \mathsf{P}^{-1}\mathsf{Q}^{-1}$, we see that $\|\mathsf{Z}\|_2 = \|\mathsf{P}\|_2$, $\|\mathsf{Z}^{-1}\|_2 = \|\mathsf{P}^{-1}\|_2$ and $|\det \mathsf{Z}| = \det \mathsf{P}$. Hence we can assume without loss of generality that the matrix Z itself is Hermitian positive definite. Let $\lambda_1 \le \cdots \le \lambda_m$ be the eigenvalues of Z, so that

$$\lambda_1 > 0, \quad \|\mathsf{Z}\|_2 = \lambda_m, \quad \|\mathsf{Z}^{-1}\|_2 = \frac{1}{\lambda_1} \quad \text{and} \quad |\det \mathsf{Z}| = \lambda_1 \cdots \lambda_m.$$

Thus

$$\|\mathsf{Z}^{-1}\|_2 = \frac{1}{\lambda_1} = \frac{\lambda_2 \cdots \lambda_m}{\lambda_1 \cdots \lambda_m} \le \frac{(\lambda_m)^{m-1}}{\lambda_1 \cdots \lambda_m} = \frac{\|\mathsf{Z}\|_2^{m-1}}{|\det \mathsf{Z}|}.$$

(b) The inequality in (a), and the inequalities $\|\mathsf{Z}^{-1}\|_1 \le \sqrt{m}\,\|\mathsf{Z}^{-1}\|_2$, $\|\mathsf{Z}\|_2 \le \sqrt{m}\,\|\mathsf{Z}\|_1$, and $|\det \mathsf{Z}| = |\lambda_1|\cdots|\lambda_m| \ge \epsilon^m$ imply

$$\|\mathsf{Z}^{-1}\|_1 \le \frac{\sqrt{m}\,\|\mathsf{Z}\|_2^{m-1}}{|\det \mathsf{Z}|} \le \frac{\sqrt{m}\,(\sqrt{m}\,\|\mathsf{Z}\|_1)^{m-1}}{\epsilon^m} = \frac{m^{m/2}}{\epsilon^m}\|\mathsf{Z}\|_1^{m-1}.$$

This completes the proof. ∎

Proposition 2.21

Under the hypotheses of Theorem 2.18, let $\underline{\varphi} := [\varphi_1, \ldots, \varphi_m]$ *and for each large n,* $\underline{\varphi_n} := [\varphi_{n,1}, \ldots, \varphi_{n,m}]$. *Define*

$$\delta := \min_{j=1,\ldots,m} \operatorname{dist}(\varphi_j, \operatorname{span}\{\varphi_i : i \ne j,\, i = 1, \ldots, m\}) > 0,$$
$$\delta_n := \min_{j=1,\ldots,m} \operatorname{dist}(\varphi_{n,j}, \operatorname{span}\{\varphi_{n,i} : i \ne j,\, i = 1, \ldots, m\}) > 0.$$

Let $\underline{\varphi_{(n)}}$ *be as in Theorem 2.18(c). Then for all large n,*

$$\frac{\|\underline{\varphi_n} - \underline{\varphi_{(n)}}\|_\infty}{\|\underline{\varphi_{(n)}}\|_\infty} \le \frac{\ell(\mathrm{C})}{2\pi}\frac{m}{\delta}\beta_1(\mathrm{C})\,\alpha_2(\mathrm{C})\,\|(\underline{T_n} - \underline{T})\,\underline{\varphi}\|_\infty,$$

which tends to zero, and

$$\|\underline{\varphi_n} - \underline{P}\,\underline{\varphi_n}\|_\infty \le \frac{\ell(\mathrm{C})}{2\pi}\alpha_1(\mathrm{C})\,\beta_{2,n}(\mathrm{C})\,\|(\underline{T_n} - \underline{T})\,\underline{\varphi_n}\|_\infty.$$

If the sequence $(\|\underline{\varphi_n}\|_\infty)$ *is bounded, then* $\|(\underline{T_n} - \underline{T})\,\underline{\varphi_n}\|_\infty$ *tends to zero. If, in addition, the sequence* (δ_n) *is bounded away from zero, then the sequence* $(\beta_{2,n}(\mathrm{C}))$ *is bounded and consequently* $\|\underline{\varphi_n} - \underline{P}\,\underline{\varphi_n}\|_\infty$ *tends to zero.*

Proof

As $\underline{P_n}\,\underline{\varphi_{(n)}} = \underline{\varphi_n}$ and $\underline{\varphi_{(n)}} = \underline{\varphi}\,\mathsf{C}_n^{-1}$, Proposition 2.15 shows that for all large n,

$$\begin{aligned}\|\underline{\varphi_n} - \underline{\varphi_{(n)}}\|_\infty &= \|(\underline{P_n} - \underline{P})\,\underline{\varphi}\,\mathsf{C}_n^{-1}\|_\infty \\ &\le \frac{\ell(\mathrm{C})}{2\pi}\beta_1(\mathrm{C})\,\alpha_2(\mathrm{C})\,\|(\underline{T_n} - \underline{T})\,\underline{\varphi}\|_\infty\,\|\mathsf{C}_n^{-1}\|_1.\end{aligned}$$

Let $\underline{\varphi_{(n)}} := [\varphi_{(n),1}, \ldots, \varphi_{(n),m}]$ and $c'_{n,i,j}$ denote the (i,j)th entry of the matrix C_n^{-1} for $1 \le i, j \le m$. Then for each $j = 1, \ldots, m$, we have

$$\varphi_{(n),j} = c'_{n,1,j}\varphi_1 + \ldots + c'_{n,m,j}\varphi_m,$$

so that for $i = 1, \ldots, m$,

$$\delta |c'_{n,i,j}| \leq \|\varphi_{(n),j}\| \leq \| \underline{\varphi_{(n)}} \|_\infty .$$

It follows that

$$\|\mathsf{C}_n^{-1}\|_1 = \max_{j=1,\ldots,m} \sum_{i=1}^{m} |c'_{n,i,j}| \leq \frac{m}{\delta} \| \underline{\varphi_{(n)}} \|_\infty .$$

Thus

$$\frac{\| \underline{\varphi_n} - \underline{\varphi_{(n)}} \|_\infty}{\| \underline{\varphi_{(n)}} \|_\infty} \leq \frac{\ell(\mathrm{C})}{2\pi} \frac{m}{\delta} \beta_1(\mathrm{C})\, \alpha_2(\mathrm{C})\, \|(\underline{T_n} - \underline{T})\, \underline{\varphi} \|_\infty ,$$

where $\|(\underline{T_n} - \underline{T})\, \underline{\varphi} \|_\infty \leq \|(T_n - T)_{|M}\| \, \| \underline{\varphi} \|_\infty \to 0$ by Proposition 2.15.

Again by Proposition 2.15, we have for all large n,

$$\begin{aligned} \| \underline{\varphi_n} - \underline{P}\, \underline{\varphi_n} \|_\infty &= \|(\underline{P_n} - \underline{P})\, \underline{\varphi_n} \|_\infty \\ &\leq \frac{\ell(\mathrm{C})}{2\pi}\, \alpha_1(\mathrm{C})\, \beta_{2,n}(\mathrm{C})\, \|(\underline{T_n} - \underline{T})\, \underline{\varphi_n} \|_\infty . \end{aligned}$$

Assume that the sequence $(\| \underline{\varphi_n} \|_\infty)$ is bounded. Then

$$\|(\underline{T_n} - \underline{T})\, \underline{\varphi_n} \|_\infty \leq \|(T_n - T)_{|M_n}\| \, \| \underline{\varphi_n} \|_\infty \to 0$$

by Proposition 2.15.

Suppose now that the sequence (δ_n) is bounded away from zero. For each large n, let $\theta_{n,i,j}$ denote the (i,j)th entry of the matrix Θ_n for $1 \leq i, j \leq m$. Since $\underline{T_n}\, \underline{\varphi_n} = \underline{\varphi_n}\, \Theta_n$, we have for each $j = 1, \ldots, m$,

$$T_n \varphi_{n,j} = \theta_{n,1,j} \varphi_{n,1} + \cdots + \theta_{n,m,j} \varphi_{n,m} ,$$

so that for $i = 1, \ldots, m$,

$$\delta_n |\theta_{n,i,j}| \leq \|T_n \varphi_{n,j}\| \leq \| \underline{T_n}\, \underline{\varphi_n} \|_\infty .$$

It follows that

$$\|\Theta_n\|_1 := \max_{j=1,\ldots,m} \sum_{i=1}^{m} |\theta_{n,i,j}| \leq \frac{m}{\delta_n} \|T_n\| \, \| \underline{\varphi_n} \|_\infty .$$

Hence the sequence $(\|\Theta_n\|_1)$ is bounded. As a consequence, the sequence $\left(\sup_{z \in \mathrm{C}} \|\Theta_n - z\mathsf{I}\|_1 \right)$ is bounded.

By Corollary 1.22, there is a Cauchy contour $\widetilde{C}$ separating Λ from $(\mathrm{sp}(T) \setminus \Lambda) \cup C$. Let $\epsilon := \mathrm{dist}(\widetilde{C}, C) > 0$. Since $\widetilde{C} \in \mathcal{C}(T, \Lambda)$, Proposition 2.9 shows that $\Lambda_n = \mathrm{sp}(T_n) \cap \mathrm{int}(\widetilde{C})$ for all sufficiently large n. Thus for each eigenvalue λ_n of Θ_n and each $z \in C$, we have

$$|\lambda_n - z| \geq \mathrm{dist}(\mathrm{sp}(\Theta_n), C) = \mathrm{dist}(\Lambda_n, C) > \epsilon > 0.$$

By Kato's result (Lemma 2.20), it follows that the sequence $(\beta_{2,n}(C))$ is bounded. Consequently, $\| \underline{\varphi_n} - \underline{P}\, \underline{\varphi_n} \|_\infty$ tends to zero. ∎

2.5 Exercises

Unless otherwise stated, X denotes a Banach space over $\mathbb{C}$, $X \neq \{0\}$ and $T \in \mathrm{BL}(X)$. Prove the following assertions.

2.1 (a) If each $T_n - T$ is compact and $T_n \xrightarrow{\text{n}} T$, then $T_n \xrightarrow{\text{cc}} T$. (Hint: Let $\|x_n\| \leq 1$ and $y_n := (T_{j(n)} - T)x_n$ with $j(n) \geq n_0$, $E := \{j(n) : n \geq n_0\}$. If the set E is finite, use the compactness of $T_m - T$ for a sufficiently large m; and if the set E is infinite, use the inequality $\|(T_{j(n)} - T)x_n\| \leq \|T_{j(n)} - T\|$.)

(b) If $T_n \xrightarrow{\text{cc}} T$ and $T_n^* \xrightarrow{\text{p}} T^*$, then $T_n \xrightarrow{\text{n}} T$. (Hint: If $x_n \in X$ with $\|x_n\| \leq 1$ and $\|(T_{k(n)} - T)x_n\| \geq \delta > 0$, then without loss of generality $(T_{k(n)} - T)x_n \to y$ in X, where $\|y\| \geq \delta$. Let $f \in X^*$ with $\langle y\,,\, f\rangle = \|y\|$. Then $\delta \leq \langle y\,,\, f\rangle \leq \limsup_{n\to\infty} \|T_{k(n)}^* f - T^* f\| = 0$.)

2.2 If $T_n \xrightarrow{\nu} T$, then $\|(T_n - T)R(z)(T_n - T)\| \to 0$ for every nonzero z in $\mathrm{re}(T)$. (Hint: $R(z) = (TR(z) - I)/z$. Compare: Resolvent operator approximation given in [54], Section 14.)

2.3 Even if we have $T_n \xrightarrow{\nu} T$ and $U_n \xrightarrow{\nu} U$, we may not have $T_n + U_n \xrightarrow{\nu} T + U$. For example, if $T_n \xrightarrow{\nu} T$ but $T_n \overset{\text{n}}{\not\to} T$, then $T_n + I \overset{\nu}{\not\to} T + I$. Also, even if we have $T_n \xrightarrow{\nu} T$ and $S_n \xrightarrow{\nu} S$, we may not have $T_n S_n \xrightarrow{\nu} TS$. For example, if T_n and T are the operators given in Example 2.3 and S is the right shift operator defined by $Sx := \sum_{k=2}^{\infty} x(k-1)e_k$ for $x := \sum_{k=1}^{\infty} x(k)e_k \in \ell^p$, then $T_n \xrightarrow{\nu} T$, but $T_n S \overset{\nu}{\not\to} TS$.

2.4 Let $T_n \xrightarrow{\text{n}} T$. Then $T_n \xrightarrow{\nu} U$ if and only if $(T-U)T = 0 = (U-T)U$. In this case, $(T-U)^2 = 0$. In case $T_n \xrightarrow{\text{n}} O$, we have $T_n \xrightarrow{\nu} U$ if and only if $U^2 = O$.

2.5 Let $\dim X = \infty$ and (T_n) be a sequence in $\text{BL}(X)$ such that $\text{rank}(T_n)$ is finite for each n and $T_n \xrightarrow{\text{p}} I$. Then $0 \in \text{sp}(T_n)$ for each n, but $0 \notin \text{sp}(I)$.

2.6 Let $T_n \xrightarrow{\nu} T$.

(a) If $\text{sp}(T)$ is contained in an open set Ω, then $\text{sp}(T_n) \subset \Omega$ for all large n. (Hint: Let $E := \mathbb{C} \setminus \Omega$ in Theorem 2.6.)

(b) Upper semicontinuity of the spectrum: $\sup\{\text{dist}(\mu, \text{sp}(T)) : \mu \in \text{sp}(T_n)\}$ tends to 0. (Hint: Let $E := \{z \in \mathbb{C} : \text{dist}(z, \text{sp}(T)) \geq \epsilon\}$ in Theorem 2.6, where $\epsilon > 0$.)

(c) Upper semicontinuity of each spectral set: Let Λ be a spectral set for T, $\delta := \text{dist}(\Lambda, \text{sp}(T) \setminus \Lambda)/2$ and $\Lambda_n := \{\mu \in \text{sp}(T_n) : \text{dist}(\mu, \Lambda) < \delta\}$. Then $\sup\{\text{dist}(\mu, \Lambda) : \mu \in \Lambda_n\} \to 0$. (Hint: Fix $0 < \epsilon < \delta$. For all large n, $\Lambda_n := \{\mu \in \text{sp}(T_n) : \text{dist}(\mu, \Lambda) < \epsilon\}$.)

2.7 Let $T_n \xrightarrow{\text{n}} T$, λ_n be an eigenvalue of T_n and $\lambda_n \to \lambda$. Then $\lambda \in \text{sp}(T)$, but λ need not be an eigenvalue of T.

2.8 Let $T_n \xrightarrow{\nu} T$ and $z \in \mathbb{C}$. Then $z \in \text{re}(T)$ if and only if $z \in \text{re}(T_n)$ for all large n and the sequence $(\|R_n(z)\|)$ is bounded. (Hint: Proposition 2.5 with T, $\widetilde{T}$ replaced by T_n, T respectively and Theorem 2.6 with $E := \{z\}$)

2.9 Lack of lower semicontinuity of the spectrum under norm convergence: Let $X := \ell^2$ and for $x := \sum_{i=1}^{\infty} \in X$, $Tx := \sum_{i=1}^{\infty} w_i x(i) e_{i+1}$, where

$$w_i := e^{-k}, \qquad \text{if } i = 2^k(2\ell+1) \text{ for some } k, \ell \geq 0.$$

For each positive integer n, let $T_n x := \sum_{i=1}^{\infty} w_{n,i} x(i) e_{i+1}$, where

$$w_{n,i} := \begin{cases} 0 & \text{if } i = 2^n(2\ell+1) \text{ for some } \ell \geq 0, \\ w_i & \text{otherwise.} \end{cases}$$

Then each T_n is nilpotent, $\text{sp}(T_n) = \{0\}$, $\|T_n - T\| = e^{-n} \to 0$, but $\text{sp}(T)$ contains $\{z \in \mathbb{C} : |z| \leq d\}$ for some $d > 0$ and $\rho(T) \geq e^{-2r}$ with $r := \sum_{j=1}^{\infty} j/2^{j+1}$.

2.10 Lack of lower semicontinuity of the spectrum under the pointwise convergence: Let $X := \ell^2$ and, for $x \in X$,

$$Tx := \sum_{k=1}^{\infty} x(k+1)e_k, \quad T_n x := \sum_{k=1}^{n} x(k+1)e_k.$$

Then $T_n \xrightarrow{\text{p}} T$, $\text{sp}(T_n) = \{0\}$, but $\text{sp}(T)$ equals $\{\lambda \in \mathbb{C} : |\lambda| \le 1\}$.

2.11 Let Λ be a spectral set for T such that $0 \notin \Lambda$, and $T_n \xrightarrow{\nu} T$. Then, with the notation given in Section 2.4, $T_n P_n - TP_n \xrightarrow{\text{p}} O$, $T_{n|M} \xrightarrow{\text{p}} T_{|M}$, $P_n - PP_n \xrightarrow{\text{p}} O$ and $P_{n|M} \xrightarrow{\text{p}} I_{|M}$. However, (T_n) may not converge pointwise to T and (P_n) may not converge pointwise to P on X. (Hint: Use Proposition 2.15 and $X := \mathbb{C}^{2\times 1}$, $\mathsf{A} := \begin{bmatrix} a & b \\ 0 & 0 \end{bmatrix}$ and $\mathsf{A}_n := \begin{bmatrix} a & c \\ 0 & 0 \end{bmatrix}$ for all positive integers n, where $a \neq 0$ and $b \neq c$.)

2.12 Let $T_n \xrightarrow{\nu} T$ as well as $T_n \xrightarrow{\nu} U$. Then $Ux = Tx$ for every $x \in \mathcal{R}(T) \cap \mathcal{R}(U)$. Also, $\text{sp}(U) = \text{sp}(T)$, and Λ is a spectral set for U if and only if Λ is a spectral set for T. For such a set Λ, let $P := P(T, \Lambda)$ and $Q := P(U, \Lambda)$. If $0 \notin \Lambda$, then $\mathcal{R}(Q) = \mathcal{R}(P) = M$ say, and $U_{|M} = T_{|M}$. The condition '$0 \notin \Lambda$' cannot be omitted from the preceding statement. (Hint: Exercise 2.8 and Exercise 2.11.)

2.13 Let E be a closed subset of $\text{re}(T)$.

(a) Let $T_n \xrightarrow{\text{p}} T$, $\|(T_n - T)T\| \to 0$ and $\|(T_n - T)T_n\| \to 0$. Then there exists some n_0 such that for all $n \ge n_0$, $E \subset \text{re}(T_n)$ and for every $x \in X$, $R_n(z)x \to R(z)x$ uniformly for $z \in E$. As a consequence, if Λ is a spectral set for T, $\mathrm{C} \in \mathcal{C}(T, \Lambda)$, $\Lambda_n := \text{sp}(T_n) \cap \text{int}(\mathrm{C})$, $P := P(T, \Lambda)$ and $P_n := P(T_n, \Lambda_n)$, then $P_n \xrightarrow{\text{p}} P$. (Hint: $[R_n(z) - R(z)]x = R_n(z)(T - T_n)R(z)x$ tends to 0 uniformly for $z \in E$. Compare [54], Proposition 14.1.)

(b) Let $T_n \xrightarrow{\text{n}} T$. Then there exists some n_0 such that for all $n \ge n_0$, $E \subset \text{re}(T_n)$ and $\|R_n(z) - R(z)\| \to 0$ uniformly for $z \in E$. As a consequence, $\|P_n - P\| \to 0$.

2.14 Let P and $\widetilde{P}$ be projections in $\text{BL}(X)$ such that $\|P - \widetilde{P}\| < 1$. Let $Y := \mathcal{R}(P)$, $\widetilde{Y} := \mathcal{R}(\widetilde{P})$. Then $\widetilde{P}_{|Y,\widetilde{Y}}$ is bijective and $\|(\widetilde{P}_{|Y,\widetilde{Y}})^{-1}\|$ is less than or equal to $1/(1 - \|P - \widetilde{P}\|)$.

2.15 Lack of lower semicontinuity of the spectrum at a nonzero isolated point: Let $X := \ell^1$ and for $x \in X$,

$$(Tx)(k) := \begin{cases} -x(1) & \text{if } k = 1, \\ x(k-1) + x(k) & \text{if } k \neq 1, \end{cases}$$

and for each positive integer n,

$$(T_n x)(k) := \begin{cases} -x(1) + x(2k) & \text{if } k = 1, \\ x(k-1) + x(k) & \text{if } 1 < k < n, \\ 3x(k) + x(k+1) & \text{if } k = n+1, \\ 9x(k-1) + x(k) & \text{if } n+1 < k \leq 2n, \\ 0 & \text{if } k > 2n. \end{cases}$$

Then T, $T_n \in \mathrm{BL}(X)$, $\lambda := -1$ is an isolated point of $\mathrm{sp}(T)$, $T_n \xrightarrow{\mathrm{p}} T$ and $\mathrm{sp}(T_n) \cap \{z \in \mathbb{C} : |z+1| < 1\} = \emptyset$, so that there is no $\lambda_n \in \mathrm{sp}(T_n)$ such that $\lambda_n \to \lambda$. (Hint: If $x(k) := (-1/2)^{k-1}$ for all k, then $Tx = -x$; and if $z \in \mathbb{C}$, $z \neq -1$ and $|z-1| > 1$, then $T - zI$ is bijective. Also, if $X_n := \mathrm{span}\{e_1, \ldots, e_{2n}\}$ and A_n is the $(2n)\times(2n)$ matrix representing $T_{n|X_n,X_n}$, then $\det(\mathsf{A}_n - z\mathsf{I}_{2n}) = -(1+z)(1-z)^{2n-1} - 3^{2n-1}$.)

2.16 Let X be a Hilbert space and $T_n \xrightarrow{\mathrm{p}} T$.

(a) If each T_n is normal and λ is an eigenvalue of T, then the lower semicontinuity of the spectrum holds at λ. (Hint: If $Tx = \lambda x$, $x \neq 0$, then there exists $\lambda_n \in \mathrm{sp}(T_n)$ such that $|\lambda_n - \lambda| \leq \|T_n x - \lambda x\|$ by the Krylov-Weinstein Inequality. See Proposition 1.41(c).)

(b) If each T_n is selfadjoint, then the lower semicontinuity of the spectrum holds at every spectral value λ of T. (Hint: T is selfadjoint, $\lambda \in \mathbb{R}$. For $\epsilon > 0$, $z := \lambda + \mathrm{i}\epsilon \in \mathrm{re}(T_n) \cap \mathrm{re}(T)$, $\|R_n(z)\| \leq 1/\epsilon = \|R(z)\|$. If $x \in X$ with $\|x\| = 1$ and $\|R(z)x\| \geq 1/2\epsilon$, then $[R_n(z) - R(z)]x = R_n(z)(T - T_n)R(z)x \to 0$.)

2.17 For $x := \sum_{k=1}^{\infty} x(k)e_k \in \ell^1$, let $Tx := x(1)e_1$ and for each positive integer n,

$$T_n x := \sum_{k=1}^{n-1} \Big(x(k) + n^{1/n}x(k+1)\Big)\, e_k + \Big(x(n) + (-1)^n n^{(1-n)/n} x(1)\Big)\, e_n.$$

Then each T_n is a finite rank operator, the sequence $(\|T_n\|)$ is bounded and $\|(T_n - T)T\| \to 0$. However, $\lambda_{2n} := 2 \in \mathrm{sp}(T_{2n})$ but $2 \notin \mathrm{sp}(T)$. Also, there is no $\mu_n \in \mathrm{sp}(T_n)$ such that $\mu_n \to 1$ although $1 \in \mathrm{sp}(T)$. (Hint: If $X_n := \mathcal{R}(T_n) = \mathrm{span}\{e_1, \ldots, e_n\}$ and A_n is the $n\times n$ matrix representing the operator $T_{n|X_n,X_n}$, then $\det(\mathsf{A}_n - z\mathsf{I}_n) = (1-z)^n - 1$.)

2.18 Let $X := \mathbb{C}^{2\times 1}$ with any norm and $w, z \in \mathbb{C}$ such that $w \neq 0$, $z \neq 0$, $w + z \neq 0$. Consider

$$\mathsf{A} := \begin{bmatrix} -z & z \\ z & -z \end{bmatrix}, \quad \mathsf{B} := \begin{bmatrix} 0 & 2z \\ 2z & 0 \end{bmatrix}, \quad \mathsf{C} := \begin{bmatrix} z & wz \\ 1 & w \end{bmatrix}, \quad \mathsf{D} := \begin{bmatrix} z & wz \\ 0 & w \end{bmatrix}.$$

(a) Let $\mathsf{A}_n := \mathsf{B}$ for all n. Then $(\|\mathsf{A}_n\|)$ is bounded, $\|(\mathsf{A}_n - \mathsf{A})\mathsf{A}\| = 0$, $\lambda_n := 2z \in \mathrm{sp}(\mathsf{A}_n)$ for all n, $\lambda_n \to \lambda := 2z$, but $\lambda \notin \mathrm{sp}(\mathsf{A})$. (Compare Corollary 2.7.) Also, $\mu := 0 \in \mathrm{sp}(\mathsf{A})$, but there is no $\mu_n \in \mathrm{sp}(\mathsf{A}_n)$ such that $\mu_n \to \mu$.

(b) Let $\mathsf{B}_n := \mathsf{A}$ for all n. Then $(\|\mathsf{B}_n\|)$ is bounded, $\|(\mathsf{B}_n - \mathsf{B})\mathsf{B}_n\| = 0$, $\lambda_n := 0 \in \mathrm{sp}(\mathsf{B}_n)$ for all n, $\lambda_n \to \lambda := 0$, but $\lambda \notin \mathrm{sp}(\mathsf{B})$. Also, $\mu := 2z \in \mathrm{sp}(\mathsf{B})$, $\mu \neq 0$, but there is no $\mu_n \in \mathrm{sp}(\mathsf{B}_n)$ such that $\mu_n \to \mu$. (Compare Corollaries 2.7 and 2.13.)

(c) Let $\mathsf{C}_n := \mathsf{D}$ for all n. Then $(\|\mathsf{C}_n\|)$ is bounded, $\|(\mathsf{C}_n - \mathsf{C})^2\| = 0$, $\lambda_n := z \in \mathrm{sp}(\mathsf{C}_n)$ for all n, $\lambda_n \to \lambda := z$, but $z \notin \mathrm{sp}(\mathsf{C})$. Also, $\mu := w + z \neq 0$, $\mu \in \mathrm{sp}(\mathsf{C})$, but there is no $\mu_n \in \mathrm{sp}(\mathsf{C}_n)$ such that $\mu_n \to \mu$. (Compare Corollaries 2.7 and 2.13.)

2.19 Let $T_n \xrightarrow{\nu} T$, where T is compact and $\mathrm{sp}(T_n) = \{0\}$ for each n. Then $\mathrm{sp}(T) = \{0\}$. (Hint: Each nonzero $\lambda \in \mathrm{sp}(T)$ is an isolated point of $\mathrm{sp}(T)$. Use Corollary 2.13.)

2.20 (Bouldin) Let Λ be a spectral set of finite type for T and $0 \notin \Lambda$. If $T_n \xrightarrow{\mathrm{p}} T$, $\|(T_n - T)T_n\| \to 0$ and $\|T_n(T_n - T)\| \to 0$, then the conclusions of Theorem 2.6, Corollary 2.7, Proposition 2.9, Corollary 2.10, Theorem 2.12 and Corollary 2.13 hold. (Hint: If $\mathrm{C} \in \mathcal{C}(T, \Lambda)$ with $0 \in \mathrm{ext}(\mathrm{C})$, then for $z \in \mathrm{C}$,

$$I + \frac{R(z)(T_n - T)T}{z} = R(z)\Big(I + \frac{T_n - T}{z}\Big)(T_n - zI),$$

$$I + \frac{T_n(T_n - T)R(z)}{z} = (T_n - zI)\Big(I + \frac{T_n - T}{z}\Big)R(z).$$

Also, $\|(T_n - T)P\| \to 0$ since $P := P(T, \Lambda)$ is compact. Compare [23].)

2.21 Let $X := \mathbb{C}^{1\times 2}$ with the ∞-norm, $Y_1 := \{[z, 0] : z \in \mathbb{C}\}$, $Y_2 := \{[z, z] : z \in \mathbb{C}\}$ and $Y_3 := \{[z, (\sqrt{2} - 1)z] : z \in \mathbb{C}\}$. Then $\mathrm{gap}(Y_1, Y_2) = 1$ and $\mathrm{gap}(Y_1, Y_3) = \sqrt{2} - 1 = \mathrm{gap}(Y_3, Y_2)$, so that $\mathrm{gap}(Y_1, Y_2) > \mathrm{gap}(Y_1, Y_3) + \mathrm{gap}(Y_3, Y_2)$.

2.22 Let P and P_n be projections in $\mathrm{BL}(X)$ with $P_n \xrightarrow{\mathrm{p}} P$ and $\mathrm{rank}(P)$ be finite. Then $P_n \xrightarrow{\mathrm{cc}} P$ if and only if $\|(P_n - P)P_n\| \to 0$ if and only if $\mathrm{rank}(P_n) = \mathrm{rank}(P)$ for all large n. In this case, $\mathrm{gap}(\mathcal{R}(P_n), \mathcal{R}(P)) \to 0$.

2.23 With the notation of Theorem 2.18 and Proposition 2.21,

$$\frac{\|\underline{P_n}\,\underline{\varphi} - \underline{\varphi}\|_\infty}{\|\underline{\varphi}\|_\infty} \leq \frac{\ell(\mathrm{C})}{2\pi}\widetilde{\alpha}_1(\mathrm{C})\alpha_2(\mathrm{C})\,\|(T_n - T)_{|M}\| \to 0,$$

$$\|\underline{P_n}\,\underline{\varphi} - \underline{\varphi}\|_\infty \leq \frac{\ell(\mathrm{C})}{2\pi}\beta_1(\mathrm{C})\alpha_2(\mathrm{C})\,\|(\underline{T_n} - \underline{T})\,\underline{\varphi}\|_\infty \to 0.$$

2.24 Estimates for eigenvectors: Let Λ be a spectral set for T, $\mathrm{C} \in \mathcal{C}(T, \Lambda)$ and $T_n \xrightarrow{\nu} T$. With the notation of Proposition 2.9, we have the following.

(a) If $\lambda_n \in \Lambda_n$, $\varphi_n \in X$ and $T_n\varphi_n = \lambda_n\varphi_n$ for all large n, then

$$\|\varphi_n - P\varphi_n\| \leq r_n := \frac{\ell(\mathrm{C})}{2\pi}\frac{\alpha_1(\mathrm{C}_\epsilon)}{\mathrm{dist}(\lambda_n, \mathrm{C})}\|(T_n - T)\varphi_n\|,$$

and if $r_n < \|\varphi_n\|$, then $\Lambda \neq \emptyset$.

(b) If $\lambda \in \Lambda$, $\varphi \in X$ and $T\varphi = \lambda\varphi$, then

$$\|P_n\varphi - \varphi\| \leq \widetilde{r}_n := \frac{\ell(\mathrm{C})}{2\pi}\frac{\alpha_2(\mathrm{C}_\epsilon)}{\mathrm{dist}(\lambda, \mathrm{C})}\|(T_n - T)\varphi\|,$$

and if $\widetilde{r}_n < \|\varphi\|$, then $\Lambda_n \neq \emptyset$.

2.25 Let λ be a spectral value of T of finite type. Suppose that g is the geometric multiplicity and ℓ is the ascent of λ. Let $T_n \xrightarrow{\mathrm{n}} T$, or $\lambda \neq 0$ and $T_n \xrightarrow{\nu} T$. For $0 < d < \mathrm{dist}(\lambda, \mathrm{sp}(T) \setminus \{\lambda\})$ and each large n, let

$$\mathrm{sp}(T_n) \cap \{z \in \mathbb{C} : |z - \lambda| < d\} = \{\lambda_{n,1}, \ldots, \lambda_{n,r(n)}\}.$$

If $g_{n,j}$ is the geometric multiplicity of $\lambda_{n,j}$ and $\ell_{n,j}$ is the ascent of $\lambda_{n,j}$, $j = 1, \ldots, r(n)$, then

(a) $g_{n,j} \leq g$, but one may not have $g_{n,1} + \cdots + g_{n,r(n)} = g$,

(b) $\ell_{n,1} + \cdots + \ell_{n,r(n)} \geq \ell$, but the equality may not hold.

2.26 Let λ be the dominant spectral value of T (that is, $\lambda \in \mathrm{sp}(T)$ and $|\lambda| > |\mu|$ for all $\mu \in \mathrm{sp}(T)$, $\mu \neq \lambda$). Assume that λ is nonzero and of finite type. Let m be the algebraic multiplicity of λ and $T_n \xrightarrow{\nu} T$. The m spectral values of T_n (counted according to their algebraic multiplicities) having largest absolute values converge to λ.

Chapter 3

Improvement of Accuracy

Let us first review the development of the subject matter so far. In Chapter 1, we have discussed the eigenvalue problem $T\varphi = \lambda\varphi$ for a bounded operator T on a complex Banach space X. In order to obtain an approximate solution to this problem, in Chapter 2 we have considered sequences (T_n) of bounded operators on X which converge to T in an appropriate manner, and have given error estimates for the solutions of the approximate eigenvalue problem $T_n\varphi_n = \lambda_n\varphi_n$.

In the next chapter, we shall point out several methods of constructing sequences of finite rank operators T_n which approximate T and show how to reduce the eigenvalue problem $T_n\varphi_n = \lambda_n\varphi_n$ to a matrix eigenvalue problem $\mathsf{A}_n\mathsf{u}_n = \lambda_n\mathsf{u}_n$. Typically, the size of the matrix A_n increases with n. The cost of the computations is significantly affected by the increase in the size of the matrix A_n.

As the applications of the eigenvalue problems become more and more varied, along with a demand for increased accuracy, the conventional methods require a finer and finer discretization. They tend to create problems of inefficiencies and instabilities of algorithms while dealing with very large nonsparse matrices. Also, one has to keep track of the accumulation of roundoff errors and of the limitations in storage and speed of available computers. As a result, one would like to deal with a matrix eigenvalue problem of optimal size in order to maximize accuracy, minimize computational errors, and respect the storage and time requirements. In many situations, one has no option but to keep the size of the matrix eigenvalue problem moderate. But then the accuracy may not be very high.

The purpose of the present chapter is to provide two ways of attaining high precision without having to solve a matrix eigenvalue problem of a very large size. The first approach is based on an iterative refinement of a not-so-precise solution of the given eigenvalue problem, and the second is based on a higher order spectral analysis which is equivalent to reformulating the given eigenvalue problem on a suitable product of the original Banach space X. The second approach is known as acceleration. When the desired eigenvalue (or the desired cluster of eigenvalues) of T is not well separated from the rest of the spectrum of T, the iterative refinement approach may cause ill-conditioning. In this situation, the acceleration approach should be preferred.

A combination of the the two approaches mentioned above is treated in [32], [15] and [16].

3.1 Iterative Refinement

3.1.1 General Remarks

Let $\widetilde{\varphi}$ be an approximate solution of the eigenvalue problem $T\varphi = \lambda\varphi$. The Power Method and its variants are the most well-known iterative methods of improving upon $\widetilde{\varphi}$. Let $\widetilde{\varphi}^{(0)} = \widetilde{\varphi}$ and for $k = 1, 2, \ldots$ define

$$\widetilde{\varphi}^{(k)} = c_k T\widetilde{\varphi}^{(k-1)},$$

where a suitable scalar c_k is chosen so as to ensure that $\|\widetilde{\varphi}^{(k)}\|$ is neither too large nor too small. For example, if the eigenvalue λ we seek is positive, then one may let $c_k = 1/\|T\widetilde{\varphi}^{(k-1)}\|$, provided $T\widetilde{\varphi}^{(k-1)} \neq 0$. In general, one may consider some $\widetilde{f} \in X^*$ and let $c_k = 1/\langle\widetilde{\varphi}^{(k-1)}, \widetilde{f}\rangle$, provided $\langle\widetilde{\varphi}^{(k-1)}, \widetilde{f}\rangle \neq 0$. The main limitation of the Power Method is that it can be used to approximate only the dominant eigenvalue of T (if it exists). If T is invertible and has a unique spectral value of the smallest modulus, then the Power Method can be applied to T^{-1}; and then it is known as the Inverse Power Method. More generally, if λ is an isolated spectral value of T and if we can find a scalar $c \in \mathrm{re}(T)$ such that $|\lambda - c| < |\mu - c|$ for every $\mu \in \mathrm{sp}(T)$ with $\mu \neq \lambda$, then the Power Method can be applied to $(T - cI)^{-1}$, which is known as the Inverse Power Method with shift c. The implementation of this method involves solving the equation $(T - cI)x = \widetilde{\varphi}^{(k-1)}$ and letting $\widetilde{\varphi}^{(k)} = x/\langle x, \widetilde{f}\rangle$, $k = 1, 2, \ldots$

We shall now consider another method for approximating an intermediate eigenvalue λ of T, which is analogous to the following iterative refinement technique of finding an approximate solution of the operator equation $Ax = y$, where $A \in \mathrm{BL}(X)$ is invertible and y is a given element of X. Suppose that there is an invertible operator $\widetilde{A}$ in $\mathrm{BL}(X)$ which is an approximation of A and for which it is easier to find $\widetilde{x} \in X$ which satisfies $\widetilde{A}\widetilde{x} = y$. Let $\widetilde{x}^{(0)} := \widetilde{x}$. Consider the residual $\widetilde{r}^{(0)} := y - A\widetilde{x}^{(0)}$, find $\widetilde{u}^{(0)} \in X$ such that $\widetilde{A}\widetilde{u}^{(0)} = \widetilde{r}^{(0)}$, and define $\widetilde{x}^{(1)} := \widetilde{x}^{(0)} + \widetilde{u}^{(0)}$. Then $\widetilde{x}^{(1)}$ may be considered as a refinement of the initial solution $\widetilde{x}^{(0)}$. This procedure can be repeated to obtain the following iterations:

$$\begin{array}{ll} \text{For} & k = 1, 2, \ldots \\ \text{set} & \widetilde{r}^{(k-1)} := y - A\widetilde{x}^{(k-1)}, \\ \text{solve} & \widetilde{A}\widetilde{u}^{(k-1)} = \widetilde{r}^{(k-1)}, \\ \text{set} & \widetilde{x}^{(k)} := \widetilde{x}^{(k-1)} + \widetilde{u}^{(k-1)}. \end{array}$$

It can be proved that if A is invertible and $\rho(A^{-1}(A - \widetilde{A})) < 1$, then $\widetilde{A}$ is also invertible; if in addition $\rho(\widetilde{A}^{-1}(\widetilde{A} - A)) < 1$, then the sequence $\widetilde{x}^{(k)}$ converges to $x \in X$, which satisfies $Ax = y$. (See Exercise 3.1.)

To find a similar iteration scheme for solving an eigenvalue problem, let us assume, for the ease of presentation, that λ is a simple eigenvalue of T with a corresponding eigenvector φ, $\widetilde{T}$ is an approximation of T for which it is easier to find a simple eigenvalue $\widetilde{\lambda}$ with a corresponding eigenvector $\widetilde{\varphi}$. Let $\widetilde{\varphi}^*$ be the eigenvector of $\widetilde{T}^*$ corresponding to the complex conjugate of $\widetilde{\lambda}$ such that $\langle \widetilde{\varphi}, \widetilde{\varphi}^* \rangle = 1$. If $\widetilde{P}$ denotes the spectral projection associated with $\widetilde{T}$ and $\widetilde{\lambda}$, then $\widetilde{P}x = \langle x, \widetilde{\varphi}^* \rangle \widetilde{\varphi}$ for $x \in X$. Let $\widetilde{\lambda}^{(0)} = \widetilde{\lambda}$, $\widetilde{\varphi}^{(0)} = \widetilde{\varphi}$ and $\widetilde{\lambda}^{(1)} := \langle T\widetilde{\varphi}^{(0)}, \widetilde{\varphi}^* \rangle$. Consider the residual $\widetilde{r}^{(0)} = \widetilde{\lambda}^{(1)}\widetilde{\varphi}^{(0)} - T\widetilde{\varphi}^{(0)}$. Clearly, $\langle \widetilde{r}^{(0)}, \widetilde{\varphi}^* \rangle = 0$, that is, $\widetilde{r}^{(0)} \in \mathcal{N}(\widetilde{P})$ and hence there is a unique $\widetilde{u}^{(0)} \in X$ such that $(\widetilde{T} - \widetilde{\lambda} I)\widetilde{u}^{(0)} = \widetilde{r}^{(0)}$ and $\langle \widetilde{u}^{(0)}, \widetilde{\varphi}^* \rangle = 0$. Define $\widetilde{\varphi}^{(1)} = \widetilde{\varphi}^{(0)} + \widetilde{u}^{(0)}$. Then $\widetilde{\lambda}^{(1)}$ and $\widetilde{\varphi}^{(1)}$ may be considered as refinements of the initial approximations $\widetilde{\lambda}^{(0)}$ and $\widetilde{\varphi}^{(0)}$ of λ and φ, respectively. This procedure can be repeated to obtain the following Elementary Iteration:

$$\begin{array}{ll} \text{For} & k = 1, 2, \ldots \\ \text{set} & \widetilde{\lambda}^{(k)} := \langle T\widetilde{\varphi}^{(k-1)}, \widetilde{\varphi}^* \rangle, \\ & \widetilde{r}^{(k-1)} := \widetilde{\lambda}^{(k)}\widetilde{\varphi}^{(k-1)} - T\widetilde{\varphi}^{(k-1)}, \\ \text{solve} & \left\{ \begin{array}{l} (\widetilde{T} - \widetilde{\lambda} I)\widetilde{u}^{(k-1)} = \widetilde{r}^{(k-1)}, \\ \quad \langle \widetilde{u}^{(k-1)}, \widetilde{\varphi}^* \rangle = 0, \end{array} \right. \\ \text{set} & \widetilde{\varphi}^{(k)} := \widetilde{\varphi}^{(k-1)} + \widetilde{u}^{(k-1)}. \end{array}$$

Note that if $\widetilde{S}$ is the reduced resolvent associated with $\widetilde{T}$ and $\widetilde{\lambda}$, then $\widetilde{u}^{(k-1)} = \widetilde{S}\widetilde{r}^{(k-1)}$ and the above scheme can be written as follows:

$$\begin{aligned}
&\text{For } k = 1, 2, \ldots\\
&\text{set } \widetilde{\lambda}^{(k)} := \langle T\widetilde{\varphi}^{(k-1)}\,,\,\widetilde{\varphi}^*\rangle,\\
&\qquad \widetilde{\varphi}^{(k)} := \widetilde{\varphi}^{(k-1)} + \widetilde{S}(\widetilde{\lambda}^{(k)}\widetilde{\varphi}^{(k-1)} - T\widetilde{\varphi}^{(k-1)}).
\end{aligned}$$

We observe that if $\widetilde{\varphi}^{(k)} \to \varphi$ in X as $k \to \infty$, then $\widetilde{\lambda}^{(k)} \to \lambda := \langle T\varphi\,,\,\widetilde{\varphi}^*\rangle$ in $\mathbb{C}$ and $\widetilde{S}(\lambda\varphi - T\varphi) = 0$. Also, since $\langle\widetilde{\varphi}^{(k)}\,,\,\widetilde{\varphi}^*\rangle = \langle\widetilde{\varphi}^{(0)}\,,\,\widetilde{\varphi}^*\rangle = 1$ for all $k = 1, 2, \ldots$ we see that $\langle\varphi\,,\,\widetilde{\varphi}^*\rangle = 1$. Hence $\langle\lambda\varphi - T\varphi\,,\,\widetilde{\varphi}^*\rangle = \lambda - \lambda = 0$, that is, $\widetilde{P}(\lambda\varphi - T\varphi) = 0$ as well. This shows that $\lambda\varphi - T\varphi = 0$. Thus λ is an eigenvalue of T with a corresponding eigenvector φ which satisfies $\langle\varphi\,,\,\widetilde{\varphi}^*\rangle = 1$.

We shall prove the convergence of the Elementary Iteration in Subsection 3.1.2 by employing mathematical induction. (See [54] and [33] for various refinement methods based on the fixed point technique.) We shall also consider the case of a multiple eigenvalue of T and, more generally, the case of a finite cluster of multiple eigenvalues of T.

The Elementary Iteration can also be viewed as a Newton-type method of finding roots of the equation $F(x) = 0$, where

$$F(x) := \langle Tx\,,\,\widetilde{\varphi}^*\rangle x - Tx, \quad x \in X.$$

If φ is an eigenvector of T corresponding to the eigenvalue λ such that $\langle\varphi\,,\,\widetilde{\varphi}^*\rangle = 1$, then $\langle T\varphi\,,\,\widetilde{\varphi}^*\rangle = \lambda\langle\varphi\,,\,\widetilde{\varphi}^*\rangle = \lambda$ and hence $F(\varphi) = 0$. Newton's iteration for finding φ starting with some $x^{(0)}$ in X is given by

$$x^{(k)} := x^{(k-1)} - (\mathrm{D}_{k-1}F)^{-1}\,F(x^{(k-1)}), \quad k = 1, 2, \ldots$$

where the operator $\mathrm{D}_{k-1}F$ denotes the **Fréchet derivative** of F at $x^{(k-1)}$:

$$(\mathrm{D}_{k-1}F)x = \langle Tx\,,\,\widetilde{\varphi}^*\rangle x^{(k-1)} + \langle Tx^{(k-1)}\,,\,\widetilde{\varphi}^*\rangle x - Tx, \quad x \in X.$$

For reducing the calculations, if we replace $\mathrm{D}_{k-1}F$ by D_0F, we obtain the following iteration, known as the **Fixed Slope Newton Method**:

$$\widetilde{x}^{(k)} := \widetilde{x}^{(k-1)} - (\mathrm{D}_0F)^{-1}\,F(\widetilde{x}^{(k-1)}), \quad k = 1, 2, \ldots$$

Now let $x^{(0)} = \widetilde{\varphi}$ and note that

$$(\mathrm{D}_0F)x = \langle Tx\,,\,\widetilde{\varphi}^*\rangle\widetilde{\varphi} + \langle T\widetilde{\varphi}\,,\,\widetilde{\varphi}^*\rangle x - Tx, \quad x \in X.$$

Replacing the operator T by the operator $\widetilde{T}$ in the expression for D_0F, we consider the following modification of the previous iteration:

$$\widetilde{\varphi}^{(k)} := \widetilde{\varphi}^{(k-1)} - (\widetilde{\mathrm{D}}_0F)^{-1}F(\widetilde{\varphi}^{(k-1)}), \quad k = 1, 2, \ldots$$

where, for $x \in X$,

$$\begin{aligned}(\widetilde{\mathrm{D}}_0 F)x &:= \langle \widetilde{T}x\,,\,\widetilde{\varphi}^*\rangle\widetilde{\varphi} + \langle \widetilde{T}\widetilde{\varphi}\,,\,\widetilde{\varphi}^*\rangle x - \widetilde{T}x \\ &= \langle \widetilde{T}x\,,\,\widetilde{\varphi}^*\rangle\widetilde{\varphi} - (\widetilde{T} - \widetilde{\lambda}I)x,\end{aligned}$$

as $\widetilde{T}\widetilde{\varphi} = \widetilde{\lambda}\widetilde{\varphi}$ and $\langle \widetilde{\varphi}\,,\,\widetilde{\varphi}^*\rangle = 1$. If $x \in \mathcal{N}(\widetilde{P})$, that is, $\langle x\,,\,\widetilde{\varphi}^*\rangle = 0$, then $\langle \widetilde{T}x\,,\,\widetilde{\varphi}^*\rangle = \langle x\,,\,\widetilde{T}^*\widetilde{\varphi}^*\rangle = \widetilde{\lambda}\langle x\,,\,\widetilde{\varphi}^*\rangle = 0$, since $\widetilde{\varphi}^*$ is an eigenvector of $\widetilde{T}^*$ corresponding to the complex conjugate of $\widetilde{\lambda}$. Thus, for $x \in \mathcal{N}(\widetilde{P})$, we have

$$(\widetilde{\mathrm{D}}_0 F)x = -(\widetilde{T} - \widetilde{\lambda}I)x \quad \text{and hence} \quad (\widetilde{\mathrm{D}}_0 F)^{-1}x = -\widetilde{S}x.$$

By mathematical induction, it can be seen that $F(\widetilde{\varphi}^{(k-1)}) \in \mathcal{N}(\widetilde{P})$ for $k = 1, 2, \ldots$ Hence for $k = 1, 2, \ldots$ we obtain

$$\begin{aligned}\widetilde{\varphi}^{(k)} &= \widetilde{\varphi}^{(k-1)} + \widetilde{S}F(\widetilde{\varphi}^{(k-1)}) = \widetilde{\varphi}^{(k-1)} + \widetilde{S}F(\widetilde{\varphi}^{(k-1)}) \\ &= \widetilde{\varphi}^{(k-1)} + \widetilde{S}(\widetilde{\lambda}^{(k)}\widetilde{\varphi}^{(k-1)} - T\widetilde{\varphi}^{(k-1)}),\end{aligned}$$

as before. (See [9] for a treatment of Newton-type refinement methods.)

We shall also consider a combination of the Power Method and the Elementary Iteration, which we call the **Double Iteration**, first for the case of a simple eigenvalue of T and then for a cluster of a finite number of multiple eigenvalues of T. (See [6] and [13].) We shall see that the Double Iteration gives improved error estimates.

Before giving the convergence results and the error analysis for the above-mentioned iterations, we mention another iterative refinement method which is known as the **Rayleigh-Schrödinger Iteration**. For s in $[0, 1]$, assume that the operator $T(s) := \widetilde{T} + s(T - \widetilde{T})$ has an eigenvalue $\lambda(s)$ and a corresponding eigenvector $\varphi(s)$ which have convergent power series expansions:

$$\lambda(s) = \widetilde{\lambda} + \sum_{j=1}^{\infty} \mu_j s^j \quad \text{and} \quad \varphi(s) = \widetilde{\varphi} + \sum_{j=1}^{\infty} \psi_j s^j.$$

(See [58] for conditions under which such expansions are valid.)

In particular, for $s = 0$, we have $\widetilde{T}\widetilde{\varphi} = \widetilde{\lambda}\widetilde{\varphi}$. Assume, for the ease of presentation, that $\widetilde{\lambda}$ is a simple eigenvalue of $\widetilde{T}$ and that $\widetilde{\varphi}^*$ is the eigenvector of $\widetilde{T}^*$ corresponding to the complex conjugate of $\widetilde{\lambda}$ such that $\langle \widetilde{\varphi}\,,\,\widetilde{\varphi}^*\rangle = 1$. Suppose further that

$$\langle \varphi(s)\,,\,\widetilde{\varphi}^*\rangle = 1 \quad \text{for all } s \in \,]0, 1].$$

Then $\langle \psi_j \, , \widetilde{\varphi}^* \rangle = 0$ for every $j = 1, 2, \ldots$ Since $T(s)\varphi(s) = \lambda(s)\varphi(s)$ for $s \in [0, 1]$, we have

$$\Big[\widetilde{T} + s(T - \widetilde{T})\Big]\Big(\widetilde{\varphi} + \sum_{j=1}^{\infty} \psi_j s^j\Big) = \Big(\widetilde{\lambda} + \sum_{j=1}^{\infty} \mu_j s^j\Big)\Big(\widetilde{\varphi} + \sum_{j=1}^{\infty} \psi_j s^j\Big).$$

Equating the coefficients of s^j for $j = 1$, we obtain

$$(\widetilde{T} - \widetilde{\lambda} I)\psi_1 = -(T - \widetilde{T})\widetilde{\varphi} + \mu_1 \widetilde{\varphi}$$

and for $j = 2, 3, \ldots$ we obtain

$$(\widetilde{T} - \widetilde{\lambda} I)\psi_j = -(T - \widetilde{T})\psi_{j-1} + \sum_{i=1}^{j-1} \mu_i \psi_{j-i} + \mu_j \widetilde{\varphi}.$$

Note that $\langle (\widetilde{T} - \widetilde{\lambda} I)\psi_j \, , \widetilde{\varphi}^* \rangle = \langle \psi_j \, , (\widetilde{T} - \widetilde{\lambda} I)^* \widetilde{\varphi}^* \rangle = \langle \psi_j \, , 0 \rangle = 0$, and $\langle \widetilde{\varphi} \, , \widetilde{\varphi}^* \rangle = 1$, $\langle \psi_j \, , \widetilde{\varphi}^* \rangle = 0$ for $j = 1, 2, \ldots$

Hence $-\langle (T - \widetilde{T})\widetilde{\varphi} \, , \widetilde{\varphi}^* \rangle + \mu_1 = 0$, that is,

$$\mu_1 = \langle (T - \widetilde{T})\widetilde{\varphi} \, , \widetilde{\varphi}^* \rangle$$

and for $j = 2, 3, \ldots$ $-\langle (T - \widetilde{T})\psi_{j-1} \, , \widetilde{\varphi}^* \rangle + \mu_j = 0$, that is,

$$\mu_j = \langle (T - \widetilde{T})\psi_{j-1} \, , \widetilde{\varphi}^* \rangle.$$

For $j = 1, 2, \ldots$ ψ_j is uniquely determined by the relation obtained by equating the coefficients of s^j stated earlier. In fact,

$$\psi_1 = \widetilde{S}[\mu_1 \widetilde{\varphi} - (T - \widetilde{T})\widetilde{\varphi}],$$

and for $j = 2, 3 \ldots$

$$\psi_j = \widetilde{S}\Big[\mu_j \widetilde{\varphi} + \sum_{i=1}^{j-1} \mu_i \psi_{j-i} - (T - \widetilde{T})\psi_{j-1}\Big].$$

Let $\widetilde{\lambda}^{(0)} := \widetilde{\lambda}$, $\widetilde{\varphi}^{(0)} := \widetilde{\varphi}$, and for $k = 1, 2, \ldots$ define

$$\widetilde{\lambda}^{(k)} := \widetilde{\lambda} + \sum_{j=1}^{k} \mu_j, \quad \widetilde{\varphi}^{(k)} := \widetilde{\varphi} + \sum_{j=1}^{k} \psi_j.$$

Then the pair $(\widetilde{\lambda}^{(k)}, \widetilde{\varphi}^{(k)})$ provides an iterative refinement of the initial eigenpair $(\widetilde{\lambda}, \widetilde{\varphi})$. This approach can also be extended to the general case of a cluster of a finite number of multiple eigenvalues of T. (See [64], [56], [57], [49], [3] and [5] for the convergence of the Rayleigh-Schrödinger Iteration along with error estimation.) Because of the necessity of computing an ever-increasing number of coefficients, the Rayleigh-Schrödinger Iteration is not preferred to the Elementary Iteration or the Double Iteration discussed before. (See [50] for a comparative performance of these three iterations.)

3.1.2 Refinement Schemes for a Simple Eigenvalue

In this subsection we give estimates for iterative refinements of initial approximations of a simple eigenvalue and of a corresponding eigenvector of $T \in \mathrm{BL}(X)$. We begin by recollecting some basic approximation results from Chapter 2. These give estimates for the initial approximations.

Lemma 3.1
Let λ be a simple eigenvalue of T and φ be a corresponding eigenvector. Assume that

$$\text{(i) } T_n \xrightarrow{\text{n}} T \quad \text{or} \quad \text{(ii) } \lambda \neq 0 \text{ and } T_n \xrightarrow{\nu} T.$$

Then for each large enough n, T_n has a unique simple eigenvalue λ_n such that $\lambda_n \to \lambda$.

Let φ_n be an eigenvector of T_n corresponding to λ_n and φ_n^* be the eigenvector of T_n^* corresponding to its (simple) eigenvalue $\overline{\lambda}_n$ such that $\langle \varphi_n , \varphi_n^* \rangle = 1$. Then $\langle \varphi , \varphi_n^* \rangle \neq 0$ for all large n. If we let

$$\varphi_{(n)} := \frac{\varphi}{\langle \varphi , \varphi_n^* \rangle},$$

then for all large n, we have

$$\max\left\{ |\lambda_n - \lambda|, \frac{||\varphi_n - \varphi_{(n)}||}{||\varphi_n||} \right\} \leq c\,||T_n - T||$$

and if $\lambda \neq 0$, then

$$\max\left\{ |\lambda_n - \lambda|, \frac{||\varphi_n - \varphi_{(n)}||}{||\varphi_n||} \right\} \leq c\,||(T_n - T)T||,$$

where c is a constant, independent of n.

Proof

If $\epsilon > 0$ is small enough, then by Theorem 2.17(a), there is a positive integer n_0 such that for each $n \geq n_0$, we have a unique $\lambda_n \in \mathrm{sp}(T_n)$ satisfying $|\lambda_n - \lambda| < \epsilon$. Further, λ_n is a simple eigenvalue of T_n and $\lambda_n \to \lambda$. If φ_n and $\varphi_{(n)}$ are as stated, then by Theorem 2.17(b), $||\varphi_{(n)}|| \leq 2||\varphi_n||$ and

$$|\lambda_n - \lambda| \leq 2\,\epsilon\,\alpha_2(\mathrm{C}_\epsilon)\frac{||(T_n - T)\varphi||}{||\varphi||},$$

$$\frac{||\varphi_n - \varphi_{(n)}||}{||\varphi_{(n)}||} \leq \alpha_2(\mathrm{C}_\epsilon)\frac{||(T_n - T)\varphi||}{||\varphi||}$$

for all large n and for some constants $\alpha_1(\mathrm{C}_\epsilon)$ and $\alpha_2(\mathrm{C}_\epsilon)$, independent of n. Since $||(T_n - T)\varphi|| \leq ||T_n - T||\;||\varphi||$, and in case $\lambda \neq 0$, $||(T_n - T)\varphi|| = ||(T_n - T)T\varphi/\lambda|| \leq ||(T_n - T)T||\;||\varphi||/|\lambda|$, the desired estimates for $|\lambda_n - \lambda|$ and $||\varphi_n - \varphi_{(n)}||/||\varphi_n||$ follow. ∎

Unless otherwise stated, we shall assume throughout this subsection that λ is a simple eigenvalue of T and φ is a corresponding eigenvector. Further, the notation λ_n, φ_n^* and $\varphi_{(n)}$ will have the meanings given in the statement of Lemma 3.1. Let P_n and S_n respectively denote the spectral projection and the reduced resolvent associated with T_n and λ_n.

Lemma 3.2

Assume that

$$\text{(i) } T_n \xrightarrow{\text{n}} T \quad \textit{or} \quad \text{(ii) } \lambda \neq 0 \textit{ and } T_n \xrightarrow{\nu} T.$$

(a) *The sequences* $(||P_n||)$ *and* $(||S_n||)$ *are bounded. Also,*

$$||(T_n - T)S_n T_n|| \leq \widetilde{c}\,||(T_n - T)T_n||$$

and

$$||(T_n - T)S_n T|| \leq \widetilde{c}\max\{||(T_n - T)T_n||,\, ||(T_n - T)T||\}$$

for some constant $\widetilde{c}$*, independent of* n.

(b) (i) *The sequence* $(||\varphi_{(n)}||)$ *is bounded if and only if the sequence* $(||\varphi_n||)$ *is bounded.*

(ii) *The sequence* $(||\varphi_n^*||)$ *is bounded if and only if the sequence* $(||\varphi_n||)$ *is bounded away from zero.*

Proof

(a) Since $T_n \xrightarrow{\nu} T$, the boundedness of the sequence $(\|P_n\|)$ follows from Proposition 2.9(b) if we let $\Lambda = \{\lambda\}$. Next, let C denote the circle with center λ and radius ϵ, where $0 < \epsilon < \mathrm{dist}(\lambda, \mathrm{sp}(T) \setminus \{\lambda\})$, drawn counterclockwise. As $\lambda_n \to \lambda$, there is a positive integer n_0 such that $|\lambda_n - \lambda| \leq \epsilon/2$ for all $n \geq n_0$, and then

$$\begin{aligned} \|S_n\| &= \Big\| -\frac{1}{2\pi \mathrm{i}} \int_{\mathrm{C}} R_n(z) \frac{dz}{\lambda_n - z} \Big\| \\ &\leq \frac{\ell(\mathrm{C})}{2\pi} \frac{\alpha_2(\mathrm{C})}{\mathrm{dist}(\lambda_n, \mathrm{C})} \leq \frac{\epsilon\, \alpha_2(\mathrm{C})}{\epsilon/2} = 2\alpha_2(\mathrm{C}), \end{aligned}$$

where $\alpha_2(\mathrm{C}) := \sup\{\|R_n(z)\| \, : \, z \in \mathrm{C}, n \geq n_0\} < \infty$ by Theorem 2.6. Thus the sequence $(\|S_n\|)$ is bounded. Since $S_n T_n = T_n S_n$ by Proposition 1.24(a), we have

$$\|(T_n - T) S_n T_n\| \leq \|(T_n - T) T_n\| \, \|S_n\|.$$

Finally, we may assume without loss of generality that the circle C does not pass through the origin, that is, $\epsilon \neq |\lambda|$. Since $T_n R_n(z) = I + z R_n(z)$ for all $z \in \mathrm{C}$ by Proposition 1.10(e),

$$\begin{aligned} S_n &= -\frac{1}{2\pi \mathrm{i}} \int_{\mathrm{C}} R_n(z) \frac{dz}{\lambda_n - z} = -\frac{1}{2\pi \mathrm{i}} \int_{\mathrm{C}} [T_n R_n(z) - I] \frac{dz}{z(\lambda_n - z)} \\ &= -\frac{1}{2\pi \mathrm{i}} \Big[T_n \int_{\mathrm{C}} R_n(z) \frac{dz}{z(\lambda_n - z)} - I \int_{\mathrm{C}} \frac{dz}{z(\lambda_n - z)} \Big]. \end{aligned}$$

Let $\delta = \min\{|z| \, : \, z \in \mathrm{C}\}$. Then $\delta = |\,|\lambda| - \epsilon| > 0$. Since $|\lambda_n - z| \geq \epsilon/2$ for all $z \in \mathrm{C}$, we see that

$$\begin{aligned} \|(T_n - T) S_n T\| &\leq \frac{\ell(\mathrm{C})}{2\pi} \frac{2}{\delta \epsilon} \left(\|(T_n - T) T_n\| \alpha_2(\mathrm{C}) \|T\| + \|(T_n - T) T\| \right) \\ &\leq \frac{2}{\delta} (\alpha_2(\mathrm{C}) \|T\| + 1) \max\{\|(T_n - T) T_n\|, \|(T_n - T) T\|\}. \end{aligned}$$

Hence the result follows if we choose $\widetilde{c}$ such that $\widetilde{c} \geq \|S_n\|$ for all $n \geq n_0$ and $\widetilde{c} \geq 2(\alpha_2(\mathrm{C})\|T\| + 1)/\delta$.

(b) (i) Since for all large n, $P_n \varphi_{(n)} = \langle \varphi_{(n)} \,,\, \varphi_n^* \rangle \varphi_n = \varphi_n$, we see that

$$\|\varphi_n\| = \|P_n \varphi_{(n)}\| \leq \|P_n\| \, \|\varphi_{(n)}\|,$$

where the sequence $(\|P_n\|)$ is bounded.

On the other hand, by Corollary 2.10,

$$\|P_n\varphi - \varphi\| = \|(P_n - P)P\varphi\| \leq \|(P_n - P)P\| \, \|\varphi\| \to 0,$$

and so $\|\langle \varphi \, , \, \varphi_n^* \rangle \varphi_n\| = \|P_n\varphi\| \to \|\varphi\| \neq 0$. Hence $\dfrac{\|\varphi\|}{2} \leq |\langle \varphi \, , \, \varphi_n^* \rangle| \|\varphi_n\|$ and

$$\|\varphi_{(n)}\| = \left\| \frac{\varphi}{\langle \varphi \, , \, \varphi_n^* \rangle} \right\| \leq 2\|\varphi_n\|$$

for all large n. Thus the sequence $(\|\varphi_{(n)}\|)$ is bounded if and only if the sequence $(\|\varphi_n\|)$ is bounded.

(ii) Note that for all large n,

$$\begin{aligned} \|P_n\| &= \sup\{\|P_n x\| \, : \, x \in X, \|x\| \leq 1\} \\ &= \sup\{\|\langle x \, , \, \varphi_n^* \rangle \varphi_n\| \, : \, x \in X, \|x\| \leq 1\} \\ &= \|\varphi_n^*\| \, \|\varphi_n\|. \end{aligned}$$

Since each P_n is a nonzero projection, we have $\|P_n\| \geq 1$ for all large n. Also, the sequence $(\|P_n\|)$ is bounded. It follows that the sequence $(\|\varphi_n^*\|)$ is bounded if and only if the sequence $(\|\varphi_n\|)$ is bounded away from zero. ∎

We shall now analyze the Elementary Iteration with the initial approximate eigenvector $\widetilde{\varphi} = \varphi_n$, where n is a fixed positive integer for which the conclusions of Lemma 3.1 hold. The successive iterates are then obtained as follows:

$$(E) \qquad \begin{cases} \varphi_n^{(0)} := \varphi_n, \text{ and for } k = 1, 2, \ldots \\ \lambda_n^{(k)} := \langle T\varphi_n^{(k-1)} \, , \, \varphi_n^* \rangle, \\ \varphi_n^{(k)} := \varphi_n^{(k-1)} + S_n \left(\lambda_n^{(k)} \varphi_n^{(k-1)} - T\varphi_n^{(k-1)} \right). \end{cases}$$

We establish three important relations about these iterates.

For $x \in X$, we have

$$\langle S_n x \, , \, \varphi_n^* \rangle = \langle S_n x \, , \, P_n^* \varphi_n^* \rangle = \langle P_n S_n x \, , \, \varphi_n^* \rangle = \langle 0 \, , \, \varphi_n^* \rangle = 0.$$

Also, since

$$\langle \varphi_n^{(0)} \, , \, \varphi_n^* \rangle = \langle \varphi_n \, , \, \varphi_n^* \rangle = 1$$

and for $k = 1, 2, \ldots$

$$\begin{aligned} \langle \varphi_n^{(k)} \, , \, \varphi_n^* \rangle &= \langle \varphi_n^{(k-1)} \, , \, \varphi_n^* \rangle + \langle S_n \left(\lambda_n^{(k)} \varphi_n^{(k-1)} - T\varphi_n^{(k-1)} \right), \, \varphi_n^* \rangle \\ &= \langle \varphi_n^{(k-1)} \, , \, \varphi_n^* \rangle, \end{aligned}$$

it follows that

$$(E)_1 \qquad \langle \varphi_n^{(k)}\,,\, \varphi_n^*\rangle = 1 \quad \text{for all } k = 0, 1, 2, \ldots$$

The condition $(E)_1$ is equivalent to '$P_n\varphi_n^{(k)} = \varphi_n$ for all $k = 0, 1, 2, \ldots$'

Next, if $x \in X$ satisfies $\langle x\,,\, \varphi_n^*\rangle = 0$, then

$$\langle T_n x\,,\, \varphi_n^*\rangle = \langle x\,,\, T_n^*\varphi_n^*\rangle = \langle x\,,\, \overline{\lambda}_n\varphi_n^*\rangle = \lambda_n\langle x\,,\, \varphi_n^*\rangle = 0.$$

Now $\langle \varphi_{(n)}\,,\, \varphi_n^*\rangle = 1$ by the very definition of $\varphi_{(n)}$. Hence for $k = 1, 2, \ldots$ we have

$$\langle \varphi_n^{(k-1)} - \varphi_{(n)}\,,\, \varphi_n^*\rangle = 1 - 1 = 0,$$

so that

$$\langle T_n(\varphi_n^{(k-1)} - \varphi_{(n)})\,,\, \varphi_n^*\rangle = 0,$$

and since

$$\langle T\varphi_{(n)}\,,\, \varphi_n^*\rangle = \frac{\langle T\varphi\,,\, \varphi_n^*\rangle}{\langle \varphi\,,\, \varphi_n^*\rangle} = \lambda,$$

we obtain

$$\lambda_n^{(k)} - \lambda = \langle T\varphi_n^{(k-1)}\,,\, \varphi_n^*\rangle - \langle T\varphi_{(n)}\,,\, \varphi_n^*\rangle = \langle T(\varphi_n^{(k-1)} - \varphi_{(n)})\,,\, \varphi_n^*\rangle.$$

Thus

$$(E)_2 \qquad \lambda_n^{(k)} - \lambda = \langle (T - T_n)(\varphi_n^{(k-1)} - \varphi_{(n)})\,,\, \varphi_n^*\rangle \quad \text{for } k = 1, 2, \ldots$$

Finally, for $k = 1, 2, \ldots$ we have

$$\begin{aligned}
\varphi_n^{(k)} &= \varphi_n^{(k-1)} + S_n\left[\lambda_n^{(k)}\varphi_n^{(k-1)} - T\varphi_n^{(k-1)}\right) \\
&= \varphi_n^{(k-1)} + S_n\left[(\lambda_n^{(k)} - \lambda_n)\varphi_n^{(k-1)} + \lambda_n\varphi_n^{(k-1)}\right. \\
&\qquad \left. + (T_n - T)\varphi_n^{(k-1)} - T_n\varphi_n^{(k-1)}\right] \\
&= \varphi_n^{(k-1)} + S_n\left[(\lambda_n^{(k)} - \lambda_n)\varphi_n^{(k-1)} + (T_n - T)\varphi_n^{(k-1)}\right] \\
&\qquad - S_n(T_n - \lambda_n I)\varphi_n^{(k-1)}.
\end{aligned}$$

Now by Proposition 1.24(b), $S_n(T_n - \lambda_n I)\varphi_n^{(k-1)} = (I - P_n)\varphi_n^{(k-1)} = \varphi_n^{(k-1)} - \langle \varphi_n^{(k-1)}\,,\, \varphi_n^*\rangle\varphi_n = \varphi_n^{(k-1)} - \varphi_n$, so that

$$\varphi_n^{(k)} = \varphi_n + S_n\left[(\lambda_n^{(k)} - \lambda_n)\varphi_n^{(k-1)} + (T_n - T)\varphi_n^{(k-1)}\right].$$

But since $T\varphi_{(n)} = \lambda\varphi_{(n)}$, we also have

$$\begin{aligned}
\varphi_{(n)} &= P_n\varphi_{(n)} + (I - P_n)\varphi_{(n)} = \varphi_n + S_n(T_n - \lambda_n I)\varphi_{(n)} \\
&= \varphi_n + S_n\left[(T_n - T)\varphi_{(n)} + (\lambda - \lambda_n)\varphi_{(n)}\right].
\end{aligned}$$

Hence

$$(E)_3\quad\left\{\begin{aligned} &\text{for } k=1,2,\dots\\ &\varphi_n^{(k)}-\varphi_{(n)} = S_n\big[(T_n-T)(\varphi_n^{(k-1)}-\varphi_{(n)})\\ &\qquad\qquad +(\lambda_n^{(k)}-\lambda_n)\varphi_n^{(k-1)}-(\lambda-\lambda_n)\varphi_{(n)}\big]\\ &\qquad = S_n\big[(T_n-T)(\varphi_n^{(k-1)}-\varphi_{(n)})\\ &\qquad\qquad +(\lambda_n^{(k)}-\lambda)\varphi_n^{(k-1)}+(\lambda-\lambda_n)(\varphi_n^{(k-1)}-\varphi_{(n)})\big]\,. \end{aligned}\right.$$

The equations $(E)_1$, $(E)_2$ and $(E)_3$ will play an important role in the proofs of convergence and of error estimation of the Elementary Iteration (E).

Theorem 3.3
Assume that

$$\text{(i) } T_n \xrightarrow{\text{n}} T \quad \textit{or} \quad \text{(ii) } \lambda \neq 0 \textit{ and } T_n \xrightarrow{\nu} T.$$

For all large n, let φ_n be so chosen that the sequence $(\|\varphi_n\|)$ is bounded and also bounded away from zero.

Let $\lambda_n^{(0)} := \lambda_n$, $\varphi_n^{(0)} := \varphi_n$ and for $k = 1, 2, \ldots$ let $\lambda_n^{(k)}$ and $\varphi_n^{(k)}$ be the iterates of the Elementary Iteration (E).

(a) *If $T_n \xrightarrow{\text{n}} T$, then there is a positive integer n_1 such that for all $n \geq n_1$ and for $k = 0, 1, 2, \ldots$*

$$\max\{|\lambda_n^{(k)}-\lambda|,\ \|\varphi_n^{(k)}-\varphi_{(n)}\|\} \leq (\beta\|T_n-T\|)^{k+1},$$

where β is a constant, independent of n and k, and if in fact $\lambda \neq 0$, then

$$\max\{|\lambda_n^{(k)}-\lambda|,\ \|\varphi_n^{(k)}-\varphi_{(n)}\|\} \leq \alpha\|(T_n-T)T\|(\beta\|T_n-T\|)^{k},$$

where α and β are constants, independent of n and k.

(b) *If $\lambda \neq 0$ and $T_n \xrightarrow{\nu} T$, then there is a positive integer n_1 such that for all $n \geq n_1$ and for all $k = 0, 1, 2, \ldots$*

$$\begin{aligned}&\max\{|\lambda_n^{(2k)}-\lambda|,\ |\lambda_n^{(2k+1)}-\lambda|,\ \|\varphi_n^{(2k)}-\varphi_{(n)}\|,\ \|\varphi_n^{(2k+1)}-\varphi_{(n)}\|\}\\ &\leq \alpha\|(T_n-T)T\|(\beta\max\{\|(T_n-T)T\|,\ \|(T_n-T)T_n\|\})^k,\end{aligned}$$

where α and β are constants, independent of n and k.

Proof
By Lemma 3.2(a), the sequence $(\|S_n\|)$ is bounded. Also, since the sequence $(\|\varphi_n\|)$ is bounded and also bounded away from zero, Lemma 3.2(b) shows that the sequences $(\|\varphi_{(n)}\|)$ and $(\|\varphi_n^*\|)$ are bounded. Further, the sequence $(\|T_n\|)$ is bounded. Hence there are constants γ, p, q, s, t such that for all large n,

$$\|\varphi_n\| \leq \gamma, \ \|\varphi_n^*\| \leq p, \ \|\varphi_{(n)}\| \leq q, \ \|S_n\| \leq s \text{ and } \|T_n - T\| \leq t.$$

Let $\gamma_1 = \max\{1, \gamma\}$.

(a) Let $T_n \xrightarrow{\text{n}} T$. By Lemma 3.1, there is a positive integer n_0, and there is a constant c such that for all $n \geq n_0$,

$$\max\{|\lambda_n - \lambda|, \ \|\varphi_n - \varphi_{(n)}\|\} \leq c\gamma_1\|T_n - T\|.$$

Let

$$\beta := \max\{c\gamma_1, \ p, \ s(1 + p + pq + c\gamma_1)\}.$$

Choose $n_1 \geq n_0$ such that $\|T_n - T\| \leq 1/\beta$ for all $n \geq n_1$.

Fix $n \geq n_1$. We show by induction on k that

$$\max\{|\lambda_n^{(k)} - \lambda|, \|\varphi_n^{(k)} - \varphi_{(n)}\|\} \leq (\beta\|T_n - T\|)^{k+1}, \quad k = 0, 1, 2, \ldots$$

Since $\lambda_n^{(0)} = \lambda_n$, $\varphi_n^{(0)} = \varphi_n$, $\max\{1, \gamma\}c = c\gamma_1 \leq \beta$ and $n_1 \geq n_0$, we see that the desired inequality holds if $k = 0$.

Now assuming that the desired inequality holds for a given $k \geq 0$, we prove that it holds with k replaced by $k + 1$.

By the equation $(E)_2$, we have

$$\begin{aligned}
|\lambda_n^{(k+1)} - \lambda| &\leq \|\varphi_n^*\| \ \|T_n - T\| \ \|\varphi_n^{(k)} - \varphi_{(n)}\| \\
&\leq p\|T_n - T\|(\beta\|T_n - T\|)^{k+1} \\
&\leq (\beta\|T_n - T\|)^{k+2},
\end{aligned}$$

since $p \leq \beta$. Next, by the equation $(E)_3$,

$$\begin{aligned}
\|\varphi_n^{(k+1)} - \varphi_{(n)}\| &= \Big\|S_n \left[(T_n - T)(\varphi_n^{(k)} - \varphi_{(n)}) + (\lambda_n^{(k+1)} - \lambda)\varphi_n^{(k)}\right. \\
&\qquad \left. +(\lambda - \lambda_n)(\varphi_n^{(k)} - \varphi_{(n)})\right]\Big\| \\
&\leq \|S_n\| \left(\|T_n - T\| \ \|\varphi_n^{(k)} - \varphi_{(n)}\| + |\lambda_n^{(k+1)} - \lambda| \ \|\varphi_n^{(k)}\|\right. \\
&\qquad \left. +|\lambda - \lambda_n| \ \|\varphi_n^{(k)} - \varphi_{(n)}\|\right).
\end{aligned}$$

Note that

$$|\lambda_n^{(k+1)} - \lambda| \leq \|T_n - T\| \ \|\varphi_n^{(k)} - \varphi_{(n)}\| \ \|\varphi_n^*\|, \quad |\lambda - \lambda_n| \leq c\gamma_1\|T_n - T\|,$$

and since $\beta\|T_n - T\| \leq 1$ for $n \geq n_1$, we have

$$\|\varphi_n^{(k)} - \varphi_{(n)}\| \leq (\beta\|T_n - T\|)^{k+1} \leq 1,$$

$$\|\varphi_n^{(k)}\| \leq \|\varphi_n^{(k)} - \varphi_{(n)}\| + \|\varphi_{(n)}\| \leq 1 + q.$$

Hence

$$\begin{aligned}\|\varphi_n^{(k+1)} - \varphi_{(n)}\| &\leq s(1 + p + pq + c\gamma_1)\|T_n - T\| \, \|\varphi_n^{(k)} - \varphi_{(n)}\| \\ &\leq \beta\|T_n - T\|(\beta\|T_n - T\|)^{k+1} \\ &\leq (\beta\|T_n - T\|)^{k+2},\end{aligned}$$

since $s(1 + p + pq + c\gamma_1) \leq \beta$.

Thus the desired inequality holds for $k + 1$ and the induction is complete.

Next, if $\lambda \neq 0$, then by Lemma 3.1 there is a positive integer n_0, and there is a constant c such that for all $n \geq n_0$;

$$\max\{|\lambda_n - \lambda|, \|\varphi_n - \varphi_{(n)}\|\} \leq c\gamma_1\|(T_n - T)T\|.$$

Hence if we let $\alpha := c\gamma_1$, $\beta := \max\{p, s(1 + p + pq + \alpha\|T\|)\}$ and choose $n_1 \geq n_0$ such that $\|T_n - T\| \leq 1/\alpha$, $\|T_n - T\| \leq 1/\beta$ for all $n \geq n_1$, then the preceding induction argument shows that for each $n \geq n_1$ and for all $k = 0, 1, 2, \ldots$

$$\max\{|\lambda_n^{(k)} - \lambda|, \|\varphi_n^{(k)} - \varphi_{(n)}\|\} \leq \alpha\|(T_n - T)T\|(\beta\|T_n - T\|)^k.$$

(b) Let $\lambda \neq 0$ and $T_n \xrightarrow{\nu} T$. By Lemma 3.1 and Lemma 3.2(a), there is a positive integer n_0, and there are constants c, $\widetilde{c}$ such that for all $n \geq n_0$,

$$\begin{aligned}&\max\{|\lambda_n - \lambda|, \|\varphi_n - \varphi_{(n)}\|\} \leq c\gamma_1\|T_n - T\|, \\ &\max\{\|(T_n - T)S_nT_n\|, \|(T_n - T)S_nT\|\} \leq \widetilde{c}\epsilon_n,\end{aligned}$$

where $\epsilon_n := \max\{\|(T_n - T)T_n\|, \|(T_n - T)T\|\}$. Let

$$\begin{aligned}\alpha &:= c\gamma_1 \max\{1, tp, s(t + q)\}, \\ \beta &:= \max\{p\beta_1, \beta_2, pt\beta_2, \beta_3\},\end{aligned}$$

where $\beta_1 := 2\widetilde{c} + 3ts\alpha$, $\beta_2 := s[\beta_1(1 + p + pq) + \alpha]$ and $\beta_3 := \beta_2 s[t(1 + p + pq) + \alpha]$.

Choose $n_1 \geq n_0$ such that $\epsilon_n \leq \min\{1/\alpha, 1/\beta, 1\}$ for all $n \geq n_1$. Fix $n \geq n_1$. We show by induction on k that for $k = 0, 1, 2, \ldots$

$$\max\{|\lambda_n^{(2k)} - \lambda|, |\lambda_n^{(2k+1)} - \lambda|, \|\varphi_n^{(2k)} - \varphi_{(n)}\|, \|\varphi_n^{(2k+1)} - \varphi_{(n)}\|\} \\ \leq \alpha\|(T_n - T)T\|(\beta\epsilon_n)^k.$$

Let $k = 0$. Clearly, since $\lambda_n^{(0)} = \lambda_n$ and $\varphi_n^{(0)} = \varphi_n$,

$$\max\{|\lambda_n^{(0)} - \lambda|, \|\varphi_n^{(0)} - \varphi_{(n)}\|\} \leq c\gamma_1\|(T_n - T)T\| \leq \alpha\|(T_n - T)T\|.$$

Also, by the equation $(E)_2$, we have

$$|\lambda_n^{(1)} - \lambda| = |\langle (T - T_n)(\varphi_n^{(0)} - \varphi_{(n)}), \varphi_n^*\rangle| \\ \leq ptc\gamma_1\|(T_n - T)T\| \leq \alpha\|(T_n - T)T\|$$

and by the equation $(E)_3$, we have

$$\varphi_n^{(1)} - \varphi_{(n)} = S_n\left[(T_n - T)(\varphi_n^{(0)} - \varphi_{(n)}) + (\lambda_n^{(1)} - \lambda_n)\varphi_n^{(0)} \right. \\ \left. -(\lambda - \lambda_n)\varphi_{(n)}\right] \\ = S_n\left[(T_n - T)(\varphi_n^{(0)} - \varphi_{(n)}) + (\lambda_n - \lambda)\varphi_{(n)}\right],$$

since $S_n\varphi_n^{(0)} = S_n\varphi_n = S_nP_n\varphi_n = 0$. Hence

$$\|\varphi_n^{(1)} - \varphi_{(n)}\| \leq s(t + q)c\gamma_1\|(T_n - T)T\| \leq \alpha(T_n - T)T\|.$$

Thus the desired inequality holds for $k = 0$.

Now assuming that the desired inequality holds for a given $k \geq 0$, we prove that it holds with k replaced by $k + 1$.

First note that the equation $(E)_3$ gives

$$\varphi_n^{(2k+1)} - \varphi_{(n)} = S_n\left[(T_n - T)(\varphi_n^{(2k)} - \varphi_{(n)}) + (\lambda_n^{(2k+1)} - \lambda)\varphi_n^{(2k)} \right. \\ \left. + (\lambda - \lambda_n)(\varphi_n^{(2k)} - \varphi_{(n)})\right] \\ = S_n\left[(T_n - T)(\varphi_n^{(2k)} - \varphi_{(n)}) \right. \\ + (\lambda_n^{(2k+1)} - \lambda)(\varphi_n^{(2k)} - \varphi_{(n)}) \\ + (\lambda_n^{(2k+1)} - \lambda)(\varphi_{(n)} - \varphi_n) \\ \left. +(\lambda - \lambda_n)(\varphi_n^{(2k)} - \varphi_{(n)})\right],$$

since $S_n\varphi_n = 0$. As $\max\{|\lambda - \lambda_n|, \|\varphi_{(n)} - \varphi_n\|\} \leq \alpha\|(T_n - T)T\| \leq \alpha\epsilon_n$, we have

$$\|(T_n - T)(\varphi_n^{(2k+1)} - \varphi_{(n)})\| \leq \|(T_n - T)S_n(T_n - T)\| \, \|\varphi_n^{(2k)} - \varphi_{(n)}\| \\ +\|T_n - T\| \, \|S_n\| \left(|\lambda_n^{(2k+1)} - \lambda| \, \|\varphi_n^{(2k)} - \varphi_{(n)}\| \right. \\ \left. +\alpha\epsilon_n|\lambda_n^{(2k+1)} - \lambda| + \alpha\epsilon_n\|\varphi_n^{(2k)} - \varphi_{(n)}\|\right).$$

Now $\|(T_n - T)S_n(T_n - T)\| \leq \|(T_n - T)S_nT_n\| + \|(T_n - T)S_nT\| \leq 2\widetilde{c}\epsilon_n$ and by the induction hypothesis, we have

$$\begin{aligned}\max\{|\lambda_n^{(2k+1)} - \lambda|, \|\varphi_n^{(2k)} - \varphi_{(n)}\|\} &\leq \alpha\|(T_n - T)T\|(\beta\epsilon_n)^k \\ &\leq \alpha\|(T_n - T)T\| \leq \alpha\epsilon_n,\end{aligned}$$

since $\beta\epsilon_n \leq 1$. Hence

$$\begin{aligned}&\|(T_n - T)(\varphi_n^{(2k+1)} - \varphi_{(n)})\| \\ &= [2\widetilde{c} + ts(\alpha + \alpha + \alpha)]\,\epsilon_n\alpha\|(T_n - T)T\|(\beta\epsilon_n)^k \\ &= \beta_1\epsilon_n\alpha\|(T_n - T)T\|(\beta\epsilon_n)^k,\end{aligned}$$

and by the equation $(E)_2$, we have

$$\begin{aligned}|\lambda_n^{(2k+2)} - \lambda| &= |\langle (T - T_n)(\varphi_n^{(2k+1)} - \varphi_{(n)})\,,\,\varphi_n^*\rangle| \\ &\leq p\beta_1\epsilon_n\alpha\|(T_n - T)T\|(\beta\epsilon_n)^k \\ &\leq \alpha\|(T_n - T)T\|(\beta\epsilon_n)^{k+1},\end{aligned}$$

as $p\beta_1 \leq \beta$. Again, by the equation $(E)_3$, we have

$$\begin{aligned}\|\varphi_n^{(2k+2)} - \varphi_{(n)}\| \leq \|S_n\|\,\big(&\|(T_n - T)(\varphi_n^{(2k+1)} - \varphi_{(n)})\| + \\ &|\lambda_n^{(2k+2)} - \lambda|\,\|\varphi_n^{(2k+1)}\| + |\lambda - \lambda_n|\,\|\varphi_n^{(2k+1)} - \varphi_{(n)}\|\big)\,.\end{aligned}$$

Now

$$\begin{aligned}\|\varphi_n^{(2k+1)}\| &\leq \|\varphi_n^{(2k+1)} - \varphi_{(n)}\| + \|\varphi_{(n)}\| \\ &\leq \alpha\|(T_n - T)T\|(\beta\epsilon_n)^k + \|\varphi_{(n)}\| \\ &\leq \alpha\epsilon_n(\beta\epsilon_n)^k + q \\ &\leq 1 + q,\end{aligned}$$

as $\alpha\epsilon_n \leq 1$ and $\beta\epsilon_n \leq 1$. Hence we obtain

$$\begin{aligned}\|\varphi_n^{(2k+2)} - \varphi_{(n)}\| &\leq s\,[\beta_1 + p\beta_1(1 + q) + \alpha]\,\epsilon_n\alpha\|(T_n - T)T\|(\beta\epsilon_n)^k \\ &= \beta_2\epsilon_n\alpha\|(T_n - T)T\|(\beta\epsilon_n)^k \\ &\leq \alpha\|(T_n - T)T\|(\beta\epsilon_n)^{k+1},\end{aligned}$$

since $\beta_2 \leq \beta$. Further, by the equation $(E)_2$, we have

$$\begin{aligned}|\lambda_n^{(2k+3)} - \lambda| &= |\langle (T - T_n)(\varphi_n^{(2k+2)} - \varphi_{(n)})\,,\,\varphi_n^*\rangle| \\ &\leq \|T - T_n\|\,\|\varphi_n^{(2k+2)} - \varphi_{(n)}\|\,\|\varphi_n^*\| \\ &\leq pt\beta_2\epsilon_n\alpha\|(T_n - T)T\|(\beta\epsilon_n)^k \\ &\leq \alpha\|(T_n - T)T\|(\beta\epsilon_n)^{k+1},\end{aligned}$$

as $pt\beta_2 \leq \beta$. Finally, since

$$\begin{aligned}||\varphi_n^{(2k+2)}|| &\leq ||\varphi_n^{(2k+2)} - \varphi_{(n)}|| + ||\varphi_{(n)}|| \\ &\leq \alpha||(T_n - T)T||(\beta\epsilon_n)^{k+1} + ||\varphi_{(n)}|| \\ &\leq 1 + q,\end{aligned}$$

the equation $(E)_3$ gives

$$\begin{aligned}||\varphi_n^{(2k+3)} - \varphi_{(n)}|| &\leq ||S_n|| \left(||T_n - T||\ ||\varphi_n^{(2k+2)} - \varphi_{(n)}||\right. \\ &\quad + |\lambda_n^{(2k+3)} - \lambda|\ ||\varphi_n^{(2k+2)}|| \\ &\quad \left. + |\lambda - \lambda_n|\ ||\varphi_n^{(2k+2)} - \varphi_{(n)}||\right) \\ &\leq s[t\beta_2 + pt\beta_2(1+q) + \alpha\beta_2]\epsilon_n\alpha||(T_n - T)T||(\beta\epsilon_n)^k \\ &= \beta_3\epsilon_n\alpha||(T_n - T)T||(\beta\epsilon_n)^k \\ &\leq \alpha||(T_n - T)T||(\beta\epsilon_n)^{k+1},\end{aligned}$$

since $||(T - T_n)T_n|| \leq \epsilon_n \leq 1$ and $\beta_3 \leq \beta$.

Thus the desired inequality holds for $k+1$ and the induction is complete. ∎

The preceding theorem shows that if $T_n \xrightarrow{n} T$, then the sequence of iterates given by the Elementary Iteration (E) converges 'geometrically', while if $\lambda \neq 0$ and $T_n \xrightarrow{\nu} T$, then it converges 'semigeometrically'.

In order to improve the rate of convergence of the iterates, especially when $T_n \xrightarrow{\nu} T$ and $\lambda \neq 0$, we modify the Elementary Iteration (E) by inserting a step of 'Power Iteration' between its successive steps. This modified iteration will be called the **Double Iteration**.

Let $\lambda \neq 0$ and n be a fixed positive integer for which the conclusions of Lemma 3.1 hold. Keeping in mind the equation $\varphi = \dfrac{T\varphi}{\lambda}$, the successive iterates are defined as follows:

$$(D) \qquad \begin{cases} \psi_n^{(0)} := \varphi_n, \text{ and for } k = 1, 2, \ldots \\[1ex] \lambda_n^{(k)} := \langle T\psi_n^{(k-1)}, \varphi_n^* \rangle, \text{ and if } \lambda_n^{(k)} \neq 0, \\[1ex] \varphi_n^{(k-1)} := \dfrac{T\psi_n^{(k-1)}}{\lambda_n^{(k)}}, \\[1ex] \mu_n^{(k)} := \langle T\varphi_n^{(k-1)}, \varphi_n^* \rangle, \\[1ex] \psi_n^{(k)} := \varphi_n^{(k-1)} + S_n(\mu_n^{(k)}\varphi_n^{(k-1)} - T\varphi_n^{(k-1)}). \end{cases}$$

Whenever these iterates are well defined, we obtain the following relations, as in the case of the Elementary Iteration:

$(D)_1$ $$\langle \psi_n^{(k)}\,,\, \varphi_n^*\rangle = 1 \quad \text{for } k = 0, 1, 2, \ldots$$

and also

$(D)_{1,1}$ $$\langle \varphi_n^{(k)}\,,\, \varphi_n^*\rangle = \frac{\langle T\psi_n^{(k)}\,,\, \varphi_n^*\rangle}{\lambda_n^{(k+1)}} = 1 \quad \text{for } k = 0, 1, 2, \ldots$$

Next,

$(D)_2$ $$\lambda_n^{(k)} - \lambda = \langle (T - T_n)(\psi_n^{(k-1)} - \varphi_{(n)})\,,\, \varphi_n^*\rangle \quad \text{for } k = 1, 2, \ldots$$

and also

$(D)_{2,2}$ $$\mu_n^{(k)} - \lambda = \langle (T - T_n)(\varphi_n^{(k-1)} - \varphi_{(n)})\,,\, \varphi_n^*\rangle \quad \text{for } k = 1, 2, \ldots$$

Further,

$(D)_3$ $$\begin{cases} \text{for } k = 1, 2, \ldots \\ \psi_n^{(k)} - \varphi_{(n)} = S_n\left[(T_n - T)(\varphi_n^{(k-1)} - \varphi_{(n)}) + (\mu_n^{(k)} - \lambda)\varphi_n^{(k-1)}\right. \\ \qquad\qquad \left. +(\lambda - \lambda_n)(\varphi_n^{(k-1)} - \varphi_{(n)})\right] \end{cases}$$

and also

$(D)_{3,3}$ $$\begin{cases} \text{for } k = 0, 1, 2, \ldots \\ \varphi_n^{(k)} - \varphi_{(n)} = T\left(\dfrac{\psi_n^{(k)}}{\lambda_n^{(k+1)}} - \dfrac{\varphi_{(n)}}{\lambda}\right) \\ \qquad = T\left(\dfrac{\psi_n^{(k)} - \varphi_{(n)}}{\lambda_n^{(k+1)}} + \dfrac{(\lambda - \lambda_n^{(k+1)})\varphi_{(n)}}{\lambda\,\lambda_n^{(k+1)}}\right), \\ \text{provided } \lambda_n^{(k+1)} \neq 0. \end{cases}$$

Theorem 3.4
Let $\lambda \neq 0$ and $T_n \xrightarrow{\nu} T$. For all large n, let φ_n be so chosen that the sequence $(\|\varphi_n\|)$ is bounded and also bounded away from zero.

Let $\mu_n^{(0)} = \lambda_n$ and $\psi_n^{(0)} = \varphi_n$. Then there is a positive integer n_1 such that for each fixed $n \geq n_1$, all the iterates $\mu_n^{(k)}$ and $\psi_n^{(k)}$ of the Double Iteration (D) are well defined, and for $k = 0, 1, 2, \ldots$ they satisfy

$$\max\{|\mu_n^{(k)} - \lambda|,\, \|\psi_n^{(k)} - \varphi_{(n)}\|\} \leq (\beta\|(T_n - T)T\|)^{k+1},$$

where β is a constant, independent of n and k.

Proof

Let γ, p, q, s, t and γ_1 be the constants stated at the beginning of the proof of Theorem 3.3. Since $\lambda \neq 0$ and $T_n \xrightarrow{\nu} T$, Lemma 3.1 shows that there is a positive integer n_0, and there is a constant c such that for all $n \geq n_0$,

$$\max\{|\lambda_n - \lambda|, \|\varphi_n - \varphi_{(n)}\|\} \leq c\gamma_1 \|(T_n - T)T\|.$$

Let

$$\beta := \max\{c\gamma_1, pc\gamma_1, \beta_1, \beta_2, p\beta_2\},$$

where $\beta_1 := \dfrac{2}{|\lambda|}\left(1 + \dfrac{tpq}{|\lambda|}\right)$ and $\beta_2 := \beta_1 s\left(1 + \dfrac{2\|T\|}{|\lambda|}(p + pq) + c\gamma_1\|T\|\right)$.

Since $\lambda \neq 0$ and $\|(T_n - T)T\| \to 0$, choose n_1 such that $\|(T_n - T)T\| \leq 1/\beta$ and $\|(T_n - T)T\| \leq \dfrac{|\lambda|}{2t\beta}$. Fix $n \geq n_1$. We show by induction on k that for $k = 0, 1, 2, \ldots$ the iterates $\psi_n^{(k)}$ and $\mu_n^{(k+1)}$ are well defined and

(i) $\|\psi_n^{(k)} - \varphi_{(n)}\| \leq (\beta\|(T_n - T)T\|)^{k+1}$,

(ii) $|\lambda_n^{(k+1)} - \lambda| \leq \|T_n - T\|(\beta\|(T_n - T)T\|)^{k+1}$,

(iii) $|\lambda_n^{(k+1)}| \geq \dfrac{|\lambda|}{2}$ and

(iv) $|\mu_n^{(k+1)} - \lambda| \leq (\beta\|(T_n - T)T\|)^{k+2}$.

Since $\psi_n^{(0)} = \varphi_n$ and $c\gamma_1 \leq \beta$, we have $\|\psi_n^{(0)} - \varphi_{(n)}\| \leq \beta\|(T_n - T)T\|$. By the equation $(D)_2$, it follows that

$$\begin{aligned}
|\lambda_n^{(1)} - \lambda| &= |\langle (T - T_n)(\psi_n^{(0)} - \varphi_{(n)}) \,,\, \varphi_n^*\rangle| \\
&\leq pc\gamma_1\|T_n - T\|\|(T_n - T)T\| \\
&\leq t\beta\|(T_n - T)T\| \leq \frac{|\lambda|}{2},
\end{aligned}$$

since $pc\gamma_1 \leq \beta$ and $\|(T_n - T)T\| \leq \dfrac{|\lambda|}{2t\beta}$. As a consequence, $|\lambda_n^{(1)}| \geq |\lambda| - |\lambda_n^{(1)} - \lambda| \geq |\lambda|/2$. In particular, $\lambda_n^{(1)} \neq 0$ and the iterates $\varphi_n^{(0)}$ and $\mu_n^{(1)}$ are well defined. By the equations $(D)_{2,2}$ and $(D)_{3,3}$, we have

$$|\mu_n^{(1)} - \lambda| = |\langle (T - T_n)(\varphi_n^{(0)} - \varphi_{(n)}) \,,\, \varphi_n^*\rangle|$$

$$= \left| \langle (T - T_n)T\Big(\frac{\psi_n^{(0)} - \varphi_{(n)}}{\lambda_n^{(1)}} + \frac{(\lambda - \lambda_n^{(1)})\varphi_{(n)}}{\lambda\,\lambda_n^{(1)}}\Big)\,,\,\varphi_n^*\rangle \right|$$
$$\leq \|(T_n - T)T\|\Big(\frac{2}{|\lambda|} + \frac{2tpq}{|\lambda|^2}\Big)pc\gamma_1\|(T_n - T)T\|$$
$$\leq (\beta\|(T_n - T)T\|)^2,$$

since $pc\gamma_1 \leq \beta$ and $\beta_1 = \dfrac{2}{|\lambda|}\left(1 + \dfrac{tpq}{|\lambda|}\right) \leq \beta$. Thus the desired inequalities hold for $k = 0$.

Now assuming that the iterates $\psi_n^{(k)}$ and $\mu_n^{(k+1)}$ are well defined (so that $\lambda_n^{(k)} \neq 0$ and $\lambda_n^{(k+1)} \neq 0$) and that the inequalities (i) to (iv) hold for a given $k \geq 0$, we prove that the iterates $\psi_n^{(k+1)}$ and $\mu_n^{(k+2)}$ are well defined and the inequalities (i) to (iv) hold with k replaced by $k+1$.

Since $\mu_n^{(k+1)}$ is well defined, so is $\psi_n^{(k+1)}$. By the equation $(D)_3$, we have

$$\psi_n^{(k+1)} - \varphi_{(n)} = S_n\left[(T_n - T)(\varphi_n^{(k)} - \varphi_{(n)}) \right.$$
$$\left. +(\mu_n^{(k+1)} - \lambda)\varphi_n^{(k)} + (\lambda - \lambda_n)(\varphi_n^{(k)} - \varphi_{(n)})\right],$$

and by the equations $(D)_{3,3}$ and $(D)_2$, we obtain

$$\|(T_n - T)(\varphi_n^{(k)} - \varphi_{(n)})\| = \Big\|(T_n - T)T\Big(\frac{\psi_n^{(k)} - \varphi_{(n)}}{\lambda_n^{(k+1)}}$$
$$+\frac{(\lambda - \lambda_n^{(k+1)})\varphi_{(n)}}{\lambda\,\lambda_n^{(k+1)}}\Big)\Big\|$$
$$\leq \|(T_n - T)T\|\left(\frac{2}{|\lambda|} + \frac{2tpq}{|\lambda|^2}\right)\|\psi_n^{(k)} - \varphi_{(n)}\|$$
$$\leq \beta_1\|(T_n - T)T\|(\beta\|(T_n - T)T\|)^{k+1}.$$

As a result, we also obtain by the equation $(D)_{2,2}$,

$$|\mu_n^{(k+1)} - \lambda| = |\langle (T - T_n)(\varphi_n^{(k)} - \varphi_{(n)})\,,\,\varphi_n^*\rangle|$$
$$\leq p\beta_1\|(T_n - T)T\|(\beta\|(T_n - T)T\|)^{k+1}.$$

Further,

$$\|\varphi_n^{(k)}\| = \frac{\|T\psi_n^{(k)}\|}{|\lambda_n^{(k+1)}|} \leq \frac{2\|T\|}{|\lambda|}\left(\|\psi_n^{(k)} - \varphi_n^*\| + \|\varphi_n^*\|\right) \leq \frac{2\|T\|}{|\lambda|}(1+q),$$

since $\|\psi_n^{(k)} - \varphi_{(n)}\| \le (\beta\|(T_n - T)T\|)^{k+1} \le 1$. Again, by the equations $(D)_{3,3}$ and $(D)_2$, we see that

$$\begin{aligned}
&|\lambda - \lambda_n|\, \|\varphi_n^{(k)} - \varphi_{(n)}\| \\
&\le c\gamma_1\|(T_n - T)T\|\, \|T\|\Big(\frac{2}{|\lambda|} + \frac{2+pq}{|\lambda|^2}\Big)\|\psi_n^{(k)} - \varphi_{(n)}\| \\
&\le c\gamma_1\|T\|\beta_1\|(T_n - T)T\|(\beta\|(T_n - T)T\|)^{k+1}.
\end{aligned}$$

From the preceding four estimates, we obtain

$$\begin{aligned}
\|\psi_n^{(k+1)} - \varphi_{(n)}\| &= \beta_2\|(T_n - T)T\|(\beta\|(T_n - T)T\|)^{k+1} \\
&\le (\beta\|(T_n - T)T\|)^{k+2},
\end{aligned}$$

since $\beta_2 \le \beta$. This proves the inequality (i) with k replaced by $k+1$.

Next, by the equation $(D)_2$,

$$\begin{aligned}
|\lambda_n^{(k+2)} - \lambda| &= |\langle (T - T_n)(\psi_n^{(k+1)} - \varphi_{(n)}) \,,\, \varphi_n^* \rangle| \\
&\le p\|T_n - T\|\beta_2\|(T_n - T)T\|(\beta\|(T_n - T)T\|)^{k+1} \\
&\le \|T_n - T\|(\beta\|(T_n - T)T\|)^{k+2},
\end{aligned}$$

since $p\beta_2 \le \beta$. This proves the inequality (ii) with k replaced by $k+1$.

Also, we have

$$|\lambda_n^{(k+2)} - \lambda| \le t\beta\|(T_n - T)T\| \le \frac{|\lambda|}{2},$$

because $\beta\|(T_n - T)T\| \le 1$ and $\|(T_n - T)T\| \le \dfrac{|\lambda|}{2t\beta}$. As a consequence, $|\lambda_n^{(k+2)}| \ge |\lambda| - |\lambda_n^{(k+2)} - \lambda| \ge |\lambda|/2$. This proves the inequality (iii) with k replaced by $k+1$. Since $\lambda_n^{(k+2)} \ne 0$, the iterate $\mu_n^{(k+2)}$ is well defined.

Finally, by the equations $(D)_{2,2}$ and $(D)_{3,3}$, we have

$$\begin{aligned}
|\mu_n^{(k+2)} - \lambda| &= |\langle (T - T_n)(\varphi_n^{(k+1)} - \varphi_{(n)}) \,,\, \varphi_n^* \rangle| \\
&\le \|(T - T_n)T\|\Big(\frac{2}{|\lambda|} + \frac{2tpq}{|\lambda|^2}\Big)(\beta\|(T_n - T)T\|)^{k+2}p \\
&= p\beta_2\|(T_n - T)T\|(\beta\|(T_n - T)T\|)^{k+2} \\
&\le (\beta\|(T_n - T)T\|)^{k+3},
\end{aligned}$$

since $p\beta_2 \le \beta$. This proves the inequality (iv) with k replaced by $k+1$. Thus the induction argument is complete.

Since the inequalities (i) and (iv) hold for all $k = 0, 1, 2, \ldots$ and since $\mu_n^{(0)} = \lambda_n$, we see that the conclusion of the theorem holds for all $k = 0, 1, 2, \ldots$ ∎

3.1.3 Refinement Schemes for a Cluster of Eigenvalues

In this subsection we extend the considerations of the previous subsection to treat the case of an eigenvalue of T of finite algebraic multiplicity and, more generally, the case of a cluster of a finite number of such eigenvalues. First, we recollect some results from Chapter 2 and develop them further. For a fixed positive integer m, we consider the Banach space $\underline{X} := X^{1\times m} = \{[x_1, \ldots, x_m] : x_j \in X \text{ for } j = 1, \ldots, m\}$ with the norm

$$\|[x_1, \ldots, x_m]\|_\infty = \max\{\|x_j\| : j = 1, \ldots, m\},$$

and use the notation and results given in Subsection 1.4.2.

Lemma 3.5
Let $\Lambda = \{\lambda_1, \ldots \lambda_r\}$ be a spectral set of finite type for T, $P := P(T, \Lambda)$, rank $P = m$, *and let $\underline{\varphi} = [\varphi_1, \ldots \varphi_m]$ form a basis for $\mathcal{R}(P)$. Assume that*

$$\text{(i) } T_n \xrightarrow{\text{n}} T, \quad \textit{or} \quad \text{(ii) } 0 \notin \Lambda \textit{ and } T_n \xrightarrow{\nu} T.$$

Then for all large n, rank $P_n = m$. *Consider*

$$\widehat{\lambda} := \frac{m_1\lambda_1 + \cdots + m_r\lambda_r}{m} \quad \textit{and} \quad \widehat{\lambda}_n := \frac{m_{n,1}\lambda_{n,1} + \cdots + m_{n,r(n)}\lambda_{n,r(n)}}{m},$$

the weighted arithmetic means of the elements in Λ and Λ_n respectively, where m_j is the algebraic multiplicity of the eigenvalue λ_j of T, $j = 1, \ldots, r$, and $m_{n,j}$ is the algebraic multiplicity of the eigenvalue $\lambda_{n,j}$ of T_n, $j = 1, \ldots, r(n)$. Then $\widehat{\lambda}_n \to \widehat{\lambda}$.

Let $\underline{\varphi_n} := [\varphi_{n,1}, \ldots \varphi_{n,m}]$ form an ordered basis for $\mathcal{R}(P_n)$, and $\underline{\varphi_n^*} = [\varphi_{n,1}^*, \ldots \varphi_{n,m}^*]$ form the corresponding adjoint basis for $\mathcal{R}(P_n^*)$. The $m\times m$ matrix

$$\mathsf{C}_n := (\underline{\varphi}, \underline{\varphi_n^*})$$

is nonsingular for all large n and

$$\|\underline{\varphi_n}\,\mathsf{C}_n - \underline{\varphi}\|_\infty \to 0.$$

Also, if $\underline{\varphi_{(n)}} := \underline{\varphi}\,\mathsf{C}_n^{-1}$, then

$$\max\left\{|\widehat{\lambda}_n - \widehat{\lambda}|, \frac{\|\underline{\varphi_n} - \underline{\varphi_{(n)}}\|_\infty}{\|\underline{\varphi_n}\|_\infty}\right\} \le c\|T_n - T\|,$$

and if $0 \notin \Lambda$, *then*

$$\max\left\{|\widehat{\lambda}_n - \widehat{\lambda}|, \frac{\|\underline{\varphi_n} - \underline{\varphi_{(n)}}\|_\infty}{\|\underline{\varphi_n}\|_\infty}\right\} \leq c\|(T_n - T)T\|,$$

where c *is a constant, independent of* n.

Proof
Let $\mathrm{C} \in \mathcal{C}(T, \Lambda)$. By Theorem 2.18(a), there is a positive integer n_0 such that for each $n \geq n_0$,

$$\Lambda_n := \mathrm{int}(\mathrm{C}) \cap \mathrm{sp}(T_n) = \{\lambda_{n,1}, \ldots \lambda_{n,r(n)}\}$$

and $\mathrm{rank}\, P_n(T_n, \Lambda_n) = \mathrm{rank}\, P = m$. Further, if $\widehat{\lambda}$ and $\widehat{\lambda}_n$ are as stated, then for all large n,

$$|\widehat{\lambda}_n - \widehat{\lambda}| \leq \frac{\ell(\mathrm{C})}{\pi}\alpha_2(\mathrm{C})\|(T_n - T)_{|\mathcal{R}(P)}\|,$$

where $\alpha_2(\mathrm{C})$ is a constant, independent of n, as we have seen in Theorem 2.18(b).

Clearly, $\|(T_n - T)_{|\mathcal{R}(P)}\| \leq \|T_n - T\|$. Also, $\|(T_n - T)_{|\mathcal{R}(P)}\| \leq \|(T_n - T)P\|$. Assume now that $0 \notin \Lambda$. Then by Corollary 1.22, we may assume that $0 \in \mathrm{ext}(\mathrm{C})$. Hence Proposition 2.9(b) shows that for all large n,

$$\|(T_n - T)P\| \leq \frac{\ell(\mathrm{C})}{2\pi}\frac{\alpha_1(\mathrm{C})}{\delta(\mathrm{C})}\|(T_n - T)T\|,$$

where $\alpha_1(\mathrm{C})$ is a constant, independent of n. Hence the desired estimates for $|\widehat{\lambda}_n - \widehat{\lambda}|$ follow.

Finally, Theorem 2.18(c) shows that the $m \times m$ matrix C_n is nonsingular for all large n, $\|\underline{\varphi_n}\,\mathsf{C}_n - \underline{\varphi}\|_\infty \to 0$ and

$$\frac{\|\underline{\varphi_n} - \underline{\varphi_{(n)}}\|_\infty}{\|\underline{\varphi_{(n)}}\|_\infty} \leq \frac{\ell(\mathrm{C})}{2\pi}\alpha_1(\mathrm{C})\alpha_2(\mathrm{C})\|(T_n - T)_{|\mathcal{R}(P)}\|,$$

where $\underline{\varphi_{(n)}} := \underline{\varphi}\,\mathsf{C}_n^{-1}$ satisfies $\|\underline{\varphi_{(n)}}\|_\infty \leq 2\|\underline{\varphi_n}\|_\infty$. Thus the desired estimates for $\|\underline{\varphi_n} - \underline{\varphi_{(n)}}\|_\infty / \|\underline{\varphi_n}\|_\infty$ follow. ∎

Unless otherwise stated, we shall assume throughout this subsection that Λ is a spectral set of finite type for T.

Let $P := P(T,\Lambda)$, $\operatorname{rank} P = m$, $\underline{\varphi} := [\varphi_1, \dots \varphi_m]$ form an ordered basis for $\mathcal{R}(P)$, and $\underline{\varphi^*} := [\varphi_1^*, \dots \varphi_m^*]$ form the corresponding adjoint basis for $\mathcal{R}(P^*)$. Then, as we have seen in Theorem 1.52,

$$\underline{P}\,\underline{x} = \underline{\varphi}\,(\underline{x}\,, \underline{\varphi^*}\,) \quad \text{and} \quad \underline{T}\,\underline{\varphi} = \underline{\varphi}\,\Theta, \text{ where } \Theta := (\underline{T}\,\underline{\varphi}\,, \underline{\varphi^*}\,).$$

Also, the notation $\widehat{\lambda}$, Λ_n, $\widehat{\lambda}_n$, P_n, $\underline{\varphi_n}$, $\underline{\varphi_n^*}$, C_n and $\underline{\varphi_{(n)}}$ will have the meanings given in the statement of Lemma 3.5.

In addition, for each large n, define

$\Theta_n := (\underline{T_n}\,\underline{\varphi_n}\,, \underline{\varphi_n^*}\,)$,

$\Theta_{(n)} := (\underline{T}\,\underline{\varphi_{(n)}}\,, \underline{\varphi_n^*}\,)$,

$\mathcal{S}_n := \mathcal{S}_{\underline{\varphi_n}}$, the block reduced resolvent corresponding to T_n, Λ_n and $\underline{\varphi_n}$,

$\delta_{n,j} := \operatorname{dist}(\varphi_{n,j}, \operatorname{span}\{\varphi_{n,i} : i \neq j,\, i = 1, \dots, m\})$ for $j = 1, \dots, m$,

$\delta_n := \min\{\delta_{n,1}, \dots \delta_{n,m}\}$.

Note that

$$\underline{T_n}\,\underline{\varphi_n} = \underline{\varphi_n}\,\Theta_n, \quad \underline{T_n^*}\,\underline{\varphi_n^*} = \underline{\varphi_n^*}\,\Theta_n^*, \quad \underline{T}\,\underline{\varphi_{(n)}} = \underline{\varphi_{(n)}}\,\Theta_{(n)}$$

and

$$(\underline{\varphi_n}\,, \underline{\varphi_n^*}\,) = \mathsf{I}_m = (\underline{\varphi_{(n)}}\,, \underline{\varphi_n^*}\,).$$

Lemma 3.6
Assume that

$$\text{(i) } T_n \xrightarrow{\text{n}} T, \quad \textit{or} \quad \text{(ii) } 0 \notin \Lambda \textit{ and } T_n \xrightarrow{\nu} T.$$

(a) (i) *The sequence* $(\|\,\underline{\varphi_{(n)}}\,\|_\infty)$ *is bounded if and only if the sequence* $(\|\,\underline{\varphi_n}\,\|_\infty)$ *is bounded.*

(ii) *The sequence* $(\|\,\underline{\varphi_n^*}\,\|_1)$ *is bounded if and only if the sequence* (δ_n) *is bounded away from zero.*

(b) *Assume that the sequence* $(\|\,\underline{\varphi_n}\,\|_\infty)$ *is bounded and that the sequence* (δ_n) *is bounded away from zero. Then*

(i) *for all large* n,

$$\|\Theta_n - \Theta_{(n)}\|_1 \leq c\|T_n - T\|$$

and if $0 \notin \Lambda$,

$$\|\Theta_n - \Theta_{(n)}\|_1 \leq c\|(T_n - T)T\|,$$

where c *is a constant, independent of* n,

(ii) *the sequence* $(\|\mathcal{S}_n\|)$ *is bounded,*

$$\|(\underline{T_n} - \underline{T})\mathcal{S}_n \underline{T_n}\| \leq \widetilde{c}\|(T_n - T)T_n\|$$

and

$$\|(\underline{T_n} - \underline{T})\mathcal{S}_n \underline{T}\| \leq \widetilde{c}\max\{\|(T_n - T)T_n\|, \|(T_n - T)T\|\},$$

where $\widetilde{c}$ *is a constant, independent of* n.

Proof

(a) (i) Since for all large n,

$$\begin{aligned}\underline{P_n}\,\underline{\varphi_{(n)}} &= \underline{\varphi_n}\,(\underline{\varphi_{(n)}}\,,\,\underline{\varphi_n^*}\,) \\ &= \underline{\varphi_n}\,(\underline{\varphi}\,\mathsf{C}_n^{-1},\,\underline{\varphi_n^*}\,) \\ &= \underline{\varphi_n}\,(\underline{\varphi},\,\underline{\varphi_n^*}\,)\mathsf{C}_n^{-1} = \underline{\varphi_n}\,,\end{aligned}$$

we see that

$$\|\,\underline{\varphi_n}\,\|_\infty \leq \|\,\underline{P_n}\,\|\;\|\,\underline{\varphi_{(n)}}\,\|_\infty = \|P_n\|\;\|\,\underline{\varphi_{(n)}}\,\|_\infty,$$

where the sequence $(\|P_n\|)$ is bounded by Proposition 2.9(b). On the other hand, as we have seen in the proof of Theorem 2.18(b),

$$\left\|\left(P_{n|M,M_n}\right)^{-1}\right\| \leq 2$$

for all large n, where $M := \mathcal{R}(P)$ and $M_n := \mathcal{R}(P_n)$. Hence

$$\|\,\underline{\varphi_{(n)}}\,\|_\infty = \left\|\left(\underline{P_{n|M,M_n}}\right)^{-1}\underline{\varphi_n}\right\|_\infty \leq 2\|\,\underline{\varphi_n}\,\|_\infty$$

for all large n. It follows that the sequence $(\|\,\underline{\varphi_{(n)}}\,\|_\infty)$ is bounded if and only if the sequence $(\|\,\underline{\varphi_n}\,\|_\infty)$ is bounded.

(ii) Let the sequence (δ_n) be bounded away from 0. For all large n, we have

$$\underline{P_n^*}\,\underline{\varphi^*} = \underline{\varphi_n^*}\,(\underline{\varphi_n}\,,\underline{\varphi^*})^*.$$

Let $\mathsf{D}_n := (\underline{\varphi_n}\,,\underline{\varphi^*})$. Then

$$\mathsf{D}_n\mathsf{C}_n = (\underline{\varphi_n}\,\mathsf{C}_n,\underline{\varphi^*}) \to (\underline{\varphi}\,,\underline{\varphi^*}) = \mathsf{I}_m,$$

since $\underline{\varphi_n}\,\mathsf{C}_n \to \underline{\varphi}$ by Lemma 3.5. Hence for all large n, the matrices D_n and C_n are nonsingular and $\|\mathsf{C}_n^{-1}\mathsf{D}_n^{-1} - \mathsf{I}\|_1 \to 0$. Now $\underline{\varphi_n^*} = (\underline{P_n^*}\,\underline{\varphi^*})(\mathsf{D}_n^*)^{-1}$ and, by Proposition 1.49,

$$\|\underline{\varphi_n^*}\|_1 \le \|\underline{P_n^*}\,\underline{\varphi^*}\|_1\|(\mathsf{D}_n^{-1})^*\|_\infty \le \|\underline{P_n^*}\|\,\|\underline{\varphi^*}\|_1\|\mathsf{D}_n^{-1}\|_1,$$

where $\|\underline{P_n^*}\| = \|P_n^*\| \le \|P_n\|$ and the sequence $(\|P_n\|)$ is bounded by Proposition 2.9(b). Also,

$$\|\mathsf{D}_n^{-1}\|_1 = \|\mathsf{C}_n\mathsf{C}_n^{-1}\mathsf{D}_n^{-1}\|_1 \le \|\mathsf{C}_n\|_1\|\mathsf{C}_n^{-1}\mathsf{D}_n^{-1}\|_1,$$

where the sequence $(\|\mathsf{C}_n^{-1}\mathsf{D}_n^{-1}\|_1)$ tends to 1. Thus, to conclude that the sequence $(\|\underline{\varphi_n^*}\|_1)$ is bounded, it is enough to show that the sequence $(\|\mathsf{C}_n\|_1)$ is bounded.

Noting that $\mathsf{C}_n = [\langle\varphi_j\,,\varphi_{n,i}^*\rangle]$ and

$$\|\mathsf{C}_n\|_1 = \max_{j=1,\ldots,m}\Big\{\sum_{i=1}^{m}|\langle\varphi_j\,,\varphi_{n,i}^*\rangle|\Big\},$$

we need only prove that for each fixed pair (i,j), $1 \le i,j \le m$, the sequence $(|\langle\varphi_i\,,\varphi_{n,j}^*\rangle|)$ is bounded. So fix i,j, $1 \le i,j \le m$. For all large n, we have

$$P_n\varphi_i = \langle\varphi_i\,,\varphi_{n,1}^*\rangle\varphi_{n,1} + \cdots + \langle\varphi_i\,,\varphi_{n,m}^*\rangle\varphi_{n,m} = \langle\varphi_i\,,\varphi_{n,j}^*\rangle\varphi_{n,j} + x_{n,j}$$

for some $x_{n,j} \in X_{n,j} := \operatorname{span}\{\varphi_{n,\ell} : \ell \ne j,\ \ell = 1,\ldots,m\}$, so that

$$\begin{aligned}\|P_n\varphi_i\| &\ge |\langle\varphi_i\,,\varphi_{n,j}^*\rangle|\operatorname{dist}(\varphi_{n,j},X_{n,j})\\ &= |\langle\varphi_i\,,\varphi_{n,j}^*\rangle|\delta_{n,j}\\ &\ge |\langle\varphi_i\,,\varphi_{n,j}^*\rangle|\delta_n.\end{aligned}$$

Hence $|\langle\varphi_i\,,\varphi_{n,j}^*\rangle| \le \|P_n\varphi_i\|/\delta_n \le \|P_n\|\,\|\varphi_i\|/\delta_n$, where the sequence $(\|P_n\|)$ is bounded and the sequence (δ_n) is bounded away from 0. Thus the sequence $(|\langle\varphi_i\,,\varphi_{n,j}^*\rangle|)$ is bounded.

Conversely, assume that the sequence $(\|\underline{\varphi_n^*}\|_1)$ is bounded. To prove that the sequence (δ_n) is bounded away from zero, it is enough to show that for each $j = 1, \ldots, m$, the sequence $(\delta_{n,j})$ is bounded away from zero. So fix j, $1 \le j \le m$ and let $X_{n,j} = \text{span}\{\varphi_{n,i} : i \ne j, i = 1, \ldots, m\}$ as before. Since $\langle \varphi_{n,j}, \varphi_{n,j}^* \rangle = 1$ and $\langle x_{n,j}, \varphi_{n,j}^* \rangle = 0$ for all $x_{n,j} \in X_{n,j}$, we have

$$1 = \langle \varphi_{n,j} - x_{n,j}, \varphi_{n,j}^* \rangle \le \|\varphi_{n,j} - x_{n,j}\| \, \|\varphi_{n,j}^*\|.$$

Taking infimum over all $x_{n,j} \in X_{n,j}$, we obtain

$$1 \le \delta_{n,j}\|\varphi_{n,j}^*\| \le \delta_{n,j}\|\underline{\varphi_n^*}\|_1.$$

As the sequence $(\|\underline{\varphi_n^*}\|_1)$ is bounded, we see that sequence $(\delta_{n,j})$ is bounded away from zero.

(b) By (a) above, the sequences $(\|\underline{\varphi_{(n)}}\|_\infty)$ and $(\|\underline{\varphi_n^*}\|_1)$ are bounded.

(i) We have by Proposition 1.46,

$$\begin{aligned}\|\Theta_n - \Theta_{(n)}\|_1 &= \|(\underline{T_n}\,\underline{\varphi_n} - \underline{T}\,\underline{\varphi_{(n)}}, \underline{\varphi_n^*})\|_1 \\ &\le \|\underline{T_n}\,\underline{\varphi_n} - \underline{T}\,\underline{\varphi_{(n)}}\|_\infty \|\underline{\varphi_n^*}\|_1.\end{aligned}$$

We note that

$$\begin{aligned}\|\underline{T_n}\,\underline{\varphi_n} - \underline{T}\,\underline{\varphi_{(n)}}\|_\infty &\le \|\underline{T_n}(\underline{\varphi_n} - \underline{\varphi_{(n)}})\|_\infty + \|(\underline{T_n} - \underline{T})\underline{P}\,\underline{\varphi_{(n)}}\|_\infty \\ &\le \|T_n\|\,\|\underline{\varphi_n} - \underline{\varphi_{(n)}}\|_\infty + \|(T_n - T)P\|\,\|\underline{\varphi_{(n)}}\|_\infty.\end{aligned}$$

Now $\|\underline{\varphi_n} - \underline{\varphi_{(n)}}\|_\infty \le c\|T_n - T\|\,\|\underline{\varphi_n}\|_\infty$ by Lemma 3.5. Also, $\|(T_n - T)P\| \le \|T_n - T\|\,\|P\|$, while if $0 \notin \Lambda$, then $\|\underline{\varphi_n} - \underline{\varphi_{(n)}}\|_\infty \le c\|(T_n - T)T\|\,\|\underline{\varphi_n}\|_\infty$ and $\|(T_n - T)P\| \le c\|(T_n - T)T\|$ as in the proof of Lemma 3.5, where c is a constant, independent of n. Hence the desired estimates for $\|\Theta_n - \Theta_{(n)}\|_1$ follow.

(ii) For all n greater than or equal to some n_0 and all $\underline{x} \in \underline{X}$, we have

$$\mathcal{S}_n\,\underline{x} = -\frac{1}{2\pi \mathrm{i}} \int_{\mathrm{C}} \underline{R_n(z)}\,\underline{x}\,(\underline{R_n(z)}\,\underline{\varphi_n}, \underline{\varphi_n^*})dz.$$

Hence by Propositions 1.49 and 1.46,

$$\|\mathcal{S}_n\| \le \frac{\ell(\mathrm{C})}{2\pi}\alpha_2(\mathrm{C})^2\|\underline{\varphi_n}\|_\infty\|\underline{\varphi_n^*}\|_1,$$

where $\alpha_2(\mathrm{C}) = \sup\{\|R_n(z)\| : z \in \mathrm{C}, n \ge n_0\} < \infty$ by Theorem 2.6. Thus the sequence $(\|\mathcal{S}_n\|)$ is bounded.

As the operators $\mathcal{S}_n$ and $\underline{T_n}$ commute, we have

$$\|(\underline{T_n} - \underline{T})\mathcal{S}_n \underline{T_n}\| \le \|(\underline{T_n} - \underline{T})\underline{T_n}\|\,\|\mathcal{S}_n\| = \|(T_n - T)T_n\|\,\|\mathcal{S}_n\|.$$

As a result, the sequence $(\|(\underline{T_n} - \underline{T})\mathcal{S}_n \underline{T_n}\|)$ is bounded.

Finally, we may assume without loss of generality that the Cauchy contour C does not pass through 0, so that

$$\delta := \min\{|z| : z \in \mathrm{C}\} > 0.$$

Now for all large n and all $\underline{x} \in \underline{X}$, we have

$$(\underline{T_n} - \underline{T})\mathcal{S}_n \underline{T}\,\underline{x} = -\frac{1}{2\pi \mathrm{i}}\int_{\mathrm{C}} (\underline{T_n} - \underline{T})\,\underline{R_n(z)}\,\underline{T}\,\underline{x}\,(\underline{R_n(z)}\,\underline{\varphi_n}, \underline{\varphi_n^*})dz.$$

By Proposition 1.10(e), we have for all $z \in \mathrm{C}$,

$$(T_n - T)R_n(z)T = \frac{1}{z}\big((T_n - T)T_nR_n(z)T - (T - T_n)T\big).$$

Hence

$$\begin{aligned}\|(\underline{T_n} - \underline{T})\mathcal{S}_n \underline{T}\| \le\ & \frac{\ell(\mathrm{C})}{2\pi\delta}(\|(T_n - T)T_n\|\alpha_2(\mathrm{C})\|T\| \\ & + \|(T - T_n)T\|)\alpha_2(\mathrm{C})\|\,\underline{\varphi_n}\,\|_\infty \|\,\underline{\varphi_n^*}\,\|_1.\end{aligned}$$

Thus the sequence $(\|(\underline{T_n} - \underline{T})\mathcal{S}_n \underline{T}\|)$ is also bounded. ∎

We remark that for each large n, $\underline{\varphi_n}$ forms an ordered basis for $M_n := \mathcal{R}(P_n)$ and $(\underline{\varphi_n}, \underline{\varphi_n^*}) = \mathsf{I}_m$, so that the matrix $\Theta_n = (\underline{T_n}\,\underline{\varphi_n}, \underline{\varphi_n^*})$ represents the operator $T_{n|M_n,M_n}$ with respect to $\underline{\varphi_n}$. Similarly, $\underline{\varphi}$ forms an ordered basis for $M := \mathcal{R}(P)$ and since the matrix $\mathsf{C}_n = (\underline{\varphi}, \underline{\varphi_n^*})$ is nonsingular for each large n, $\underline{\varphi_{(n)}} = \underline{\varphi}\,\mathsf{C}_n^{-1}$ forms an ordered basis for M. As $(\underline{\varphi_{(n)}}, \underline{\varphi_n^*}) = (\underline{\varphi}, \underline{\varphi_n^*})\mathsf{C}_n^{-1} = \mathsf{I}_m$, we see that the matrix $\Theta_{(n)} = (\underline{T}\,\underline{\varphi_{(n)}}, \underline{\varphi_n^*})$ represents the operator $T_{|M,M}$ with respect to $\underline{\varphi_{(n)}}$. Hence

$$\widehat{\lambda}_n = \frac{1}{m}\operatorname{tr}\Theta_n, \quad \widehat{\lambda} = \frac{1}{m}\operatorname{tr}\Theta_{(n)},$$
$$|\widehat{\lambda}_n - \widehat{\lambda}| = \frac{1}{m}|\operatorname{tr}(\Theta_n - \Theta_{(n)})| \le \frac{1}{m}\|\Theta_n - \Theta_{(n)}\|_1.$$

In the iteration schemes for the cluster Λ of eigenvalues of T, we shall use the bases $\underline{\varphi_n}$, $\underline{\varphi_{(n)}}$ and the matrices Θ_n, $\Theta_{(n)}$ for a fixed n, as in the case of a simple eigenvalue treated in Subsection 3.1.2. If $\Theta_n^{(k)}$ denotes the kth iterate of Θ_n, then $\widehat{\lambda}_n^{(k)} := \frac{1}{m} \operatorname{tr} \Theta_n^{(k)}$ will be the approximation of $\widehat{\lambda}$ that we seek.

Fix a positive integer n for which the conclusions of Lemma 3.5 hold. In analogy with the Elementary Iteration (E) for a simple eigenvalue, we consider the **Elementary Iteration** $(\underline{E})$ for a cluster of eigenvalues as follows:

$$(\underline{E}) \qquad \begin{cases} \underline{\varphi_n}^{(0)} := \underline{\varphi_n}\,, \text{ and for } k = 1, 2, \ldots \\[2ex] \Theta_n^{(k)} := \left(\underline{T}\, \underline{\varphi_n}^{(k-1)}, \underline{\varphi_n^*} \right), \\[2ex] \underline{\varphi_n}^{(k)} := \underline{\varphi_n}^{(k-1)} + \mathcal{S}_n \left(\underline{\varphi_n}^{(k-1)} \Theta_n^{(k)} - \underline{T}\, \underline{\varphi_n}^{(k-1)} \right). \end{cases}$$

We recall that by Lemma 1.51,

$$\underline{P_n}\, \underline{x} = \underline{\varphi_n} \left(\underline{x}, \underline{\varphi_n^*} \right), \quad \underline{x} \in \underline{X},$$

$$\underline{P_n^*}\, \underline{f} = \underline{\varphi_n^*} \left(\underline{\varphi_n}, \underline{f} \right)^*, \quad \underline{f} \in \underline{X^*}.$$

Let $\mathcal{Z}_n : \underline{X} \to \underline{X}$ be defined by

$$\mathcal{Z}_n\, \underline{x} = \underline{x}\, \Theta_n, \quad \underline{x} \in \underline{X}.$$

Then by Proposition 1.56, the operators $\underline{T_n}$, $\underline{P_n}$, $\mathcal{S}_n$ and $\mathcal{Z}_n$ commute, and for all $\underline{x} \in \underline{X}$,

$$\mathcal{S}_n(\underline{T_n}\, \underline{x} - \underline{x}\, \Theta_n) = \underline{x} - \underline{P_n}\, \underline{x}, \quad \mathcal{S}_n\, \underline{P_n}\, \underline{x} = \underline{0}.$$

These considerations allow us to establish the following three relations in exactly the same manner as we established the relations $(E)_1, (E)_2, (E)_3$ in the previous subsection. We merely replace T, T_n, P_n, S_n, λ, λ_n, $\lambda_n^{(k)}$, $\varphi_{(n)}$, φ_n and φ_n^* by $\underline{T}$, $\underline{T_n}$, $\underline{P_n}$, $\mathcal{S}_n$, $\Theta_{(n)}$, Θ_n, $\Theta_n^{(k)}$, $\underline{\varphi_{(n)}}$, $\underline{\varphi_n}$, and $\underline{\varphi_n^*}$ respectively, and the scalar product $\langle \cdot\,, \cdot \rangle$ by the Gram matrix $(\cdot, \cdot)$. We note that for $z \in \mathbb{C}$ and $x \in X$, the element zx of X is to be replaced by the element $\underline{x}\,\mathsf{Z}$ of $\underline{X}$, where $\underline{x} \in \underline{X}$ and $\mathsf{Z} \in \mathbb{C}^{m \times m}$. Thus we have

$(\underline{E})_1$ $\left(\underline{\varphi_n}^{(k)}, \underline{\varphi_n^*} \right) = \mathsf{I}_m$ and hence $\underline{P_n}\, \underline{\varphi_n}^{(k)} = \underline{\varphi_n}$ for $k = 0, 1, 2, \ldots$

$(\underline{E})_2$ $\Theta_n^{(k)} - \Theta_{(n)} = \left((\underline{T} - \underline{T_n})(\underline{\varphi_n}^{(k-1)} - \underline{\varphi_{(n)}}), \underline{\varphi_n^*} \right)$ for $k = 1, 2, \ldots$

and

$$(\underline{E})_3 \quad \begin{cases} \text{for } k = 1, 2, \ldots \\ \underline{\varphi_n}^{(k)} - \underline{\varphi_{(n)}} = \mathcal{S}_n \Big[(\underline{T_n} - \underline{T})(\underline{\varphi_n}^{(k-1)} - \underline{\varphi_{(n)}}) \\ \qquad + \underline{\varphi_n}^{(k-1)} (\Theta_n^{(k)} - \Theta_n) - \underline{\varphi_{(n)}} (\Theta_{(n)} - \Theta_n) \Big] \\ \qquad = \mathcal{S}_n \Big[(\underline{T_n} - \underline{T})(\underline{\varphi_n}^{(k-1)} - \underline{\varphi_{(n)}}) \\ \qquad + \underline{\varphi_n}^{(k-1)} (\Theta_n^{(k)} - \Theta_{(n)}) \\ \qquad + (\underline{\varphi_n}^{(k-1)} - \underline{\varphi_{(n)}})(\Theta_{(n)} - \Theta_n) \Big]. \end{cases}$$

Using the above-mentioned relations, we obtain the following analogue of Theorem 3.3.

Theorem 3.7
Assume that

$$\text{(i) } T_n \xrightarrow{\text{n}} T, \quad \textit{or} \quad \text{(ii) } 0 \notin \Lambda \textit{ and } T_n \xrightarrow{\nu} T.$$

For each large n, let $\underline{\varphi_n}$ be so chosen that the sequence $(\| \underline{\varphi_n} \|_\infty)$ is bounded and the sequence (δ_n) is bounded away from zero.
Let $\Theta_n^{(0)} := \Theta_n$, $\underline{\varphi_n}^{(0)} := \underline{\varphi_n}$ and for $k = 1, 2, \ldots$ let $\Theta_n^{(k)}$ and $\underline{\varphi_n}^{(k)}$ be the iterates of the Elementary Iteration $(\underline{E})$. Further, let

$$\widehat{\lambda}_n^{(k)} = \frac{1}{m} \operatorname{tr} \Theta_n^{(k)} \quad \textit{for } k = 1, 2, \ldots$$

(a) *If $T_n \xrightarrow{\text{n}} T$, then there is a positive integer n_1 such that for all $n \geq n_1$ and for $k = 0, 1, 2, \ldots$*

$$\max\{|\widehat{\lambda}_n^{(k)} - \widehat{\lambda}|, \| \underline{\varphi_n}^{(k)} - \underline{\varphi_{(n)}} \|\} \leq (\beta \|T_n - T\|)^{k+1},$$

where β is a constant, independent of n and k, and if in fact $0 \notin \Lambda$, then

$$\max\{|\widehat{\lambda}_n^{(k)} - \widehat{\lambda}|, \| \underline{\varphi_n}^{(k)} - \underline{\varphi_{(n)}} \|\} \leq \alpha \|(T_n - T)T\| (\beta \|T_n - T\|)^k,$$

where α and β are constants, independent of n and k.

(b) *If $0 \notin \Lambda$ and $T_n \xrightarrow{\nu} T$, then there is a positive integer n_1 such that for all $n \geq n_1$ and for all $k = 0, 1, 2, \ldots$*

$$\max\{|\widehat{\lambda}_n^{(2k)} - \widehat{\lambda}|, |\widehat{\lambda}_n^{(2k+1)} - \widehat{\lambda}|, \| \underline{\varphi_n}^{(2k)} - \underline{\varphi_{(n)}} \|, \| \underline{\varphi_n}^{(2k+1)} - \underline{\varphi_{(n)}} \|\}$$
$$\leq \alpha \|(T_n - T)T\| (\beta \max\{\|(T_n - T)T\|, \|(T_n - T)T_n\|\})^k,$$

where α and β are constants, independent of n and k.

Proof
As $|\widehat{\lambda}_n^{(k)} - \widehat{\lambda}| \leq \|\Theta_n^{(k)} - \Theta_{(n)}\|_1$ for all large n and all $k = 0, 1, 2, \ldots$ it is enough to obtain the desired bounds for $\|\Theta_n^{(k)} - \Theta_{(n)}\|_1$ in place of $|\widehat{\lambda}_n^{(k)} - \widehat{\lambda}|$. By Lemma 3.6(b)(ii), the sequence $(\|\mathcal{S}_n\|)$ is bounded. Also, since the sequence $(\| \underline{\varphi_n} \|_\infty)$ is bounded and the sequence (δ_n) is bounded away from zero, Lemma 3.6(a) shows that the sequences $(\| \underline{\varphi_{(n)}} \|_\infty)$ and $(\| \underline{\varphi_n^*} \|_1)$ are bounded. Further, the sequence $(\| \underline{T_n} \|)$ is bounded. Hence there are constants γ, p, q, s, t such that for all large n,

$$\| \underline{\varphi_n} \|_\infty \leq \gamma, \ \| \underline{\varphi_n^*} \|_1 \leq p, \ \| \underline{\varphi_{(n)}} \|_\infty \leq q, \ \|\mathcal{S}_n\| \leq s \text{ and } \| \underline{T_n} - \underline{T} \| \leq t.$$

Let $\gamma_1 = \max\{1, \gamma\}$.

(a) Let $T_n \xrightarrow{\text{n}} T$. By Lemma 3.5 and by Lemma 3.6(b)(i), there is a positive integer n_0 and there is a constant c such that

$$\max\{\|\Theta_n - \Theta_{(n)}\|_1, \| \underline{\varphi_n} - \underline{\varphi_{(n)}} \|_\infty\} \leq c\gamma_1 \|T_n - T\|.$$

Let $\beta := \max\{c\gamma_1, p, s(1 + p + pq + c\gamma_1)\}$ and choose $n_1 \geq n_0$ such that $\|T_n - T\| \leq 1/\beta$ for all $n \geq n_1$.

Fix $n \geq n_1$. It can be shown by induction on k that

$$\max\{\|\Theta_n^{(k)} - \Theta_{(n)}\|_1, \| \underline{\varphi_n}^{(k)} - \underline{\varphi_{(n)}} \|_\infty\} \leq (\beta \|T_n - T\|)^{k+1}, \ k = 0, 1, 2, \ldots$$

exactly as in the proof of Theorem 3.3(a) by employing the relations $(\underline{E})_2$ and $(\underline{E})_3$. For achieving this, we merely replace the norm $\| \cdot \|$ on X and the absolute value $| \cdot |$ on $\mathbb{C}$ by the norm $\| \cdot \|_\infty$ on $\underline{X}$ and the norm $\| \cdot \|_1$ on $\mathbb{C}^{m \times m}$ respectively.

Next, if $0 \notin \Lambda$, then by Lemma 3.5 and by Lemma 3.6(b)(i), there is a positive integer n_0 and there is a constant c such that

$$\max\{\|\Theta_n - \Theta_{(n)}\|_1, \| \underline{\varphi_n} - \underline{\varphi_{(n)}} \|_\infty\} \leq c\gamma_1 \|(T_n - T)T\|.$$

Hence if we let $\alpha := c\gamma_1$, $\beta := \max\{p, s(1+p+pq+\alpha\|T\|)\}$ and choose $n_1 \geq n_0$ such that $\|T_n - T\| \leq 1/\alpha$, $\|T_n - T\| \leq 1/\beta$ for all $n \geq n_1$, then for each fixed $n \geq n_1$ and all $k = 0, 1, 2, \ldots$ we obtain

$$\max\left\{\|\Theta_n^{(k)} - \Theta_{(n)}\|_1, \|\underline{\varphi_n}^{(k)} - \underline{\varphi_{(n)}}\|_\infty\right\} \leq \alpha\|(T_n - T)T\|(\beta\|T_n - T\|)^k,$$

as before.

(b) Let $0 \notin \Lambda$ and $T_n \xrightarrow{\nu} T$. By Lemma 3.5 and Lemma 3.6(b), there is a positive integer n_0 and there are constants c, $\widetilde{c}$ such that for all $n \geq n_0$,

$$\max\{\|\Theta_n - \Theta_{(n)}\|, \|\underline{\varphi_n} - \underline{\varphi_{(n)}}\|_\infty\} \leq c\gamma_1\|(T_n - T)T\|,$$
$$\max\{\|(\underline{T_n} - \underline{T})\mathcal{S}_n \underline{T_n}\|, \|(\underline{T_n} - \underline{T})\mathcal{S}_n \underline{T}\|\} \leq \widetilde{c}\epsilon_n,$$

where $\epsilon_n := \max\{\|(T_n - T)T_n\|, \|(T_n - T)T\|\}$. Then there are constants α and β such that if $n \geq n_0$ and if $\epsilon_n \leq \min\{1/\alpha, 1/\beta, 1\}$, then for $k = 0, 1, 2, \ldots$

$$\max\left\{\|\Theta_n^{(2k)} - \Theta_{(n)}\|_1, \|\Theta_n^{(2k+1)} - \Theta_{(n)}\|_1, \|\underline{\varphi_n}^{(2k)} - \underline{\varphi_{(n)}}\|_\infty,\right.$$
$$\left.\|\underline{\varphi_n}^{(2k+1)} - \underline{\varphi_{(n)}}\|_\infty\right\} \leq \alpha\|(T_n - T)T\|(\beta\epsilon_n)^k,$$

exactly as in the proof of Theorem 3.3(b). ∎

We now consider an analogue of the Double Iteration (D). Let Λ be a spectral set for T such that $0 \notin \Lambda$ and $\operatorname{rank} P(T, \Lambda) = m < \infty$. Fix a positive integer n for which the conclusions of Lemma 3.5 hold. Keeping in mind the equation $\underline{\varphi_{(n)}} = \underline{T}\,\underline{\varphi_{(n)}}(\Theta_{(n)})^{-1}$, the successive iterates are defined as follows:

$$(\underline{D}) \quad \begin{cases} \underline{\psi_n}^{(0)} := \underline{\varphi_n}, \text{ and for } k = 1, 2, \ldots \\ \Theta_n^{(k)} := (\underline{T}\,\underline{\psi_n}^{(k-1)}, \underline{\varphi_n^*}), \text{ and if } \Theta_n^{(k)} \text{ is nonsingular,} \\ \underline{\varphi_n}^{(k-1)} := \underline{T}\,\underline{\psi_n}^{(k-1)}(\Theta_n^{(k)})^{-1}, \\ \Upsilon_n^{(k)} := (\underline{T}\,\underline{\varphi_n}^{(k-1)}, \underline{\varphi_n^*}), \\ \underline{\psi_n}^{(k)} := \underline{\varphi_n}^{(k-1)} + \mathcal{S}_n(\underline{\varphi_n}^{(k-1)}\Upsilon_n^{(k)} - \underline{T}\,\underline{\varphi_n}^{(k-1)}). \end{cases}$$

Whenever the iterates are well defined, we obtain the following relations as in the case of the Elementary Iteration $(\underline{E})$.

$$(\underline{D})_1 \qquad (\underline{\psi_n}^{(k)}, \underline{\varphi_n^*}) = \mathsf{I}_m \quad \text{for } k = 0, 1, 2, \ldots$$

and also

$(\underline{D})_{1,1}$

$$\langle \underline{\varphi_n}^{(k)}, \underline{\varphi_n^*} \rangle = \langle \underline{T}\, \underline{\psi_n}^{(k)}, \underline{\varphi_n^*} \rangle (\Theta_n^{(k+1)})^{-1} = \mathsf{I}_m \quad \text{for } k = 0, 1, 2, \ldots$$

Next,

$$(\underline{D})_2 \;\; \Theta_n^{(k)} - \Theta_{(n)} = \langle (\underline{T} - \underline{T_n})(\underline{\psi_n}^{(k-1)} - \varphi_{(n)}), \underline{\varphi_n^*} \rangle \quad \text{for } k = 1, 2, \ldots$$

and also

$(\underline{D})_{2,2}$

$$\Upsilon_n^{(k)} - \Theta_{(n)} = \langle (\underline{T} - \underline{T_n})(\underline{\varphi_n}^{(k-1)} - \underline{\varphi_{(n)}}), \underline{\varphi_n^*} \rangle \quad \text{for } k = 1, 2, \ldots$$

Further,

$$(\underline{D})_3 \quad \begin{cases} \text{for } k = 1, 2, \ldots \\ \underline{\psi_n}^{(k)} - \underline{\varphi_{(n)}} = \mathcal{S}_n \Big[(\underline{T_n} - T)(\underline{\varphi_n}^{(k-1)} - \underline{\varphi_{(n)}}) \\ \qquad + \underline{\varphi_n}^{(k-1)}(\Upsilon_n^{(k)} - \Theta_{(n)}) \\ \qquad + (\underline{\varphi_n}^{(k-1)} - \underline{\varphi_{(n)}})(\Theta_{(n)} - \Theta_n) \Big] \end{cases}$$

and also

$$(\underline{D})_{3,3} \quad \begin{cases} \text{for } k = 0, 1, 2, \ldots \\ \underline{\varphi_n}^{(k)} - \underline{\varphi_{(n)}} = \underline{T} \left[\underline{\psi_n}^{(k)} (\Theta_n^{(k+1)})^{-1} - \underline{\varphi_{(n)}}\, \Theta_{(n)}^{-1} \right] \\ \qquad = \underline{T} \Big[(\underline{\psi_n}^{(k)} - \underline{\varphi_{(n)}})(\Theta_n^{(k+1)})^{-1} \\ \qquad + \underline{\varphi_{(n)}}\, (\Theta_n^{(k+1)})^{-1} (\Theta_{(n)} - \Theta_n^{(k+1)}) \Theta_{(n)}^{-1} \Big], \\ \text{provided } \Theta_n^{(k+1)} \text{ is nonsingular.} \end{cases}$$

Theorem 3.8
Let $0 \notin \Lambda$ and $T_n \stackrel{\nu}{\to} T$. For each large n, let $\underline{\varphi_n}$ be so chosen that the sequence $(\|\varphi_n\|_\infty)$ is bounded and the sequence (δ_n) is bounded away from zero.

Let $\Upsilon_n^{(0)} = \Theta_n$ and $\underline{\psi_n}^{(0)} = \underline{\varphi_n}$. Then there is a positive integer n_1 such that for each fixed $n \geq n_1$, all the iterates $\Upsilon_n^{(k)}$ and $\underline{\psi_n}^{(k)}$ of the

Double Iteration ($\underline{D}$) *are well defined; and if we let*

$$\widehat{\mu}_n^{(k)} := \frac{1}{m} \operatorname{tr} \Upsilon_n^{(k)},$$

then for all $k = 0, 1, 2, \ldots$

$$\max\{|\widehat{\mu}_n^{(k)} - \widehat{\lambda}|, \|\underline{\psi_n}^{(k)} - \underline{\varphi_{(n)}}\|_\infty\} \le (\beta\|(T_n - T)T\|)^{k+1},$$

where β *is a constant, independent of* n *and* k.

Proof
As $|\widehat{\mu}_n^{(k)} - \lambda| \le \|\Upsilon_n^{(k)} - \Theta_{(n)}\|_1$ for all large n and all $k = 1, 2, \ldots$ it is enough to obtain the desired bounds for $\|\Upsilon_n^{(k)} - \Theta_{(n)}\|_1$ in place of $|\widehat{\mu}_n^{(k)} - \lambda|$.

Let $\|\underline{\varphi_n}\|_\infty \le \gamma$ for all large n and $\gamma_1 := \max\{1, \gamma\}$.

Since $0 \notin \Lambda$ and $T_n \xrightarrow{\nu} T$, Lemma 3.5 and Lemma 3.6(b)(i) show that there is a positive integer n_0 and there is a positive constant c such that for all $n \ge n_0$,

$$\max\{\|\Theta_n - \Theta_{(n)}\|_1, \|\underline{\varphi_n} - \underline{\varphi_{(n)}}\|_\infty\} \le c\gamma_1\|(T_n - T)T\|.$$

By Lemma 3.6(a), there are constants p and q such that

$$\|\underline{\varphi_n^*}\|_1 \le p \quad \text{and} \quad \|\underline{\varphi_{(n)}}\|_\infty \le q.$$

Let M denote the spectral subspace associated with T and Λ. Since $\operatorname{sp}(T_{|M,M}) = \Lambda$ and $0 \notin \Lambda$, we note that the operator $T_{|M,M}$ is invertible in $\operatorname{BL}(M)$ and hence the operator $\underline{T_M} := \underline{T_{|M,M}}$ is invertible in $\operatorname{BL}(\underline{M})$. Now for all large n,

$$\underline{T_M}\,\underline{\varphi_{(n)}} = \underline{T}\,\underline{\varphi_{(n)}} = \underline{\varphi_{(n)}}\,\Theta_{(n)},$$

so that $\underline{\varphi_{(n)}} = (\underline{T_M})^{-1}\underline{\varphi_{(n)}}\,\Theta_{(n)}$ and

$$\begin{aligned} \mathsf{I}_m = \langle \underline{\varphi_{(n)}}, \underline{\varphi_n^*} \rangle &= \langle (\underline{T_M})^{-1}\underline{\varphi_{(n)}}\,\Theta_{(n)}, \underline{\varphi_n^*} \rangle \\ &= \langle (\underline{T_M})^{-1}\underline{\varphi_{(n)}}, \underline{\varphi_n^*} \rangle \Theta_{(n)}. \end{aligned}$$

Thus for all large n, the $m \times m$ matrix $\Theta_{(n)}$ is nonsingular and

$$\begin{aligned} \|(\Theta_{(n)})^{-1}\|_1 &= \|\langle (\underline{T_M})^{-1}\underline{\varphi_{(n)}}, \underline{\varphi_n^*} \rangle\|_1 \\ &\le \|(\underline{T_M})^{-1}\|\,\|\underline{\varphi_{(n)}}\|_\infty \|\underline{\varphi_n^*}\|_1 \\ &\le pq\|(T_M)^{-1}\|. \end{aligned}$$

Let $\alpha := pq\|(T_M)^{-1}\|$. Then for all large n,

$$\|(\Theta_{(n)})^{-1}\| \leq \alpha.$$

By Lemma 3.6(b)(ii), there is a constant s such that $\|\mathcal{S}_n\| \leq s$ for all large n. Also, let $\|T_n - T\| \leq t$ for all n.

Consider

$$\beta := \max\{c\gamma_1, pc\gamma_1, \beta_1, \beta_2, p\beta_2\},$$

where $\beta_1 := 2\alpha(1 + \alpha tpq)$ and $\beta_2 := \beta_1 s(1 + 2\alpha\|T\|(p + pq) + c\gamma_1\|T\|)$.

Choose $n_1 \geq n_0$ such that $\|(T_n - T)T\| \leq 1/\beta$, $\|(T_n - T)T\| \leq \dfrac{1}{2\alpha t\beta}$, and fix $n \geq n_1$. It can be shown by induction on k and by employing the relations $(\underline{D})_2$, $(\underline{D})_{2,2}$, $(\underline{D})_3$ and $(\underline{D})_{3,3}$ that for $k = 0, 1, 2, \ldots$ the iterates $\underline{\psi_n}^{(k)}$ and $\Upsilon_n^{(k+1)}$ are well defined and

(i) $\|\underline{\psi_n}^{(k)} - \underline{\varphi_{(n)}}\|_\infty \leq (\beta\|(T_n - T)T\|)^{k+1}$,

(ii) $\|\Theta_n^{(k+1)} - \Theta_{(n)}\|_1 \leq \|T_n - T\|(\beta\|(T_n - T)T\|)^{k+1}$,

(iii) $\Theta_n^{(k+1)}$ is nonsingular and $\|(\Theta_n^{(k+1)})^{-1}\| \leq 2\alpha$,

(iv) $\|\Upsilon_n^{(k+1)} - \Theta_{(n)}\|_1 \leq (\beta\|(T_n - T)T\|)^{k+2}$.

The induction argument is very similar to the one we gave in the proof of Theorem 3.4. For achieving this, we merely replace T, T_n, S_n, λ, λ_n, $\lambda_n^{(k)}$, $\mu_n^{(k)}$, $\varphi_{(n)}$, $\varphi_n^{(k)}$, $\psi_n^{(k)}$ and φ_n^* by $\underline{T}$, $\underline{T_n}$, $\mathcal{S}_n$, $\Theta_{(n)}$, Θ_n, $\Theta_n^{(k)}$, $\Upsilon_n^{(k)}$, $\underline{\varphi_{(n)}}$, $\underline{\varphi_n}^{(k)}$, $\underline{\psi_n}^{(k)}$ and $\underline{\varphi_n}^{(*)}$ respectively, and the scalar product $\langle \cdot\,, \cdot \rangle$ by the Gram matrix $(\cdot, \cdot)$. Also, we note that if for some $k = 1, 2, \ldots$

$$\|\Theta_n^{(k)} - \Theta_{(n)}\|_1 \leq \frac{1}{2\alpha} \leq \frac{1}{2\|(\Theta_{(n)})^{-1}\|},$$

then the matrix $\Theta_n^{(k)}$ is nonsingular and $\|(\Theta_n^{(k)})^{-1}\| \leq 2\alpha$. As we remarked in the proof of Theorem 3.7(a), we replace the norm $\|\cdot\|$ and X and the absolute value $|\cdot|$ on $\mathbb{C}$ by the norm $\|\cdot\|_\infty$ on $\underline{X}$ and the norm $\|\cdot\|_1$ on $\mathbb{C}^{m\times m}$ respectively.

Since the inequalities (i) and (iv) hold for all $k = 0, 1, 2, \ldots$ and since $\Upsilon_n^{(0)} = \Theta_n$, the desired estimates hold for all $k = 0, 1, 2, \ldots$ ∎

In Subsection 5.1.2 we shall show how the sequence $(\underline{\varphi_n})$ mentioned in Theorems 3.7 and 3.8 can be chosen in practice in such a way that $(\|\underline{\varphi_n}\|_\infty)$ is bounded and (δ_n) is bounded away from zero.

3.2 Acceleration

3.2.1 Motivation

In this section, we consider another approach toward improving the accuracy of an approximate solution of the eigenvalue problem. Let $T \in \mathrm{BL}(X)$. Suppose we wish to find $\varphi \neq 0$ in X and $\lambda \neq 0$ in $\mathbb{C}$ such that $T\varphi = \lambda\varphi$, but instead we have found $\widetilde{\varphi} \neq 0$ in X and $\widetilde{\lambda} \neq 0$ in $\mathbb{C}$ such that $\widetilde{T}\widetilde{\varphi} = \widetilde{\lambda}\widetilde{\varphi}$, where $\widetilde{T} \in \mathrm{BL}(X)$ is 'near' T.

Let us add $\widetilde{T}\varphi$ to both sides of the equation $\lambda\varphi = T\varphi$ and rewrite it as follows:

$$\lambda\varphi - (T - \widetilde{T})\varphi = \widetilde{T}\varphi.$$

If $\rho(T - \widetilde{T}) < |\lambda|$, then by Theorem 1.13(d), the operator $\lambda I - (T - \widetilde{T})$ is invertible in $\mathrm{BL}(X)$ and so

$$\begin{aligned}\varphi = [\lambda I - (T - \widetilde{T})]^{-1}\widetilde{T}\varphi &= \frac{1}{\lambda}\left[I - \left(\frac{T - \widetilde{T}}{\lambda}\right)\right]^{-1}\widetilde{T}\varphi \\ &= \frac{1}{\lambda}\left[\sum_{k=0}^{\infty}\left(\frac{T - \widetilde{T}}{\lambda}\right)^k\right]\widetilde{T}\varphi.\end{aligned}$$

The idea is to truncate the infinite series $\displaystyle\sum_{k=0}^{\infty}\left(\frac{T - \widetilde{T}}{\lambda}\right)^k$ after a finite number of terms, say q terms, and attempt to find $\widetilde{\varphi}^{[q]} \neq 0$ in X and $\widetilde{\lambda}^{[q]} \neq 0$ in $\mathbb{C}$ such that

$$\widetilde{\varphi}^{[q]} = \frac{1}{\widetilde{\lambda}^{[q]}}\left[\sum_{k=0}^{q-1}\left(\frac{T - \widetilde{T}}{\widetilde{\lambda}^{[q]}}\right)^k\right]\widetilde{T}\widetilde{\varphi}^{[q]}.$$

The preceding equation can be rewritten as follows:

$$(\widetilde{\lambda}^{[q]})^q\widetilde{\varphi}^{[q]} = \sum_{k=0}^{q-1}(\widetilde{\lambda}^{[q]})^{q-1-k}(T - \widetilde{T})^k\widetilde{T}\widetilde{\varphi}^{[q]}.$$

If $q = 1$ and we let $\widetilde{\lambda} := \widetilde{\lambda}^{[q]}$, $\widetilde{\varphi} := \widetilde{\varphi}^{[q]}$, then the preceding equation reduces to the equation $\widetilde{\lambda}\widetilde{\varphi} = \widetilde{T}\widetilde{\varphi}$, that is, to the ordinary eigenequation for the approximation $\widetilde{T}$ of T. This equation is linear in the scalar $\widetilde{\lambda}$.

For a general positive integer q, the above equation leads to a **polynomial eigenvalue problem** of order q. Solving this problem is a part of **higher order spectral approximation**. It was first considered in detail by Dellwo and Friedman in [31], and it was systematically developed in the framework of a product space by Alam, Kulkarni and Limaye in [13]. We observe that the polynomial eigenequation can be recast as follows:

$$\widetilde{\lambda}^{[q]}\begin{bmatrix} \widetilde{\varphi}^{[q]} \\ \dfrac{\widetilde{\varphi}^{[q]}}{\widetilde{\lambda}^{[q]}} \\ \vdots \\ \dfrac{\widetilde{\varphi}^{[q]}}{(\widetilde{\lambda}^{[q]})^{q-1}} \end{bmatrix} = \begin{bmatrix} \widetilde{T} & (T-\widetilde{T})\widetilde{T} & \cdots & \cdots & (T-\widetilde{T})^{q-1}\widetilde{T} \\ I & O & \cdots & \cdots & O \\ O & I & O & \cdots & O \\ \vdots & \ddots & \ddots & \ddots & \vdots \\ O & \cdots & O & I & O \end{bmatrix}\begin{bmatrix} \widetilde{\varphi}^{[q]} \\ \dfrac{\widetilde{\varphi}^{[q]}}{\widetilde{\lambda}^{[q]}} \\ \vdots \\ \dfrac{\widetilde{\varphi}^{[q]}}{(\widetilde{\lambda}^{[q]})^{q-1}} \end{bmatrix}.$$

In a similar manner, the ordinary eigenequation $\lambda\varphi = T\varphi$ with $\lambda \neq 0$ can be recast as follows:

$$\lambda\begin{bmatrix} \varphi \\ \dfrac{\varphi}{\lambda} \\ \vdots \\ \dfrac{\varphi}{\lambda^{q-1}} \end{bmatrix} = \begin{bmatrix} T & O & \cdots & \cdots & O \\ I & O & \cdots & \cdots & O \\ O & I & O & \cdots & O \\ \vdots & \ddots & \ddots & \ddots & \vdots \\ O & \cdots & O & I & O \end{bmatrix}\begin{bmatrix} \varphi \\ \dfrac{\varphi}{\lambda} \\ \vdots \\ \dfrac{\varphi}{\lambda^{q-1}} \end{bmatrix}.$$

These representations lead us to consider the linear space of all columns of length q with entries in X (as against the linear space of all rows of a given length with entries in X, which we have studied in Subsection 1.4.2).

Let X be a complex Banach space, $q \geq 2$ an integer and

$$\boldsymbol{X}^{[q]} := X^{q\times 1} = \left\{[x_1,\ldots,x_q]^\top : x_i \in X \quad \text{for } 1 \leq i \leq q\right\},$$

where

$$[x_1,\ldots,x_q]^\top := \begin{bmatrix} x_1 \\ \vdots \\ x_q \end{bmatrix}.$$

For $\boldsymbol{x} := [x_1, \ldots, x_q]^\top \in \boldsymbol{X}^{[q]}$, define

$$\|\boldsymbol{x}\|_\infty := \max\{\|x_1\|, \ldots, \|x_q\|\}.$$

Then $\|\cdot\|_\infty$ is a complete norm on $\boldsymbol{X}^{[q]}$.

Given T, $\widetilde{T} \in \mathrm{BL}(X)$, we define $\boldsymbol{T}^{[q]}$ and $\widetilde{\boldsymbol{T}}^{[q]}$ in $\mathrm{BL}(\boldsymbol{X}^{[q]})$ as follows: For $\boldsymbol{x} = [x_1, \ldots, x_q]^\top$, let

$$\boldsymbol{T}^{[q]}\boldsymbol{x} := [Tx_1, x_1, \ldots, x_{q-1}]^\top$$

and

$$\widetilde{\boldsymbol{T}}^{[q]}\boldsymbol{x} = \left[\sum_{k=0}^{q-1}(T-\widetilde{T})^k\widetilde{T}x_{k+1}, x_1, \ldots, x_{q-1}\right]^\top.$$

The operators $\boldsymbol{T}^{[q]}$ and $\widetilde{\boldsymbol{T}}^{[q]}$ can be represented by $q\times q$ matrices as follows:

$$\boldsymbol{T}^{[q]} := \begin{bmatrix} T & O & \cdots & \cdots & O \\ I & O & \cdots & \cdots & O \\ O & I & O & \cdots & O \\ \vdots & \ddots & \ddots & \ddots & \vdots \\ O & \cdots & O & I & O \end{bmatrix},$$

$$\widetilde{\boldsymbol{T}}^{[q]} := \begin{bmatrix} \widetilde{T} & (T-\widetilde{T})\widetilde{T} & \cdots & \cdots & (T-\widetilde{T})^{q-1}\widetilde{T} \\ I & O & \cdots & \cdots & O \\ O & I & O & \cdots & O \\ \vdots & \ddots & \ddots & \ddots & \vdots \\ O & \cdots & O & I & O \end{bmatrix}.$$

Note that $\boldsymbol{T}^{[q]}$ depends only on T, while $\widetilde{\boldsymbol{T}}^{[q]}$ depends on both T and $\widetilde{T}$. When the integer q is fixed, we write $\boldsymbol{X}$, $\boldsymbol{T}$ and $\widetilde{\boldsymbol{T}}$ for $\boldsymbol{X}^{[q]}$, $\boldsymbol{T}^{[q]}$ and $\widetilde{\boldsymbol{T}}^{[q]}$, respectively.

Fix an integer $q \geq 2$. For $\boldsymbol{x} = [x_1, \ldots, x_q]^\top \in \boldsymbol{X}$ and $0 \neq \lambda \in \mathbb{C}$, we have $\boldsymbol{T}\boldsymbol{x} = \lambda\boldsymbol{x}$ if and only if $Tx_1 = \lambda x_1$ and $x_{k+1} = x_1/\lambda^k$ for

$k = 1, \ldots, q-1$. Also, for $\boldsymbol{x} = [x_1, \ldots, x_q]^\top \in \boldsymbol{X}$ and $0 \neq \widetilde{\lambda} \in \mathbb{C}$, we have $\widetilde{\boldsymbol{T}}\boldsymbol{x} = \widetilde{\lambda}\boldsymbol{x}$ if and only if $\sum\limits_{k=0}^{q-1} \widetilde{\lambda}^{q-1-k}(T - \widetilde{T})^k \widetilde{T} x_1 = \widetilde{\lambda}^q x_1$ and $x_{k+1} = x_1/\widetilde{\lambda}^k$ for $k = 1, \ldots, q-1$.

For $\varphi \neq 0$ in X and $\lambda \neq 0$ in $\mathbb{C}$, let

$$\boldsymbol{\varphi} := \left[\varphi, \frac{\varphi}{\lambda}, \ldots, \frac{\varphi}{\lambda^{q-1}}\right]^\top \in \boldsymbol{X}.$$

Then, as we have remarked above,

$$\boldsymbol{T}\boldsymbol{\varphi} = \lambda\boldsymbol{\varphi} \quad \text{if and only if} \quad T\varphi = \lambda\varphi.$$

Similarly, for $\widetilde{\varphi} \neq 0$ in X and $\widetilde{\lambda} \neq 0$ in $\mathbb{C}$, let

$$\widetilde{\boldsymbol{\varphi}} := \left[\widetilde{\varphi}, \frac{\widetilde{\varphi}}{\widetilde{\lambda}}, \ldots, \frac{\widetilde{\varphi}}{\widetilde{\lambda}^{q-1}}\right]^\top \in \boldsymbol{X}.$$

Then, as we have mentioned earlier,

$$\widetilde{\boldsymbol{T}}\widetilde{\boldsymbol{\varphi}} = \widetilde{\lambda}\widetilde{\boldsymbol{\varphi}} \text{ if and only if } \widetilde{\lambda}^q \widetilde{\varphi} = \sum_{k=0}^{q-1} \widetilde{\lambda}^{q-1-k}(T - \widetilde{T})^k \widetilde{T}\widetilde{\varphi}.$$

Thus our polynomial eigenvalue problem in X can be lifted to an ordinary (that is, linear) eigenvalue problem in the column space $\boldsymbol{X}$.

It is easy to see that $\boldsymbol{T}, \widetilde{\boldsymbol{T}} \in \mathrm{BL}(\boldsymbol{X})$, and if $\|\cdot\|$ denotes the operator norm on $\mathrm{BL}(\boldsymbol{X})$ as well, then

$$\|\boldsymbol{T}\| = \max\{1, \|T\|\}, \quad \|\widetilde{\boldsymbol{T}}\| \leq \max\Big\{1, \sum_{k=0}^{q-1} \|(T - \widetilde{T})^k \widetilde{T}\|\Big\}.$$

We shall now show that the nonzero spectral values of $\boldsymbol{T}$ are the same as the nonzero spectral value of T, and their multiplicities are not changed. On the other hand, the nonzero spectral values of $\widetilde{\boldsymbol{T}}$ can be very different from those of $\widetilde{T}$. As we shall see later, the nonzero spectral values of $\widetilde{\boldsymbol{T}}$ would, in fact, provide a better approximation of the nonzero spectral values of T. Also, even if $\widetilde{T}$ is of finite rank, $\widetilde{\boldsymbol{T}}$ may not be of finite rank. However, the computation of the nonzero spectral values of $\widetilde{\boldsymbol{T}}$ can be reduced to a matrix eigenvalue problem, as we shall see in Section 5.3.

Let $\mathbf{I}$ denote the identity operator on $\boldsymbol{X}$. Consider $z \neq 0$ in $\mathbb{C}$. It can be easily seen that $\boldsymbol{T} - z\mathbf{I}$ is invertible in $\mathrm{BL}(\boldsymbol{X})$ if and only if $T - zI$ is invertible in $\mathrm{BL}(X)$, and in that case

$$R(\boldsymbol{T}, z) = \begin{bmatrix} R(T,z) & O & \cdots & \cdots & O \\ \dfrac{R(T,z)}{z} & -\dfrac{I}{z} & O & \cdots & O \\ \vdots & \vdots & \ddots & \ddots & \vdots \\ \vdots & \vdots & \ddots & \ddots & O \\ \dfrac{R(T,z)}{z^{q-1}} & -\dfrac{I}{z^{q-1}} & -\dfrac{I}{z^{q-2}} & \cdots & -\dfrac{I}{z} \end{bmatrix}.$$

In order to obtain a representation for a spectral projection associated with $\boldsymbol{T}$, we introduce the following notation.

Let Λ be a spectral set for T such that $0 \notin \Lambda$, and let $\mathrm{C} \in \mathcal{C}(T, \Lambda)$ be such that $0 \in \mathrm{ext}(\mathrm{C})$. Define

$$J_k(T, \Lambda) := -\frac{1}{2\pi \mathrm{i}} \int_{\mathrm{C}} R(T, z) \frac{dz}{z^k} \quad \text{for } k = 0, 1, \ldots, q-1.$$

It is easy to see that $J_k(T, \Lambda)$ does not depend on $\mathrm{C} \in \mathcal{C}(T, \Lambda)$, provided $0 \in \mathrm{ext}(\mathrm{C})$. Clearly $J_0(T, \Lambda) = P(T, \Lambda)$. We shall denote $J_k(T, \Lambda)$, simply by J_k when the context is clear just as we denote $P(T, \Lambda)$ by P.

By the standard techniques of contour integration, it follows that

$$PJ_k = J_k P = J_k \quad \text{for } k = 1, \ldots, q-1.$$

(Compare the proof of Proposition 1.24(c).) Also, by employing the identity $TR(T, z) = I + zR(T, z) = R(T, z)T$ for $z \in \mathrm{re}(T)$ (1.10(e)), it can be seen that

$$TJ_k = J_k T = J_{k-1} \quad \text{for } k = 1, \ldots, q-1.$$

We consider a map $J(T, \Lambda) : X \to \boldsymbol{X}$ defined by

$$J(T, \Lambda)x = [J_0 x, J_1 x, \ldots, J_{q-1} x]^\top.$$

Again, we shall denote $J(T, \Lambda)$ simply by J when the context is clear.

Proposition 3.9

(a) *If $0 \neq \lambda \in \mathbb{C}$, then λ is an eigenvalue of $\boldsymbol{T}$ if and only if λ is an eigenvalue of T, and then the geometric multiplicities of λ are the same. In fact, the map* $\varphi \longmapsto \left[\varphi, \frac{\varphi}{\lambda}, \cdots, \frac{\varphi}{\lambda^{q-1}}\right]^\top$ *from $\mathcal{N}(T - \lambda I)$ to $\mathcal{N}(\boldsymbol{T} - \lambda \mathbf{I})$ is a bijection.*

(b) $\mathrm{sp}(\boldsymbol{T}) \setminus \{0\} = \mathrm{sp}(T) \setminus \{0\}$. *Also, if $0 \notin \Lambda \subset \mathbb{C}$, then Λ is a spectral set for $\boldsymbol{T}$ if and only if Λ is a spectral set for T. In that case, there is* $\mathrm{C} \in \mathcal{C}(T, \Lambda)$ *such that* $0 \in \mathrm{ext}(\mathrm{C})$, *and every such* C *belongs to* $\mathcal{C}(\boldsymbol{T}, \Lambda)$. *If $P := P(T, \Lambda)$ and $\boldsymbol{P} = P(\boldsymbol{T}, \Lambda)$, then*

$$\boldsymbol{P} = \begin{bmatrix} P & O & \cdots & O \\ J_1 & O & \cdots & O \\ \vdots & \vdots & & \vdots \\ J_{q-1} & O & \cdots & O \end{bmatrix}.$$

Also, $\mathrm{rank}\, \boldsymbol{P} = \mathrm{rank}\, P$. *In fact, the map J gives a bijection from $\mathcal{R}(P)$ to $\mathcal{R}(\boldsymbol{P})$.*

In particular, if $\lambda \neq 0$, then λ is an isolated point of $\mathrm{sp}(\boldsymbol{T})$ *if and only if λ is an isolated point of* $\mathrm{sp}(T)$*; and then its algebraic multiplicities are the same.*

Proof

(a) The statement is easy to verify.

(b) Let $0 \neq z \in \mathbb{C}$. As we have noted before, $z \in \mathrm{re}(\boldsymbol{T})$ if and only if $z \in \mathrm{re}(T)$. Hence $\mathrm{sp}(\boldsymbol{T}) \setminus \{0\} = \mathrm{sp}(T) \setminus \{0\}$. Letting $Y = \boldsymbol{X}$, $A = \boldsymbol{T}$ and $E = \{0\}$ in Proposition 1.29, we see that Λ is a spectral set for $\boldsymbol{T}$ if and only if Λ is a spectral set for T. In that case, there is $\mathrm{C} \in \mathcal{C}(T, \Lambda)$ such that $0 \in \mathrm{ext}(\mathrm{C})$ and every such C belongs to $\mathcal{C}(\boldsymbol{T}, \Lambda)$. Now the matrix representation of $R(\boldsymbol{T}, z)$, $z \in \mathrm{C}$, gives the following matrix representation of $\boldsymbol{P} = -\dfrac{1}{2\pi \mathrm{i}} \displaystyle\int_{\mathrm{C}} R(\boldsymbol{T}, z)dz$:

$$\begin{bmatrix} \dfrac{-1}{2\pi\mathrm{i}} \displaystyle\int_{\mathrm{C}} R(T,z)dz & O & \cdots & \cdots & O \\ \dfrac{-1}{2\pi\mathrm{i}} \displaystyle\int_{\mathrm{C}} R(T,z)\frac{dz}{z} & \dfrac{I}{2\pi\mathrm{i}} \displaystyle\int_{\mathrm{C}} \frac{dz}{z} & O & \cdots & O \\ \vdots & \vdots & \ddots & \ddots & \vdots \\ \vdots & \vdots & \ddots & \ddots & O \\ \dfrac{-1}{2\pi\mathrm{i}} \displaystyle\int_{\mathrm{C}} R(T,z)\frac{dz}{z^{q-1}} & \dfrac{I}{2\pi\mathrm{i}} \displaystyle\int_{\mathrm{C}} \frac{dz}{z^{q-1}} & \dfrac{I}{2\pi i} \displaystyle\int_{\mathrm{C}} \frac{dz}{z^{q-2}} & \cdots & \dfrac{I}{2\pi\mathrm{i}} \displaystyle\int_{\mathrm{C}} \frac{dz}{z} \end{bmatrix}.$$

But $\dfrac{-1}{2\pi\mathrm{i}} \displaystyle\int_{\mathrm{C}} R(T,z)dz = P$, $\dfrac{-1}{2\pi\mathrm{i}} \displaystyle\int_{\mathrm{C}} R(T,z)\frac{dz}{z^k} = J_k$ and $\displaystyle\int_{\mathrm{C}} \frac{dz}{z^k} = 0$ for $1 \leq k \leq q-1$ as $0 \in \mathrm{ext}(\mathrm{C})$. Thus for $\boldsymbol{x} = [x_1, \ldots, x_q]^\top \in \boldsymbol{X}$, we have

$$\boldsymbol{P}\boldsymbol{x} = [Px_1, J_1x_1, \ldots, J_{q-1}x_1]^\top,$$

so that $\boldsymbol{x} \in \mathcal{R}(\boldsymbol{P})$ if and only if $x_1 \in \mathcal{R}(P)$ and $x_2 = J_1x_1, \ldots, x_q = J_{q-1}x_1$. The desired results now follow easily. ▮

The preceding proposition shows that the spectral considerations for $\boldsymbol{T}$ and for T are the same, if we exclude 0 from these considerations. (See also Exercise 3.6.) As a result, given an approximation $\widetilde{T}$ of T, we may approximate the nonzero spectral values of $\boldsymbol{T}$ by the nonzero spectral values of $\widetilde{\boldsymbol{T}}$. In doing so, the accuracy of approximation is improved. The following result is crucial in this respect.

Proposition 3.10
Let Λ be a spectral set for T such that $0 \notin \Lambda$, $\mathrm{C} \in \mathcal{C}(T, \Lambda)$ with $0 \in \mathrm{ext}(\mathrm{C})$

and $\boldsymbol{P} := P(\boldsymbol{T}, \Lambda)$. *Then*

$$(\widetilde{\boldsymbol{T}} - \boldsymbol{T})\boldsymbol{P} = \begin{bmatrix} -(T-\widetilde{T})^q J_{q-1} & O & \cdots & O \\ O & O & \cdots & O \\ \vdots & \vdots & & \vdots \\ O & O & \cdots & O \end{bmatrix}.$$

In particular, if $0 \neq \lambda \in \Lambda$, φ *is an eigenvector of* T *corresponding to* λ *and* $\boldsymbol{\varphi} = \left[\varphi, \frac{\varphi}{\lambda}, \cdots, \frac{\varphi}{\lambda^{q-1}}\right]^\top$, *then*

$$(\widetilde{\boldsymbol{T}} - \boldsymbol{T})\boldsymbol{\varphi} = \left[-\frac{(T-\widetilde{T})^q \varphi}{\lambda^{q-1}}, 0, \cdots, 0\right]^\top.$$

Proof

By Proposition 3.9(b), we have

$$(\widetilde{\boldsymbol{T}} - \boldsymbol{T})\boldsymbol{P} = \begin{bmatrix} (\widetilde{T}-T)P + \sum_{k=1}^{q-1}(T-\widetilde{T})^k \widetilde{T} J_k & O & \cdots & O \\ O & O & \cdots & O \\ \vdots & \vdots & & \vdots \\ O & O & \cdots & O \end{bmatrix}.$$

Now since $TJ_k = J_{k-1}$ for $k = 1, \ldots, q-1$ and $J_0 = P$, we have

$$\begin{aligned} \sum_{k=1}^{q-1}(T-\widetilde{T})^k \widetilde{T} J_k &= \sum_{k=1}^{q-1}(T-\widetilde{T})^k [T - (T-\widetilde{T})] J_k \\ &= \sum_{k=1}^{q-1}(T-\widetilde{T})^k T J_k - \sum_{k=1}^{q-1}(T-\widetilde{T})^{k+1} J_k \\ &= \sum_{k=1}^{q-1}(T-\widetilde{T})^k J_{k-1} - \sum_{k=1}^{q-1}(T-\widetilde{T})^{k+1} J_k \\ &= (T-\widetilde{T})P - (T-\widetilde{T})^q J_{q-1}. \end{aligned}$$

Hence the desired expression for $(\widetilde{\boldsymbol{T}} - \boldsymbol{T})\boldsymbol{P}$ follows.

Next, let $0 \neq \lambda \in \Lambda$, $T\varphi = \lambda\varphi$ and $\boldsymbol{\varphi} = \left[\varphi, \frac{\varphi}{\lambda}, \dots, \frac{\varphi}{\lambda^{q-1}}\right]^{\top}$. Then $\boldsymbol{\varphi}$ is an eigenvector of $\boldsymbol{T}$ corresponding to λ and

$$(\widetilde{\boldsymbol{T}} - \boldsymbol{T})\boldsymbol{\varphi} = (\widetilde{\boldsymbol{T}} - \boldsymbol{T})\boldsymbol{P}\boldsymbol{\varphi} = \left[-(T - \widetilde{T})^q J_{q-1}\varphi, 0, \dots, 0\right]^{\top}.$$

Since $R(T, z)\varphi = \varphi/(\lambda - z)$ for each $z \in \mathrm{C}$, we have

$$J_{q-1}\varphi = -\frac{1}{2\pi \mathrm{i}} \int_{\mathrm{C}} \frac{\varphi}{\lambda - z} \frac{dz}{z^{q-1}} = \frac{\varphi}{\lambda^{q-1}}$$

by Cauchy's Integral Formula. Hence the desired expression for $(\widetilde{\boldsymbol{T}} - \boldsymbol{T})\boldsymbol{\varphi}$ follows. ∎

3.2.2 Higher Order Spectral Approximation

For a fixed integer $q \geq 2$, we consider a sequence (T_n) in $\mathrm{BL}(X)$ which approximates $T \in \mathrm{BL}(X)$ in a manner which we shall presently specify. This mode of convergence is even weaker than ν-convergence. For each pair (T, T_n), $n = 1, 2, \dots$ we consider the operators $\boldsymbol{T}$, $\boldsymbol{T}_n$ defined on the linear space $\boldsymbol{X}$ of columns of length q with entries in X. We shall show that the nonzero spectral values of finite type of T are approximated by the nonzero spectral values of finite type of $\boldsymbol{T}_n$ with sharper error estimates as compared to the error estimates obtained in Section 2.4. A similar statement holds for a basis of a spectral subspace corresponding to a spectral set of finite type for T, provided such a set does not contain 0.

Throughout this subsection we assume that $T \in \mathrm{BL}(X)$ and (T_n) is a sequence in $\mathrm{BL}(X)$ such that $(T_n - T) \xrightarrow{\nu} O$, that is,

$$(\|T_n\|) \text{ is a bounded sequence and } \|(T_n - T)^2\| \to 0 \text{ as } n \to \infty.$$

This assumption is in general weaker than the assumption: $T_n \xrightarrow{\nu} T$. For example, consider the 1-norm on $X := \mathbb{C}^{2\times 1}$, and let $T \in \mathrm{BL}(X)$ be defined by the matrix $\mathsf{A} := \begin{bmatrix} 1 & 1 \\ 1 & 1 \end{bmatrix}$ with respect to the standard basis for X. For $n = 1, 2, \dots$ let $T_n \in \mathrm{BL}(X)$ be defined by the matrix $\mathsf{A}_n := \begin{bmatrix} 1 & 1 \\ 0 & 1 \end{bmatrix}$ with respect to the standard basis for X. Then

$$(\mathsf{A}_n - \mathsf{A})^2 = \mathsf{O} \quad \text{but} \quad (\mathsf{A}_n - \mathsf{A})\mathsf{A} = \begin{bmatrix} 0 & 0 \\ 1 & 1 \end{bmatrix} \overset{\mathrm{n}}{\not\to} \mathsf{O}.$$

For $n = 1, 2, \ldots$ replace $\widetilde{T}$ by T_n in the definition of $\widetilde{\boldsymbol{T}}$ and let

$$\boldsymbol{T}_n := \begin{bmatrix} T_n & (T-T_n)T_n & \cdots & \cdots & (T-T_n)^{q-1}T_n \\ I & O & \cdots & \cdots & O \\ O & I & O & \cdots & O \\ \vdots & \ddots & \ddots & \ddots & \vdots \\ O & \cdots & O & I & O \end{bmatrix}.$$

Then it follows that

$$\boldsymbol{T}_n - \boldsymbol{T} = \begin{bmatrix} -(T-T_n) & (T-T_n)T_n & \cdots & (T-T_n)^{q-1}T_n \\ O & O & \cdots & O \\ \vdots & \vdots & & \vdots \\ O & O & \cdots & O \end{bmatrix},$$

$$(\boldsymbol{T}_n - \boldsymbol{T})\boldsymbol{T} = \begin{bmatrix} -(T-T_n)^2 & (T-T_n)^2T_n & \cdots & (T-T_n)^{q-1}T_n & O \\ O & O & \cdots & O & O \\ \vdots & \vdots & & \vdots & \vdots \\ O & O & \cdots & O & O \end{bmatrix},$$

and

$$(\boldsymbol{T}_n - \boldsymbol{T})\boldsymbol{T}_n = \begin{bmatrix} O & \cdots & O & -(T-T_n)^qT_n \\ O & \cdots & O & O \\ \vdots & & \vdots & \vdots \\ O & \cdots & O & O \end{bmatrix}.$$

Proposition 3.11
Let $(T_n - T) \xrightarrow{\nu} O$. Then $\boldsymbol{T}_n \xrightarrow{\nu} \boldsymbol{T}$. In fact, there is a positive integer n_0 such that for $n \geq n_0$, $\|(T_n - T)^2\| \leq 1/2$ and then

$$\begin{aligned}
\|\boldsymbol{T}_n\| &\leq \max\{1, 2(\|T_n\| + \|(T - T_n)T_n\|)\} \\
\|(\boldsymbol{T}_n - \boldsymbol{T})\boldsymbol{T}\| &\leq [1 + 2(\|T_n\| + \|(T - T_n)T_n\|)]\, \|(T_n - T)^2\|, \\
\|(\boldsymbol{T}_n - \boldsymbol{T})\boldsymbol{T}_n\| &= \|(T_n - T)^q T_n\| \\
&\leq \max\{\|T_n\|, \|(T_n - T)T_n\|\}\, \|(T_n - T)^2\|.
\end{aligned}$$

Proof
Since $\|(T_n - T)^2\| \to 0$, there is a positive integer n_0 such that for $n \geq n_0$, $\|(T_n - T)^2\| \leq 1/2$. For positive integers n and k, we have

$$\|(T - T_n)^k T_n\| \leq \begin{cases} \|(T - T_n)^2\|^{k/2} \|T_n\| & \text{if } k \text{ is even,} \\ \\ \|(T - T_n)^2\|^{(k-1)/2} \|(T - T_n)T_n\| & \text{if } k \text{ is odd.} \end{cases}$$

Hence if $n \geq n_0$, we obtain

$$\begin{aligned}
\sum_{k=0}^{\infty} \|(T - T_n)^k T_n\| &= \sum_{j=0}^{\infty} \|(T - T_n)^{2j} T_n\| + \sum_{j=0}^{\infty} \|(T - T_n)^{2j+1} T_n\| \\
&\leq \Big(\sum_{j=0}^{\infty} \|(T - T_n)^2\|^j \Big) (\|T_n\| + \|(T - T_n)T_n\|) \\
&\leq 2(\|T_n\| + \|(T - T_n)T_n\|).
\end{aligned}$$

Hence for $n \geq n_0$, we have

$$\begin{aligned}
\|\boldsymbol{T}_n\| &\leq \max\Big\{1, \sum_{k=0}^{q-1} \|(T - T_n)^k T_n\|\Big\} \\
&\leq \max\{1, 2(\|T_n\| + \|(T - T_n)T_n\|)\}.
\end{aligned}$$

Since the sequence $(\|T_n\|)$ is bounded, it follows that the sequence $(\|\boldsymbol{T}_n\|)$ is also bounded.

Next, let $n \geq n_0$. We note that

$$\begin{aligned}
\|(\boldsymbol{T}_n - \boldsymbol{T})\boldsymbol{T}\| &\leq \|(T - T_n)^2\| + \sum_{k=2}^{q-1} \|(T - T_n)^k T_n\| \\
&\leq \|(T - T_n)^2\| + \|(T - T_n)^2\| \sum_{k=0}^{\infty} \|(T - T_n)^k T_n\| \\
&\leq \big[1 + 2(\|T_n\| + \|(T - T_n)T_n\|)\big] \|(T - T_n)^2\|.
\end{aligned}$$

Also, since $\|(T - T_n)^2\| \leq 1$, we have

$$\|(\boldsymbol{T}_n - \boldsymbol{T})\boldsymbol{T}_n\| = \|(T - T_n)^q T_n\|$$

$$\leq \begin{cases} \|(T - T_n)^2\|^{q/2}\|T_n\| & \text{if } q \text{ is even,} \\ \|(T - T_n)^2\|^{(q-1)/2}\|(T - T_n)T_n\| & \text{if } q \text{ is odd} \end{cases}$$

$$\leq \max\{\|T_n\|, \|(T_n - T)T_n\|\}\|(T - T_n)^2\|.$$

Since the sequence $(\|T_n\|)$ is bounded and $\|(T - T_n)^2\| \to 0$, it follows that $\|(\boldsymbol{T}_n - \boldsymbol{T})\boldsymbol{T}\| \to 0$ and $\|(\boldsymbol{T}_n - \boldsymbol{T})\boldsymbol{T}_n\| \to 0$. Thus $\boldsymbol{T}_n \xrightarrow{\nu} \boldsymbol{T}$. ∎

Consider now a spectral set Λ for T such that $0 \notin \Lambda$. By Proposition 3.9(b), Λ is a spectral set for $\boldsymbol{T}$ as well. We shall use the following notation throughout this section:

$$P := P(T, \Lambda), \quad M := \mathcal{R}(P), \quad \boldsymbol{P} = P(\boldsymbol{T}, \Lambda), \quad \boldsymbol{M} = \mathcal{R}(\boldsymbol{P}).$$

For $\mathrm{C} \in \mathcal{C}(T, \Lambda)$, we let

$$\ell(\mathrm{C}) := \text{ the length of C}, \qquad \delta(\mathrm{C}) := \min\{|z| : z \in \mathrm{C}\},$$

$$\alpha_1(\mathrm{C}) := \sup_{z \in \mathrm{C}} \|R(T, z)\|, \qquad \alpha_{1,q}(\mathrm{C}) := \sup_{z \in \mathrm{C}} \|R(\boldsymbol{T}, z)\|.$$

Since $\boldsymbol{T}_n \xrightarrow{\nu} \boldsymbol{T}$ (Proposition 3.11), the theory of spectral approximation developed in Chapter 2 becomes applicable for the operator $\boldsymbol{T}$ and the sequence $(\boldsymbol{T}_n)$.

Theorem 3.12
Let Λ be a spectral set for T such that $0 \notin \Lambda$, and consider $\mathrm{C} \in \mathcal{C}(T, \Lambda)$ with $0 \in \mathrm{ext}(\mathrm{C})$. Fix an integer $q \geq 2$. Assume that $(T_n - T) \xrightarrow{\nu} O$.

(a) *There is a positive integer $n_{0,q}$ such that $\mathrm{C} \subset \mathrm{re}(\boldsymbol{T}_n)$ for all $n \geq n_{0,q}$ and*

$$\alpha_{2,q}(\mathrm{C}) := \sup\{\|R(\boldsymbol{T}_n, z)\| : z \in \mathrm{C}, n \geq n_{0,q}\} < \infty.$$

Further, for each large n, if we let

$$\Lambda_n^{[q]} := \mathrm{sp}(\boldsymbol{T}_n) \cap \mathrm{int}(\mathrm{C}) \quad \textit{and} \quad \boldsymbol{P}_n := -\frac{1}{2\pi \mathrm{i}} \int_{\mathrm{C}} R(\boldsymbol{T}_n, z)dz,$$

then $\Lambda_n^{[q]}$ *and* $\boldsymbol{P}_n$ *do not depend on* $\mathrm{C} \in \mathcal{C}(T, \Lambda)$, *provided* $0 \in \mathrm{ext}(\mathrm{C})$.

(b) *Let* $\boldsymbol{M}_n = \mathcal{R}(\boldsymbol{P}_n)$ *for all large* n. *Then*

$$\|(\boldsymbol{T}_n - \boldsymbol{T})_{|\boldsymbol{M}_n}\| \leq \frac{\ell(\mathrm{C})\alpha_{2,q}(\mathrm{C})}{2\pi\delta(\mathrm{C})}\|(T_n - T)^q T_n\|$$

and

$$\begin{aligned}\|(\boldsymbol{T}_n - \boldsymbol{T})_{|\boldsymbol{M}}\| &\leq \frac{\ell(\mathrm{C})\alpha_1(\mathrm{C})}{2\pi[\delta(\mathrm{C})]^{q-1}}\|(T_n - T)^q_{\ |M}\| \\ &\leq \frac{[\ell(\mathrm{C})\alpha_1(\mathrm{C})]^2}{4\pi^2[\delta(\mathrm{C})]^q}\|(T_n - T)^q T\|.\end{aligned}$$

(c) *Suppose that* $\Lambda := \{\lambda_1, \ldots, \lambda_r\}$, *where each* λ_j *is a spectral value of* T *of finite type. If* m_j *is the algebraic multiplicity of* λ_j *and* $m := m_1 + \cdots + m_r$, *then for all large* n, $\operatorname{rank} \boldsymbol{P}_n = m$ *and*

$$\Lambda_n^{[q]} = \{\lambda_{n,1}^{[q]}, \ldots, \lambda_{n,r(n,q)}^{[q]}\},$$

where each $\lambda_{n,j}^{[q]}$ *is a spectral value of* $\boldsymbol{T}_n$ *of finite type. Further, if* $m_{n,j}^{[q]}$ *is the algebraic multiplicity of* $\lambda_{n,j}^{[q]}$ *for* $j = 1, \ldots, r(n,q)$, *then* $m_{n,1}^{[q]} + \cdots + m_{n,r(n,q)}^{[q]} = m$, *and the weighted arithmetic means*

$$\widehat{\lambda} := \frac{m_1\lambda_1 + \cdots + m_n\lambda_n}{m} \qquad \text{and}$$

$$\widehat{\lambda}_n^{[q]} := \frac{m_{n,1}^{[q]}\lambda_{n,1}^{[q]} + \cdots + m_{n,r(n,q)}^{[q]}\lambda_{n,r(n,q)}^{[q]}}{m}$$

satisfy

$$|\widehat{\lambda}_n^{[q]} - \widehat{\lambda}| \leq \frac{\ell(\mathrm{C})}{\pi}\min\{\alpha_{1,q}(\mathrm{C})\|(\boldsymbol{T}_n - \boldsymbol{T})_{|\boldsymbol{M}_n}\|, \alpha_{2,q}(\mathrm{C})\|(\boldsymbol{T}_n - \boldsymbol{T})_{|\boldsymbol{M}}\|\}.$$

Proof

(a) By Proposition 3.9(b), Λ is a spectral set for $\boldsymbol{T}$ and $\mathrm{C} \in \mathcal{C}(\boldsymbol{T}, \Lambda)$. Also, by Proposition 3.11, $\boldsymbol{T}_n \stackrel{\nu}{\to} \boldsymbol{T}$. Hence the desired conclusions follow from Theorem 2.6 and Proposition 2.9(a).

(b) By Proposition 2.9(b), we have for all large n,

$$\|(\boldsymbol{T}_n - \boldsymbol{T})_{|\boldsymbol{M}_n}\| \leq \|(\boldsymbol{T}_n - \boldsymbol{T})\boldsymbol{P}_n\| \leq \frac{\ell(\mathrm{C})\alpha_{2,q}(\mathrm{C})}{2\pi\delta(\mathrm{C})}\|(\boldsymbol{T}_n - \boldsymbol{T})\boldsymbol{T}_n\|,$$

where $\|(\boldsymbol{T}_n - \boldsymbol{T})\boldsymbol{T}_n\| = \|(T_n - T)^q T_n\|$, as stated in Proposition 3.11.

Let now $\boldsymbol{x} = [x_1, \ldots, x_q]^\top \in \boldsymbol{M}$. Then by Proposition 3.9(b), $x_1 \in M$, that is, $Px_1 = x_1$. Since P commutes with $R(z)$ for each $z \in \mathrm{C}$, we have

$$PJ_{q-1}x_1 = \left(-\frac{1}{2\pi \mathrm{i}} \int_{\mathrm{C}} R(z) \frac{dz}{z^{q-1}}\right) Px_1 = J_{q-1}x_1,$$

so that $J_{q-1}x_1 \in M$. Hence letting $\widetilde{T} := T_n$ in Proposition 3.10, we have

$$\begin{aligned} \|(\boldsymbol{T}_n - \boldsymbol{T})\boldsymbol{x}\|_\infty &= \|(\boldsymbol{T}_n - \boldsymbol{T})\boldsymbol{P}\boldsymbol{x}\|_\infty = \|(T - T_n)^q J_{q-1}x_1\| \\ &\le \|(T - T_n)^q_{|M}\| \, \|J_{q-1}x_1\|, \end{aligned}$$

where $\|J_{q-1}x_1\| \le \dfrac{\ell(\mathrm{C})\alpha_1(\mathrm{C})}{2\pi[\delta(\mathrm{C})]^{q-1}} \|x_1\|$. As $\|x_1\| \le \|\boldsymbol{x}\|_\infty$, we obtain

$$\|(\boldsymbol{T}_n - \boldsymbol{T})_{|\boldsymbol{M}}\| \le \frac{\ell(\mathrm{C})\alpha_1(\mathrm{C})}{2\pi[\delta(\mathrm{C})]^{q-1}} \|(T - T_n)^q_{|M}\|,$$

where

$$\|(T - T_n)^q_{|M}\| \le \|(T - T_n)^q P\| \le \frac{\ell(\mathrm{C})\alpha_1(\mathrm{C})}{2\pi\delta(\mathrm{C})} \|(T_n - T)^q T\|,$$

since $P = -\dfrac{T}{2\pi \mathrm{i}} \displaystyle\int_{\mathrm{C}} R(z) \frac{dz}{z}$.

(c) By Proposition 3.9(b), $\operatorname{rank} \boldsymbol{P} = \operatorname{rank} P = m$. Hence the desired conclusion follows from Theorem 2.18(a) and (b). ∎

The preceding theorem shows that by employing a spectral analysis of order $q \ge 2$, we may obtain error estimates for a spectral approximation in terms of $\|(\boldsymbol{T}_n - \boldsymbol{T})_{|\boldsymbol{M}}\| = O(\|(T_n - T)^q T\|)$ and $\|(\boldsymbol{T}_n - \boldsymbol{T})_{|\boldsymbol{M}_n}\| = O(\|(T_n - T)^q T_n\|)$.

If we increase the order of spectral analysis to $q + 1$ or to $q + 2$, then the error estimates improve in the following manner.

Let $T_n \xrightarrow{\mathrm{n}} T$. Since $\|(T_n - T)^{q+1}T\| \le \|T_n - T\| \, \|(T_n - T)^q T\|$ and $\|(T_n - T)^{q+1}T_n\| \le \|T_n - T\| \, \|(T_n - T)^q T_n\|$, we have $\|(T_n - T)^{q+1}T\| = o(\|(T_n - T)^q T\|)$ and $\|(T_n - T)^{q+1}T_n\| = o(\|(T_n - T)^q T_n\|)$. Thus the error estimates improve in a 'geometric' manner as q increases.

Let $(T_n - T) \xrightarrow{\nu} O$. Since $\|(T_n - T)^{q+2}T\| \le \|(T_n - T)^2\| \, \|(T_n - T)^q T\|$ and $\|(T_n - T)^{q+2}T_n\| \le \|(T_n - T)^2\| \, \|(T_n - T)^q T_n\|)$, we have $\|(T_n - T)^{q+2}T\| = o(\|(T_n - T)^q T\|)$ and $\|(T_n - T)^{q+2}T_n\| = o(\|(T_n -$

$T)^q T_n\|$). Thus the error estimates improve in a 'semigeometric' manner as q increases.

Theorem 4.12(b) describes a situation in which we can estimate the rate at which the norms $\|(T_n - T)^q T\|$ and $\|(T_n - T)^q T_n\|$ tend to zero.

3.2.3 Simple Eigenvalue and Cluster of Eigenvalues

We first prove an accelerated approximation result for a nonzero simple eigenvalue of T which is analogous to Theorem 2.17. Recall the notation $\alpha_{1,q}(\mathrm{C})$ and $\alpha_{2,q}(\mathrm{C})$ employed in Theorem 3.12.

Theorem 3.13
Let λ be a nonzero simple eigenvalue of T, ϵ such that $0 < \epsilon < \min\{|\lambda|, \mathrm{dist}(\lambda, \mathrm{sp}(T) \setminus \{\lambda\})\}$ and C_ϵ the positively oriented circle with center λ and radius ϵ. Assume that $(T_n - T) \xrightarrow{\nu} O$ and fix an integer $q \geq 2$.

(a) *There is a positive integer $n_{0,q}$ such that for each $n \geq n_{0,q}$, we have a unique $\lambda_n^{[q]} \in \mathrm{sp}(\boldsymbol{T}_n)$ satisfying $|\lambda_n^{[q]} - \lambda| < \epsilon$. Also, $\lambda_n^{[q]}$ is a simple eigenvalue of $\boldsymbol{T}_n$ for each $n \geq n_{0,q}$ and $\lambda_n^{[q]} \to \lambda$ as $n \to \infty$.*

(b) *Let φ be an eigenvector of T corresponding to λ and for $n \geq n_{0,q}$, let $\varphi_n^{[q]} \in X$ be the first component of an eigenvector $\boldsymbol{\varphi}_n$ of $\boldsymbol{T}_n$ corresponding to $\lambda_n^{[q]}$. Then for each large n, $P\varphi_n^{[q]}$ is an eigenvector of T corresponding to λ,*

$$|\lambda_n^{[q]} - \lambda| \leq 2\epsilon \min\left\{\frac{\alpha_{1,q}(\mathrm{C}_\epsilon)}{|\lambda| - \epsilon}\frac{\|(T - T_n)^q T_n \varphi_n^{[q]}\|}{\|\varphi_n^{[q]}\|}, \alpha_{2,q}(\mathrm{C}_\epsilon)\frac{\|(T - T_n)^q \varphi\|}{\|\varphi\|}\right\}$$

and

$$\begin{aligned}\|\varphi_n^{[q]} - P\varphi_n^{[q]}\| &\leq \frac{\epsilon\, \alpha_{1,q}(\mathrm{C}_\epsilon)}{\mathrm{dist}(\lambda_n^{[q]}, \mathrm{C}_\epsilon)}\frac{1}{|\lambda_n^{[q]}|^q}\|(T - T_n)^q T_n \varphi_n^{[q]}\| \\ &\leq \frac{2\,\alpha_{1,q}(\mathrm{C}_\epsilon)}{(|\lambda| - \epsilon)^q}\|(T - T_n)^q T_n \varphi_n^{[q]}\|.\end{aligned}$$

(c) *Let $\boldsymbol{\varphi} := \left[\varphi, \frac{\varphi}{\lambda}, \ldots, \frac{\varphi}{\lambda^{q-1}}\right]^\top$, $\boldsymbol{\varphi}_n := \left[\varphi_n^{[q]}, \frac{\varphi_n^{[q]}}{\lambda_n^{[q]}}, \ldots \frac{\varphi_n^{[q]}}{(\lambda_n^{[q]})^{q-1}}\right]^\top$ and $\boldsymbol{P}_n := P(\boldsymbol{T}_n, \{\lambda_n^{[q]}\})$ for $n \geq n_{0,q}$. Let $c_{n,q}$ be the complex number such that $\boldsymbol{P}_n\boldsymbol{\varphi} = c_{n,q}\boldsymbol{\varphi}_n$ and $\varphi_{(n)}^{[q]} := \varphi/c_{n,q}$. Then $\varphi_{(n)}^{[q]}$ is an eigenvector of T corresponding to λ and*

$$\frac{\|\varphi_n^{[q]} - \varphi_{(n)}^{[q]}\|}{\|\varphi_{(n)}^{[q]}\|} \leq \frac{\alpha_{2,q}(\mathrm{C}_\epsilon)}{|\lambda|^{q-1}}\frac{\|(T_n - T)^q \varphi\|}{\|\varphi\|} \leq \frac{\alpha_{2,q}(\mathrm{C}_\epsilon)}{|\lambda|^q}\|(T_n - T)^q T\|.$$

Proof

(a) Let $\Lambda = \{\lambda\}$ in Theorem 3.12. Since λ is a nonzero simple eigenvalue of T, that is, $m := \operatorname{rank} P(T, \Lambda) = 1$, the results follow immediately.

(b) Since $\boldsymbol{T}_n \xrightarrow{\nu} \boldsymbol{T}$ by Proposition 3.11 and since λ is a nonzero simple eigenvalue of $\boldsymbol{T}$ by Proposition 3.9(b), Theorem 2.17(b) shows that if $\boldsymbol{\varphi}_n$ is an eigenvector of $\boldsymbol{T}_n$ corresponding to its simple eigenvalue $\lambda_n^{[q]}$, then for each large n, $\boldsymbol{P}\boldsymbol{\varphi}_n$ is an eigenvector of $\boldsymbol{T}$ corresponding to λ,

$$|\lambda_n^{[q]} - \lambda| \leq 2\epsilon \min\left\{\alpha_{1,q}(\mathrm{C}_\epsilon)\frac{\|(\boldsymbol{T}_n - \boldsymbol{T})\boldsymbol{\varphi}_n\|_\infty}{\|\boldsymbol{\varphi}_n\|_\infty}, \alpha_{2,q}(\mathrm{C}_\epsilon)\frac{\|(\boldsymbol{T}_n - \boldsymbol{T})\boldsymbol{\varphi}\|_\infty}{\|\boldsymbol{\varphi}\|_\infty}\right\}$$

and

$$\|\boldsymbol{\varphi}_n - \boldsymbol{P}\boldsymbol{\varphi}_n\|_\infty \leq \frac{\epsilon\,\alpha_{1,q}(\mathrm{C}_\epsilon)}{\operatorname{dist}(\lambda_n^{[q]}, \mathrm{C}_\epsilon)}\|(\boldsymbol{T}_n - \boldsymbol{T})\boldsymbol{\varphi}_n\|_\infty.$$

Now $\boldsymbol{\varphi}_n = \left[\varphi_n^{[q]}, \frac{\varphi_n^{[q]}}{\lambda_n^{[q]}}, \ldots, \frac{\varphi_n^{[q]}}{(\lambda_n^{[q]})^{q-1}}\right]^\top$ for some $\varphi_n^{[q]} \in X$ and

$$\boldsymbol{P}\boldsymbol{\varphi}_n = [P\varphi_n^{[q]}, J_1\varphi_n^{[q]}, \ldots, J_{q-1}\varphi_n^{[q]}]^\top \neq \boldsymbol{0}.$$

Since $J_k x = J_k P x$ for all $x \in X$ and $k = 1, \ldots, q-1$, we see that for each large n, $P\varphi_n^{[q]} \neq 0$, and as $\mathcal{R}(P) = \mathcal{N}(T - \lambda I)$, $P\varphi_n^{[q]}$ is an eigenvector of T corresponding to λ.

Next,

$$(\boldsymbol{T}_n - \boldsymbol{T})\boldsymbol{\varphi}_n = \frac{(\boldsymbol{T}_n - \boldsymbol{T})\boldsymbol{T}_n\boldsymbol{\varphi}_n}{\lambda_n^{[q]}} = \frac{1}{\lambda_n^{[q]}}\left[-\frac{(T - T_n)^q T_n\varphi_n^{[q]}}{(\lambda_n^{[q]})^{q-1}}, 0, \ldots, 0\right]^\top,$$

so that $\|(\boldsymbol{T}_n - \boldsymbol{T})\boldsymbol{\varphi}_n\|_\infty = \|(T - T_n)^q T_n\varphi_n^{[q]}\|/|\lambda_n^{[q]}|^q$. Also, $\|\boldsymbol{\varphi}_n\|_\infty = \|\varphi_n^{[q]}\| \max\{1, 1/|\lambda_n^{[q]}|^{q-1}\}$. Hence

$$\frac{\|(\boldsymbol{T}_n - \boldsymbol{T})\boldsymbol{\varphi}_n\|_\infty}{\|\boldsymbol{\varphi}_n\|_\infty} = \frac{\|(T - T_n)^q T_n\varphi_n^{[q]}\|}{|\lambda_n^{[q]}| \max\{|\lambda_n^{[q]}|^{q-1}, 1\}\|\varphi_n^{[q]}\|} \leq \frac{\|(T - T_n)^q T_n\varphi_n^{[q]}\|}{|\lambda_n^{[q]}|\ \|\varphi_n^{[q]}\|}.$$

Since $|\lambda_n^{[q]}| \geq |\lambda| - |\lambda_n^{[q]} - \lambda| > |\lambda| - \epsilon$, it follows that

$$|\lambda_n^{[q]} - \lambda| \leq \frac{2\epsilon\,\alpha_{1,q}(\mathrm{C}_\epsilon)}{|\lambda| - \epsilon}\frac{\|(T - T_n)^q T_n\varphi_n^{[q]}\|}{\|\varphi_n^{[q]}\|}.$$

Also, for all large n, since $\lambda_n^{[q]} \in \operatorname{int}(\mathrm{C}_{\epsilon/2})$ by Theorem 3.12(a), we have $\operatorname{dist}(\lambda_n^{[q]}, \mathrm{C}_\epsilon) \geq \epsilon/2$ and hence

$$\begin{aligned}\|\varphi_n^{[q]} - P\varphi_n^{[q]}\| \leq \|\boldsymbol{\varphi}_n - \boldsymbol{P}\boldsymbol{\varphi}_n\|_\infty &\leq \frac{\epsilon\,\alpha_{1,q}(\mathrm{C}_\epsilon)}{\operatorname{dist}(\lambda_n^{[q]}, \mathrm{C}_\epsilon)}\frac{\|(T - T_n)^q T_n\varphi_n^{[q]}\|}{|\lambda_n^{[q]}|^q}\\ &\leq \frac{2\,\alpha_{1,q}(\mathrm{C}_\epsilon)}{(|\lambda| - \epsilon)^q}\|(T - T_n)^q T_n\varphi_n^{[q]}\|,\end{aligned}$$

as desired. Finally, letting $\widetilde{T} = T_n$ in Proposition 3.10, we have

$$(\boldsymbol{T}_n - \boldsymbol{T})\boldsymbol{\varphi} = \left[-\frac{(T - T_n)^q\varphi}{\lambda^{q-1}}, 0, \ldots, 0\right]^\top .$$

Hence

$$\begin{aligned}\frac{\|(\boldsymbol{T}_n - \boldsymbol{T})\boldsymbol{\varphi}\|_\infty}{\|\boldsymbol{\varphi}\|_\infty} &= \frac{\|(T - T_n)^q\varphi\|}{|\lambda|^{q-1}\|\varphi\| \max\{1, 1/|\lambda|^{q-1}\}} \\ &= \frac{\|(T - T_n)^q\varphi\|}{\max\{|\lambda|^{q-1}, 1\}\|\varphi\|} \leq \frac{\|(T - T_n)^q\varphi\|}{\|\varphi\|}\end{aligned}$$

and

$$|\lambda_n^{[q]} - \lambda| \leq 2\epsilon\, \alpha_{2,q}(\mathrm{C}_\epsilon) \frac{\|(T - T_n)^q\varphi\|}{\|\varphi\|}.$$

(c) As in Theorem 2.17(b), for each large n, $\boldsymbol{P}_n\boldsymbol{\varphi}$ is an eigenvector of $\boldsymbol{T}_n$ corresponding to $\lambda_n^{[q]}$; and since $\mathcal{R}(\boldsymbol{P}_n) = \mathcal{N}(\boldsymbol{T} - \lambda_n^{[q]}\mathbf{I})$, there is a nonzero complex number $c_{n,q}$ such that $\boldsymbol{P}_n\boldsymbol{\varphi}_n = c_{n,q}\boldsymbol{\varphi}_n$. Let $\boldsymbol{\varphi}_{(n)} := \boldsymbol{\varphi}/c_{n,q}$. Then since $\varphi_n^{[q]} - \varphi_{(n)}^{[q]}$ is the first component of $\boldsymbol{\varphi}_n - \boldsymbol{\varphi}_{(n)}$, we have

$$\|\varphi_n^{[q]} - \varphi_{(n)}^{[q]}\| \leq \|\boldsymbol{\varphi}_n - \boldsymbol{\varphi}_{(n)}\|_\infty .$$

Also,

$$\|\boldsymbol{\varphi}_{(n)}\|_\infty = \frac{\|\boldsymbol{\varphi}\|_\infty}{|c_{n,q}|} = \frac{\|\varphi\|}{|c_{n,q}|} \max\left\{1, \frac{1}{|\lambda|^{q-1}}\right\} = \|\varphi_{(n)}^{[q]}\| \max\left\{1, \frac{1}{|\lambda|^{q-1}}\right\}.$$

Hence

$$\frac{\|\varphi_n^{[q]} - \varphi_{(n)}^{[q]}\|}{\|\varphi_n^{[q]}\|} \leq \frac{\|\boldsymbol{\varphi}_n - \boldsymbol{\varphi}_{(n)}\|_\infty}{\|\boldsymbol{\varphi}_{(n)}\|_\infty |\lambda|^{q-1}} \max\left\{|\lambda|^{q-1}, 1\right\}.$$

Now by Theorem 2.17(c),

$$\begin{aligned}\frac{\|\boldsymbol{\varphi}_n - \boldsymbol{\varphi}_{(n)}\|_\infty}{\|\boldsymbol{\varphi}_{(n)}\|_\infty} &\leq \alpha_{2,q}(\mathrm{C}_\epsilon) \frac{\|(\boldsymbol{T}_n - \boldsymbol{T})\boldsymbol{\varphi}\|_\infty}{\|\boldsymbol{\varphi}\|_\infty} \\ &= \alpha_{2,q}(\mathrm{C}_\epsilon) \frac{\|(T - T_n)^q\varphi\|}{\max\{|\lambda|^{q-1}, 1\}\|\varphi\|},\end{aligned}$$

as we have seen in (b) above. Thus

$$\begin{aligned}\frac{\|\varphi_n^{[q]} - \varphi_{(n)}^{[q]}\|}{\|\varphi_{(n)}^{[q]}\|} &\leq \alpha_{2,q}(\mathrm{C}_\epsilon) \frac{\|(T - T_n)^q\varphi\|}{|\lambda|^{q-1}\|\varphi\|} = \alpha_{2,q}(\mathrm{C}_\epsilon) \frac{\|(T - T_n)^q T\varphi\|}{|\lambda|^q\|\varphi\|} \\ &\leq \frac{\alpha_{2,q}(\mathrm{C}_\epsilon)}{|\lambda|^q} \|(T - T_n)^q T\|,\end{aligned}$$

as desired. ∎

In order to extend the preceding result to the case of a nonzero multiple eigenvalue of T (and more generally, to the case of a finite number of such eigenvalues), we consider the linear space of all rows of a given length with entries in $\boldsymbol{X}^{[q]}$, where q is an integer ≥ 2. We shall generalize Theorem 3.13 and obtain an accelerated approximation result for a spectral set of finite type for T which can be compared with Proposition 2.21.

Let m be a fixed positive integer. In Subsection 1.4.2, we have introduced the product space $X^{1\times m}$. If $\underline{X} := X^{1\times m}$ and $\underline{x} = [x_1, \ldots, x_m] \in \underline{X}$, we had considered $\|\underline{x}\|_\infty = \max\{\|x_j\|_\infty : 1 \leq j \leq m\}$. In analogy with these considerations, let

$$\underline{\boldsymbol{X}}^{[q]} := (\boldsymbol{X}^{[q]})^{1\times m} = \{[\boldsymbol{x}_1, \ldots, \boldsymbol{x}_m] : \boldsymbol{x}_j \in \boldsymbol{X}^{[q]}, 1 \leq j \leq m\},$$

and for $\underline{\boldsymbol{x}} = [\boldsymbol{x}_1, \ldots, \boldsymbol{x}_m] \in \underline{\boldsymbol{X}}^{[q]}$,

$$\|\underline{\boldsymbol{x}}\|_\infty := \max\{\|\boldsymbol{x}_j\|_\infty : 1 \leq j \leq m\}.$$

Then $\underline{\boldsymbol{X}}^{[q]}$ is a Banach space over $\mathbb{C}$. Note that $\underline{\boldsymbol{X}}^{[q]} = X^{q\times m}$. Whenever we fix an integer q, we write $\underline{\boldsymbol{X}}$ for $\underline{\boldsymbol{X}}^{[q]}$, just as we wrote $\boldsymbol{X}$ for $\boldsymbol{X}^{[q]}$.

If $\boldsymbol{F} : \boldsymbol{X} \to \boldsymbol{X}$, then its natural extension to $\underline{\boldsymbol{X}}$ is denoted by $\underline{\boldsymbol{F}}$. Thus $\underline{\boldsymbol{F}}[\boldsymbol{x}_1, \ldots, \boldsymbol{x}_m] = [\boldsymbol{F}\boldsymbol{x}_1, \ldots, \boldsymbol{F}\boldsymbol{x}_m]$ for $\underline{\boldsymbol{x}} = [\boldsymbol{x}_1, \ldots, \boldsymbol{x}_m] \in \underline{\boldsymbol{X}}$.

Identifying $\boldsymbol{X}^{1\times m} = (X^{q\times 1})^{1\times m}$ with $(X^{1\times m})^{q\times 1}$, we note that $\underline{\boldsymbol{X}}$ consists of all $\underline{\boldsymbol{x}} = \begin{bmatrix} \underline{x}_1 \\ \vdots \\ \underline{x}_q \end{bmatrix}$, where $\underline{x}_i \in \underline{X}$ for $1 \leq i \leq q$, and

$$\|\underline{\boldsymbol{x}}\|_\infty = \max\{\|\underline{x}_i\|_\infty : 1 \leq i \leq q\}.$$

Let $\boldsymbol{F} : \boldsymbol{X} \to \boldsymbol{X}$ be given by

$$\boldsymbol{F} = \begin{bmatrix} F_{1,1} & \cdots & F_{1,q} \\ \vdots & & \vdots \\ F_{q,1} & \cdots & F_{q,q} \end{bmatrix},$$

where $F_{i,k} : X \to X$ for $i, k = 1, \ldots, q$. Consider $\underline{\boldsymbol{x}} = [\boldsymbol{x}_1, \ldots, \boldsymbol{x}_m] = \begin{bmatrix} \underline{x}_1 \\ \vdots \\ \underline{x}_q \end{bmatrix} = [x_{k,\ell}]$, $1 \leq k \leq q$, $1 \leq \ell \leq m$, where $\boldsymbol{x}_j = [x_{1,j}, \ldots, x_{q,j}]^\top$ for

$j = 1, \ldots, m$ and $\underline{x}_i = [x_{i,1}, \ldots, x_{i,m}]$ for $i = 1, \ldots, q$. Then we have

$$\underline{\boldsymbol{F}}\,\underline{\boldsymbol{x}} = \begin{bmatrix} \sum_{k=1}^{q} F_{1,k}x_{k,1} & \cdots & \sum_{k=1}^{q} F_{1,k}x_{k,m} \\ \vdots & & \vdots \\ \sum_{k=1}^{q} F_{q,k}x_{k,1} & \cdots & \sum_{k=1}^{q} F_{q,k}x_{k,m} \end{bmatrix} = \begin{bmatrix} \underline{F_{1,1}} & \cdots & \underline{F_{1,q}} \\ \vdots & & \vdots \\ \underline{F_{q,1}} & \cdots & \underline{F_{q,q}} \end{bmatrix} \begin{bmatrix} \underline{x}_1 \\ \vdots \\ \underline{x}_q \end{bmatrix},$$

where $\underline{F_{i,k}} : \underline{X} \to \underline{X}$ is the natural extension of $F_{i,k}$.

In particular, if $T \in \mathrm{BL}(X)$, and

$$\boldsymbol{T} := \begin{bmatrix} T & O & \cdots & \cdots & O \\ I & O & \cdots & \cdots & O \\ O & I & O & \cdots & O \\ \vdots & \ddots & \ddots & \ddots & \vdots \\ O & \cdots & O & I & O \end{bmatrix}$$

as before, then

$$\underline{\boldsymbol{T}}\,\underline{\boldsymbol{x}} = \begin{bmatrix} \underline{T} & \underline{O} & \cdots & \cdots & \underline{O} \\ \underline{I} & \underline{O} & \cdots & \cdots & \underline{O} \\ \underline{O} & \underline{I} & \underline{O} & \cdots & \underline{O} \\ \vdots & \ddots & \ddots & \ddots & \vdots \\ \underline{O} & \cdots & \underline{O} & \underline{I} & \underline{O} \end{bmatrix} \begin{bmatrix} \underline{x}_1 \\ \vdots \\ \vdots \\ \underline{x}_q \end{bmatrix} = \begin{bmatrix} \underline{T}\,\underline{x}_1 \\ \underline{x}_1 \\ \vdots \\ \vdots \\ \underline{x}_{q-1} \end{bmatrix}.$$

If $\mathsf{Z} \in \mathbb{C}^{m \times m}$, we define

$$\underline{\boldsymbol{x}}\,\mathsf{Z} := \begin{bmatrix} \underline{x}_1\mathsf{Z} \\ \vdots \\ \underline{x}_q\mathsf{Z} \end{bmatrix}.$$

Let Λ be a spectral set for T such that $0 \notin \Lambda$, and let $\mathrm{C} \in \mathcal{C}(T, \Lambda)$ such that $0 \in \mathrm{ext}(\mathrm{C})$. Consider the map $J = J(T, \Lambda)$ from X to $\boldsymbol{X}$ introduced just before stating Proposition 3.9: For $x \in X$,

$$Jx = [J_0x, J_1x, \ldots, J_{q-1}x]^\top,$$

where $J_0 = P := P(T, \Lambda)$ and $J_k x = \dfrac{-1}{2\pi \mathrm{i}} \displaystyle\int_C R(T, z) \dfrac{dz}{z^k}$, $k = 0, \ldots, q-1$. Let $\underline{J} : \underline{X} \to \underline{\boldsymbol{X}}$ denote the natural extension of J to $\underline{X}$. Thus $\underline{J}\,\underline{x} = [Jx_1, \ldots, Jx_m]$ for $\underline{x} = [x_1, \ldots, x_m] \in \underline{X}$. Clearly,

$$\underline{J}\,\underline{x} = \begin{bmatrix} Px_1 & \cdots & Px_m \\ \vdots & & \vdots \\ J_{q-1}x_1 & \cdots & J_{q-1}x_m \end{bmatrix} = \begin{bmatrix} \underline{P}\,\underline{x} \\ \vdots \\ \underline{J_{q-1}}\,\underline{x} \end{bmatrix}.$$

For $\mathsf{Z} \in \mathbb{C}^{m \times m}$, we have

$$\underline{J}(\underline{x}\mathsf{Z}) = \begin{bmatrix} \underline{P}(\underline{x}\mathsf{Z}) \\ \vdots \\ \underline{J_{q-1}}(\underline{x}\mathsf{Z}) \end{bmatrix} = \begin{bmatrix} (\underline{P}\,\underline{x})\mathsf{Z} \\ \vdots \\ (\underline{J_{q-1}}\,\underline{x})\mathsf{Z} \end{bmatrix} = (\underline{J}\,\underline{x})\mathsf{Z}.$$

Proposition 3.14
Consider $T \in \mathrm{BL}(X)$ and $\mathsf{Z} \in \mathbb{C}^{m \times m}$.

(a) *Let Λ be a spectral set for T such that $0 \notin \Lambda$ and $J := J(T, \Lambda)$. Then*

$$\underline{\boldsymbol{T}}\,\underline{J} = \underline{J}\,\underline{T}.$$

In particular, if $\underline{x} \in \underline{X}$ and $\underline{T}\,\underline{x} = \underline{x}\mathsf{Z}$, then

$$\underline{\boldsymbol{T}}(\underline{J}\,\underline{x}) = (\underline{J}\,\underline{x})\mathsf{Z}.$$

(b) *Let $\underline{\boldsymbol{x}} = \begin{bmatrix} \underline{x}_1 \\ \vdots \\ \underline{x}_q \end{bmatrix} \in \underline{\boldsymbol{X}}$. Then $\underline{\boldsymbol{P}}\,\underline{\boldsymbol{x}} = \underline{J}\,\underline{x}_1$. Also, $\underline{\boldsymbol{T}}\,\underline{\boldsymbol{x}} = \underline{\boldsymbol{x}}\mathsf{Z}$ if and only if $\underline{T}\,\underline{x}_1 = \underline{x}_1\mathsf{Z}$ and $\underline{x}_{k+1}\mathsf{Z}^k = \underline{x}_1$ for $k = 1, \ldots, q-1$.*

Proof

(a) Let $\underline{x} \in \underline{X}$. Since $TP = PT$ and $J_k T = J_{k-1}$ for $k = 1, \ldots, q-1$, we obtain

$$\underline{\boldsymbol{T}}\,\underline{J}\,\underline{x} = \underline{\boldsymbol{T}} \begin{bmatrix} \underline{P}\,\underline{x} \\ \underline{J_1}\,\underline{x} \\ \vdots \\ \underline{J_{q-1}}\,\underline{x} \end{bmatrix} = \begin{bmatrix} \underline{T}\,\underline{P}\,\underline{x} \\ \underline{P}\,\underline{x} \\ \vdots \\ \underline{J_{q-2}}\,\underline{x} \end{bmatrix} = \begin{bmatrix} \underline{P}\,\underline{T}\,\underline{x} \\ \underline{J_1}\,\underline{T}\,\underline{x} \\ \vdots \\ \underline{J_{q-1}}\,\underline{T}\,\underline{x} \end{bmatrix} = \underline{J}\,\underline{T}\,\underline{x}.$$

Thus $\underline{\boldsymbol{T}}\,\underline{J} = \underline{J}\,\underline{T}$. If $\underline{T}\,\underline{x} = \underline{x}\,\mathsf{Z}$, then

$$\underline{\boldsymbol{T}}(\underline{J}\,\underline{x}) = \underline{J}(\underline{T}\,\underline{x}) = \underline{J}(\underline{x}\,\mathsf{Z}) = (\underline{J}\,\underline{x})\mathsf{Z}.$$

(b) Recalling the matrix representation of $\boldsymbol{P}$ given in Proposition 3.9(b), we have

$$\underline{\boldsymbol{P}}\,\underline{\boldsymbol{x}} = \begin{bmatrix} \underline{P} & \underline{O} \cdots \underline{O} \\ \underline{J_1} & \underline{O} \cdots \underline{O} \\ \vdots & \vdots \quad\; \vdots \\ \underline{J_{q-1}} & \underline{O} \cdots \underline{O} \end{bmatrix} \begin{bmatrix} \underline{x}_1 \\ \vdots \\ \underline{x}_q \end{bmatrix} = \begin{bmatrix} \underline{P}\,\underline{x}_1 \\ \underline{J_1}\,\underline{x}_1 \\ \vdots \\ \underline{J_{q-1}}\,\underline{x}_1 \end{bmatrix} = \underline{J}\,\underline{x}_1.$$

Also, we note that

$$\underline{\boldsymbol{T}}\,\underline{\boldsymbol{x}} = \begin{bmatrix} \underline{T}\,\underline{x}_1 \\ \underline{x}_1 \\ \vdots \\ \underline{x}_{q-1} \end{bmatrix} \quad \text{and} \quad \underline{\boldsymbol{x}}\,\mathsf{Z} = \begin{bmatrix} \underline{x}_1\mathsf{Z} \\ \underline{x}_2\mathsf{Z} \\ \vdots \\ \underline{x}_q\mathsf{Z} \end{bmatrix}.$$

Hence $\underline{\boldsymbol{T}}\,\underline{\boldsymbol{x}} = \underline{\boldsymbol{x}}\,\mathsf{Z}$ if and only if $\underline{T}\,\underline{x}_1 = \underline{x}_1\mathsf{Z}$, $\underline{x}_1 = \underline{x}_2\mathsf{Z}, \ldots, \underline{x}_{q-1} = \underline{x}_q\mathsf{Z}$, that is, $\underline{T}\,\underline{x}_1 = \underline{x}_1\mathsf{Z}$ and $\underline{x}_{k+1}\mathsf{Z}^k = \underline{x}_k\mathsf{Z}^{k-1} = \cdots = \underline{x}_1$ for $k = 1, \ldots, q-1$. ∎

Let Λ be a spectral set of finite type for T such that $0 \notin \Lambda$, and let the rank of the associated spectral projection P be m. Let $\underline{\varphi} := [\varphi_1, \ldots, \varphi_m]$ form an ordered basis for the spectral subspace $M := \mathcal{R}(P)$. Then $\underline{T}\,\underline{\varphi} = \underline{\varphi}\,\Theta$ for some $\Theta \in \mathbb{C}^{m\times m}$. Since $\mathrm{sp}(\Theta) = \Lambda$ and $0 \notin \Lambda$, we see that the matrix Θ is nonsingular.

Fix an integer $q \geq 2$. By Proposition 3.9(b), Λ is a spectral set for $\boldsymbol{T}$, the rank of the associated spectral projection $\boldsymbol{P}$ is m, and

$$\underline{\boldsymbol{\varphi}} := \underline{J}\,\underline{\varphi} = [J\varphi_1, \ldots, J\varphi_m] = \begin{bmatrix} \underline{P}\,\underline{\varphi} \\ \underline{J_1}\,\underline{\varphi} \\ \vdots \\ \underline{J_{q-1}}\,\underline{\varphi} \end{bmatrix}$$

forms an ordered basis for the associated spectral subspace $\boldsymbol{M} := \mathcal{R}(\boldsymbol{P})$. Also, by Proposition 3.14(a),

$$\underline{\boldsymbol{T}}\,\underline{\boldsymbol{\varphi}} = \underline{\boldsymbol{T}}(\underline{J}\,\underline{\varphi}) = (\underline{J}\,\underline{\varphi})\Theta = \underline{\boldsymbol{\varphi}}\,\Theta,$$

and by Proposition 3.14(b), $\underline{J_k}\ \underline{\varphi}\,\Theta^k = \underline{P}\ \underline{\varphi} = \underline{\varphi}$ for $k = 1, \ldots, q-1$. Thus

$$\underline{\boldsymbol{\varphi}} = \begin{bmatrix} \underline{\varphi} \\ \underline{\varphi}\,\Theta^{-1} \\ \vdots \\ \underline{\varphi}\,\Theta^{-q+1} \end{bmatrix}.$$

Assume that $(T_n - T) \overset{\nu}{\to} O$ and consider $\mathrm{C} \in \mathcal{C}(T, \Lambda)$ such that $0 \in \mathrm{ext}(\mathrm{C})$. Then by Theorem 3.12(c), there is a positive integer $n_{0,q}$ such that for each $n \geq n_{0,q}$,

$$\Lambda_n^{[q]} := \mathrm{sp}(\boldsymbol{T}_n) \cap \mathrm{int}(\mathrm{C})$$

is a spectral set for $\boldsymbol{T}_n$ and the associated spectral projection $\boldsymbol{P}_n$ is of rank m. Let $\underline{\boldsymbol{\varphi}_n}$ form a basis for the associated spectral subspace $\boldsymbol{M}_n := \mathcal{R}(\boldsymbol{P}_n)$. Then

$$\underline{\boldsymbol{T}_n}\ \underline{\boldsymbol{\varphi}_n} = \underline{\boldsymbol{\varphi}_n}\,\Theta_n^{[q]}$$

for some $\Theta_n^{[q]} \in \mathbb{C}^{m \times m}$. Since $\mathrm{sp}(\Theta_n^{[q]}) = \Lambda_n^{[q]} \subset \mathrm{int}(\mathrm{C})$ and $0 \in \mathrm{ext}(\mathrm{C})$, we see that the matrix $\Theta_n^{[q]}$ is nonsingular. Let $\underline{\boldsymbol{\varphi}_n} = \begin{bmatrix} \underline{\varphi_{n,1}^{[q]}} \\ \vdots \\ \underline{\varphi_{n,q}^{[q]}} \end{bmatrix}$.

Since

$$\underline{\boldsymbol{T}_n}\ \underline{\boldsymbol{\varphi}_n} = \begin{bmatrix} \underline{T_n} & (\underline{T} - \underline{T_n})\underline{T_n} & \cdots & \cdots & (\underline{T} - \underline{T_n})^{q-1}\underline{T_n} \\ \underline{I} & \underline{O} & \cdots & \cdots & \underline{O} \\ \underline{O} & \underline{I} & \underline{O} & \cdots & \underline{O} \\ \vdots & \ddots & \ddots & \ddots & \vdots \\ \underline{O} & \cdots & \underline{O} & \underline{I} & \underline{O} \end{bmatrix} \begin{bmatrix} \underline{\varphi_{n,1}^{[q]}} \\ \vdots \\ \vdots \\ \vdots \\ \underline{\varphi_{n,q}^{[q]}} \end{bmatrix}$$

and

$$\underline{\boldsymbol{\varphi}_n}\,\Theta_n^{[q]} = \begin{bmatrix} \underline{\varphi_{n,1}^{[q]}}\,\Theta_n^{[q]} \\ \vdots \\ \underline{\varphi_{n,q}^{[q]}}\,\Theta_n^{[q]} \end{bmatrix},$$

we have $\underline{\varphi_{n,1}^{[q]}} = \underline{\varphi_{n,2}^{[q]}}\,\Theta_n^{[q]}$, $\underline{\varphi_{n,2}^{[q]}} = \underline{\varphi_{n,3}^{[q]}}\,\Theta_n^{[q]}, \ldots, \underline{\varphi_{n,q-1}^{[q]}} = \underline{\varphi_{n,q}^{[q]}}\,\Theta_n^{[q]}$. Hence $\underline{\varphi_{n,k+1}^{[q]}} = \underline{\varphi_{n,1}^{[q]}}\,(\Theta_n^{[q]})^{-k}$ for $k = 1, \ldots, q-1$. Write $\underline{\varphi_n^{[q]}} := \underline{\varphi_{n,1}^{[q]}}$ in order to simplify the notation slightly. Then

$$\underline{\boldsymbol{\varphi}_n} = \begin{bmatrix} \underline{\varphi_n^{[q]}} \\ \underline{\varphi_n^{[q]}}\,(\Theta_n^{[q]})^{-1} \\ \vdots \\ \underline{\varphi_n^{[q]}}\,(\Theta_n^{[q]})^{-q+1} \end{bmatrix}.$$

As in Section 2.4, consider

$$\beta_1(\mathrm{C}) := \sup\{\|R(\Theta, z)\| : z \in \mathrm{C}\},$$
$$\beta_{2,n,q}(\mathrm{C}) := \sup\{\|R(\Theta_n^{[q]}, z)\| : z \in \mathrm{C}\}.$$

We now obtain improved error estimates for approximations of the ordered basis $\underline{\varphi}$ for M.

Theorem 3.15

Under the hypotheses of Theorem 3.12(c) and with the notation stated above, we have the following:

(a) *For each large n, $\underline{P}\,\underline{\varphi_n^{[q]}}$ forms an ordered basis for the spectral subspace M of T and*

$$\|\,\underline{\varphi_n^{[q]}} - \underline{P}\,\underline{\varphi_n^{[q]}}\,\|_\infty \leq \frac{\ell(\mathrm{C})}{2\pi}\alpha_{1,q}(\mathrm{C})\beta_{2,n,q}(\mathrm{C})\|(\Theta_n^{[q]})^{-1}\|_1^{q-1}\|(\underline{T} - \underline{T_n})^q\,\underline{T_n}\,\underline{\varphi_n^{[q]}}\,(\Theta_n^{[q]})^{-1}\|_\infty.$$

If the sequence $\|\,\underline{\varphi_n^{[q]}}\,\|_\infty$ is bounded, then

$$\|(\underline{T} - \underline{T_n})^q\,\underline{T_n}\,\underline{\varphi_n^{[q]}}\,\|_\infty \to 0.$$

Let $\underline{\boldsymbol{\varphi}_n} = [\boldsymbol{\varphi}_{n,1}, \ldots, \boldsymbol{\varphi}_{n,m}]$ and

$$\delta_{n,q} := \min_{j=1,\ldots,m} \operatorname{dist}(\boldsymbol{\varphi}_{n,j}, \operatorname{span}\{\boldsymbol{\varphi}_{n,i} : i \neq j, i = 1, \ldots, m\}).$$

If the sequence $\|\,\underline{\boldsymbol{\varphi}_n}\,\|_\infty$ is bounded and the sequence $(\delta_{n,q})$ is bounded away from zero, then the sequences $(\beta_{2,n,q}(\mathrm{C}))$ and $(\|(\Theta_n^{[q]})^{-1}\|_1)$ are bounded, and in particular, $\|\,\underline{\varphi_n^{[q]}} - \underline{P}\,\underline{\varphi_n^{[q]}}\,\|_\infty = O(\|(T - T_n)^q T_n\|)$ as $n \to \infty$.

(b) *For each large n, $\boldsymbol{P}_n\boldsymbol{\varphi}$ forms an ordered basis for $\boldsymbol{M}_n$ and $\underline{\boldsymbol{P}_n}\ \underline{\boldsymbol{\varphi}} = \underline{\boldsymbol{\varphi}_n}\,\mathsf{C}_{n,q}$ for some nonsingular $m\times m$ matrix $\mathsf{C}_{n,q}$. If*

$$\underline{\varphi^{[q]}_{(n)}} := \underline{\varphi}\,\mathsf{C}^{-1}_{n,q},$$

then $\underline{\varphi^{[q]}_{(n)}}$ forms an ordered basis for M and

$$\frac{\|\,\underline{\varphi^{[q]}_n} - \underline{\varphi^{[q]}_{(n)}}\,\|_\infty}{\|\,\underline{\varphi^{[q]}_{(n)}}\,\|_\infty} \le \frac{\ell(\mathrm{C})}{2\pi}\frac{m}{\delta_q}\beta_1(\mathrm{C})\alpha_{2,q}(\mathrm{C})\gamma_{n,q}\|(\underline{T} - \underline{T_n})^q\,\underline{\varphi}\,\Theta^{-q+1}\|_\infty,$$

where

$$\delta_q := \min_{j=1,\dots,m} \operatorname{dist}(J\varphi_j, \operatorname{span}\{J\varphi_i\ :\ i \ne j, i = 1,\dots,m\}),$$

$$\gamma_{n,q} := \max\left\{1, \frac{\|\,\underline{J_1}\ \underline{\varphi^{[q]}_{(n)}}\,\|_\infty}{\|\,\underline{\varphi^{[q]}_{(n)}}\,\|_\infty}, \dots, \frac{\|\,\underline{J_{q-1}}\ \underline{\varphi^{[q]}_{(n)}}\,\|_\infty}{\|\,\underline{\varphi^{[q]}_{(n)}}\,\|_\infty}\right\}$$
$$\le \max\{\|\mathrm{C}_{n,q}\Theta^{-k}\mathrm{C}^{-1}_{n,q}\|_1\ :\ k = 0,\dots,q-1\}.$$

In particular,

$$\frac{\|\,\underline{\varphi^{[q]}_n} - \underline{\varphi^{[q]}_{(n)}}\,\|_\infty}{\|\,\underline{\varphi^{[q]}_{(n)}}\,\|_\infty} = O(\|(T - T_n)^q T\|) \text{ as } n \to \infty.$$

Proof

(a) Since $\boldsymbol{T}_n \overset{\nu}{\to} \boldsymbol{T}$ by Proposition 3.11, Theorem 2.18(c) shows that for each large n, $\underline{\boldsymbol{P}\ \boldsymbol{\varphi}_n}$ forms an ordered basis for $\boldsymbol{M}$. Now by Proposition 3.14(b), we have

$$\underline{\boldsymbol{P}}\ \underline{\boldsymbol{\varphi}_n} = \underline{J}\ \underline{\varphi^{[q]}_n} = \begin{bmatrix} \underline{P}\ \underline{\varphi^{[q]}_n} \\ \underline{J_1}\ \underline{\varphi^{[q]}_n} \\ \vdots \\ \underline{J_{q-1}}\ \underline{\varphi^{[q]}_n} \end{bmatrix}.$$

We show that $\underline{P}\ \underline{\varphi^{[q]}_n}$ forms an ordered basis for the m-dimensional subspace $M := \mathcal{R}(P)$. Let $\underline{P}\ \underline{\varphi^{[q]}_n} = [x_1,\dots,x_m]$ and $c_1x_1+\dots+c_mx_m = 0$ for some $c_1,\dots,c_m$ in $\mathbb{C}$. Then for each $k = 1,\dots,q-1$, $\underline{J_k}\ \underline{\varphi^{[q]}_n} = \underline{J_k}\ \underline{P}\ \underline{\varphi^{[q]}_n} = [J_kx_1,\dots,J_kx_m]$, so that $\underline{\boldsymbol{P}}\ \underline{\boldsymbol{\varphi}_n} = \begin{bmatrix} x_1 & \cdots & x_m \\ J_1x_1 & \cdots & J_1x_m \\ \vdots & & \vdots \\ J_{q-1}x_1 & \cdots & J_{q-1}x_m \end{bmatrix}.$

Let $\boldsymbol{x}_j := [x_j, J_1x_j, \ldots, J_{q-1}x_j]^\top$ for $j = 1, \ldots, m$. Then $c_1\boldsymbol{x}_1 + \cdots + c_m\boldsymbol{x}_m = [0, J_10, \ldots, J_{q-1}0]^\top = \mathbf{0}$. Since $[\boldsymbol{x}_1, \ldots, \boldsymbol{x}_m]$ forms a basis for $\boldsymbol{M}$, we have $c_1 = \cdots = c_m = 0$, as required.

Applying Proposition 2.21 to $\boldsymbol{T}$ and $\boldsymbol{T}_n$ in place of T and T_n, we obtain

$$\|\underline{\boldsymbol{\varphi}_n} - \underline{\boldsymbol{P}}\,\underline{\boldsymbol{\varphi}_n}\|_\infty \leq \frac{\ell(\mathrm{C})}{2\pi}\alpha_{1,q}(\mathrm{C})\beta_{2,n,q}(\mathrm{C})\|(\underline{\boldsymbol{T}_n} - \underline{\boldsymbol{T}})\,\underline{\boldsymbol{\varphi}_n}\|_\infty.$$

Clearly,

$$\|\underline{\varphi_n^{[q]}} - \underline{P}\,\underline{\varphi_n^{[q]}}\|_\infty \leq \|\underline{\boldsymbol{\varphi}_n} - \underline{\boldsymbol{P}}\,\underline{\boldsymbol{\varphi}_n}\|_\infty.$$

Also, since $\underline{\boldsymbol{T}_n}\,\underline{\boldsymbol{\varphi}_n} = \underline{\boldsymbol{\varphi}_n}\,\Theta_n^{[q]}$, we have

$$(\underline{\boldsymbol{T}_n} - \underline{\boldsymbol{T}})\,\underline{\boldsymbol{\varphi}_n} = \begin{bmatrix} \underline{O} & \cdots & \underline{O} & -(\underline{T} - \underline{T_n})^q\,\underline{T_n} \\ \underline{O} & \cdots & \underline{O} & \underline{O} \\ \vdots & & \vdots & \vdots \\ \underline{O} & \cdots & \underline{O} & \underline{O} \end{bmatrix} \begin{bmatrix} \underline{\varphi_n^{[q]}} \\ \underline{\varphi_n^{[q]}}\,(\Theta_n^{[q]})^{-1} \\ \vdots \\ \underline{\varphi_n^{[q]}}\,(\Theta_n^{[q]})^{-q+1} \end{bmatrix} (\Theta_n^{[q]})^{-1}$$

$$= \begin{bmatrix} -(\underline{T} - \underline{T_n})^q\,\underline{T_n}\,\underline{\varphi_n^{[q]}}\,(\Theta_n^{[q]})^{-q} \\ \underline{O} \\ \vdots \\ \underline{O} \end{bmatrix},$$

so that

$$\|(\underline{\boldsymbol{T}_n} - \underline{\boldsymbol{T}})\,\underline{\boldsymbol{\varphi}_n}\|_\infty \leq \|(\underline{T} - \underline{T_n})^q\,\underline{T_n}\,\underline{\varphi_n^{[q]}}\,(\Theta_n^{[q]})^{-1}\|_\infty \|(\Theta_n^{[q]})^{-1}\|_1^{q-1}.$$

Hence

$$\|\underline{\varphi_n^{[q]}} - \underline{P}\,\underline{\varphi_n^{[q]}}\|_\infty \leq \frac{\ell(\mathrm{C})}{2\pi}\alpha_{1,q}(\mathrm{C})\beta_{2,n,q}(\mathrm{C})\|(\Theta_n^{[q]})^{-1}\|_1^{q-1}\|(\underline{T} - \underline{T_n})^q\,\underline{T_n}\,\underline{\varphi_n^{[q]}}\,(\Theta_n^{[q]})^{-1}\|_\infty,$$

as desired. If $\|\underline{\varphi_n^{[q]}}\|_\infty \leq \gamma$, then

$$\|(\underline{T} - \underline{T_n})^q\,\underline{T_n}\,\underline{\varphi_n^{[q]}}\|_\infty \leq \gamma\|(T - T_n)^q T_n\| \to 0.$$

Let $\theta_{n,i,j}^{[q]}$ denote the (i, j)th entry of the matrix $\Theta_n^{[q]}$ for $1 \leq i, j \leq m$. Since $\underline{\boldsymbol{T}_n}\,\underline{\boldsymbol{\varphi}_n} = \underline{\boldsymbol{\varphi}_n}\,\Theta_n^{[q]}$, we have for each $j = 1, \ldots, m$,

$$\boldsymbol{T}_n\boldsymbol{\varphi}_{n,j} = \theta_{n,1,j}^{[q]}\boldsymbol{\varphi}_{n,1} + \cdots + \theta_{n,m,j}^{[q]}\boldsymbol{\varphi}_{n,m},$$

so that for $i = 1, \ldots, m$,

$$\delta_{n,q}|\theta^{[q]}_{n,i,j}| \leq ||\boldsymbol{T}_n \boldsymbol{\varphi}_{n,j}||_\infty \leq ||\, \underline{\boldsymbol{T}_n}\ \underline{\boldsymbol{\varphi}_n}\, ||_\infty .$$

It follows that

$$||\Theta^{[q]}_n||_1 = \max_{j=1,\ldots,m} \sum_{i=1}^{m} |\theta^{[q]}_{n,i,j}| \leq \frac{m}{\delta_{n,q}} ||\boldsymbol{T}_n||_\infty ||\, \underline{\boldsymbol{\varphi}_n}\, ||_\infty ,$$

where $(||\boldsymbol{T}_n||_\infty)$ is a bounded sequence. Assume now that the sequence $(||\, \underline{\boldsymbol{\varphi}_n}\, ||_\infty)$ is bounded and the sequence $(\delta_{n,q})$ is bounded away from zero. Then the sequence $(||\Theta^{[q]}||_1)$ is bounded. Also, since $\mathrm{sp}(\Theta^{[q]}_n) = \Lambda^{[q]}_n \subset \mathrm{int}(\mathrm{C})$ and $0 \in \mathrm{ext}(\mathrm{C})$, we see that

$$|\lambda^{[q]}_n| \geq \delta(\mathrm{C}) > 0$$

for every eigenvalue $\lambda^{[q]}_n$ of $\Theta^{[q]}_n$. Hence by Kato's result (2.20), the sequence $(||(\Theta^{[q]}_n)^{-1}||_1)$ is bounded. Further, we note that the sequence $\left(\sup_{z \in \mathrm{C}} ||\Theta^{[q]}_n - z\mathsf{I}||_1\right)$ is bounded. By Corollary 1.22, there is a Cauchy contour $\widetilde{\mathrm{C}}$ separating Λ from $(\mathrm{sp}(T) \setminus \Lambda)$ and also from $\mathrm{C} \cup \{0\}$. Let $\epsilon := \mathrm{dist}(\widetilde{\mathrm{C}}, \mathrm{C}) > 0$. Since $\widetilde{\mathrm{C}} \in \mathcal{C}(T, \Lambda)$ and $0 \in \mathrm{ext}(\widetilde{\mathrm{C}})$, Theorem 3.12(a) shows that $\Lambda^{[q]}_n = \mathrm{sp}(\boldsymbol{T}_n) \cap \mathrm{int}(\widetilde{\mathrm{C}})$ for all sufficiently large n. Thus for each eigenvalue $\lambda^{[q]}_n$ of $\Theta^{[q]}_n$ and each $z \in \mathrm{C}$, we have

$$|\lambda^{[q]}_n - z| \geq \mathrm{dist}(\mathrm{sp}(\Theta^{[q]}_n), \mathrm{C}) = \mathrm{dist}(\Lambda^{[q]}_n, \mathrm{C}) > \epsilon > 0.$$

Again by Kato's result, it follows that the sequence $(\beta_{2,n,q}(\mathrm{C}))$ is bounded. Consequently,

$$||\, \underline{\varphi^{[q]}_n} - \underline{P}\ \underline{\varphi^{[q]}_n}\, ||_\infty = O(||(T - T_n)^q T_n||).$$

(b) As in (a) above, Theorem 2.18 shows that for each large n, $\underline{\boldsymbol{P}_n}\ \underline{\boldsymbol{\varphi}}$ forms an ordered basis for $\boldsymbol{M}_n$. Since $\underline{\boldsymbol{\varphi}_n}$ also forms an ordered basis for $\boldsymbol{M}_n$, there is a nonsingular $m \times m$ matrix $\mathrm{C}_{n,q}$ such that $\underline{\boldsymbol{P}_n}\ \underline{\boldsymbol{\varphi}_n} = \underline{\boldsymbol{\varphi}_n}\, \mathrm{C}_{n,q}$. Define

$$\underline{\boldsymbol{\varphi}_{(n)}} := \underline{\boldsymbol{\varphi}}\, \mathrm{C}^{-1}_{n,q} = \begin{bmatrix} \underline{\varphi}\, \mathrm{C}^{-1}_{n,q} \\ \underline{J_1}\ \underline{\varphi}\, \mathrm{C}^{-1}_{n,q} \\ \vdots \\ \underline{J_{q-1}}\ \underline{\varphi}\, \mathrm{C}^{-1}_{n,q} \end{bmatrix}$$

and $\underline{\varphi}^{[q]}_{(n)} := \underline{\varphi}\,\mathsf{C}^{-1}_{n,q}$. As $\underline{\varphi}$ forms an ordered basis for M and the matrix $\mathsf{C}_{n,q}$ is nonsingular, it follows that $\underline{\varphi}^{[q]}_{(n)}$ also forms an ordered basis for M.

Applying Proposition 2.21 to $\boldsymbol{T}$ and $\boldsymbol{T}_n$ in place of T and T_n, and noting that $\underline{\boldsymbol{T}}\,\underline{\boldsymbol{\varphi}} = \underline{\boldsymbol{\varphi}}\,\Theta$ with $\underline{\boldsymbol{\varphi}} = [J\varphi_1, \ldots, J\varphi_m]$, we obtain

$$\frac{\|\,\underline{\boldsymbol{\varphi}_n} - \underline{\boldsymbol{\varphi}_{(n)}}\,\|_\infty}{\|\,\underline{\boldsymbol{\varphi}_{(n)}}\,\|_\infty} \leq \frac{\ell(\mathrm{C})}{2\pi}\frac{m}{\delta_q}\beta_1(\mathrm{C})\alpha_{2,q}(\mathrm{C})\|(\,\underline{\boldsymbol{T}_n} - \underline{\boldsymbol{T}}\,)\,\underline{\boldsymbol{\varphi}}\,\|_\infty,$$

where $\delta_q = \min\limits_{j=1,\ldots,m} \operatorname{dist}(J\varphi_j, \operatorname{span}\{J\varphi_i \,:\, i \neq j,\ i = 1, \ldots, m\})$.
Clearly,

$$\|\,\underline{\varphi_n^{[q]}} - \underline{\varphi^{[q]}_{(n)}}\,\|_\infty \leq \|\,\underline{\boldsymbol{\varphi}_n} - \underline{\boldsymbol{\varphi}_{(n)}}\,\|_\infty.$$

Also, since $(\,\underline{J_k}\,\underline{\varphi})\mathrm{C}^{-1}_{n,q} = \underline{J_k}\,(\,\underline{\varphi}\,\mathrm{C}^{-1}_{n,q}) = \underline{J_k}\,\underline{\varphi^{[q]}_{(n)}}$ for $k = 1, \ldots, q-1$, we have

$$\begin{aligned}\|\,\underline{\boldsymbol{\varphi}_{(n)}}\,\|_\infty &= \max\left\{\|\,\underline{\varphi^{[q]}_{(n)}}\,\|_\infty, \|\,\underline{J_1}\,\underline{\varphi^{[q]}_{(n)}}\,\|_\infty, \ldots, \|\,\underline{J_{q-1}}\,\underline{\varphi^{[q]}_{(n)}}\,\|_\infty\right\}\\ &= \|\,\underline{\varphi^{[q]}_{(n)}}\,\|_\infty \gamma_{n,q},\end{aligned}$$

where $\gamma_{n,q} := \max\left\{1, \dfrac{\|\,\underline{J_1}\,\underline{\varphi^{[q]}_{(n)}}\,\|_\infty}{\|\,\underline{\varphi^{[q]}_{(n)}}\,\|_\infty}, \ldots, \dfrac{\|\,\underline{J_{q-1}}\,\underline{\varphi^{[q]}_{(n)}}\,\|_\infty}{\|\,\underline{\varphi^{[q]}_{(n)}}\,\|_\infty}\right\}$. Further, since for $k = 0, \ldots, q-1$,

$$\begin{aligned}\underline{J_k}\,\underline{\varphi^{[q]}_{(n)}} &= (\,\underline{J_k}\,\underline{\varphi})\mathrm{C}^{-1}_{n,q} = \underline{\varphi}\,\underline{\Theta}^{-k}\mathrm{C}^{-1}_{n,q}\\ &= \underline{\varphi}\,\mathrm{C}^{-1}_{n,q}(\mathsf{C}_{n,q}\Theta^{-k}\mathrm{C}^{-1}_{n,q}) = \underline{\varphi^{[q]}_{(n)}}\,(\mathsf{C}_{n,q}\,\underline{\Theta}^{-k}\mathrm{C}^{-1}_{n,q}),\end{aligned}$$

we have

$$\gamma_{n,q} \leq \max\left\{\|\mathsf{C}_{n,q}\Theta^{-k}\mathrm{C}^{-1}_{n,q}\|_1 \,:\, k = 0, \ldots, q-1\right\}.$$

Next,

$$\begin{aligned}(\,\underline{\boldsymbol{T}_n} - \underline{\boldsymbol{T}}\,)\,\underline{\boldsymbol{\varphi}} &= (\,\underline{\boldsymbol{T}_n} - \underline{\boldsymbol{T}}\,)\,\underline{\boldsymbol{P}}\,\underline{\boldsymbol{\varphi}}\\ &= \begin{bmatrix} -(\,\underline{T} - \underline{T_n}\,)^q\,\underline{J_{q-1}} & \underline{O} & \cdots & \underline{O}\\ \underline{O} & \underline{O} & \cdots & \underline{O}\\ \vdots & \vdots & & \vdots\\ \underline{O} & \underline{O} & \cdots & \underline{O}\end{bmatrix}\begin{bmatrix}\underline{\varphi}\\ \underline{J_1}\,\underline{\varphi}\\ \vdots\\ \underline{J_{q-1}}\,\underline{\varphi}\end{bmatrix}\\ &= \begin{bmatrix} -(\,\underline{T} - \underline{T_n}\,)^q\,\underline{J_{q-1}}\,\underline{\varphi}\\ \underline{O}\\ \vdots\\ \underline{O}\end{bmatrix}.\end{aligned}$$

Hence $\|(\underline{\boldsymbol{T}_n} - \underline{\boldsymbol{T}})\underline{\boldsymbol{\varphi}}\|_\infty = \|(\underline{T} - \underline{T_n})^q \underline{\varphi}\,\Theta^{-q+1}\|_\infty$ and we obtain

$$\frac{\|\underline{\varphi_n^{[q]}} - \underline{\varphi_{(n)}^{[q]}}\|_\infty}{\|\underline{\varphi_{(n)}^{[q]}}\|_\infty} \leq \frac{\ell(\mathrm{C})}{2\pi}\frac{m}{\delta_q}\beta_1(\mathrm{C})\alpha_{2,q}(\mathrm{C})\gamma_{n,q}\|(\underline{T} - \underline{T_n})^q \underline{\varphi}\,\Theta^{-q+1}\|_\infty.$$

But $\gamma_{n,q} \leq \max\{1, \|J_1\|, \ldots, \|J_{q-1}\|\}$ and

$$\begin{aligned}\|(\underline{T} - \underline{T_n})^q \underline{\varphi}\,\Theta^{-q+1}\|_\infty &= \|(\underline{T} - \underline{T_n})^q \underline{T}\,\underline{\varphi}\,\Theta^{-q}\|_\infty \\ &\leq \|(T - T_n)^q T\|\,\|\underline{\varphi}\|_\infty \|\Theta^{-1}\|_1^q,\end{aligned}$$

so that

$$\frac{\|\underline{\varphi_n^{[q]}} - \underline{\varphi_{(n)}^{[q]}}\|_\infty}{\|\underline{\varphi_{(n)}^{[q]}}\|_\infty} = O(\|(T - T_n)^q T\|) \quad \text{as } n \to \infty,$$

as desired. ∎

Remark 3.16
We observe that if $m = 1$, $\Lambda = \{\lambda\}$ and C is the circle C_ϵ of radius ϵ with center λ in the preceding theorem, then we obtain Theorem 3.13. To see this, we note the following: $\ell(\mathrm{C})/2\pi = \epsilon$, $\beta_{2,n,q} = \sup\{1/|\lambda_n^{[q]} - z| : z \in \mathrm{C}_\epsilon\} = 1/\operatorname{dist}(\lambda_n^{[q]}, \mathrm{C}_\epsilon)$, $\Theta_n^{[q]} = [\lambda_n^{[q]}]$, $\|(\Theta_n^{[q]})^{-1}\|_1 = 1/|\lambda_n^{[q]}|$, $\Theta = [\lambda]$, $\beta_1(\mathrm{C}) = \sup\{1/|\lambda - z| : z \in \mathrm{C}_\epsilon\} = 1/\epsilon$, $\delta_q = \|J\varphi\|_\infty = \|\varphi\| \max\{1, 1/|\lambda|, \ldots, 1/|\lambda|^{q-1}\}$, $\mathsf{C}_{n,q} = [c_{n,q}]$, $J_k\varphi_{(n)}^{[q]} = \varphi/c_{n,q}\lambda^k$ for $k = 0, \ldots, q-1$ and $\gamma_{n,q} = \max\{1, 1/|\lambda|, \ldots, 1/|\lambda|^{q-1}\}$.

Also, if $q = 1$ in the preceding theorem, then we obtain Proposition 2.9. To see this we note the following: $\|(\underline{T} - \underline{T_n})\underline{T_n}\,\underline{\varphi_n}\,\Theta_n^{-1}\|_\infty = \|(\underline{T} - \underline{T_n})\underline{\varphi_n}\|_\infty$, $\delta_q = \delta$ and $\gamma_{n,q} = 1$. ∎

Remark 3.17
In Remark 2.16, we have noted that if $T_n \stackrel{\nu}{\to} T$, then

$$\begin{aligned}\operatorname{gap}(M, M_n) &:= \max\{\delta(M, M_n), \delta(M_n, M)\} \\ &\leq \max\{\|(P_n - P)_{|M}\|, \|(P_n - P)_{|M_n}\|\},\end{aligned}$$

where, by Proposition 2.15, we have for all large n,

$$\begin{aligned}\|(P_n - P)_{|M}\| &\leq \frac{\ell(\mathrm{C})}{2\pi}\alpha_1(\mathrm{C})\alpha_2(\mathrm{C})\|(T_n - T)_{|M}\|, \\ \|(P_n - P)_{|M_n}\| &\leq \frac{\ell(\mathrm{C})}{2\pi}\alpha_1(\mathrm{C})\alpha_2(\mathrm{C})\|(T_n - T)\|_{|M_n}\|.\end{aligned}$$

Similarly, it follows that if $(T_n - T) \xrightarrow{\nu} O$, then $\boldsymbol{T}_n \xrightarrow{\nu} \boldsymbol{T}$ and if

$$\text{gap}(\boldsymbol{M}, \boldsymbol{M}_n) := \max\{\delta(\boldsymbol{M}, \boldsymbol{M}_n), \delta(\boldsymbol{M}_n, \boldsymbol{M})\},$$

then for all large n,

$$\delta(\boldsymbol{M}, \boldsymbol{M}_n) \leq \|(\boldsymbol{P}_n - \boldsymbol{P})_{|\boldsymbol{M}}\| \leq \frac{\ell(\mathrm{C})}{2\pi}\alpha_{1,q}(\mathrm{C})\alpha_{2,q}(\mathrm{C})\|(\boldsymbol{T}_n - \boldsymbol{T})_{|\boldsymbol{M}}\|,$$

$$\delta(\boldsymbol{M}_n, \boldsymbol{M}) \leq \|(\boldsymbol{P}_n - \boldsymbol{P})_{|\boldsymbol{M}_n}\| \leq \frac{\ell(\mathrm{C})}{2\pi}\alpha_{1,q}(\mathrm{C})\alpha_{2,q}(\mathrm{C})\|(\boldsymbol{T}_n - \boldsymbol{T})_{|\boldsymbol{M}_n}\|.$$

We have seen in Theorem 3.12 that

$$\|(\boldsymbol{T}_n - \boldsymbol{T})_{|\boldsymbol{M}}\| = O(\|(T_n - T)^q T\|) \text{ and } \|(\boldsymbol{T}_n - \boldsymbol{T})_{|\boldsymbol{M}_n}\| = O(\|(T_n - T)^q T_n\|).$$

Let us define $M_{n,q}$ to be the subspace of X consisting of all the first components of the elements of $\boldsymbol{M}_n$, and investigate

$$\text{gap}(M, M_{n,q}) := \max\{\delta(M, M_{n,q}), \delta(M_{n,q}, M)\}.$$

To relate $\delta(M, M_{n,q})$ with $\delta(\boldsymbol{M}, \boldsymbol{M}_n)$, consider $x \in M$ with $\|x\| = 1$. Then $Jx = [x, J_1 x, \ldots, J_{q-1}x]^\top \in \boldsymbol{M}$. Now

$$\begin{aligned} \text{dist}(x, M_{n,q}) &= \inf\{\|x - x_{n,q}\| \,:\, x_{n,q} \in M_{n,q}\} \\ &\leq \inf\{\|Jx - \boldsymbol{x}_n\|_\infty \,:\, \boldsymbol{x}_n \in \boldsymbol{M}_n\} \\ &\leq \|Jx\|_\infty \,\text{dist}\left(\frac{Jx}{\|Jx\|_\infty}, \boldsymbol{M}_n\right). \end{aligned}$$

Hence

$$\begin{aligned} \delta(M, M_{n,q}) &:= \sup\{\text{dist}(x, M_{n,q}) \,:\, x \in M, \|x\| = 1\} \\ &\leq \|J_{|M}\| \sup\{\text{dist}(\boldsymbol{x}, \boldsymbol{M}_n) \,:\, \boldsymbol{x} \in \boldsymbol{M}, \|\boldsymbol{x}\|_\infty = 1\} \\ &= \|J_{|M}\|\delta(\boldsymbol{M}, \boldsymbol{M}_n). \end{aligned}$$

It is not easy to find a bound for $\delta(M_{n,q}, M)$ in terms of $\delta(\boldsymbol{M}_n, \boldsymbol{M})$. We refer to Theorem 3.5 of [14] for a way out. ∎

3.2.4 Dependence on the Order of the Spectral Analysis

In our development of the acceleration technique, we have so far fixed an integer $q \geq 2$ and considered spectral analysis of order q. It is interesting to investigate whether this analysis depends on the choice of q and, if so, in what manner.

Let $T \in \mathrm{BL}(X)$ and $(T_n - T) \xrightarrow{\nu} O$. It follows from Proposition 3.11 that $\boldsymbol{T}_n^{[q]} \xrightarrow{\nu} \boldsymbol{T}^{[q]}$ as $n \to \infty$, uniformly in $q \in \{2, 3, \ldots, \}$, that is, (i) $(\|\boldsymbol{T}_n^{[q]}\|)$ is bounded in n and q, and (ii) $\|(\boldsymbol{T}_n^{[q]} - \boldsymbol{T}^{[q]})\boldsymbol{T}^{[q]}\|$ as well as $\|(\boldsymbol{T}_n^{[q]} - \boldsymbol{T}^{[q]})\boldsymbol{T}_n^{[q]}\|$ tend to zero as $n \to \infty$, uniformly in $q \in \{2, 3, \ldots\}$.

Let E be a closed subset of $\mathrm{re}(T)$ such $0 \notin E$. Define

$$\delta(E) := \min\{|z| : z \in E\}, \quad \alpha_1(E) := \sup\{\|R(T, z)\| : z \in E\}$$

and for $q = 2, 3, \ldots$

$$\alpha_{1,q}(E) := \sup\{\|R(\boldsymbol{T}^{[q]}, z)\| : z \in E\}.$$

Let us first consider the case when $\delta(E) > 1$. The reason for this restriction will soon be clear. The matrix representation of $R(\boldsymbol{T}^{[q]}, z)$ given earlier shows that for all $z \in E$,

$$\begin{aligned}
&\|R(\boldsymbol{T}^{[q]}, z)\| \\
&\le \max\left\{\|R(T, z)\|, \frac{\|R(T, z)\|}{|z|} + \frac{1}{|z|}, \ldots, \frac{\|R(T, z)\|}{|z|^{q-1}} + \frac{1}{|z|^{q-1}} + \cdots + \frac{1}{|z|}\right\} \\
&\le \max\left\{\alpha_1(E), \frac{\alpha_1(E)}{\delta(E)} + \frac{1}{\delta(E)}, \ldots, \frac{\alpha_1(E)}{[\delta(E)]^{q-1}} + \frac{1}{[\delta(E)]^{q-1}} + \cdots + \frac{1}{\delta(E)}\right\} \\
&\le \alpha_1(E) + \frac{1}{[\delta(E)]^{q-1}} + \cdots + \frac{1}{\delta(E)} \\
&< \alpha_1(E) + \frac{1}{\delta(E) - 1}.
\end{aligned}$$

Let $\alpha_{1,\infty}(E) := \alpha_1(E) + 1/[\delta(E) - 1]$. Then the sequence $(\alpha_{1,q}(E))$ is bounded by $\alpha_{1,\infty}(E)$. [On the other hand, if $\delta(E) \le 1$, then the sequence $(\alpha_{1,q}(E))$ may not be bounded, as the simple example in Exercise 3.11 shows.]

The proof of Theorem 2.6 shows that there is a positive integer n_0 (independent of q) such that for each $n \ge n_0$ and each $q \in \{2, 3, \ldots, \}$, $E \subset \mathrm{re}(\boldsymbol{T}_n^{[q]})$.

As a consequence, we note that if $\lambda_n \in \cup\{\mathrm{sp}(\boldsymbol{T}_n^{[q]}) : q = 2, 3, \ldots\}$ and $\lambda_n \to \lambda$ as $n \to \infty$, where $|\lambda| > 1$, then $\lambda \in \mathrm{sp}(T)$. (See the proof of Corollary 2.7.)

Further, define

$$\alpha_{2,q}(E) := \sup\{\|R(\boldsymbol{T}_n^{[q]}, z)\| : z \in E, \ n \ge n_0\}.$$

As in the proof of Theorem 2.6, we have

$$\alpha_{2,q}(E) \le 2\alpha_{1,\infty}(E)[1 + t\,\alpha_1(E)],$$

where $\|\boldsymbol{T}^{[q]} - \boldsymbol{T}_n^{[q]}\| \leq t$ for all n and q. Let $\alpha_{2,\infty}(E) := 2\alpha_{1,\infty}(E)[1 + t\,\alpha_1(E)]$. Then the sequence $(\alpha_{2,q}(E))$ is bounded by $\alpha_{2,\infty}(E)$.

Let now Λ be a nonempty spectral set for T such that $0 \notin \Lambda$, $P := P(T, \Lambda)$ and $\boldsymbol{P}^{[q]} := P(\boldsymbol{T}^{[q]}, \Lambda)$.

Assume first that

$$\min\{|\lambda| : \lambda \in \Lambda\} > 1.$$

By Proposition 1.29, find $\mathrm{C} \in \mathcal{C}(T, \Lambda)$ such that $\{z \in \mathbb{C} : |z| \leq 1\} \subset \mathrm{ext}(\mathrm{C})$. Then $\delta(\mathrm{C}) > 1$ and there is a positive integer n_0 such that $\mathrm{C} \subset \mathrm{re}(\boldsymbol{T}_n^{[q]})$ for all $n \geq n_0$ and all $q = 2, 3, \ldots$ As we proved in Proposition 2.9, if $n \geq n_0$ and $q = 2, 3, \ldots$ then

$$\Lambda_n^{[q]} := \mathrm{sp}(\boldsymbol{T}_n^{[q]}) \cap \mathrm{int}(\mathrm{C}) \quad \text{and} \quad \boldsymbol{P}_n^{[q]} := P(\boldsymbol{T}_n^{[q]}, \Lambda_n^{[q]})$$

do not depend on $\mathrm{C} \in \mathcal{C}(T, \Lambda)$ provided $\delta(\mathrm{C}) > 1$, and $\|(\boldsymbol{P}_n^{[q]} - \boldsymbol{P}^{[q]})\boldsymbol{P}^{[q]}\|$ as well as $\|(\boldsymbol{P}_n^{[q]} - \boldsymbol{P}^{[q]})\boldsymbol{P}_n^{[q]}\|$ tend to zero as $n \to \infty$, uniformly in $q \in \{2, 3, \ldots\}$. As a consequence, there is a positive integer n_1 such that for all $n \geq n_1$ and all $q = 2, 3, \ldots, \rho(\boldsymbol{P}_n^{[q]} - \boldsymbol{P}^{[q]}) < 1$, so that $\mathrm{rank}\, \boldsymbol{P}_n^{[q]} = \mathrm{rank}\, \boldsymbol{P}^{[q]} = \mathrm{rank}\, P$ and, in particular, $\Lambda_n^{[q]} \neq \emptyset$.

Since for all $q = 2, 3, \ldots$

$$\alpha_{1,q}(\mathrm{C}) \leq \alpha_{1,\infty}(\mathrm{C}) < \infty \quad \text{and} \quad \alpha_{2,q}(\mathrm{C}) \leq \alpha_{2,\infty}(\mathrm{C}) < \infty,$$

we see that the conclusions of Theorem 3.12 hold true if we replace $n_{0,q}$ by n_0, $\alpha_{1,q}(\mathrm{C})$ by $\alpha_{1,\infty}(\mathrm{C})$ and $\alpha_{2,q}(\mathrm{C})$ by $\alpha_{2,\infty}(\mathrm{C})$. In case the spectral set Λ is of finite type, then for all large n and all $q = 2, 3, \ldots$ we obtain

$$|\widehat{\lambda}_n^{[q]} - \widehat{\lambda}| \leq \frac{\ell(\mathrm{C})}{\pi}\alpha_{1,\infty}(\mathrm{C})\frac{\ell(\mathrm{C})\alpha_{2,\infty}(\mathrm{C})}{2\pi\delta(\mathrm{C})}\|(T_n - T)^q T_n\|,$$

so that $|\widehat{\lambda}_n^{[q]} - \widehat{\lambda}| = O(\|(T_n - T)^q T_n\|)$ as $n, q \to \infty$, and also

$$|\widehat{\lambda}_n^{[q]} - \widehat{\lambda}| \leq \frac{\ell(\mathrm{C})}{\pi}\alpha_{2,\infty}(\mathrm{C})\frac{[\ell(\mathrm{C})\alpha_1(\mathrm{C})]^2}{4\pi^2[\delta(\mathrm{C})]^q}\|(T_n - T)^q T_n\|,$$

so that $|\widehat{\lambda}_n^{[q]} - \widehat{\lambda}| = O(\|(T_n - T)^q T\|/[\delta(\mathrm{C})]^q$ as $n, q \to \infty$.

Let us now turn to the general case of a spectral set Λ of T such that $0 \notin \Lambda$. Since Λ is closed, there is a positive number s such that

$$\min\{|\lambda| : \lambda \in \Lambda\} > s.$$

Again by Proposition 1.29, find $\mathrm{C} \in \mathcal{C}(T, \Lambda)$ such that $\delta(\mathrm{C}) > s$. As we have mentioned before, the sequences $(\alpha_{1,q}(\mathrm{C}))$ and $(\alpha_{2,q}(\mathrm{C}))$ may not be bounded if $s \le 1$. We may, however, multiply the operators T and T_n by the positive number $r := 1/s$ and apply our earlier considerations to the operators rT and rT_n. Now $(rT_n - rT) \stackrel{\nu}{\to} O$, $r\Lambda := \{r\lambda : \lambda \in \Lambda\}$ is a spectral set for T and $\min\{|\mu| : \mu \in r\Lambda\} = r \min\{|\lambda| : \lambda \in \Lambda\} > rs = 1$. Also, $r\mathrm{C} \in \mathcal{C}(rT, r\Lambda), 0 \in \mathrm{ext}(r\mathrm{C})$ and

$$P(rT, r\Lambda) = -\frac{1}{2\pi \mathrm{i}} \int_{r\mathrm{C}} R(rT, w) dw = -\frac{1}{2\pi \mathrm{i}} \int_{\mathrm{C}} R(T, z) dz = P(T, \Lambda).$$

For $\boldsymbol{x} := [x_1, \ldots, x_q]^\top \in \boldsymbol{X}^{[q]}$, let

$$\mathrm{diag}[1, r, \ldots, r^{q-1}]\boldsymbol{x} := [x_1, rx_2, \ldots, r^{q-1}x_q]^\top.$$

Consider the operators

$$r\star \boldsymbol{T}^{[q]} := \left(\mathrm{diag}[1, r, \ldots, r^{q-1}]\right)^{-1} \left(r\boldsymbol{T}^{[q]}\right) \mathrm{diag}[1, r, \ldots, r^{q-1}]$$

and

$$r\star \boldsymbol{T}_n^{[q]} := \left(\mathrm{diag}[1, r, \ldots, r^{q-1}]\right)^{-1} \left(r\boldsymbol{T}_n^{[q]}\right) \mathrm{diag}[1, r, \ldots, r^{q-1}]$$

from $\boldsymbol{X}^{[q]}$ to $\boldsymbol{X}^{[q]}$. It can be easily seen that if we replace T by rT, and T_n by rT_n in the definitions of the operators $\boldsymbol{T}^{[q]}$ and $\boldsymbol{T}_n^{[q]}$, then we obtain the operators $r\star \boldsymbol{T}^{[q]}$ and $r\star \boldsymbol{T}_n^{[q]}$. Since similar operators have the same spectrum, we have

$$\mathrm{sp}(r\star \boldsymbol{T}^{[q]}) = \{r\lambda : \lambda \in \mathrm{sp}(\boldsymbol{T}^{[q]})\} \text{ and } \mathrm{sp}(r\star \boldsymbol{T}_n^{[q]}) = \{r\mu : \mu \in \mathrm{sp}(\boldsymbol{T}_n^{[q]})\}.$$

Also, we see that

$$\begin{aligned} P(r\star \boldsymbol{T}^{[q]}, r\Lambda) &= \frac{-1}{2\pi \mathrm{i}} \int_{r\mathrm{C}} R(r\star \boldsymbol{T}^{[q]}, w) dw \\ &= \frac{-1}{2\pi \mathrm{i}} \int_{r\mathrm{C}} \left(\mathrm{diag}[1, r, \ldots, r^{q-1}]\right)^{-1} R(r\boldsymbol{T}^{[q]}, w) \,\mathrm{diag}[1, r, \ldots, r^{q-1}] dw \\ &= \left(\mathrm{diag}[1, r, \ldots, r^{q-1}]\right)^{-1} \left(\frac{-1}{2\pi \mathrm{i}} \int_{\mathrm{C}} R(\boldsymbol{T}^{[q]}, z) dz\right) \mathrm{diag}[1, r, \ldots, r^{q-1}] dz. \end{aligned}$$

Thus

$$P(r\star \boldsymbol{T}^{[q]}, r\Lambda) = \left(\mathrm{diag}[1, r, \ldots, r^{q-1}]\right)^{-1} P(\boldsymbol{T}^{[q]}, \Lambda) \,\mathrm{diag}[1, r, \ldots, r^{q-1}].$$

Similarly, there is an integer $n_0(r)$ such that for all $n \ge n_0(r)$, we have

$$P(r\star \boldsymbol{T}_n^{[q]}, r\Lambda_n^{[q]}) = \left(\mathrm{diag}[1, r, \ldots, r^{q-1}]\right)^{-1} P(\boldsymbol{T}_n^{[q]}, \Lambda_n^{[q]}) \,\mathrm{diag}[1, r, \ldots, r^{q-1}].$$

As a result, an element $\boldsymbol{x} = [x_1, \dots, x_q]^\top$ in $\boldsymbol{X}^{[q]}$ belongs to the range of $P(r\star \boldsymbol{T}_n^{[q]}, r\Lambda_n^{[q]})$ if and only if $P(\boldsymbol{T}_n^{[q]}, \Lambda_n^{[q]})[x_1, rx_2, \dots, r^{q-1}x_q]^\top = [x_1, rx_2, \dots, r^{q-1}x_q]^\top$, that is, the element $[x_1, rx_2, \dots, r^{q-1}x_q]^\top$ belongs to the range of $P(\boldsymbol{T}_n^{[q]}, \Lambda_n^{[q]})$. In particular, $\lambda_n^{[q]}$ is a nonzero spectral value of $\boldsymbol{T}_n^{[q]}$ of algebraic multiplicity m if and only if $r\lambda_n^{[q]}$ is a nonzero spectral value of $\mathbf{r}\boldsymbol{T}_n^{[q]}$ of algebraic multiplicity m.

As in Theorem 3.12(c), let Λ be a spectral for T of finite type, $0 \notin \Lambda$ and $\widehat{\lambda}$ denote the weighted arithmetic mean of the spectral values of T belonging to Λ. Then $r\widehat{\lambda}$ is the weighted arithmetic mean of the spectral values of rT belonging to $r\Lambda$. Also, it follows from the considerations given above, that if $r\widehat{\lambda}_n^{[q]}$ is the weighted arithmetic mean of the spectral values of $r\star \boldsymbol{T}_n^{[q]}$ belonging to $r\Lambda_n^{[q]} := \mathrm{sp}(r\star \boldsymbol{T}_n^{[q]}]) \cap \mathrm{int}(r\mathrm{C})$, then $\widehat{\lambda}_n^{[q]}$ is the weighted arithmetic mean of the spectral values of $\boldsymbol{T}_n^{[q]}$ belonging to $\Lambda_n^{[q]} := \mathrm{sp}(\boldsymbol{T}_n^{[q]}) \cap \mathrm{int}(\mathrm{C})$. Applying Theorem 3.12(c) to rT and rT_n, we obtain for all large n and $q = 2, 3, \dots$

$$|r\widehat{\lambda}_n^{[q]} - r\widehat{\lambda}| \le \frac{\ell(r\mathrm{C})}{\pi} a_q(r\mathrm{C}) \frac{\ell(r\mathrm{C}) b_q(r\mathrm{C})}{2\pi\delta(r\mathrm{C})} \|(rT_n - rT)^q rT_n\|$$

as well as

$$|r\widehat{\lambda}_n^{[q]} - r\widehat{\lambda}| \le \frac{\ell(r\mathrm{C})}{\pi} b_q(r\mathrm{C}) \frac{[\ell(r\mathrm{C}) a_1(r\mathrm{C})]^2}{4\pi^2[\delta(r\mathrm{C})]^q} \|(rT_n - rT)^q rT\|,$$

where

$$\begin{aligned}
a_q(r\mathrm{C}) &:= \sup\{\|R(r\star \boldsymbol{T}^{[q]}, w)\| \,:\, w \in r\mathrm{C}\},\\
b_q(r\mathrm{C}) &:= \sup\{\|R(r\star \boldsymbol{T}_n^{[q]}, w)\| \,:\, n \ge n_0(r), w \in r\mathrm{C}\},\\
a_1(r\mathrm{C}) &:= \sup\{\|R(rT, w)\| \,:\, w \in r\mathrm{C}\}.
\end{aligned}$$

Note that $\ell(r\mathrm{C}) = r\ell(\mathrm{C})$, $\delta(r\mathrm{C}) = r\delta(\mathrm{C}) > rs = 1$ and $a_1(r\mathrm{C}) = \alpha_1(\mathrm{C})/r$.

As the sequences $(a_q(r\mathrm{C}))$ and $(b_q(r\mathrm{C}))$ are bounded in q, there is an integer $n_1(r)$ such that for all $n \ge n_1(r)$ and all $q = 2, 3, \dots$

$$|\widehat{\lambda}_n^{[q]} - \widehat{\lambda}| \le \frac{\alpha}{s^{q-1}} \min\{\|(T_n - T)^q T_n\|, \|(T_n - T)^q T\|\},$$

where α is a constant independent of n and q.

We observe that if $s > 1$, then the estimate $|\widehat{\lambda}_n^{[q]} - \widehat{\lambda}| = O(\|(T_n - T)^q T_n\|/s^{q-1})$ (as $n, q \to \infty$) obtained above is an improvement over the estimate $|\widehat{\lambda}_n^{[q]} - \widehat{\lambda}| = O(\|(T_n - T)^q T_n\|)$ (as $n, q \to \infty$) stated earlier, while the estimate $|\widehat{\lambda}_n^{[q]} - \widehat{\lambda}| = O(\|(T_n - T)^q T\|/s^{q-1})$(as $n, q \to$

∞) obtained above matches with the estimate $|\widehat{\lambda}_n^{[q]} - \widehat{\lambda}| = O(\|T_n - T)^q T\|/[\delta(\text{C})]^q)$(as $n, q \to \infty$) stated earlier.

For uniform estimates for the accelerated approximation of an eigenvector of T corresponding to a nonzero simple eigenvalue, see Exercises 3.12 and 3.13. Similar considerations hold for uniform estimates for the accelerated approximation of an ordered basis of the spectral subspace of T corresponding to a spectral set of finite type not containing zero.

3.3 Exercises

Unless otherwise stated, X denotes a Banach space over $\mathbb{C}$, $X \neq \{0\}$ and $T \in \text{BL}(X)$. Prove the following assertions.

3.1 Let $A \in \text{BL}(X)$ be invertible and $\widetilde{A} \in \text{BL}(X)$ satisfy $\rho(A^{-1}(A - \widetilde{A})) < 1$. Then $\widetilde{A}$ is invertible. If, in addition, $\rho(\widetilde{A}^{-1}(\widetilde{A} - A)) < 1$, then

$$A^{-1} = \sum_{j=0}^{\infty} \left[\widetilde{A}^{-1}(\widetilde{A} - A)\right]^j \widetilde{A}^{-1}.$$

Let $y \in X$. If $\widetilde{A}\widetilde{x} = y$, $\widetilde{x}^{(0)} := \widetilde{x}$ and $\widetilde{x}^{(k)} := \widetilde{x}^{(k-1)} + \widetilde{A}^{-1}\left(y - A\widetilde{x}^{(k-1)}\right)$ for $k = 1, 2, \dots$ then $\widetilde{x}^{(k)} = \left[\widetilde{A}^{-1}(\widetilde{A}^{-1} - A)\right]^k \widetilde{A}^{-1}\widetilde{x}$, and hence $\widetilde{x}^{(k)} \to x$ in X and $Ax = y$.

3.2 Let X be a Hilbert space with an orthonormal basis $\{\tilde{e}, e_1, e_2, \dots, \}$ and consider a diagonal operator $\widetilde{T} \in \text{BL}(X)$ given by

$$\widetilde{T}x := \widetilde{\lambda}\langle x\,,\,\widetilde{e}\rangle\widetilde{e} + \sum_{j=1}^{\infty} \ell_j \langle x\,,\,e_j\rangle e_j, \quad x \in X.$$

Let $\widetilde{V} := T - \widetilde{T}$. Then the coefficients μ_1, ψ_1, μ_2 in the Rayleigh-Schrödinger series associated with T, $\widetilde{T}$ and $\widetilde{\lambda}$ are given by

$$\mu_1 = \langle \widetilde{V}\widetilde{e}, \widetilde{e}\rangle, \quad \psi_1 = -\sum_{j=1}^{\infty} \frac{\langle \widetilde{V}\widetilde{e}\,,\,e_j\rangle}{\ell_j - \widetilde{\lambda}} e_j, \quad \mu_2 = -\sum_{j=1}^{\infty} \frac{\langle \widetilde{V}\widetilde{e}\,,\,e_j\rangle\langle \widetilde{V}e_j\,,\,\widetilde{e}\rangle}{\ell_j - \widetilde{\lambda}}.$$

(Hint: $\widetilde{P}x = \langle x\,,\,\widetilde{e}\rangle\widetilde{e}$ and $\widetilde{S}x = \sum_{j=1}^{\infty} \frac{\langle x\,,\,e_j\rangle}{\ell_j - \widetilde{\lambda}} e_j$ for $x \in X$.)

3.3 The first iterate $\varphi_n^{(1)}$ in the Elementary Iteration is given by

$$\varphi_n^{(1)} = \varphi_n - S_n T \varphi_n,$$

and the first iterate $\psi_n^{(1)}$ in the Double Iteration is given by

$$\psi_n^{(1)} = \frac{T\varphi_n}{\langle T\varphi_n \,,\, \varphi_n^* \rangle} + \frac{S_n T}{\langle T\varphi_n \,,\, \varphi_n^* \rangle} \left[\frac{\langle T^2\varphi_n \,,\, \varphi_n^* \rangle}{\langle T\varphi_n \,,\, \varphi_n^* \rangle} \varphi_n - T\varphi_n \right],$$

provided $\langle T\varphi_n \,,\, \varphi_n^* \rangle \neq 0$.

3.4 Let $T_n \in \mathrm{BL}(X)$ for $n = 1, 2, \ldots$

(a) Let λ be a nonzero simple eigenvalue of T, λ_n and φ_n be as in Lemma 3.1, and $\lambda_n^{(1)}$ and $\varphi_n^{(1)}$ be the first iterates in the Elementary iteration (E). If $T_n(T - T_n)\varphi_n = 0$, then $\lambda_n^{(1)} = \lambda_n$ and $\varphi_n^{(1)} = T\varphi_n/\lambda_n$.

(b) Let Λ be a spectral set for T such that $0 \notin \Lambda$, Θ_n and $\underline{\varphi_n}$ be as in Lemma 3.5, and $\Theta_n^{(1)}$ and $\underline{\varphi_n}^{(1)}$ be the first iterates in the Elementary Iteration $(\underline{E})$. If $\underline{T_n}\,(\underline{T} - \underline{T_n})\,\underline{\varphi_n} = \underline{0}$, then $\Theta_n^{(1)} = \Theta_n$ and $\underline{\varphi_n}^{(1)} = (\underline{T}\,\underline{\varphi_n})\Theta_n^{-1}$. (Hint: Remarks 1.25, 1.58. Note: These results hold if T_n is the Galerkin approximation of T as defined in Section 4.1.)

3.5 Let $\underline{\varphi}$, $\underline{\varphi}^*$, $\underline{\varphi_n}$, $\underline{\varphi_n^*}$ and C_n be as in Lemma 3.5 and Lemma 3.6. Let $\mathsf{D}_n = (\underline{\varphi_n} \,,\, \underline{\varphi}^*)$. Then the sequence $(\|\mathsf{C}_n^{-1}\mathsf{D}_n^{-1}\|_1)$ is bounded. Further, the sequence $(\|\,\underline{\varphi_n}\,\|_\infty)$ is bounded if and only if the sequence $(\|\mathsf{C}_n^{-1}\|_1)$ is bounded, while the sequence $(\|\,\underline{\varphi_n^*}\,\|_1)$ is bounded if and only if the sequence $(\|\mathsf{D}_n^{-1}\|_1)$ is bounded.

3.6 Let $0 \neq \lambda \in \mathbb{C}$. Then the ascent of λ as an eigenvalue of T is equal to the ascent of λ as an eigenvalue of $\boldsymbol{T} \in \mathrm{BL}(\boldsymbol{X})$. (Hint: If $P := P(T, \{\lambda\})$, $\boldsymbol{P} := P(\boldsymbol{T}, \{\lambda\})$, $D = P(T - \lambda I)$ and $\boldsymbol{D} = \boldsymbol{P}(\boldsymbol{T} - \lambda \mathbf{I})$, then for $k = 1, 2, \ldots D^k = O$ if and only if $\boldsymbol{D}^k = \boldsymbol{O}$.)

3.7 Let $q \geq 2$. Then $\mathcal{N}(\boldsymbol{T}) = \{[0, \ldots, 0, x]^\top \in \boldsymbol{X} \,:\, x \in X\}$, so that 0 is an eigenvalue of $\boldsymbol{T}$ even if 0 may not be an eigenvalue of T. If 0 is an eigenvalue of T and φ is a corresponding eigenvector, then $[\varphi, 0, \ldots, 0]^\top$ is not an eigenvector of $\boldsymbol{T}$ corresponding to 0, but $[0, \ldots, 0, \varphi]^\top$ is an eigenvector of $\boldsymbol{T}$ corresponding to 0.

3.8 If $T_n \xrightarrow{\mathrm{n}} T$, then $\boldsymbol{T}_n^{[q]} \xrightarrow{\mathrm{n}} \boldsymbol{T}^{[q]}$, uniformly for $q = 2, 3, \ldots$

3.9 Let C be a Cauchy contour in $\mathrm{re}(T)$ with $0 \in \mathrm{ext}(\mathrm{C})$. For $k = 0, 1, 2, \ldots$ consider $J_k := -\dfrac{1}{2\pi\mathrm{i}} \displaystyle\int_{\mathrm{C}} R(T, z) \frac{dz}{z^k}$. Let $\underline{\varphi} \in X^{1\times m}$ and $\Theta \in$

$\mathbb{C}^{m\times m}$ satisfy $\underline{T}\,\underline{\varphi} = \underline{\varphi}\Theta$. Then $(\underline{J_k}\,\underline{\varphi})\Theta^k = \underline{J_0}\,\underline{\varphi}$, where $\underline{J_0}\,\underline{\varphi} = \underline{\varphi}$ if $\mathrm{sp}(\Theta) \subset \mathrm{int}(\mathrm{C})$ and $\underline{J_0}\,\underline{\varphi} = \underline{0}$ if $\mathrm{csp}(\Theta) \subset \mathrm{ext}(\mathrm{C})$.

In particular, if $m = 1$, $\underline{\varphi} = \varphi$, $\Theta = [\lambda]$ and $\lambda \neq 0$, then $J_k\varphi = \varphi/\lambda^k$ if $\lambda \in \mathrm{int}(\mathrm{C})$ and $J_k\varphi = 0$ if $\lambda \in \mathrm{ext}(\mathrm{C})$.

3.10 Under the hypotheses and with the notation of Theorem 3.15, let

$$\begin{aligned}\widetilde{\alpha}_{1,q}(\mathrm{C}) &:= \sup\{\|R(\boldsymbol{T}^{[q]}, z)_{|\boldsymbol{M}}\| \,:\, z \in \mathrm{C}\}\\ \widetilde{\alpha}_{2,q}(\mathrm{C}) &:= \sup\{\|R(\boldsymbol{T}_n^{[q]}, z)_{|\boldsymbol{M}_n}\| \,:\, z \in \mathrm{C}, n \geq n_{0,q}\}.\end{aligned}$$

Then

$$\begin{aligned}\frac{\|\,\underline{\varphi_n^{[q]}} - \underline{P}\,\underline{\varphi_n^{[q]}}\,\|_\infty}{\|\,\underline{\varphi_n^{[q]}}\,\|_\infty} &\leq \frac{\ell(\mathrm{C})}{2\pi}\alpha_{1,q}(\mathrm{C})\widetilde{\alpha}_{2,q}(\mathrm{C}) \max\left\{1, \|(\Theta_n^{[q]})^{-q+1}\right\}\\ &\quad\times\left[\frac{\ell(\mathrm{C})}{2\pi}\frac{\alpha_{2,q}(\mathrm{C})}{\delta(\mathrm{C})}\|(T_n - T)^q T_n\|\right],\\ \frac{\|\,\underline{\varphi_n^{[q]}} - \underline{\varphi_{(n)}^{[q]}}\,\|_\infty}{\|\,\underline{\varphi_{(n)}^{[q]}}\,\|_\infty} &\leq \frac{\ell(\mathrm{C})}{2\pi}\widetilde{\alpha}_{1,q}(\mathrm{C})\alpha_{2,q}(\mathrm{C})\gamma_{n,q}\\ &\quad\times\left[\frac{\ell(\mathrm{C})\alpha_1(\mathrm{C})}{2\pi}\right]^2 \frac{1}{[\delta(\mathrm{C})]^q}\|(T_n - T)^q T\|.\end{aligned}$$

(Hint: Proposition 2.21, Theorem 3.12 (b), Theorem 3.15.)

3.11 Fix $c \in \mathbb{C}$, let $X := \mathbb{C}$ and define $Tx := cx$ for $x \in X$. For $q = 1, 2, \ldots \boldsymbol{x} = [x_1, \ldots, x_q]^\top \in \boldsymbol{X}^{[q]}$ and $z \in \mathbb{C} \setminus \{c, 0\}$, we have

$$R(\boldsymbol{T}^{[q]}, z)\boldsymbol{x} = \left[\frac{x_1}{c-z}, \frac{x_1}{z(c-z)} - \frac{x_2}{z}, \ldots, \frac{x_1}{z^{q-1}(c-z)} - \frac{x_2}{z^{q-1}} - \cdots \frac{x_q}{z}\right]^\top.$$

If $0 < |z| < 1$ and $z \neq c$, then $\|R(\boldsymbol{T}^{[q]}, z)\| \geq 1/|z|^{q-1}|c-z|$, which tends to ∞ as $q \to \infty$. Also, if $c \neq 1$, then $\|R(\boldsymbol{T}^{[q]}, 1)\| \geq (q-1) - 1/|c-1|$, which tends to ∞ as $q \to \infty$.

3.12 Let λ be a nonzero simple eigenvalue of T, and $\lambda_n^{[q]}$, $\varphi_n^{[q]}$ be as in Theorem 3.13. Then there is some n_0 such that for all $n \geq n_0$ and all $q = 2, 3, \ldots P\varphi_n^{[q]}$ is an eigenvector of T corresponding to λ, and

$$\|\varphi_n^{[q]} - P\varphi_n^{[q]}\| \leq \gamma\|(T - \lambda_n^{[q]} I)\varphi_n^{[q]}\|,$$

where γ is a constant independent of n and q.

3.13 Let λ be a nonzero simple eigenvalue of T and $|\lambda| > s > 0$. Assume that $(T_n - T) \xrightarrow{\nu} O$. Let $0 < \epsilon < \min\{|\lambda| - s, \operatorname{dist}(\lambda, \operatorname{sp}(T) \setminus \{\lambda\})\}$. Then there is n_0 such that for each $n \geq n_0$ and $q = 2, 3, \ldots$ there is a unique $\lambda_n^{[q]} \in \operatorname{sp}(\boldsymbol{T}_n^{[q]})$ satisfying $|\lambda_n^{[q]} - \lambda| < \epsilon$. Also, $\lambda_n^{[q]}$ is a simple eigenvalue of $\boldsymbol{T}_n^{[q]}$ and $\lambda_n^{[q]} \to \lambda$ as $n \to \infty$, uniformly for $q = 2, 3, \ldots$ With φ, $\varphi_n^{[q]}$ and $\varphi_{(n)}^{[q]}$ as in 3.13, we have for each large n,

$$|\lambda_n^{[q]} - \lambda| \leq \frac{\alpha}{s^{q-1}} \min\left\{ \frac{\|(T - T_n)^q T_n \varphi_n^{[q]}\|}{\|\varphi_n^{[q]}\|}, \frac{\|(T - T_n)^q \varphi\|}{\|\varphi\|} \right\},$$

$$\|\varphi_n^{[q]} - P\varphi_n^{[q]}\| \leq \frac{\alpha}{(|\lambda| - \epsilon)^q} \|(T - T_n)^q T_n \varphi_n^{[q]}\|,$$

$$\frac{\|\varphi_n^{[q]} - \varphi_{(n)}^{[q]}\|}{\|\varphi_{(n)}^{[q]}\|} \leq \frac{\alpha}{|\lambda|^q} \frac{\|(T_n - T)^q T \varphi\|}{\|\varphi\|},$$

where α is a constant, independent of n and q. (Hint: Let $r = 1/s$ and apply Theorem 3.13 to rT, rT_n.)

3.14 Let $0 \neq \lambda \in \mathbb{C}$, $0 \neq \lambda_n^{[q]} \in \mathbb{C}$, $\boldsymbol{x} = [x_1, \ldots, x_q]^\top \in \boldsymbol{X}^{[q]}$ and $r > 0$.

(a) $\boldsymbol{x}$ is an eigenvector of $r \star \boldsymbol{T}^{[q]}$ corresponding to $r\lambda$ if and only if $Tx_1 = \lambda x_1$ and $x_{k+1} = x_1/(r\lambda)^k$ for $k = 1, \ldots, q-1$.

(b) $\boldsymbol{x}$ is an eigenvector of $r \star \boldsymbol{T}_n^{[q]}$ corresponding to $r\lambda_n^{[q]}$ if and only if $\sum_{k=0}^{q-1} (\lambda_n^{[q]})^{q-1-k} (T - T_n)^k T_n x_1 = (\lambda_n^{[q]})^q x_1$ and $x_{k+1} = x_1/(r\lambda_n^{[q]})^k$ for $k = 1, \ldots, q-1$.

In particular, $x_1 \in X$ is the first component of an eigenvector of $r \star \boldsymbol{T}^{[q]}$ (resp., $r \star \boldsymbol{T}_n^{[q]}$) corresponding to $r\lambda$ (resp., $r\lambda_n^{[q]}$) if and only if x_1 is the first component of an eigenvector of $\boldsymbol{T}^{[q]}$ (resp., $\boldsymbol{T}_n^{[q]}$) corresponding to λ (resp., $\lambda_n^{[q]}$).

3.15 Let $(T_n - T) \xrightarrow{\nu} O$. If E is a closed subset of $\operatorname{re}(T)$ such that $0 \notin E$, then there is a positive integer n_0 such that $E \subset \operatorname{re}(\boldsymbol{T}_n^{[q]})$ for all $n \geq n_0$ and all $q = 2, 3, \ldots$ For $n = 1, 2, \ldots$ consider $(SP)_n = \bigcup_{q=2}^{\infty} \operatorname{sp}(\boldsymbol{T}_n^{[q]})$.

(a) Property U: If $\lambda_n \in (SP)_n$ and $\lambda_n \to \lambda \neq 0$, then $\lambda \in \operatorname{sp}(T)$.

(b) Property L at a nonzero isolated point of $\operatorname{sp}(T)$: Let λ be a nonzero isolated point of $\operatorname{sp}(T)$ and $0 < \epsilon < \min\{|\lambda|, \operatorname{dist}(\lambda, \operatorname{sp}(T) \setminus \{\lambda\})\}$. If $\lambda_n \in (SP)_n$ and $|\lambda_n - \lambda| < \epsilon$, then $\lambda_n \to \lambda$. Also, there is an integer n_1 such that for each $n \geq n_1$ and each $q = 2, 3, \ldots$ the set $\{\lambda_n^{[q]} \in \operatorname{sp}(\boldsymbol{T}_n^{[q]}) : |\lambda - \lambda_n^{[q]}| < \epsilon\}$ is nonempty.

(c) Let Λ be a spectral set for T, $0 \notin \Lambda$ and $\mathrm{C} \in \mathcal{C}(T, \Lambda)$ with $0 \in \operatorname{ext}(\mathrm{C})$. Then there is some n_1 such that for all $n \geq n_1$ and all $q = 2, 3, \ldots$ $\Lambda_n^{[q]} := \operatorname{sp}(\boldsymbol{T}_n^{[q]}) \cap \operatorname{int}(\mathrm{C})$ and $\boldsymbol{P}_n^{[q]}$ do not depend on C. (Compare Proposition 2.9 (a).)

Chapter 4

Finite Rank Approximations

Consider a complex Banach space X and a bounded linear operator T on X, that is, $T \in \mathrm{BL}(X)$. In Chapter 2, we saw how spectral values of finite type and the associated spectral subspaces of T are approximated if a sequence (T_n) in $\mathrm{BL}(X)$ 'converges' to T; specifically if $T_n \xrightarrow{\mathrm{n}} T$, and more generally if $T_n \xrightarrow{\nu} T$. We shall see in Chapter 5 that if the rank of T_n is finite, then the spectral computations for T_n can be reduced to solving a matrix eigenvalue problem in a canonical way. For this reason, we now consider various situations in which it is possible to find a sequence (T_n) of bounded finite rank operators such that $T_n \xrightarrow{\mathrm{n}} T$, or at least, $T_n \xrightarrow{\nu} T$. First we treat some finite rank approximations which employ bounded finite rank projections. Next we treat the case of Fredholm integral operators, for which degenerate kernel approximations as well as approximations based on quadrature rules are described. For a weakly singular integral operator T, we present an approximation (T_n^K) such that $T_n^K = T_n + U_n$, where T_n is of finite rank, $T_n^K \xrightarrow{\nu} T$ but $T_n^K \stackrel{\mathrm{cc}}{\not\to} T$ and $T_n^K \stackrel{\mathrm{n}}{\not\to} T$. In the last section we give localization results along the lines of a result of Brakhage.

4.1 Approximations Based on Projections

Let (π_n) be a sequence of bounded projections defined on X, that is, each π_n is in $\mathrm{BL}(X)$ and $\pi_n^2 = \pi_n$. Define

$$T_n^P := \pi_n T, \quad T_n^S := T\pi_n \quad \text{and} \quad T_n^G := \pi_n T \pi_n.$$

The bounded operators T_n^P, T_n^S and T_n^G are known as the **projection approximation** of T, the **Sloan approximation** of T (cf. [67]) and the **Galerkin approximation** of T, respectively. All three are finite rank operators if the rank of the projection π_n is finite. We prove a basic convergence result for these approximations. Recall that we denote the identity operator on X by I.

Theorem 4.1
Let $T \in \mathrm{BL}(X)$ and $\pi_n \xrightarrow{\text{p}} I$. Then

(a) $T_n^P \xrightarrow{\text{p}} T$, $T_n^S \xrightarrow{\text{p}} T$ *and* $T_n^G \xrightarrow{\text{p}} T$.

(b) *If T is a compact operator, then* $T_n^P \xrightarrow{\text{n}} T$, $T_n^S \xrightarrow{\nu} T$, $T_n^G \xrightarrow{\nu} T$.

(c) *If T is a compact operator and $\pi_n^* \xrightarrow{\text{p}} I^*$, then* $T_n^S \xrightarrow{\text{n}} T$, $T_n^G \xrightarrow{\text{n}} T$.

Proof
The Uniform Boundedness Principle (Theorem 9.1 of [55] or Theorem 4.7-3 of [48]) shows that the sequence $(\|\pi_n\|)$ is bounded.

(a) Since $\pi_n \xrightarrow{\text{p}} I$, $T_n^P x = \pi_n T x \to Tx$ for each $x \in X$, that is, $T_n^P \xrightarrow{\text{p}} T$. Also, since T is continuous and

$$T_n^S - T = T(\pi_n - I), \quad T_n^G - T = (T_n^P - T)\pi_n + T_n^S - T,$$

we see that $T_n^S \xrightarrow{\text{p}} T$ and $T_n^G \xrightarrow{\text{p}} T$.

(b) Let the operator T be compact. The set $E := \{Tx : x \in X, \|x\| \le 1\}$ is relatively compact in the Banach space X. By the Banach-Steinhaus Theorem (Corollary 9.2(a) of [55]), the pointwise convergence of the sequence (π_n) to I is uniform on the set E, that is,

$$\|T_n^P - T\| = \|\pi_n T - T\| = \sup\{\|(\pi_n - I)y\| : y \in E\} \to 0.$$

Thus $T_n^P \xrightarrow{\text{n}} T$.

Since $\|T_n^S\| \le \|\pi_n\| \, \|T\|$ and $\|T_n^G\| \le \|\pi_n\|^2 \|T\|$ for all n, the sequences (T_n^S) and (T_n^G) are bounded in $\mathrm{BL}(X)$. Further, we note that

$$(T_n^S - T)T = T(T_n^P - T), \quad (T_n^G - T)T = (T_n^P - T)T_n^P + (T_n^S - T)T,$$

$$(T_n^S - T)T_n^S = (T_n^S - T)T\pi_n, \quad (T_n^G - T)T_n^G = (T_n^P - T)T_n^G.$$

Hence $T_n^S \xrightarrow{\nu} T$ and $T_n^G \xrightarrow{\nu} T$.

(c) Let T be compact and $\pi_n^* \xrightarrow{\text{p}} I^*$ in addition to $\pi_n \xrightarrow{\text{p}} I$. Since T^* is a compact operator on X^*, we see that

$$\|T_n^S - T\| = \|(T_n^S - T)^*\| = \|\pi_n^* T^* - T^*\| \to 0$$

by (a) above. Also,

$$\begin{aligned}\|T_n^G - T\| &= \|\pi_n(T_n^S - T) + T_n^P - T\| \\ &\le \alpha\|T_n^S - T\| + \|T_n^P - T\| \to 0.\end{aligned}$$

Thus $T_n^S \xrightarrow{\text{n}} T$ and $T_n^G \xrightarrow{\text{n}} T$. ∎

We remark that if T is compact, (T_n^S) and (T_n^G) in fact converge to T in a collectively compact manner. (See Theorem 4.5 of [25] or Theorem 15.1 of [54].) However, they may not in general converge to T in the norm. (See the comment before Proposition 4.6.)

We point out that the Hahn-Banach Extension Theorem (Theorem 7.8 of [55] or Theorem 4.3-2 of [48]) is used while asserting $\|(T_n^S - T)^*\| = \|T_n^S - T\|$ in the proof of (c) above.

For an estimation of the rate at which $\|T_n^P - T\|$, $\|(T_n^S - T)T\|$, $\|(T_n^S - T)T_n^S\|$, $\|(T_n^G - T)T\|$ and $\|(T_n^G - T)T_n^G\|$ may tend to zero, see Exercises 4.2 and 4.12.

We now give several ways of constructing a sequence (π_n) of bounded projections on X such that $\pi_n \xrightarrow{\text{p}} I$ and the rank of each π_n is finite.

4.1.1 Truncation of a Schauder Expansion

Assume that the Banach space X has a Schauder basis, that is, there are $e_1, e_2, \ldots$ in X such that $\|e_j\| = 1$ for each positive integer j and for every $x \in X$, there are unique scalars $c_1(x), c_2(x), \ldots$ for which

$$x = \sum_{j=1}^{\infty} c_j(x) e_j := \lim_{n\to\infty} \sum_{j=1}^{n} c_j(x) e_j.$$

As a consequence of the Bounded Inverse Theorem (Theorem 11.1 of [55]), each linear functional $x \mapsto c_j(x)$, $x \in X$, is continuous on X (Theorem 11.4 of [55]).

For each positive integer n, define

$$\pi_n x := \sum_{j=1}^{n} c_j(x) e_j, \quad x \in X.$$

Clearly, $\pi_n \in \mathrm{BL}(X)$, $\pi_n^2 = \pi_n$, $\pi_n \xrightarrow{\mathrm{p}} I$ and $\operatorname{rank} \pi_n = n$.

If an operator $T \in \mathrm{BL}(X)$ is represented by the infinite matrix $(t_{i,j})$ with respect to a Schauder basis $e_1, e_2, \ldots$ that is,

$$Te_j = \sum_{i=1}^{\infty} t_{i,j} e_i, \quad j = 1, 2, \ldots$$

then

$$T_n^P e_j = \pi_n T e_j \quad = \sum_{i=1}^{n} t_{i,j} e_i, \quad j = 1, 2, \ldots$$

$$T_n^S e_j = T \pi_n e_j \quad = \begin{cases} \sum_{i=1}^{\infty} t_{i,j} e_i & \text{if } j = 1, \ldots, n, \\ 0 & \text{if } j > n, \end{cases}$$

$$T_n^G e_j = \pi_n T \pi_n e_j = \begin{cases} \sum_{i=1}^{n} t_{i,j} e_i & \text{if } j = 1, \ldots, n, \\ 0 & \text{if } j > n. \end{cases}$$

Hence the matrix representing the projection approximation T_n^P of T is obtained by truncating each column of the matrix $(t_{i,j})$ at the nth entry and putting all zeros thereafter, while the matrix representing the Sloan approximation T_n^S of T is obtained by replacing every column after the nth column of the matrix $(t_{i,j})$ by a column of all zeros. The matrix representing the Galerkin approximation T_n^G of T is obtained by carrying out both these operations, so that the entries in the top left $n \times n$ corner of this matrix are the same as the corresponding entries of the matrix $(t_{i,j})$ and all other entries are equal to zero.

We now consider a special case. Let X be a separable Hilbert space and let $e_1, e_2, \ldots$ form an orthonormal basis for X. If $\langle \cdot\, , \cdot \rangle$ denotes the inner product on X, then by the Fourier Expansion Theorem (Theorem 22.7 of [55] or Theorem 3.5-2 of [48]), we have

$$x = \sum_{j=1}^{\infty} \langle x\, , e_j \rangle e_j, \quad x \in X.$$

Thus for each positive integer n,

$$\pi_n x = \sum_{j=1}^{n} \langle x\, , e_j \rangle e_j, \quad x \in X.$$

Since $\langle \pi_n x\,,\, y\rangle = \sum_{j=1}^{n} \langle x\,,\, e_j\rangle\langle e_j\,,\, y\rangle = \langle x\,,\, \pi_n y\rangle$ for all $x, y \in X$, we see that $\pi_n^* = \pi_n$, that is, the projection π_n is orthogonal. Hence if T is a compact operator on X, then we obtain not only $T_n^P \overset{\text{n}}{\to} T$ but also $T_n^S \overset{\text{n}}{\to} T$ and $T_n^G \overset{\text{n}}{\to} T$ by Theorem 4.1(b),(c).

Example 4.2 Some classical Schauder bases and orthonormal bases:
(a) Standard Schauder basis for ℓ^p, $1 \le p < \infty$.
For each positive integer j, let

$$e_k := (0, \ldots, 0, 1, 0, 0, \ldots),$$

where only the kth entry is 1. Then (e_k) forms a Schauder basis for ℓ^p. If $p = 2$, this is an orthonormal basis.

(b) Schauder basis of saw-tooth functions for $C^0([0,1])$.
For $t \in [0,1]$, let $e_0(t) := t$, $e_1(t) := 1 - t$,

$$e_2(t) := \begin{cases} 2e_0(t) & \text{if } 0 \le t \le \dfrac{1}{2}, \\ 2e_1(t) & \text{if } \dfrac{1}{2} < t \le 1. \end{cases}$$

For each positive integer m and $j = 1, \ldots, 2^m$, let

$$e_{2^m + j}(t) := e_2(2^m t - j + 1),$$

where we let $e_2(t) := 0$ for $t \notin [0,1]$. The sequence of these saw-tooth functions forms a Schauder basis for $C^0([0,1])$.

(c) Orthonormal basis of Haar functions for $L^2([0,1])$.
For $t \in [0,1]$, let $e_{0,0}(t) := 1$,

$$e_{1,0}(t) := \begin{cases} 1 & \text{if } 0 \le t < \dfrac{1}{2}, \\ -1 & \text{if } \dfrac{1}{2} < t < 1, \\ 0 & \text{if } t = \dfrac{1}{2}. \end{cases}$$

For each positive integer m and $j = 1, \ldots, 2^m$, let

$$e_{m,j}(t) := \begin{cases} 2^{m/2} & \text{if } \dfrac{j-1}{2^m} \le t < \dfrac{2j-1}{2^{m+1}}, \\ -2^{m/2} & \text{if } \dfrac{2j-1}{2^{m+1}} \le t < \dfrac{j}{2^m}, \\ 0 & \text{otherwise.} \end{cases}$$

The sequence of these **Haar functions** forms an orthonormal basis for the space $L^2([0,1])$.

(d) **Orthonormal basis of trigonometric functions for** $L^2([-\pi,\pi])$.
For $t \in [-\pi,\pi]$, let $e_j(t) := \dfrac{1}{\sqrt{2\pi}}(\cos jt + \mathrm{i}\,\sin jt)$, $j = 0, \pm1, \pm2, \ldots$ The sequence of these **trigonometric functions** forms an orthonormal basis for the space $L^2([-\pi,\pi])$.

For $t \in [0,\pi]$, let $e_0(t) := \dfrac{1}{\sqrt{\pi}}$, $e_j(t) := \dfrac{\sqrt{2}}{\sqrt{\pi}}\cos jt$ and $\widetilde{e}_j(t) := \dfrac{\sqrt{2}}{\sqrt{\pi}}\sin jt$. Then each of the sequences (e_j) and $(\widetilde{e}_j)$ forms an orthonormal basis for the space $L^2([0,\pi])$.

(e) **Orthonormal bases of polynomials for** $L^2_w(\,]a,b[\,)$.
For $t \in \,]a,b[$, let $y_j(t) := t^j$ for each nonnegative integer j. Applying the Gram-Schmidt Process described in Proposition 1.39 to the linearly independent set $\{y_0, y_1, y_2, \ldots\}$, we obtain a sequence of orthonormal polynomials (e_n). If $a := -1$ and $b := 1$, then they are known as the **Legendre polynomials**. Note that $e_0(t) = 1/\sqrt{2}$, $e_1(t) = \sqrt{3}\,t/\sqrt{2}$ and $e_2(t) = \sqrt{10}(3t^2 - 1)/4$ for $t \in \,]-1,1[$. The Legendre polynomials of higher degrees are often found by employing a three-term recurrence relation.

If w is a positive continuous function on $]a,b[$, one can orthonormalize $y_0, y_1, y_2, \ldots$ with respect to the inner product

$$\langle x\,,\,y\rangle_w := \int_a^b x(t)\overline{y(t)}w(t)\,dt$$

defined on the set $L^2_w(\,]a,b[\,)$ of all equivalence classes of Lebesgue measurable complex-valued functions x on $]a,b[$ satisfying $\int_a^b |x(t)|^2 w(t)\,dt < \infty$. Then the resulting sequence (e_j) of polynomials forms an orthonormal basis for $L^2_w(\,]a,b[\,)$ and the sequence $(\sqrt{w}\,e_j)$ forms an orthonormal basis for $L^2(\,]a,b[\,)$. ∎

4.1.2 Interpolatory Projections

Let $X := C^0([a,b])$ with the sup norm $\|\cdot\|_\infty$. For positive integers n and $r(n)$, consider the **nodes** $t_{n,1}, \ldots, t_{n,r(n)}$ in $[a,b]$:

$$a \le t_{n,1} < t_{n,2} < \cdots < t_{n,r(n)-1} < t_{n,r(n)} \le b$$

and let $t_{n,0} := a$ and $t_{n,r(n)+1} := b$. Consider functions $e_{n,1}, \ldots, e_{n,r(n)}$ in $C^0([a,b])$ such that

$$e_{n,j}(t_{n,k}) = \delta_{j,k}, \quad j,\, k = 1, \ldots, r(n).$$

We Define

$$(\pi_n x)(t) := \sum_{j=1}^{r(n)} x(t_{n,j}) e_{n,j}(t), \quad x \in X,\, t \in [a,b].$$

Then $\pi_n : X \to X$, and clearly $\pi_n^2 = \pi_n$. Also,

$$\mathcal{R}(\pi_n) = \operatorname{span}\{e_{n,1}, \ldots, e_{n,r(n)}\}$$

and the set $\{e_{n,1}, \ldots, e_{n,r(n)}\}$ is linearly independent. Hence $\operatorname{rank}(\pi_n) = r(n)$. Since for each $x \in X$,

$$(\pi_n x)(t_{n,j}) = x(t_{n,j}), \quad j = 1, \ldots, r(n),$$

that is, $\pi_n x$ interpolates x at $t_{n,1}, \ldots, t_{n,r(n)}$, we say that π_n is an **interpolatory projection**. An approximation based on an interpolatory projection is also known as a **collocation approximation**.

We show that $\|\pi_n\| = \left\| \sum_{j=1}^{r(n)} |e_{n,j}| \right\|_\infty$. For $x \in X$ and $t \in [a,b]$, we have

$$\begin{aligned} |(\pi_n x)(t)| &\le \sum_{j=1}^{r(n)} |x(t_{n,j})|\, |e_{n,j}(t)| \le \|x\|_\infty \sum_{j=1}^{r(n)} |e_{n,j}(t)| \\ &\le \|x\|_\infty \left\| \sum_{j=1}^{r(n)} |e_{n,j}| \right\|_\infty. \end{aligned}$$

Hence $\|\pi_n\| \le \left\| \sum_{j=1}^{r(n)} |e_{n,j}| \right\|_\infty$. On the other hand, choose $t_0 \in [a,b]$ such that

$$\Big(\sum_{j=1}^{r(n)} |e_{n,j}| \Big)(t_0) = \left\| \sum_{j=1}^{r(n)} |e_{n,j}| \right\|_\infty$$

and define $x_0 \in X$ as follows:

$$x_0(t_{n,j}) := \operatorname{sgn} e_{n,j}(t_0), \quad j = 1, \ldots, r(n),$$

$x_0(a) := x(t_{n,1})$, $x_0(b) := x(t_{n,r(n)})$ and x_0 is a polynomial of degree ≤ 1 on each of the subintervals $[a, t_{n,1}], [t_{n,1}, t_{n,2}], \ldots, [t_{r(n)-1}, t_{r(n)}], [t_{r(n)}, b]$. Then $\|x_0\|_\infty = 1$ and

$$(\pi_n x_0)(t_0) = \sum_{j=1}^{r(n)} x_0(t_{n,j}) e_{n,j}(t_0) = \sum_{j=1}^{r(n)} |e_{n,j}(t_0)| = \left\| \sum_{j=1}^{r(n)} |e_{n,j}| \right\|_\infty .$$

Hence $\|\pi_n\| \ge \left\| \sum_{j=1}^{r(n)} |e_{n,j}| \right\|_\infty$ and we are through.

As the following examples will show, we often have $r(n) = n$.

Example 4.3 Some classical nodes:
Some classical choices of nodes $t_{n,1}, \ldots, t_{n,n}$ in $[0,1]$ for which the mesh $h_n := \max\{t_{n,j} - t_{n,j-1} : j = 1, \ldots, n+1\} \to 0$ are as follows.

(a) $t_{n,j} := \dfrac{j}{n}$, $j = 1, \ldots, n$

(b) $t_{n,j} := \dfrac{j-1}{n}$, $j = 1, \ldots, n$

(c) $t_{n,j} := \dfrac{2j-1}{2n}$, $j = 1, \ldots, n$

(d) $t_{n,j} := \dfrac{j-1}{n-1}$, $j = 1, \ldots, n$ for $n > 1$

(e) $t_{n,j} := \begin{cases} \dfrac{j - (1/\sqrt{3})}{n} & \text{if } j = 1, 3, \ldots, n-1, \\ \dfrac{j - 1 + (1/\sqrt{3})}{n} & \text{if } j = 2, 4, \ldots, n, \end{cases}$ for $n = 2, 4, \ldots$

The nodes in (e) above are obtained by considering the roots $\pm 1/\sqrt{3}$ of the second degree Legendre polynomial in $[-1, 1]$ and transfering them to each of the $n/2$ intervals $[2(k-1)/n,\ 2k/n]$, $k = 1, \ldots, n/2$, where n is an even positive integer, by a linear change of variable. ∎

We now give some specific choices of functions $e_{n,j} \in C^0([a,b])$, $j = 1, \ldots, n$, which yield important sequences of interpolatory projections.

(i) Piecewise Linear Interpolation: For each positive integer n, define $e_{n,j}$ in $C^0([a,b])$ as follows:

$$e_{n,j}(t_{n,k}) := \delta_{j,k}, \quad j, k = 1, \ldots, n,$$

$e_{n,1}(a) := 1$, $e_{n,n}(b) := 1$, $e_{n,j}(a) := 0$ for $j = 2, \ldots, n$, $e_{n,j}(b) := 0$ for $j = 1, \ldots, n-1$ and $e_{n,j}$ is a polynomial of degree ≤ 1 on each of the subintervals $[a, t_{n,1}], [t_{n,1}, t_{n,2}], \ldots, [t_{n,n-1}, t_{n,n}]$ and $[t_{n,n}, b]$. The functions $e_{n,1}, \ldots, e_{n,n}$ are known as the **hat functions** because of the shape of their graphs.

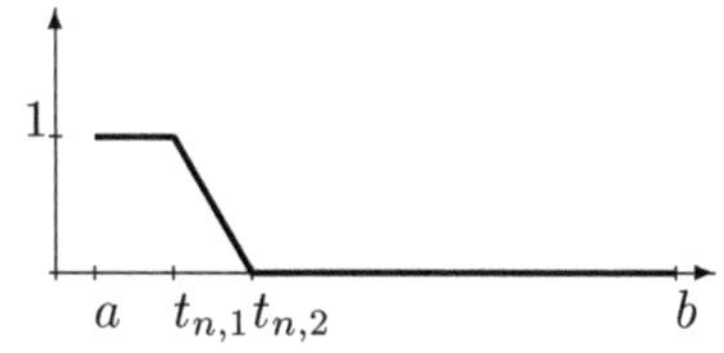

the hat function $e_{n,1}$

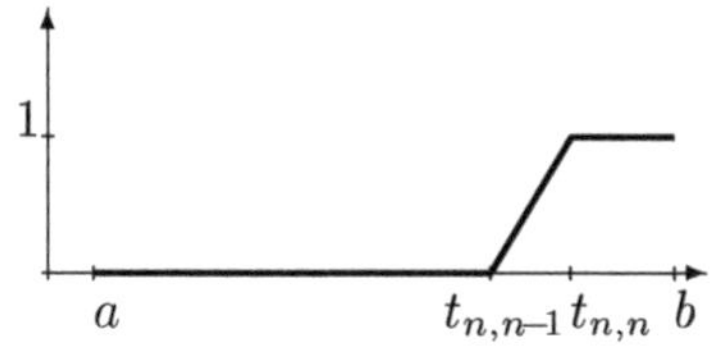

the hat function $e_{n,n}$

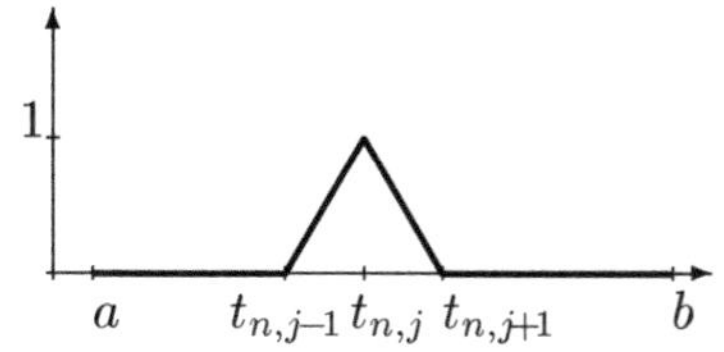

the hat function $e_{n,j}$, $1 < j < n$

Note that $e_{n,j}(t) \geq 0$ for all positive integers n, $j = 1, \ldots, n$ and $t \in [a, b]$. If $t \in [t_{n,j-1}, t_{n,j}]$ for some $j = 1, \ldots, n+1$, then $e_{n,j-1}(t) + e_{n,j}(t) = 1$, while $e_{n,k}(t) = 0$ for all $k \neq j-1, j$. In particular,

$$e_{n,1}(t) + \cdots + e_{n,n}(t) = 1 \quad \text{for all } t \in [a, b].$$

Let $t_{n,0} := a$ and $t_{n,n+1} := b$. We show that if the mesh

$$h_n := \max\{t_{n,j} - t_{n,j-1} \,:\, j = 1, \ldots, n+1\} \to 0,$$

then $\pi_n \xrightarrow{\text{p}} I$. Fix $x \in C^0([a, b])$ and let $\epsilon > 0$. By the uniform continuity of x on $[a, b]$, there is some $\delta > 0$ such that $|x(s) - x(t)| < \epsilon$, whenever $s, t \in [a, b]$ and $|s - t| < \delta$. Since $h_n \to 0$, choose n_0 such that $h_n < \delta$ for all $n \geq n_0$. Let $n \geq n_0$ and $t \in [a, b]$. If $t \in [t_{n,j-1}, t_{n,j}]$, then $|t_{n,j} - t| \leq h_n < \delta$ and hence $|x(t_{n,j}) - x(t)| < \epsilon$. Thus

$$\begin{aligned}
|(\pi_n x)(t) - x(t)| &= \Big| \sum_{i=1}^{n} [x(t_{n,i}) - x(t)] e_{n,i}(t) \Big| \\
&\leq |x(t_{n,j-1}) - x(t)| e_{n,j-1}(t) + |x(t_{n,j}) - x(t)| e_{n,j}(t) \\
&< \epsilon \left[e_{n,j-1}(t) + e_{n,j}(t) \right] = \epsilon.
\end{aligned}$$

Thus $\|\pi_n x - x\|_\infty \to 0$ for every $x \in C^0([a, b])$, that is, $\pi_n \xrightarrow{\text{p}} I$.

If in fact $x \in C^1([a,b])$ and $t \in [t_{n,j-1}, t_{n,j}]$ for some $j = 1, \ldots, n+1$, then by the Mean Value Theorem, $|x(t_{n,j-1}) - x(t)| = |x'(s_{n,j})|(t - t_{n,j-1})$ and $|x(t_{n,j}) - x(t)| = |x'(u_{n,j})|(t_{n,j} - t)$ for some $s_{n,j} \in [t_{n,j-1}, t]$ and $u_{n,j} \in [t, t_{n,j}]$, so that

$$\begin{aligned}
|\pi_n x(t) - x(t)| &\le |x'(s_{n,j})|(t - t_{n,j-1})e_{n,j-1}(t) \\
&\quad + |x'(u_{n,j})|(t_{n,j} - t)e_{n,j}(t) \\
&\le \|x'\|_\infty [(t - t_{n,j-1}) + (t_{n,j} - t)] \\
&\le \|x'\|_\infty h_n.
\end{aligned}$$

Thus

$$\|\pi_n x - x\|_\infty \le \|x'\|_\infty h_n \quad \text{for all } x \in C^1([a,b]).$$

It can also be proved that if $t_{n,1} = a$ and $t_{n,n} = b$ for all n, then

$$\|\pi_n x - x\|_\infty \le \frac{1}{8}\|x''\|_\infty (h_n)^2 \quad \text{for all } x \in C^2([a,b]).$$

(See Exercise 4.8.)

Piecewise quadratic interpolation is treated in Exercise 4.9.

(ii) Cubic Spline Interpolation: For each positive integer n and nodes $a = t_{n,1} < \cdots < t_{n,n} = b$ in $[a,b]$, let X_n denote the subspace of $X := C^0([a,b])$ consisting of all $x \in C^2([a,b])$ such that x is a polynomial of degree ≤ 3 on each of the $n-1$ subintervals $[t_{n,1}, t_{n,2}], \ldots, [t_{n,n-1}, t_{n,n}]$. Then $\dim X_n = n+2$. [Note: A polynomial of degree ≤ 3 on each of the $n-1$ subintervals has 4 degrees of freedom, which are constrained by 3 continuity conditions at each of the $n-2$ points $t_{n,2}, \ldots, t_{n,n-1}$ and $4(n-1) - 3(n-2) = n+2$. See Theorem 4.4 of [65].] An element of X_n is known as a cubic spline function with knots at $t_{n,1}, \ldots, t_{n,n}$. It can be shown that for each $j = 1, \ldots, n$, there is a unique function $e_{n,j}$ in X_n such that

$$e_{n,j}(t_{n,k}) = \delta_{j,k}, \quad k = 1, \ldots, n,$$

and the third derivative of $e_{n,j}$ exists at $t_{n,2}$ and $t_{n,n-1}$. The last requirement says that for each $e_{n,j}$, the two points $t_{n,2}$ and $t_{n,n-1}$ are not really knots. For $x \in X$, let $\pi_n x := \sum_{j=1}^{n} x(t_{n,j})e_j$, $x \in X$, as usual. Let h_n and $\widetilde{h}_n$ denote the maximum and the minimum of $\{t_{n,j} - t_{n,j-1} : j = 2, \ldots, n\}$, respectively. If there is a constant α such that $(h_n/\widetilde{h}_n) \le \alpha$ for all n (a so-called sequence of quasiuniform partitions) and $h_n \to 0$, then it is known that $\pi_n \xrightarrow{\text{p}} I$. Note that all the classical choices of nodes yield quasiuniform partitions. Further, it can be shown that

$$\|\pi_n x - x\|_\infty \leq \alpha \|x^{(4)}\|_\infty (h_n)^4 \quad \text{for all } x \in C^4([a,b]),$$

where α is a constant. (See Corollary 6.21 of [65] and pages 55, 56 of [29].)

Cubic spline interpolation for the case where $a < t_{n,1}$ and $b > t_{n,n}$ is considered in Exercise 4.11(b).

It is possible to choose a basis for the space X_n of cubic splines such that each function in the basis is nonzero on at most four of the subintervals $[a, t_{n,1}], \ldots, [t_{n,n-1}, b]$. Such functions are known as B-splines and have proved to be well suited for numerical computations. (See [29] for a discussion of B-splines and [70] for spectral computations using cubic splines.)

(iii) Lagrange Interpolation: Let $a \leq t_{n,1} < \cdots < t_{n,n} \leq b$. The Lagrange polynomials $p_{n,1}, \ldots, p_{n,n}$ with nodes at $t_{n,1}, \ldots, t_{n,n}$ are defined by

$$p_{n,j}(t) := \prod_{\substack{k=1 \\ k \neq j}}^{n} \frac{t - t_{n,k}}{t_{n,j} - t_{n,k}}, \quad t \in [a,b].$$

Note that each $p_{n,j}$ is a polynomial of degree $n-1$ and $p_{n,j}(t_{n,k}) = \delta_{j,k}$ for $j,\ k = 1, \ldots, n$. The interpolatory projection π_n, obtained by letting $e_{n,j} := p_{n,j}$, $j = 1, \ldots, n$, is known as the Lagrange interpolatory projection. A result of Kharshiladze and Lozinski implies that no matter how the nodes $t_{n,1}, \ldots, t_{n,n}$ are chosen, there is some $x \in C^0([a,b])$ for which the sequence $(\|\pi_n x - x\|_\infty)$ is unbounded. (See page 214 of [28].) Thus $\pi_n \overset{p}{\not\to} I$. If, however, $a := -1$, $b := 1$ and the nodes $t_{n,1}, \ldots, t_{n,n}$ are the roots of the nth Legendre polynomial in $[-1,1]$, then it can be proved that $\|\pi_n x - x\|_2 \to 0$ for each $x \in C^0([-1,1])$. (See page 137 of [28].) Using this result of Erdös and Turan, it can be seen that if T is a compact operator on $L^2([-1,1])$ with $\mathcal{R}(T) \subset C^0([-1,1])$, then $T_n^P \overset{n}{\to} T$; and if T is a Fredholm integral operator on $C^0([-1,1])$ with a continuous kernel, then $T_n^S \overset{\nu}{\to} T$. (See Exercise 4.13.)

4.1.3 Orthogonal Projections on Subspaces of Piecewise Constant Functions

Let $X := L^2([a,b])$. For each positive integer n, consider

$$a = t_{n,0} < t_{n,1} < \cdots < t_{n,n-1} < t_{n,n} = b$$

and the subspace $X_n := \{x \in X : x \text{ is constant on } [t_{n,j-1}, t_{n,j}[,\ j = 1, \ldots, n\}$ of X. For $x \in X$, let

$$(\pi_n x)(t) := \frac{1}{t_{n,j} - t_{n,j-1}} \int_{t_{n,j-1}}^{t_{n,j}} x(s)\, ds, \ t \in [t_{n,j-1}, t_{n,j}[, \ j = 1, \ldots, n,$$

$$(\pi_n x)(b) := (\pi_n x)(t_{n,n-1}).$$

It is clear that $\pi_n : X \to X$ is linear, $\mathcal{R}(\pi_n) = X_n$ and if $x \in X_n$, then $\pi_n x = x$. Hence $\pi_n^2 = \pi_n$ and $\operatorname{rank} \pi_n = \dim X_n = n$. Let $x \in X$. Then for $j = 1, \ldots, n$,

$$\begin{aligned}
\int_{t_{n,j-1}}^{t_{n,j}} |(\pi_n x)(t)|^2\, dt &= \int_{t_{n,j-1}}^{t_{n,j}} \left| \frac{1}{t_{n,j} - t_{n,j-1}} \int_{t_{n,j-1}}^{t_{n,j}} x(s)\, ds \right|^2 dt \\
&\leq \int_{t_{n,j-1}}^{t_{n,j}} \left[\frac{1}{t_{n,j} - t_{n,j-1}} \int_{t_{n,j-1}}^{t_{n,j}} |x(s)|^2\, ds \right] dt \\
&= \int_{t_{n,j-1}}^{t_{n,j}} \left[|x(s)|^2 \left(\int_{t_{n,j-1}}^{t_{n,j}} \frac{1}{t_{n,j} - t_{n,j-1}}\, dt \right) \right] ds \\
&= \int_{t_{n,j-1}}^{t_{n,j}} |x(s)|^2\, ds.
\end{aligned}$$

Hence

$$\|\pi_n x\|_2^2 = \sum_{j=1}^{n} \int_{t_{n,j-1}}^{t_{n,j}} |(\pi_n x)(t)|^2\, dt \leq \sum_{j=1}^{n} \int_{t_{n,j-1}}^{t_{n,j}} |x(s)|^2\, ds = \|x\|_2^2.$$

Thus $\|\pi_n\|_2 \leq 1$. It follows that π_n is an orthogonal projection defined on the Hilbert space $L^2([a, b])$ and that the range X_n of π_n is contained in the set of all piecewise constant functions.

We now show that if $h_n := \max\{t_{n,j} - t_{n,j-1} : j = 1, \ldots, n\} \to 0$, then $\|\pi_n x - x\|_2 \to 0$ for every $x \in L^2([a, b])$. Since the sequence $(\|\pi_n\|_2)$ is bounded and $C^0([a, b])$ is a dense subset of $L^2([a, b])$, it is enough to show that

$$\|\pi_n x - x\|_2 \to 0 \quad \text{for every } x \in C^0([a, b]).$$

Consider then $x \in C^0([a, b])$. Let $\epsilon > 0$. As x is uniformly continuous on $[a, b]$, there is $\delta > 0$ such that $|x(s) - x(t)| < \epsilon$ whenever $s, t \in [a, b]$ and $|s - t| < \delta$. Since $h_n \to 0$, choose n_0 such that $h_n < \delta$ for all $n \geq n_0$.

Then for all $n \geq n_0$ and $j = 1, \ldots, n$, we have

$$\begin{aligned}
&\int_{t_{n,j-1}}^{t_{n,j}} |(\pi_n x)(t) - x(t)|^2 \, dt \\
&= \int_{t_{n,j-1}}^{t_{n,j}} \left| \frac{1}{t_{n,j} - t_{n,j-1}} \int_{t_{n,j-1}}^{t_{n,j}} [x(t) - x(s)] \, ds \right|^2 dt \\
&\leq \int_{t_{n,j-1}}^{t_{n,j}} \left[\frac{1}{t_{n,j} - t_{n,j-1}} \int_{t_{n,j-1}}^{t_{n,j}} |x(t) - x(s)|^2 \, ds \right] dt \\
&\leq \int_{t_{n,j-1}}^{t_{n,j}} \epsilon^2 \, ds = \epsilon^2 (t_{n,j} - t_{n,j-1}).
\end{aligned}$$

Hence

$$\begin{aligned}
\|\pi_n x - x\|_2^2 &= \sum_{j=1}^{n} \int_{t_{n,j-1}}^{t_{n,j}} |(\pi_n x)(t) - x(t)|^2 \, dt \\
&\leq \sum_{j=1}^{n} \epsilon^2 (t_{n,j} - t_{n,j-1}) = \epsilon^2 (b - a)
\end{aligned}$$

for all $n \geq n_0$. Thus $\|\pi_n x - x\|_2 \to 0$ for every $x \in C^0([a, b])$. We conclude that $\pi_n \xrightarrow{\text{p}} I$.

One can similarly consider orthogonal projections defined on $L^2([a, b])$ whose ranges are contained in the set of all piecewise polynomials of degree $\leq d$, where d is a given positive integer. (See Exercise 4.14.)

4.1.4 Finite Element Approximation

If A and B are operators on a Banach space X, then the problem of finding a nonzero element φ of X and a scalar λ such that $A\varphi = \lambda B\varphi$ is known as the generalized eigenvalue problem.

A 'weak formulation' of the generalized eigenvalue problem which is valid for several differential operators can be given as follows.

Let X be a Hilbert space with an inner product $\langle \cdot\, , \cdot \rangle$. A sesquilinear functional $a(\cdot\, , \cdot)$ on $X \times X$ is a complex-valued function on $X \times X$ which is linear in the first variable and conjugate-linear in the second variable. It is said to be bounded if $|a(x, y)| \leq \alpha \|x\| \, \|y\|$ for some $\alpha > 0$ and all x, $y \in X$. Consider bounded sesquilinear functionals $a(\cdot\, , \cdot)$ and $b(\cdot\, , \cdot)$ on $X \times X$, where $a(\cdot\, , \cdot)$ is strongly coercive, that is,

$$\Re\, a(x, x) \geq \alpha \|x\|^2 \quad \text{for some } \alpha > 0 \text{ and all } x \in X.$$

Consider the problem of finding a nonzero element φ of X and a scalar λ such that

$$a(\varphi, y) = \lambda b(\varphi, y) \quad \text{for all } y \in X.$$

This is known as a **weakly posed generalized eigenvalue problem.** By the Riesz Representation Theorem for a Hilbert space (Theorem 24.3 of [55] or Theorem 3.8-1 of [48]), there are $A \in \mathrm{BL}(X)$ and $B \in \mathrm{BL}(X)$ such that

$$a(x, y) = \langle Ax\,,\, y\rangle \quad \text{and} \quad b(x, y) = \langle Bx\,,\, y\rangle \quad \text{for all } x,\, y \in X.$$

The strong coercivity of the sesquilinear functional $a(\cdot\,,\cdot)$ shows that the linear operator A is bounded below and its adjoint A^* is injective. It follows that the bounded operator A is invertible in $\mathrm{BL}(X)$. Let $T := A^{-1}B$.

A **finite element approximation** of the weakly posed generalized eigenvalue problem is obtained by considering, for each positive integer n, a finite dimensional subspace X_n of X and by requiring to find a nonzero element φ_n of X_n and a scalar λ_n such that

$$a(\varphi_n, y) = \lambda_n b(\varphi_n, y) \quad \text{for all } y \in X_n.$$

Let π_n denote the orthogonal projection defined on X with $\mathcal{R}(\pi_n) = X_n$. Then the preceding equation can be written as

$$\pi_n A\varphi_n = \lambda_n \pi_n B\varphi_n, \quad 0 \neq \varphi_n \in X_n.$$

Then for every $x \in X_n$, we have

$$\begin{aligned} \alpha\|x\|^2 &\leq \Re\, a(x, x) \leq |a(x, x)| = |\langle Ax\,,\, x\rangle| \\ &= |\langle Ax\,,\, \pi_n x\rangle| = |\langle \pi_n Ax\,,\, x\rangle| \\ &\leq \|\pi_n Ax\|\,\|x\|. \end{aligned}$$

Define $A_n := \pi_n A_{|X_n, X_n}$. Then $A_n \in \mathrm{BL}(X_n)$ is injective. Since the linear space X_n is finite dimensional, A_n is invertible and $\|A_n^{-1}\| \leq 1/\alpha$. Now the earlier equation can be written as

$$\varphi_n = \lambda_n A_n^{-1} \pi_n B\varphi_n = \lambda_n A_n^{-1} \pi_n AT\varphi_n, \quad 0 \neq \varphi_n \in X_n.$$

Define $\breve{\pi}_n x := A_n^{-1} \pi_n Ax$ for $x \in X$. Then $\breve{\pi}_n \in \mathrm{BL}(X)$, $\mathcal{R}(\breve{\pi}_n) = X_n$ and

$$\breve{\pi}_n^2 = A_n^{-1}(\pi_n A_{|X_n} A_n^{-1})\pi_n A = A_n^{-1}\pi_n A = \breve{\pi}_n.$$

Thus $\breve{\pi}_n$ is a bounded finite rank projection; and if we let

$$\breve{T}_n^P := \breve{\pi}_n T,$$

then our equation becomes

$$\breve{T}_n^P \varphi_n = \frac{1}{\lambda_n}\varphi_n, \quad 0 \neq \varphi_n \in X_n,$$

provided $\lambda_n \neq 0$. This shows that a finite element approximation of a weakly posed generalized eigenvalue problem can be realized as a projection approximation of a bounded operator T. In most applications, the operator T is compact and the finite dimensional subspaces $X_1, X_2, \ldots$ of X are so chosen that $\pi_n \xrightarrow{\text{p}} I$. Since $\breve{\pi}_n = \breve{\pi}_n - \pi_n + \pi_n = A_n^{-1}\pi_n A(I - \pi_n) + \pi_n$, we see that $\breve{\pi}_n \xrightarrow{\text{p}} I$ and so $\breve{T}_n^P \xrightarrow{\text{n}} T$.

4.2 Approximations of Integral Operators

The approximation procedures discussed in the last section are applicable to any compact operator T on a complex Banach space X. In this section we consider the special case where X is a suitable function space and T is a Fredholm integral operator on X. We develop several approximation procedures which are peculiar to this case.

For the sake of simplicity, we consider either $X := L^2([a,b])$ with the 2-norm and a function $k(\cdot\,,\cdot) \in L^2([a,b]\times[a,b])$, or $X := C^0([a,b])$ with the sup norm and a function $k(\cdot\,,\cdot) \in C^0([a,b]\times[a,b])$. For $x \in X$, let

$$(Tx)(s) := \int_a^b k(s,t)x(t)\,dt, \quad s \in [a,b].$$

The operator T is known as a Fredholm integral operator with kernel $k(\cdot\,,\cdot)$. It is easy to see that $T \in \mathrm{BL}(X)$. In fact, if $X := L^2([a,b])$, then

$$\|T\| \leq \Big(\int\int_{[a,b]\times[a,b]} |k(s,t)|^2\,dm(s,t)\Big)^{1/2} = \|k(\cdot\,,\cdot)\|_2,$$

and, if $X := C^0([a,b])$, then

$$\|T\| \leq (b-a)\sup\{|k(s,t)| \,:\, s,t \in [a,b]\} = (b-a)\|k(\cdot\,,\cdot)\|_\infty.$$

Further, T is a compact operator. (See Example 17.4(b) of [55] or Theorem 8.7-5 of [48].)

One can also consider integral operators on $L^p(S)$ or $C^0(S)$, where S is a suitable subset of $\mathbb{R}^k$, $k \geq 1$ and $1 \leq p \leq \infty$. However, we shall restrict ourselves to the two cases mentioned above.

We shall denote the 2-norm on $L^2([a,b]\times[a,b])$ as well as the sup norm on $C^0([a,b]\times[a,b])$ simply by $\|\cdot\|$.

4.2.1 Degenerate Kernel Approximation

A kernel $\widetilde{k}(\cdot\,,\cdot)$ is said to be degenerate if there are $x_1,\ldots,x_r$ and $y_1,\ldots,y_r$ in X such that

$$\widetilde{k}(s,t) := \sum_{j=1}^{r} x_j(s)y_j(t), \quad s,t \in [a,b].$$

If $\widetilde{T}$ is a Fredholm integral operator with a degenerate kernel $\widetilde{k}(\cdot\,,\cdot)$, then for $x \in X$, we have

$$(\widetilde{T}x)(s) = \sum_{j=1}^{r} x_j(s) \int_a^b y_j(t)x(t)\,dt, \quad s \in [a,b],$$

so that $\mathcal{R}(\widetilde{T}) \subset \operatorname{span}\{x_1,\ldots,x_r\}$ and hence $\widetilde{T}$ is a finite rank operator.

Theorem 4.4

Let T be a Fredholm integral operator on $X := L^2([a,b])$ or $X := C^0([a,b])$ with kernel $k(\cdot\,,\cdot)$; and for each positive integer n, let $k_n(\cdot\,,\cdot)$ be a degenerate kernel such that $\|k_n(\cdot\,,\cdot) - k(\cdot\,,\cdot)\| \to 0$. Consider the degenerate kernel approximation *of T given by:*

$$(T_n^D x)(s) := \int_a^b k_n(s,t)x(t)\,dt, \quad x \in X, \ s \in [a,b].$$

Then (T_n^D) is a sequence of bounded finite rank operators on X and $T_n^D \xrightarrow{\text{n}} T$.

Proof

Note that $T_n^D - T$ is a Fredholm integral operator on X with kernel $k_n(\cdot\,,\cdot) - k(\cdot\,,\cdot)$. If $X := L^2([a,b])$, then

$$\|T_n^D - T\|_2 \leq \|k_n(\cdot\,,\cdot) - k(\cdot\,,\cdot)\|_2 \to 0$$

and if $X := C^0([a,b])$, then

$$\|T_n^D - T\|_\infty \leq (b-a)\|k_n(\cdot\,,\cdot) - k(\cdot\,,\cdot)\|_\infty \to 0.$$

Hence the result follows. ∎

We now describe various methods for constructing a sequence of degenerate kernels which converges to a given kernel in the norm.

(i) Piecewise Linear Interpolation in the Second Variable: Let $k(\cdot\,,\cdot) \in C^0([a,b]\times[a,b])$, and for a fixed $s \in [a,b]$, consider $k_s(t) := k(s,t)$, $t \in [a,b]$. For each positive integer n, let $a = t_{n,0} \leq t_{n,1} < \cdots < t_{n,n} \leq t_{n,n+1} = b$ and let π_n denote the piecewise linear interpolatory projection described in part (i) of Subsection 4.1.2. Define

$$k_n(s,t) := (\pi_n k_s)(t) = \sum_{j=1}^{n} k(s, t_{n,j}) e_{n,j}(t), \quad s,\, t \in [a,b],$$

where $e_{n,j}$, $j = 1, \ldots, n$, are the corresponding hat functions. Thus $k_n(\cdot\,,\cdot)$ is a degenerate kernel obtained by interpolating the kernel $k(\cdot\,,\cdot)$ in the second variable. We show that $\|k_n(\cdot\,,\cdot) - k(\cdot\,,\cdot)\|_\infty \to 0$ if $h_n := \max\{t_{n,j} - t_{n,j-1} : j = 1, \ldots, n+1\} \to 0$.

Let $\epsilon > 0$. By the uniform continuity of the function $k(\cdot\,,\cdot)$ on $[a,b]\times[a,b]$, there exists $\delta > 0$ such that $|k(s,t) - k(s,u)| < \epsilon$ whenever $s \in [a,b]$ and $|t-u| < \delta$. Since $h_n \to 0$, choose n_0 such that $h_n < \delta$ for all $n \geq n_0$. If $n \geq n_0$, we have

$$|k_n(s,t) - k(s,t)| = |(\pi_n k_s)(t) - k_s(t)| < \epsilon$$

for all $s, t \in [a,b]$. (See the proof of $\pi_n \xrightarrow{\text{p}} I$ given in part (i) of Subsection 4.1.2.) Thus $\|k_n(\cdot\,,\cdot) - k(\cdot\,,\cdot)\|_\infty \to 0$.

If we interpolate the kernel $k(\cdot\,,\cdot)$ in both the variables, we obtain the degenerate kernel

$$\widetilde{k}_n(s,t) = \sum_{i,j=1}^{n} k(t_{n,i}, t_{n,j}) e_{n,i}(s) e_{n,j}(t), \quad s,t \in [a,b].$$

As before, it can be seen that $\|\widetilde{k}_n(\cdot\,,\cdot) - k(\cdot\,,\cdot)\|_\infty \to 0$ if $h_n \to 0$.

(ii) Bernstein Polynomials in Two Variables: Let $k(\cdot\,,\cdot)$ belong to $C^0([0,1]\times[0,1])$. For each nonnegative integer n, consider the nth

Bernstein polynomial in two variables given by

$$k_n(s,t) := \sum_{i,j=0}^{n} k\Big(\frac{i}{n},\frac{j}{n}\Big)\binom{n}{i}\binom{n}{j} s^i(1-s)^{n-i}t^j(1-t)^{n-j}, \; s,t \in [a,b].$$

Then $k_n(\cdot\,,\cdot)$ is a degenerate kernel and $\|k_n(\cdot\,,\cdot) - k(\cdot\,,\cdot)\|_\infty \to 0$. (See page 10 of [35].)

(iii) Truncation of a Taylor Expansion: Suppose that $k(\cdot\,,\cdot)$ belongs to $C^0([a,b]\times[a,b])$ and has a uniformly and absolutely convergent Taylor series expansion about some point $(s_0,t_0) \in \mathbb{R}^2$ given by

$$k(s,t) := \sum_{i,j=0}^{\infty} c_{i,j}(s-s_0)^i(t-t_0)^j, \quad s,\, t \in [a,b],$$

where $c_{i,j} \in \mathbb{C}$ for $i,j = 0,1,\ldots$ For each positive integer n, let

$$k_n(s,t) := \sum_{i,j=0}^{n} c_{i,j}(s-s_0)^i(t-t_0)^j, \quad s,t \in [a,b].$$

It follows that $\|k_n(\cdot\,,\cdot) - k(\cdot\,,\cdot)\|_\infty \to 0$. A simple example of this kind is given by

$$e^{st} := \sum_{j=0}^{\infty} \frac{s^j t^j}{j!}, \quad s,\, t \in [a,b],$$

where $(s_0,t_0) := (0,0)$, $c_{i,j} := 0$ if $i \neq j$ and $c_{j,j} := 1/j!$ for $i,j = 0,1,\ldots$

(iv) Truncation of a Fourier Expansion: Suppose that $k(\cdot\,,\cdot)$ belongs to $L^2([a,b]\times[a,b])$. Consider an orthonormal basis (e_n) for $L^2([a,b])$, and for positive integers i and j, let

$$k_{i,j}(s,t) := e_i(s)\overline{e_j(t)}, \quad s,\, t \in [a,b].$$

Then $(k_{i,j})$ is an orthonormal basis for $L^2([a,b]\times[a,b])$. (See 22.8(d) of [55].) By the Fourier Expansion Theorem for $L^2([a,b]\times[a,b])$ (Theorem 22.7 of [55] or Theorems 3.5-2, 3.6-2 and 3.6-3 of [48]), we have

$$k(\cdot\,,\cdot) = \sum_{i,j=1}^{\infty} c_{i,j}k_{i,j}(\cdot\,,\cdot),$$

where

$$c_{i,j} := \int_a^b \int_a^b k(s,t)\overline{k_{i,j}(s,t)}\, ds\, dt,$$

and the series converges in $L^2([a,b]\times[a,b])$. For each positive integer n, let

$$k_n(s,t) := \sum_{i,j=1}^{n} c_{i,j}k_{i,j}(s,t) = \sum_{i,j=1}^{n} c_{i,j}e_i(s)\overline{e_j(t)}, \quad s,\, t \in [a,b].$$

Clearly, $\|k_n(\cdot\,,\cdot) - k(\cdot\,,\cdot)\|_2 \to 0$.

If we let $\pi_n x := \sum_{j=1}^{n} \langle x\,,\, e_j\rangle e_j$ for $x \in L^2([a,b])$, then it is easy to see that $T_n^D = \pi_n T \pi_n = T_n^G$.

Several other degenerate kernel approximations are considered in [66].

4.2.2 Approximations Based on Numerical Integration

Let $X := C^0([a,b])$ with the sup norm and let $Q : X \to \mathbb{C}$ be defined by $Q(x) := \int_a^b x(t)\,dt$, $x \in X$. It is clear that Q is a continuous linear functional on X and $\|Q\| = b - a$. A quadrature formula is a linear functional $\widetilde{Q} : X \to \mathbb{C}$ given by

$$\widetilde{Q}(x) := \sum_{j=1}^{r} w_j x(t_j), \quad x \in X,$$

where the nodes $t_1, \ldots, t_r$ satisfy $a \le t_1 < \cdots < t_r \le b$ and the weights $w_1, \ldots, w_r$ are complex numbers. It is easy to see that the functional $\widetilde{Q}$ is continuous on X and in fact we have

$$\|\widetilde{Q}\| = \sum_{j=1}^{r} |w_j|.$$

We say that (Q_n) is a convergent sequence of quadrature formulæ if

$$Q_n(x) \to Q(x) \quad \text{for every } x \in X.$$

The following result, known as Polya's Theorem, gives a criterion for a sequence of quadrature formulæ to be convergent.

A sequence of quadrature formulæ given by

$$Q_n(x) := \sum_{j=1}^{r(n)} w_{n,j} x(t_{n,j}),\ x \in X,$$

is convergent if and only if

(i) $Q_n(y) \to Q(y)$ for every y in a subset E whose span is dense in $C^0([a,b])$,

(ii) $\sum_{j=1}^{r(n)} |w_{n,j}| \le \alpha$ for some constant α and all positive integers n.

The proof depends on the Uniform Boundedness Principle. A convenient choice of a subset E whose span is dense in $C^0([a,b])$ is $\{y_0, y_1, y_2, \ldots\}$, where $y_k(t) := t^k$, $t \in [a,b]$. In any case, if the weights $w_{n,j}$are all nonnegative and if the constant function y_0 is in the set E, the condition (ii) given above is automatically satisfied since

$$\sum_{j=1}^{r(n)} |w_{n,j}| = \sum_{j=1}^{r(n)} w_{n,j} = Q_n(y_0) \to Q(y_0) = b - a.$$

We now describe a natural way of approximating a Fredholm integral operator T by employing a sequence (Q_n) of quadrature formulæ. For a fixed $s \in [a,b]$, consider the function $k_s(t) := k(s,t)$, $t \in [a,b]$. Let $x \in X$. Then $k_s\, x \in X$ for every fixed $s \in [a,b]$ and

$$(Tx)(s) = \int_a^b (k_s\, x)(t)\, dt = Q(k_s\, x).$$

We define an approximating operator T_n^N by replacing the functional Q by the quadrature formula Q_n in the equation given above. Thus for each positive integer n and all $x \in X$, let

$$\begin{aligned}(T_n^N x)(s) := Q_n(k_s\, x) &= \sum_{j=1}^{r(n)} w_{n,j}(k_s\, x)(t_{n,j}) \\ &= \sum_{j=1}^{r(n)} w_{n,j} k(s, t_{n,j}) x(t_{n,j}), \quad s \in [a,b].\end{aligned}$$

The operator T_n^N is known as the **Nyström approximation** of T based on the quadrature formula Q_n.

Let (π_n) be a sequence of bounded projections defined on X and let

$$T_n^F := \pi_n T_n^N$$

for each positive integer n. The operator T_n^F is known as the **Fredholm approximation** of T based on the quadrature formula Q_n and the bounded projection π_n.

Theorem 4.5
Let $X := C^0([a,b])$ and T be a Fredholm integral operator on X with a continuous kernel.

(a) *Let (T_n^N) be a Nyström approximation of T based on a convergent sequence of quadrature formulæ. Then $T_n^N \xrightarrow{\text{p}} T$ and $T_n^N \xrightarrow{\nu} T$.*

(b) *Let, in addition, (π_n) be a sequence of bounded projections such that $\pi_n \xrightarrow{\text{p}} I$. Then $T_n^F \xrightarrow{\text{p}} T$ and $T_n^F \xrightarrow{\nu} T$.*

Proof

(a) Let $x \in X$. Since the sequence (Q_n) of quadrature formulæ employed to define the Nyström approximation (T_n^N) is convergent, we see that

$$(T_n^N x)(s) = Q_n(k_s\, x) \to Q(k_s\, x) = (Tx)(s)$$

for each $s \in [a,b]$. We show that this convergence is uniform for $s \in [a,b]$. The subset $S := \{k_s\, x \,:\, s \in [a,b]\}$ of X is uniformly bounded since for all $s \in [a,b]$,

$$\|k_s\, x\|_\infty \le \|k_s\|_\infty \|x\|_\infty \le \|k(\cdot\,,\cdot)\|_\infty \|x\|_\infty.$$

Also, it is uniformly equicontinuous, since the functions x and $k(\cdot\,,\cdot)$ are uniformly continuous; and for all t, $u \in [a,b]$, we have

$$\begin{aligned}|k_s(t)x(t) - k_s(u)x(u)| &\le |k_s(t)x(t) - k_s(t)x(u)| \\ &\quad + |k_s(t)x(u) - k_s(u)x(u)| \\ &\le \|k(\cdot\,,\cdot)\|_\infty |x(t) - x(u)| \\ &\quad + \|x\|_\infty \sup_{s\in[a,b]} |k(s,t) - k(s,u)|.\end{aligned}$$

By Ascoli's Theorem (Theorem 3.10(a) of [55]), the set S is relatively compact; and by the Banach-Steinhaus Theorem (Theorem 9.2(a) of [55]), the pointwise convergence of the sequence (Q_n) of continuous functionals is uniform on S. This means that $\|T_n^N x - Tx\|_\infty \to 0$. Thus $T_n^N \xrightarrow{\text{p}} T$. In particular, $(\|T_n^N\|)$ is bounded.

Let $E := \{Tx \,:\, x \in X,\, \|x\|_\infty \le 1\}$. Since T is a compact operator, the set E is relatively compact in X. Again by the Banach-Steinhaus Theorem, the pointwise convergence of (T_n^N) to T is uniform on E, that is,

$$\|(T_n^N - T)T\| = \sup\{\|(T_n^N - T)y\|_\infty \,:\, y \in E\} \to 0.$$

Next, the subset

$$\widetilde{E} := \bigcup_{n=1}^{\infty} \{T_n^N x \,:\, x \in X,\ \|x\|_\infty \leq 1\}$$

of X is uniformly bounded since for all $x \in X$ with $\|x\|_\infty \leq 1$, $\|T_n^N x\| \leq \sup_{n\geq 1} \|T_n^N\| < \infty$. Also, it is uniformly equicontinuous since for all n, all $x \in X$ with $\|x\|_\infty \leq 1$ and all s, $u \in [a,b]$, we have

$$\begin{aligned}|(T_n^N x)(s) - (T_n^N x)(u)| &\leq \sum_{j=1}^{r(n)} |w_{n,j}|\,|k(s,t_{n,j}) - k(u,t_{n,j})||x(t_{n,j})| \\ &\leq \Big(\sup_{n\geq 1} \sum_{j=1}^{r(n)} |w_{n,j}|\Big)\Big(\sup_{t\in[a,b]} |k(s,t) - k(u,t)|\Big).\end{aligned}$$

Note that the function $k(\cdot\,,\cdot)$ is uniformly continuous on $[a,b]\times[a,b]$ and

$$\sup_{n\geq 1} \sum_{j=1}^{r(n)} |w_{n,j}| < \infty$$

by Polya's Theorem. Again by Ascoli's Theorem, the set $\widetilde{E}$ is relatively compact; and by the Banach-Steinhaus Theorem, the pointwise convergence of (T_n^N) to T is uniform on $\widetilde{E}$, so that

$$\|(T_n^N - T)T_n^N\| \leq \sup\{\|(T_n^N - T)\widetilde{y}\|_\infty \,:\, \widetilde{y} \in \widetilde{E}\} \to 0.$$

Thus $T_n^N \xrightarrow{\nu} T$.

(b) Let (π_n) be a sequence of bounded projections defined on X such that $\pi_n \xrightarrow{\text{p}} I$. By the Uniform Boundedness Principle, $\|\pi_n\| \leq \alpha$ for some $\alpha > 0$ and all n. Since

$$T_n^F - T = \pi_n(T_n^N - T) + \pi_n T - T = \pi_n(T_n^N - T) + T_n^P - T,$$

we see that $T_n^F \xrightarrow{\text{p}} T$. In particular, $(\|T_n^F\|)$ is bounded.

Since the sets E and $\widetilde{E}$, introduced in the proof of (a) above, are relatively compact, we have

$$\|(T_n^F - T)T\| = \sup\{\|(T_n^F - T)y\|_\infty \,:\, y \in E\} \to 0,$$

$$\|(T_n^F - T)T_n^F\| = \sup\{\|(T_n^F - T)\pi_n\widetilde{y}\|_\infty \,:\, \widetilde{y} \in \widetilde{E}\} \to 0$$

again by the Banach-Steinhaus Theorem, since $(T_n^F - T)\pi_n \xrightarrow{\text{p}} O$. Thus $T_n^F \xrightarrow{\nu} T$. ▮

We remark that in fact $T_n^N \xrightarrow{\text{cc}} T$ and $T_n^F \xrightarrow{\text{cc}} T$. (See Propositions 2.1 and 2.2 of [17], Theorem 4.11 and Corollary 4.12 of [25], or Theorem 16.2 of [54].) However, $T_n^N \overset{\text{n}}{\nrightarrow} T$ and $T_n^F \overset{\text{n}}{\nrightarrow} T$, unless, of course, $T := O$. This follows from the following result which also implies that if π_n is an interpolatory projection, then $T_n^S := T\pi_n \overset{\text{n}}{\nrightarrow} T$ and $T_n^G := \pi_n T \pi_n \overset{\text{n}}{\nrightarrow} T$, unless $T := O$.

Proposition 4.6
Let $X := C^0([a,b])$, $T \in \text{BL}(X)$ be a Fredholm integral operator with a kernel $k(\cdot,\cdot)$, and (T_n) be a sequence in $\text{BL}(X)$. Suppose that the following conditions are satisfied.

(i) *For each $x \in X$ and each $s \in [a,b]$, $(T_n x)(s) \to (Tx)(s)$ as $n \to \infty$.*

(ii) *For each n, there are $t_{n,1}, \ldots, t_{n,r(n)}$ in $[a,b]$ such that*

1. *$T_n x = 0$ whenever $x \in X$ and $x(t_{n,1}) = \cdots = x(t_{n,r(n)}) = 0$,*
2. *For each $s \in [a,b]$, there is $\delta_n(s) > 0$ such that*

$$\sum_{j=1}^{r(n)} \int_{|t_{n,j}-t|<\delta_n(s)} k(s,t)\,dt \to 0 \qquad \text{as } n \to \infty.$$

Then

$$\liminf_{n\to\infty} \|T_n - T\| \geq 2\|T\|.$$

Proof
Let $\epsilon > 0$. Then there exist $x_0 \in X$ and $s_0 \in [a,b]$ such that $\|x_0\|_\infty \leq 1$ and

$$|(Tx_0)(s_0)| > \|T\| - \epsilon.$$

Since, by condition (i), $(T_n x_0)(s_0) \to (Tx_0)(s_0)$, choose n_0 such that for all $n \geq n_0$, we have

$$|(T_n x_0)(s_0) - (Tx_0)(s_0)| < \epsilon.$$

For each $n = 1, 2, \ldots$ consider points $t_{n,1}, \ldots, t_{n,r(n)}$ and let $\delta_n := \delta_n(s_0)$, as stated in condition (ii). By altering the continuous function x_0 on the intervals $\mathcal{I}_{n,j} :=]t_{n,j} - \delta_n, t_{n,j} + \delta_n[\cap [a,b]$, $j = 1, \ldots, r(n)$, construct a function $x_n \in C^0([a,b])$ such that

$$\|x_n\|_\infty \leq 1, \ x_n(t_{n,j}) = -x_0(t_{n,j}) \text{ for } j = 1, \ldots, r(n).$$

Then

$$\begin{aligned} |(Tx_n)(s_0) - (Tx_0)(s_0)| &= \Big| \sum_{j=1}^{r(n)} \int_{\mathcal{I}_{n,j}} k(s_0,t)[x_n(t) - x_0(t)]\,dt \Big| \\ &\leq 2 \sum_{j=1}^{r(n)} \int_{\mathcal{I}_{n,j}} |k(s_0,t)|\,dt, \end{aligned}$$

so that

$$(Tx_n)(s_0) \to (Tx_0)(s_0).$$

Then for each n, we have $(x_0 + x_n)(t_{n,j}) = 0$, $j = 1, \ldots, r(n)$, so that $T_n(x_0 + x_n) = 0$ by condition (ii). In particular, $(T_n x_n)(s_0) = -(T_n x_0)(s_0)$. Hence

$$|(T_n x_n - Tx_n)(s_0)| = |-(T_n x_0)(s_0) - (Tx_n)(s_0)| \to 2|(Tx_0)(s_0)|.$$

Since $\|x_n\|_\infty \leq 1$, we see that

$$\|T_n - T\| \geq \|(T_n - T)x_n\|_\infty \geq |(T_n x_n - Tx_n)(s_0)|.$$

Thus

$$\liminf_{n\to\infty} \|T_n - T\| \geq \lim_{n\to\infty} |(T_n x_n - Tx_n)(s_0)| = 2|(Tx_0)(s_0)| \geq 2\|T\| - 2\epsilon.$$

As $\epsilon > 0$ is arbitrary, we have

$$\liminf_{n\to\infty} \|T_n - T\| \geq 2\|T\|.$$

The proof is complete. ∎

The special case of Proposition 4.6 (where $k(\cdot\,,\cdot)$ is a continuous kernel and $T_n := T_n^N$) seems to have prompted Anselone to develop the theory of collectively compact operator approximation and to extend several classical results about norm convergence. (See [17].) As we have seen in Lemma 2.2, if T is compact, then $T_n \overset{\text{cc}}{\to} T$ implies that $T_n \overset{\nu}{\to} T$. Hence our results for ν-convergence in Chapter 2 further extend the classical results.

We now give a number of convergent sequences of quadrature formulæ, which can be employed for constructing Nyström and Fredholm approximations of an integral operator.

Several quadrature formulæ arise as Riemann sums: For $n = 1, 2, \ldots$ consider a partition $a = \tau_{n,0} < \tau_{n,1} < \cdots < \tau_{n,r(n)} = b$ of the interval $[a, b]$ and let $\widetilde{h}_n := \max\{\tau_{n,j} - \tau_{n,j-1} : j = 1, \ldots, r(n)\}$. Choose $t_{n,j}$ in $[\tau_{n,j-1}, \tau_{n,j}]$ for $j = 1, \ldots, r(n)$. Letting $t_{n,0} := a$, $t_{n,r(n)+1} := b$ and $h_n := \max\{t_{n,j} - t_{n,j-1} : j = 1, \ldots, r(n) + 1\}$, we note that $h_n \leq 2\widetilde{h}_n$ as well as $\widetilde{h}_n \leq 2h_n$ since $\tau_{n,j} \in [t_{n,j}, t_{n,j+1}]$ for $j = 0, \ldots, r(n)$.

Let $w_{n,j} := \tau_{n,j} - \tau_{n,j-1}$ for $j = 1, \ldots, r(n)$, and for $x : [a, b] \to \mathbb{C}$,

$$Q_n(x) := \sum_{j=1}^{r(n)} w_{n,j} x(t_{n,j}) = \sum_{j=1}^{r(n)} x(t_{n,j})(\tau_{n,j} - \tau_{n,j-1}).$$

Then $Q_n(x)$ is a **Riemann sum** for x. Hence $Q_n(x) \to \int_a^b x(t)\, dt$ for every Riemann integrable function x on $[a, b]$, provided $\widetilde{h}_n \to 0$, or equivalently $h_n \to 0$, as $n \to \infty$.

If $x \in C^0([a, b])$, then we can describe the rate of convergence of $(Q_n(x))$ to $\int_a^b x(t)\, dt$ in terms of the **modulus of continuity** of x: For $\delta > 0$, let

$$\omega(x, \delta) := \max\{|x(s) - x(t)| : s, t \in [a, b], |s - t| \leq \delta\}.$$

By the Mean Value Theorem for Integrals, for each $j = 1, \ldots, r(n)$, there exists $s_{n,j} \in [\tau_{n,j-1}, \tau_{n,j}]$ such that

$$\begin{aligned}\left|Q_n(x) - \int_a^b x(t)\, dt\right| &= \left|\sum_{j=1}^{r(n)} \left(x(t_{n,j})(\tau_{n,j} - \tau_{n,j-1}) - \int_{\tau_{n,j-1}}^{\tau_{n,j}} x(t)\, dt\right)\right| \\ &= \left|\sum_{j=1}^{r(n)} \left[x(t_{n,j}) - x(s_{n,j})\right](\tau_{n,j} - \tau_{n,j-1})\right| \\ &\leq \omega(x, \widetilde{h}_n) \sum_{j=1}^{r(n)} (\tau_{n,j} - \tau_{n,j-1}) = (b - a)\omega(x, \widetilde{h}_n).\end{aligned}$$

The most simple examples of this kind are the **Rectangular Rules**: Consider $a := 0$, $b := 1$ and $\tau_{n,j} := j/n$ for $j = 0, \ldots, n$. For $j = 1, \ldots, n$, we have $w_{n,j} := \tau_{n,j} - \tau_{n,j-1} = 1/n$. Next, for $j = 1, \ldots, n$, let $t_{n,j} := j/n$ as in Example 4.3(a), or $t_{n,j} := (j-1)/n$ as in Example 4.3(b). Then for $x \in C^0([0,1])$,

$$Q_n(x) := \frac{1}{n}\sum_{j=1}^{n} x\Big(\frac{j}{n}\Big) \quad \text{or} \quad Q_n(x) := \frac{1}{n}\sum_{j=1}^{n} x\Big(\frac{j-1}{n}\Big)$$

and

$$\Big|Q_n(x) - \int_0^1 x(t)\,dt\Big| \le \omega(x, 1/n).$$

If in fact $x \in C^1([0,1])$, then $\omega(x, 1/n) \le \|x'\|_\infty/n$ and hence

$$\Big|Q_n(x) - \int_0^1 x(t)\,dt\Big| \le \frac{\|x'\|_\infty}{n} \quad \text{for } n = 1, 2, \ldots$$

Other classical quadrature formulæ can also be interpreted as Riemann sums, as indicated in Example 4.9.

Proposition 4.7

Let T be a Fredholm integral operator on $X := C^0([a,b])$ with a continuous kernel $k(\cdot\,,\cdot)$. Consider a convergent sequence (Q_n) of quadrature formulæ, and let T_n^N be the Nyström approximation of T based on Q_n.

Assume that for each n, the quadrature formula Q_n gives a Riemann sum, that is, there is a partition $a = \tau_{n,0} < \tau_{n,1} < \cdots < \tau_{n,r(n)} = b$ of $[a,b]$ such that $t_{n,j} \in [\tau_{n,j-1}, \tau_{n,j}]$ for $j = 1, \ldots, r(n)$ and $Q_n(x) := \sum_{j=1}^{r(n)} w_{n,j} x(t_{n,j})$, where $w_{n,j} := \tau_{n,j} - \tau_{n,j-1}$ for $j = 1, \ldots, r(n)$. For $s, t \in [a,b]$, define

$$k_{s,t}(u) := k(s,u)k(u,t) \quad \text{for } u \in [a,b].$$

Then for $n = 1, 2, \ldots$ we have

$$\|(T_n^N - T)T\|,\ \|(T_n^N - T)T_n^N\| \le (b-a) \sup_{s,t \in [a,b]} \omega(k_{s,t}, \widetilde{h}_n),$$

where $\omega(\cdot\,,\cdot)$ denotes the modulus of continuity and $\widetilde{h}_n := \max\{\tau_{n,j} - \tau_{n,j-1} : j = 1, \ldots, r(n)\}$.

Proof
Let $x \in X$ and $s \in [a,b]$. We have

$$(T_n^N - T)Tx(s) = \sum_{j=1}^{r(n)} w_{n,j} k(s, t_{n,j}) Tx(t_{n,j}) - \int_a^b k(s,u) Tx(u)\, du$$

$$= \sum_{j=1}^{r(n)} w_{n,j} k(s, t_{n,j}) \int_a^b k(t_{n,j}, t) x(t)\, dt - \int_a^b k(s,u) \Big[\int_a^b k(u,t) x(t)\, dt \Big] du$$

$$= \int_a^b \Big[\sum_{j=1}^{r(n)} w_{n,j} k(s, t_{n,j}) k(t_{n,j}, t) - \int_a^b k(s,u) k(u,t)\, du \Big] x(t)\, dt$$

$$= \int_a^b \Big[\sum_{j=1}^{r(n)} w_{n,j} k_{s,t}(t_{n,j}) - \int_a^b k_{s,t}(u)\, du \Big] x(t)\, dt.$$

Since $w_{n,j} = \tau_{n,j} - \tau_{n,j-1}$ for $j = 1, \ldots, r(n)$, we obtain

$$|(T_n^N - T)Tx(s)| \leq (b-a) \|x\|_\infty \sup_{t \in [a,b]} \omega(k_{s,t}, \widetilde{h}_n),$$

as noted earlier. Hence

$$\|(T_n^N - T)T\| \leq (b-a) \sup_{s,t \in [a,b]} \omega(k_{s,t}, \widetilde{h}_n).$$

Again, we have

$$\begin{aligned}
(T_n^N - T)T_n^N x(s) &= \sum_{j=1}^{r(n)} w_{n,j} k(s, t_{n,j}) \sum_{\ell=1}^{r(n)} w_{n,\ell} k(t_{n,j}, t_{n,\ell}) x(t_{n,\ell}) \\
&\quad - \int_a^b k(s,u) \Big[\sum_{\ell=1}^{r(n)} w_{n,\ell} k(u, t_{n,\ell}) x(t_{n,\ell}) \Big] du \\
&= \sum_{\ell=1}^{r(n)} w_{n,\ell} \Big[\sum_{j=1}^{r(n)} w_{n,j} k(s, t_{n,j}) k(t_{n,j}, t_{n,\ell}) \\
&\quad - \int_a^b k(s,u) k(u, t_{n,\ell})\, du \Big] x(t_{n,\ell}) \\
&= \sum_{\ell=1}^{r(n)} w_{n,\ell} \Big[\sum_{j=1}^{r(n)} w_{n,j} k_{s,t_{n,\ell}}(t_{n,j}) \\
&\quad - \int_a^b k_{s,t_{n,\ell}}(u)\, du \Big] x(t_{n,\ell}).
\end{aligned}$$

As above, we obtain

$$|(T_n^N - T)T_n^N x(s)| \leq \Big(\sum_{\ell=1}^{r(n)} w_{n,\ell}\Big)\|x\|_\infty \sup_{\ell=1,\ldots,r(n)} \omega(k_{s,t_{n,\ell}}, \widetilde{h}_n).$$

But $\sum_{\ell=1}^{r(n)} w_{n,\ell} = \sum_{\ell=1}^{r(n)} (\tau_{n,\ell} - \tau_{n,\ell-1}) = b - a$. Hence

$$\|(T_n^N - T)T_n^N\| \leq (b-a) \sup_{s,t\in[a,b]} \omega(k_{s,t}, \widetilde{h}_n),$$

as desired. ∎

Example 4.8
Consider the kernel

$$k(s,t) := \begin{cases} s(1-t) & \text{if } 0 \leq s \leq t \leq 1, \\ (1-s)t & \text{if } 0 \leq t < s \leq 1. \end{cases}$$

It is continuous on $[0,1]\times[0,1]$. Fix s, $t \in [0,1]$ such that $s \leq t$. Then for $u \in [0,1]$, we have

$$k_{s,t}(u) := k(s,u)k(u,t) = \begin{cases} (1-s)(1-t)u^2 & \text{if } u \leq s, \\ s(1-t)u(1-u) & \text{if } s < u < t, \\ st(1-u)^2 & \text{if } t \leq u. \end{cases}$$

It can be seen that for $\delta > 0$, we have

$$\omega(k_{s,t}, \delta) \leq 4\delta.$$

This inequality holds for $t \leq s$ as well, since the kernel $k(\cdot\,,\cdot)$ is symmetric in s and t. Proposition 4.7 now shows that for the Fredholm integral operator T with this kernel, we have

$$\|(T_n^N - T)T\|\,,\, \|(T_n^N - T)T_n^N\| \leq 4(b-a)\widetilde{h}_n,$$

where $\widetilde{h}_n$ is as defined in the proposition. ∎

Many quadrature formulæ arise from interpolatory projections considered in Subsection 4.1.2. Let $a \leq t_{n,1} < t_{n,2} < \cdots < t_{n,r(n)} \leq b$

and $e_{n,1}, \ldots, e_{n,r(n)}$ be in $C^0([a,b])$ such that $e_{n,j}(t_{n,k}) = \delta_{j,k}$ for j, $k = 1, \ldots, r(n)$. Let

$$\pi_n x := \sum_{j=1}^{r(n)} x(t_{n,j}) e_{n,j}, \quad x \in C^0([a,b]),$$

and consider the induced quadrature formula Q_n given by

$$Q_n(x) := \int_a^b (\pi_n x)(t)\, dt = \sum_{j=1}^{r(n)} w_{n,j} x(t_{n,j}), \quad x \in C^0([a,b]),$$

where $w_{n,j} := \displaystyle\int_a^b e_{n,j}(t)\, dt$, $j = 1, \ldots, r(n)$. Since for $j = 1, \ldots, r(n)$,

$$Q_n(e_{n,j}) = \int_a^b (\pi_n e_{n,j})(t)\, dt = \int_a^b e_{n,j}(t)\, dt,$$

the quadrature formula Q_n is exact on $\operatorname{span}\{e_{n,1}, \ldots, e_{n,r(n)}\}$.

Also, for $x \in C^0([a,b])$ and $s \in [a,b]$, we have

$$(T_n^N \pi_n x)(s) = \sum_{j=1}^{r(n)} w_{n,j} k(s, t_{n,j}) (\pi_n x)(t_{n,j}) = (T_n^N x)(s),$$

since $(\pi_n x)(t_{n,j}) = x(t_{n,j})$ for $j = 1, \ldots, r(n)$. Thus

$$T_n^N \pi_n = T_n^N,$$

if the quadrature formula Q_n is induced by an interpolatory projection π_n.

Assume now that $\breve{\pi}_n$ is another interpolatory projection given by

$$\breve{\pi}_n(x) := \sum_{i=1}^{\breve{r}(n)} x(\breve{t}_{n,i}) \breve{e}_{n,i}, \quad x \in C^0([a,b]).$$

If we employ $\breve{\pi}_n$ to define the Fredholm approximation $T_n^F := \breve{\pi}_n T_n^N$, then for $x \in C^0([a,b])$ and $s \in [a,b]$, we have

$$\begin{aligned}
(T_n^F x)(s) &= \sum_{i=1}^{\breve{r}(n)} (T_n^N x)(\breve{t}_{n,i}) \breve{e}_{n,i}(s) \\
&= \sum_{i=1}^{\breve{r}(n)} \Big[\sum_{j=1}^{r(n)} w_{n,j} k(\breve{t}_{n,i}, t_{n,j}) x(t_{n,j}) \Big] \breve{e}_{n,i}(s).
\end{aligned}$$

We note that in using T_n^N, the kernel $k(\cdot\,,\cdot)$ of the integral operator T is discretized in only the second variable, while in using $T_n^F := \breve{\pi}_n T_n^N$ the kernel is discretized in both the variables.

If the sequence (π_n) of interpolatory projection satisfies $\pi_n \xrightarrow{\text{p}} I$, then clearly

$$Q_n(x) := \int_a^b (\pi_n x)(t)\,dt \to \int_a^b x(t)\,dt = Q(x)$$

for all $x \in C^0([a,b])$, that is, the induced sequence of quadrature formulæ is convergent. It is possible, however, for a sequence of quadrature formulæ induced by a sequence (π_n) of interpolatory projections to be convergent even if $\pi_n \overset{\text{p}}{\not\to} I$, as Example 4.10 shows.

Example 4.9 Quadrature formulæ induced by piecewise linear interpolatory projections:
Consider nodes $t_{n,1}, \ldots, t_{n,r(n)}$ in $[a,b]$. In this case, the weights are given by

$$\begin{aligned} w_{n,j} &:= \int_a^b e_{n,j}(t)\,dt \\ &= \begin{cases} t_{n,1} - a + \dfrac{t_{n,2} - t_{n,1}}{2} & \text{if } j = 1, \\ \dfrac{t_{n,j+1} - t_{n,j-1}}{2} & \text{if } j = 2, \ldots, r(n) - 1, \\ b - t_{n,r(n)} + \dfrac{t_{n,r(n)} - t_{n,r(n)-1}}{2} & \text{if } j = r(n). \end{cases} \end{aligned}$$

Observe that the quadrature formula $Q_n(x) := \sum_{j=1}^{r(n)} w_{n,j} x(t_{n,j})$ can be considered as a Riemann sum for x: If $\tau_{n,0} := 0$, $\tau_{n,r(n)} := 1$ and $\tau_{n,j} := \dfrac{t_{n,j} + t_{n,j+1}}{2}$ for $j = 1, \ldots, r(n) - 1$, then $Q_n(x) = \sum_{j=1}^{r(n)} x(t_{n,j})(\tau_{n,j} - \tau_{n,j-1})$, where $t_{n,j} \in [\tau_{n,j-1}, \tau_{n,j}]$ for $j = 1, \ldots, r(n)$. Hence

$$\left| Q_n(x) - \int_0^1 x(t)\,dt \right| \le \omega(x, \widetilde{h}_n),$$

where $\widetilde{h}_n := \max\{\tau_{n,j} - \tau_{n,j-1} \,:\, j = 1, \ldots, r(n)\}$.

We shall now consider some special cases.

(a) Compound Mid-Point Rule:
Let $a := 0$, $b := 1$, $t_{n,j} := \dfrac{2j-1}{2n}$, $j = 1, \ldots, n$, as in Example 4.3(c). Then $w_{n,j} := \dfrac{1}{n}$ for all $j = 1, \ldots, n$ and we obtain

$$Q_n(x) := \frac{1}{n} \sum_{j=1}^{n} x\Big(\frac{2j-1}{2n}\Big), \quad x \in C^0([0,1]).$$

We also have

$$\Big|Q_n(x) - \int_0^1 x(t)\, dt\Big| \le \frac{\|x''\|_\infty}{24n^2} \text{ for } n = 1, 2 \ldots \text{ and } x \in C^2([0,1]).$$

(b) Compound Trapezoidal Rule:
Let $a := 0$, $b := 1$, $t_{n,j} := \dfrac{j-1}{n-1}$, $n > 1$, $j = 1, \ldots, n$, as in Example 4.3(d). Then

$$w_{n,1} := \frac{1}{2(n-1)} = w_{n,n}, \quad w_{n,j} := \frac{1}{n-1} \quad \text{for } j = 2, \ldots, n-1$$

and we obtain

$$Q_n(x) := \frac{1}{n-1}\left[\frac{x(0)}{2} + \sum_{j=2}^{n-1} x\Big(\frac{j-1}{n-1}\Big) + \frac{x(1)}{2}\right], \quad x \in C^0([0,1]).$$

We also have

$$\Big|Q_n(x) - \int_0^1 x(t)\, dt\Big| \le \frac{\|x''\|_\infty}{12(n-1)^2} \text{ for } n = 1, 2 \ldots \text{ and } x \in C^2([0,1]).$$

(c) Compound Gauss Two-Point Rule:
Let $a := 0$, $b := 1$, n an even positive integer and let the nodes be as in Example 4.3(e). Then $w_{n,j} := 1/n$ for all $j = 1, \ldots, n$ and we obtain for $x \in C^0([0,1])$,

$$Q_n(x) := \frac{1}{n}\left[\sum_{\substack{j=1 \\ j \text{ odd}}}^{n} x\Big(\frac{j-(1/\sqrt{3})}{n}\Big) + \sum_{\substack{j=2 \\ j \text{ even}}}^{n} x\Big(\frac{j-1+(1/\sqrt{3})}{n}\Big)\right].$$

We also have

$$\Big|Q_n(x) - \int_0^1 x(t)\, dt\Big| \le \frac{\|x^{(4)}\|_\infty}{270n^4} \text{ for } n = 1, 2 \ldots \text{ and } x \in C^4([0,1]).$$

Some of these quadrature formulæ will be used in Section 5.4 to illustrate numerical approximations of integral operators. ∎

The Compound Simpson Rule is an example of a quadrature formula induced by a piecewise quadratic interpolatory projection. (See Exercise 4.9.)

For error estimates of various compound quadrature rules mentioned above, we refer the reader to Section 7.4 of [27].

Example 4.10 Gauss-Legendre Rule:
Let $a := -1$, $b := 1$ and let $t_{n,1} \ldots, t_{n,n}$ be the roots of the Legendre polynomial of degree n. For $j = 1, \ldots, n$, let $p_{n,j}$ denote the Lagrange polynomial with nodes at $t_{n,1}, \cdots, t_{n,n}$ and define

$$(\pi_n x)(t) := \sum_{j=1}^{n} x(t_{n,j}) p_{n,j}(t), \quad x \in C^0([-1,1]).$$

Let Q_n be the induced quadrature formula. It can be shown that if $y_k(t) := t^k$, $t \in [-1,1]$, then

$$Q_n(y_k) = \int_{-1}^{1} t^k \, dt \quad \text{for } k = 0, 1, \ldots, 2n-1$$

and

$$w_{n,j} := \int_{-1}^{1} p_{n,j}(t) \, dt = \int_{-1}^{1} p_{n,j}^2(t) \, dt > 0, \quad j = 1, \ldots, n.$$

(See Theorem 9.6(b) of [55].) By Polya's Theorem, we see that (Q_n) is a convergent sequence of quadrature formulæ, although $\pi_n \overset{p}{\not\to} I$, as we have noted in part (iii) of Subsection 4.1.2. We also have

$$\left| Q_n(x) - \int_{-1}^{1} x(t) \, dt \right| \le \frac{(n!)^3 2^{2n+1} \|x^{(2n)}\|_\infty}{(2n!)^4 (2n+1)}$$

for $n = 1, 2 \ldots$ and $x \in C^{2n}([-1,1])$. (See Example 2.12 on page 108 of [22].) ∎

Finally, we consider a situation where one can estimate the rate at which $\|(T_n^N - T)^q T\|$ and $\|(T_n^N - T)^q T_n^N\|$ tend to zero, where q is a positive integer. For this purpose, we prove a preliminary result.

Lemma 4.11
Let T be a Fredholm integral operator on $X := C^0([a,b])$ with a continuous kernel $k(\cdot\,,\cdot)$. Consider a convergent sequence (Q_n) of quadrature formulæ given by

$$Q_n(x) = \sum_{j=1}^{r(n)} w_{n,j}x(t_{n,j}) \quad \text{for } n = 1,2,\ldots \text{ and } x \in X,$$

and

$$\alpha := \sup\Big\{ \sum_{j=1}^{r(n)} |w_{n,j}| : n = 1,2\ldots \Big\}.$$

Let T_n^N be the Nyström approximation of T based on Q_n. If for some nonnegative integers i and j, the (i,j)th partial derivative $\dfrac{\partial^{i+j}k}{\partial t^j \partial s^i}$ exists at every $(s,t) \in [a,b]\times[a,b]$, define

$$\alpha_{i,j} := \sup\Big\{ \Big| \frac{\partial^{i+j}k}{\partial t^j \partial s^i}(s,t) \Big| : s,t \in [a,b] \Big\}.$$

Let p be a positive integer.

(a) *Assume that for $j = 1,\ldots,p$, the jth partial derivative of $k(\cdot\,,\cdot)$ with respect to the first variable exists and is a continuous function on $[a,b]\times[a,b]$. Then for $j = 1,\ldots,p$ and for all $x \in C^0([a,b])$, Tx, $T_n^N x \in C^j([a,b])$, and*

$$\|(Tx)^{(j)}\|_\infty \le \alpha_{j,0}(b-a)\|x\|_\infty, \quad \|(T_n^N x)^{(j)}\|_\infty \le \alpha_{j,0}\,\alpha\,\|x\|_\infty.$$

(b) *Assume that for $n = 1,2,\ldots$ and $x \in C^p([a,b])$,*

$$\Big| Q_n(x) - \int_a^b x(t)dt \Big| \le \frac{c_p}{n^p}\|x^{(p)}\|_\infty,$$

where c_p is a constant, independent of n and x. Further, assume that for $j = 1,\ldots,p$, the jth partial derivative of $k(\cdot\,,\cdot)$ with respect to the second variable exists and is continuous at every $(s,t) \in [a,b]\times[a,b]$. Then for all $x \in C^p([a,b])$,

$$\|(T_n^N - T)x\|_\infty \le \frac{c_p}{n^p} \sum_{j=0}^{p} \binom{p}{j} \alpha_{0,p-j} \|x^{(j)}\|_\infty.$$

If in fact for a positive integer i and each $j = 1, \ldots, p$, the (i,j)th partial derivative $\dfrac{\partial^{i+j}k}{\partial t^j \partial s^i}$ of $k(\cdot\,,\cdot)$ exists and is continuous at every (s,t) in $[a,b]\times[a,b]$, then Tx, $T_n^N x$ belong to $C^i([a,b])$ for all $x \in C^p([a,b])$, and

$$\| \left[(T_n^N - T)x\right]^{(i)} \|_\infty \le \frac{c_p}{n^p} \sum_{j=0}^{p} \binom{p}{j} \alpha_{i,p-j} \|x^{(j)}\|_\infty .$$

Proof

(a) Let $x \in X$ and $s \in [a,b]$. Fix j, $1 \le j \le p$. Differentiating

$$Tx(s) = \int_a^b k(s,t)x(t)\,dt \quad \text{and} \quad T_n^N x(s) = \sum_{j=1}^{r(n)} w_{n,j} k(s, t_{n,j}) x(t_{n,j})$$

j times with respect to s, we obtain

$$(Tx)^{(j)}(s) = \int_a^b \frac{\partial^j k}{\partial s^j}(s,t)x(t)dt,$$

$$(T_n^N x)^{(j)}(s) = \sum_{\ell=1}^{r(n)} w_{n,\ell} \frac{\partial^j k}{\partial s^j}(s, t_{n,\ell}) x(t_{n,\ell}).$$

Hence the desired bounds for $\|(Tx)^{(j)}\|_\infty$ and $\|(T_n^N x)^{(j)}\|_\infty$ follow easily. Note that $\alpha_{j,0}$ is finite by the continuity of the jth partial derivative of $k(\cdot\,,\cdot)$ with respect to the first variable, and α is finite by Polya's Theorem.

(b) For a fixed $s \in [a,b]$, let $k_s(t) = k(s,t), t \in [a,b]$. By assumption, $k_s \in C^p([a,b])$ for each $s \in [a,b]$. Let $x \in C^p([a,b])$. Then by Leibnitz's Rule,

$$(k_s x)^{(p)} = \sum_{j=0}^{p} \binom{p}{j} k_s^{(p-j)} x^{(j)} \quad \text{for each } s \in [a,b].$$

Thus for all $s \in [a,b]$,

$$|T_n^N x(s) - Tx(s)| = \left| Q_n(k_s x) - \int_a^b k_s(t)x(t)dt \right| \le \frac{c_p}{n^p} \|(k_s x)^{(p)}\|_\infty$$

$$\le \frac{c_p}{n^p} \sum_{j=0}^{p} \binom{p}{j} \alpha_{0,p-j} \|x^{(j)}\|_\infty$$

by our assumption on the sequence (Q_n) of the quadrature formulæ. Hence the desired bound for $\|(T_n^N - T)x\|_\infty$ follows easily.

If for a positive integer i and $j = 1, \ldots, p$, the (i, j)th partial derivative of $k(\cdot\,, \cdot)$ exists and is continuous on $[a, b]\times[a, b]$, then by considering the kernel $\dfrac{\partial^i k}{\partial s^i}(\cdot\,,\cdot)$ in place of the kernel $k(\cdot\,,\cdot)$, we obtain the desired bound for $\|[(T_n^N - T)x]^{(i)}\|_\infty$ for each $x \in C^p([a, b])$. ∎

Theorem 4.12

Let T be a Fredholm integral operator on $X := C^0([a, b])$ with a continuous kernel $k(\cdot\,,\cdot)$. Consider a convergent sequence (Q_n) of quadrature formulæ and T_n^N be the Nyström approximation of T based on Q_n.

Let p be a positive integer and assume that the sequence (Q_n) satisfies

$$\left|Q_n(x) - \int_a^b x(t)dt\right| \leq \frac{c_p}{n^p}\|x^{(p)}\|_\infty \text{ for } n = 1, 2, \ldots \text{ and } x \in C^p([a, b]),$$

where c_p is constant, independent of n and x.

(a) *Let for each $i = 1, \ldots, p$, the ith partial derivatives of $k(\cdot\,,\cdot)$ with respect to the first variable as well as the second variable exist and be continuous on $[a, b]\times[a, b]$. Then*

$$\|(T_n^N - T)T\| = O\Big(\frac{1}{n^p}\Big) = \|(T_n^N - T)T_n^N\|.$$

(b) *Let in fact for each $i, j = 0, 1, \ldots, p$, the (i, j)th partial derivative $\dfrac{\partial^{i+j} k}{\partial t^j \partial s^i}$ of $k(\cdot\,,\cdot)$ exist and be continuous at every $(s, t) \in [a, b]\times[a, b]$. Then for n, $q = 1, 2, \ldots$*

$$\|(T_n^N - T)^q T\| \leq (b-a)\left(\frac{d_p}{n^p}\right)^q \text{ and } \|(T_n^N - T)^q T_n^N\| \leq \alpha\left(\frac{d_p}{n^p}\right)^q,$$

where d_p is a constant, independent of n and q, while α is a constant, independent of n, q and p.

Proof

Let $x \in X$.

(a) By Lemma 4.11(a), Tx and $T_n^N x$ belong to $C^p([a, b])$, and for $j = 1, \ldots, p$,

$$\|(Tx)^{(j)}\| \leq \alpha_{j,0}(b - a)\|x\|_\infty, \ \|(T_n^N x)^{(j)}\| \leq \alpha_{j,0}\alpha\|x\|_\infty,$$

where the constants $\alpha_{j,0}$ and α are as defined in Lemma 4.11. Hence by Lemma 4.11(b),

$$\begin{aligned}\|(T_n^N - T)Tx\|_\infty &\leq \frac{c_p}{n^p}\sum_{j=0}^{p}\binom{p}{j}\,\alpha_{0,p-j}\|(Tx)^{(j)}\|_\infty \\ &\leq (b-a)\frac{c_p}{n^p}\sum_{j=0}^{p}\binom{p}{j}\,\alpha_{0,p-j}\alpha_{j,0}\|x\|_\infty\end{aligned}$$

and

$$\begin{aligned}\|(T_n^N - T)T_n^N x\|_\infty &\leq \frac{c_p}{n^p}\sum_{j=0}^{p}\binom{p}{j}\,\alpha_{0,p-j}\|(T_n^N x)^{(j)}\|_\infty \\ &\leq \alpha\frac{c_p}{n^p}\sum_{j=0}^{p}\binom{p}{j}\,\alpha_{0,p-j}\alpha_{j,0}\|x\|_\infty.\end{aligned}$$

Hence both $\|(T_n^N - T)T\|$ and $\|(T_n^N - T)T_n^N\|$ are less than or equal to a constant times $1/n^p$, as desired.

(b) By Lemma 4.11(a), $(T_n - T)^q Tx \in C^p([a,b])$ for all $q = 1, 2, \ldots$ We claim that for $q = 1, 2, \ldots$, $n = 1, 2, \ldots$ and $i = 0, 1, \ldots, p$,

$$\|[(T_n^N - T)^q Tx]^{(i)}\|_\infty \leq (b-a)\frac{c_p\gamma_p(c_p\widetilde{\gamma}_p)^{q-1}}{n^{pq}}\|x\|_\infty,$$

where

$$\gamma_p = \max_{i=0,1,\ldots,p}\sum_{j=0}^{p}\binom{p}{j}\,\alpha_{i,p-j}\alpha_{j,0} \text{ and } \widetilde{\gamma}_p = \max_{i=0,1,\ldots,p}\sum_{j=0}^{p}\binom{p}{j}\,\alpha_{i,p-j},$$

$\alpha_{i,j}$ being the constant introduced in Lemma 4.11, $i, j = 0, \ldots, p$. To prove our claim, we use mathematical induction on q.

Let $q = 1$. By Lemma 4.11(a), we have

$$\|(Tx)^{(j)}\|_\infty \leq \alpha_{j,0}(b-a)\|x\|_\infty \quad \text{for } j = 0, \ldots, p,$$

and hence by Lemma 4.11(b),

$$\begin{aligned}\|[(T_n^N - T)Tx]^{(i)}\|_\infty &\leq \frac{c_p}{n^p}\sum_{j=0}^{p}\binom{p}{j}\,\alpha_{i,p-j}\|(Tx)^{(j)}\|_\infty \\ &\leq (b-a)\frac{c_p\gamma_p}{n^p}\|x\|_\infty\end{aligned}$$

for $n = 1, 2, \ldots$ and $i = 0, \ldots, p$. Thus our claim holds for $q = 1$.

Assume that our claim holds for some $q \geq 1$ and let $y := (T_n^N - T)^q Tx$. Again, by Lemma 4.11(b),

$$\begin{aligned}
\|[(T_n^N - T)^{q+1}Tx]^{(i)}\|_\infty &= \|[(T_n^N - T)y]^{(i)}\|_\infty \\
&\leq \frac{c_p}{n^p} \sum_{j=0}^{p} \binom{p}{j} \alpha_{i,p-j} \|y^{(j)}\|_\infty \\
&\leq \frac{c_p}{n^p} \sum_{j=0}^{p} \binom{p}{j} \alpha_{i,p-j} (b-a) \frac{c_p \gamma_p (c_p \widetilde{\gamma}_p)^{q-1}}{n^{pq}} \|x\|_\infty \\
&\leq (b-a) \frac{c_p \gamma_p (c_p \widetilde{\gamma}_p)^q}{n^{p(q+1)}} \|x\|_\infty
\end{aligned}$$

for $n = 1, 2, \ldots$ and $i = 0, \ldots, p$. Thus our claim holds for $q + 1$ and the induction is over.

In particular, letting $i = 0$, we have for $q = 1, 2, \ldots$ and $n = 1, 2, \ldots$

$$\|(T_n^N - T)^q Tx\|_\infty \leq (b-a) \frac{c_p \gamma_p (c_p \widetilde{\gamma}_p)^{q-1}}{n^{pq}} \|x\|_\infty.$$

We let $d_p := c_p \max\{\gamma_p, \widetilde{\gamma}_p\}$ and obtain for each fixed $q = 1, 2, \ldots$

$$\|(T_n^N - T)^q T\| \leq (b-a) \left(\frac{d_p}{n^p}\right)^q \quad \text{for } n = 1, 2, \ldots$$

Analogous estimate for $\|(T_n^N - T)^q T_n^N\|$ can be proved similarly upon replacing $b - a$ by the constant α introduced in Lemma 4.11. ∎

Example 4.13 Quadrature formulæ satisfying the hypothesis in Theorem 4.12:

Several well-known sequences of quadrature formulæ satisfy the condition

$$\left|Q_n(x) - \int_a^b x(t)dt\right| \leq \frac{c_p}{n^p} \|x^{(p)}\|_\infty \quad \text{for } n = 1, 2, \ldots \text{ and } x \in C^p([a, b]),$$

mentioned in Theorem 4.12: the Compound Rectangular Rule satisfies it for $p = 1$, the Compound Mid-Point and the Compound Trapezoidal Rules satisfy it for $p = 2$, the Compound Simpson and the Compound Gauss Two-Point Rules satisfy it for $p = 4$. (See Exercise 4.10 and Example 4.9(c).) Further, for any positive integer r, the Compound

Gauss r Point Rule satisfies the above-mentioned condition for $p = 2r$. (See Section 7.4 of [27].) ∎

We note that if p is a positive integer and $X := C^p([a,b])$ with the norm given by

$$\|x\| = \sum_{j=0}^{p} \|x^{(j)}\|_\infty, \quad x \in X,$$

and if T is a Fredholm integral operator with a continuous kernel $k(\cdot\,,\cdot)$ such that all the partial derivatives up to the order $2p$ exist and are continuous on $[a,b]\times[a,b]$, then $\|T_n^N - T\| = O\Big(\dfrac{1}{n^p}\Big)$, where T_n^N is a Nyström approximation of T based on a quadrature formula satisfying the condition given in Theorem 4.12. This follows easily from Lemma 4.11.

4.2.3 Weakly Singular Integral Operators

In this subsection we extend the use of the Nyström approximation to a class of Fredholm integral operators having discontinuous kernels.

Let $\kappa \in C^0(]0,1])$ be a nonnegative steadily decreasing function, that is, $\kappa(r_1) \geq \kappa(r_2)$ whenever $0 < r_1 < r_2 \leq 1$, such that $\lim_{t \to 0^+} \kappa(t) = \infty$ and $\int_0^1 \kappa(t)\,dt < \infty$. We say that κ is weakly singular at 0.

Example 4.14 Weakly singular functions:
The functions defined by

$$\begin{aligned}
\kappa(r) &:= r^{-\alpha}, \quad r \in\,]0,1], \qquad \text{where } \alpha \in\,]0,1[\quad \text{is fixed},\\
\kappa(r) &:= -\ln r, \quad r \in\,]0,1],
\end{aligned}$$

are weakly singular at 0. ∎

Lemma 4.15
Let a function κ be a weakly singular at 0. Then

(a) *for any $\epsilon > 0$, there exists $\delta > 0$ such that $\int_0^\delta \kappa(t)\,dt < \epsilon$ and*

(b) *for any $\delta \in]0,1[$, the function $s \in [\delta, 1-\delta] \mapsto \int_{s-\delta}^{s+\delta} \kappa(t)\,dt$ is steadily decreasing.*

Proof

(a) The measure μ defined by $\mu(E) := \int_E \kappa(t)\,dt$, where E is a Lebesgue measurable subset of $[0,1]$, is absolutely continuous with respect to the Lebesgue measure. Hence for all $\epsilon > 0$, there exists $\delta > 0$ such that for any interval $\mathcal{I}$ in $[0,1]$ whose length is less than or equal to δ, $\mu(\mathcal{I})$ is less than ϵ.

(b) For $s \in [\delta, 1-\delta]$, define $g(s) := \int_{s-\delta}^{s+\delta} \kappa(t)\,dt$. Then g is a continuously differentiable function, and for $s \in]\delta, 1-\delta[$, $g'(s) = \kappa(s+\delta) - \kappa(s-\delta) \leq 0$, since κ is steadily decreasing. ∎

Let $\kappa :]0,1] \to I\!R$ be a weakly singular function at 0. Throughout this section, we let

$$k(s,t) := \kappa(\,|s-t|\,), \quad s,\, t \in [0,1],\, s \neq t,$$

and consider the integral operator T defined by

$$(Tx)(s) := \int_0^1 k(s,t)x(t)\,dt, \quad x \in X,\, s \in [0,1].$$

Let $\delta \in]0,1]$ and define the truncated function $\kappa_\delta : [0,1] \to I\!R$ by

$$\kappa_\delta(t) := \begin{cases} \kappa(\delta) & \text{if } 0 \leq t \leq \delta, \\ \kappa(t) & \text{otherwise,} \end{cases}$$

This truncated function κ_δ induces a kernel $k_\delta(\cdot\,,\cdot)$ defined by

$$k_\delta(s,t) := \kappa_\delta(\,|s-t|\,), \quad s,\, t \in [0,1].$$

If $\delta := 1/n$, we write κ_n and k_n in place of $\kappa_{1/n}$ and $k_{1/n}$, respectively.

Let $X := C^0([0,1])$ and n be a positive integer. We define the auxiliary Fredholm integral operator $\widetilde{T}_n$ on X by

$$(\widetilde{T}_n x)(s) := \int_0^1 k_n(s,t)x(t)\,dt, \quad x \in X,\, s \in [0,1].$$

Since $k_n(\cdot,\cdot)$ is a continuous function on $[0,1]\times[0,1]$, $\widetilde{T}_n$ is a compact operator on X.

For each positive integer n, we define the set

$$\mathcal{I}(s,1/n) := \{t \in [0,1] : 0 < |t-s| \leq 1/n\}.$$

By Lemma 4.15(a), given $\epsilon > 0$, there exists an integer n_0 such that for $n > n_0$,

$$\int\limits_{\mathcal{I}(s,1/n)} \kappa(\,|s-t|\,)\,dt = 2\int_0^{1/n} \kappa(t)\,dt < \frac{\epsilon}{2}$$

for all $s \in [0,1]$. Hence, for $n > n_0$, $x \in X$ such that $\|x\|_\infty \leq 1$ and $s \in [0,1]$,

$$\begin{aligned}|(\widetilde{T}_n x)(s) - (Tx)(s)| &= \Big|\int\limits_{\mathcal{I}(s,1/n)} [\kappa(1/n) - \kappa(\,|s-t|\,)]x(t)\,dt\Big| \\ &\leq 2\|x\|_\infty \int\limits_{\mathcal{I}(s,1/n)} \kappa(\,|s-t|\,)\,dt < \epsilon,\end{aligned}$$

since κ is nonnegative and steadily decreasing on $]0,1]$. In particular, $Tx \in X$ for every $x \in X$, and $\widetilde{T}_n \xrightarrow{\text{n}} T$. Hence T is compact. (Compare Exercise 4.6.)

The approximation of T we want to present is motivated by the following way of rewriting $(Tx)(s)$: For $s \in [0,1]$,

$$(Tx)(s) = \int_0^1 \kappa(\,|s-t|\,)[x(t) - x(s)]\,dt + x(s)\int_0^1 \kappa(\,|s-t|\,)\,dt.$$

The continuity of x at s is supposed to mitigate the effect of the weak singularity of κ at 0 in the first integral and to make this integral more amenable for numerical integration than the integral $\displaystyle\int_0^1 \kappa(\,|s-t|\,)x(t)\,dt$.

This approach is called the Singularity Subtraction Technique and has been proposed by Kantorovich and Krylov. (See [45] and [18].) The basic idea is to truncate κ near 0 and then use a Nyström approximation of the resulting operator, rewritten as in the singularity subtraction technique.

More precisely, let (Q_n) be a convergent sequence of quadrature formulæ with nodes $t_{n,j}$ such that $0 \leq t_{n,1} < \cdots < t_{n,n} \leq 1$ and weights $w_{n,j} \geq 0$. We define the Kantorovich-Krylov approximation of T as follows.

For $x \in X$ and $s \in [0,1]$, let

$$(T_n^K x)(s) := \sum_{j=1}^{n} w_{n,j}\kappa_n(\,|s-t_{n,j}|\,)[x(t_{n,j})-x(s)]+x(s)\int_0^1 \kappa(\,|s-t|\,)\,dt.$$

We consider the following additional hypothesis on the nodes and the weights of Q_n:

$$(H) \qquad \begin{cases} \text{There exists } c>0 \text{ such that } \displaystyle\sum_{t_{n,j}\in\mathcal{I}} w_{n,j} \le c(b-a) \\ \text{whenever } 0\le a<b\le 1, \text{ and } \mathcal{I} :=]a,b] \text{ or } \mathcal{I} := [a,b[. \end{cases}$$

Note that the nodes and the weights for most of the quadrature formulæ given in Example 4.9 satisfy hypothesis (H).

Lemma 4.16
Let a and b be real numbers such that $0 \le a < b \le 1$. Let $y :]a,b] \to \mathbb{R}$ be a nonnegative steadily increasing function. Then under the hypothesis (H), for each $n = 1, 2, \dots$

$$\sum_{a<t_{n,j}\le b} w_{n,j}y(t_{n,j}) \le \frac{c(b-a)}{n}\,y(b) + c\int_a^b y(t)\,dt.$$

Similarly, let $y : [a,b[\to \mathbb{R}$ be a nonnegative steadily decreasing function. Then under the hypothesis (H), for each $n = 1, 2, \dots$

$$\sum_{a\le t_{n,j}< b} w_{n,j}y(t_{n,j}) \le \frac{c(b-a)}{n}\,y(a) + c\int_a^b y(t)\,dt.$$

Proof
Define $\mathcal{I}_{n,i} :=]a+(i-1)(b-a)/n, a+i(b-a)/n]$ for $i = 1,\dots,n$. Then

$$\begin{aligned}\sum_{a<t_{n,j}\le b} w_{n,j}y(t_{n,j}) &= \sum_{i=1}^{n}\sum_{t_{n,j}\in\mathcal{I}_{n,i}} w_{n,j}y(t_{n,j}) \\ &\le \frac{c(b-a)}{n}\sum_{i=1}^{n} y\left(a+\frac{i(b-a)}{n}\right) \\ &= \frac{c(b-a)}{n}y(b) + c\frac{b-a}{n}\sum_{i=1}^{n-1} y\left(a+\frac{i(b-a)}{n}\right) \\ &\le \frac{c(b-a)}{n}y(b) + c\int_a^b y(t)\,dt,\end{aligned}$$

since y is steadily increasing.

If y is steadily decreasing, the proof is similar. ∎

Lemma 4.17
Let $\delta \in]0,1]$, $s \in [0,1]$ and $n > 1/\delta$. Under the hypothesis (H), we have

$$\sum_{|s-t_{n,j}|<\delta} w_{n,j}\kappa_n(|s-t_{n,j}|) \le 4c\int_0^\delta \kappa(t)\,dt.$$

Proof
Let us write

$$\sum_{|s-t_{n,j}|<\delta} w_{n,j}\kappa_n(|s-t_{n,j}|) = \sum_{0\le s-t_{n,j}<\delta} w_{n,j}\kappa_n(s-t_{n,j}) + \sum_{0\le t_{n,j}-s<\delta} w_{n,j}\kappa_n(t_{n,j}-s).$$

To find an upper bound for the first sum on the right side, consider a function y defined by $y(t) := \kappa_n(s-t)$ for $t \in]a,s]$, where $a := \max\{0, s-\delta\}$. Then y is steadily increasing and Lemma 4.16 gives

$$\sum_{0\le s-t_{n,j}<\delta} w_{n,j}\kappa_n(s-t_{n,j}) \le \frac{c\delta}{n}\kappa(1/n) + c\int_0^\delta \kappa_n(t)\,dt.$$

But

$$\frac{c\delta}{n}\kappa(1/n) \le \frac{c}{n}\kappa(1/n) = c\int_0^{1/n} \kappa_n(t)\,dt$$

and

$$\int_0^{1/n} \kappa_n(t)\,dt \le \int_0^\delta \kappa_n(t)\,dt \le \int_0^\delta \kappa(t)\,dt,$$

so that

$$\sum_{0\le s-t_{n,j}<\delta} w_{n,j}\kappa_n(s-t_{n,j}) \le 2c\int_0^\delta \kappa(t)\,dt.$$

In a completely analogous manner, consider a function y defined by $y(t) := \kappa_n(t-s)$ for $t \in [s,b[$, where $b := \min\{s+\delta, 1\}$. Then y is steadily decreasing and Lemma 4.16 gives exactly the same upper bound for the second sum. ∎

For $x \in X$, $s \in [0,1]$ and $n = 1, 2, \dots$ define

$$(T_n x)(s) := \sum_{i=1}^{n} w_{n,i} \kappa_n(\,|s - t_{n,i}|\,) x(t_{n,i}).$$

The following result was proved in [52].

Proposition 4.18
Under the hypothesis (H), $T_n \overset{\text{cc}}{\to} T$.

Proof
First we prove that $T_n \overset{\text{p}}{\to} T$. Let $x \in X$, $\|x\|_\infty = 1$ and $\epsilon > 0$. There exists $\delta \in]0,1]$ such that

$$\int_0^\delta \kappa(t)\,dt < \frac{\epsilon}{9} \min\Big\{1, \frac{1}{3c}\Big\}.$$

For $n = 1, 2, \dots$ we decompose

$$(T_n - T)x = A_1 + A_{2,n} + A_{3,n},$$

where, for $s \in [0,1]$,

$$\begin{aligned} A_1(s) &:= \int_0^1 \kappa_\delta(\,|s-t|\,) x(t)\,dt - \int_0^1 \kappa(\,|s-t|\,) x(t)\,dt \\ &= \int_0^\delta [\kappa_\delta(\,|s-t|\,) - \kappa(\,|s-t|\,)] x(t)\,dt, \\ A_{2,n}(s) &:= \sum_{j=1}^{n} w_{n,j} \kappa_\delta(\,|s - t_{n,j}|\,) x(t_{n,j}) - \int_0^1 \kappa_\delta(\,|s-t|\,) x(t)\,dt \quad \text{and} \\ A_{3,n}(s) &:= \sum_{j=1}^{n} w_{n,j} \kappa_n(\,|s - t_{n,j}|\,) x(t_{n,j}) - \sum_{j=1}^{n} w_{n,j} \kappa_\delta(\,|s - t_{n,j}|\,) x(t_{n,j}). \end{aligned}$$

Clearly, for $s \in [0,1]$,

$$|A_1(s)| \le \delta \kappa(\delta) + \int_0^\delta \kappa(\,|s-t|\,)\,dt \le 3 \int_0^\delta \kappa(\,t\,)\,dt < \frac{\epsilon}{3}.$$

Let $\widetilde{T}$ denote the Fredholm integral operator with continuous kernel $k_\delta(\cdot\,,\cdot)$ and let $\widetilde{T}_n^N$ denote the Nyström approximation of $\widetilde{T}$ based on the sequence (Q_n) of (convergent) quadrature formulæ. Then by Theorem 4.5, $\widetilde{T}_n^N \xrightarrow{\text{p}} \widetilde{T}$. Hence there exists some n_0 depending on x such that for all $n > n_0$ and all $s \in [0,1]$,

$$|A_{2,n}(s)| \leq \frac{\epsilon}{3}.$$

Also, by Lemma 4.15, for $n > [1/\delta] + 1$ and all $s \in [0,1]$,

$$\begin{aligned}
|A_{3,n}(s)| &\leq \sum_{|s-t_{n,j}|<\delta} w_{n,j} |\kappa_n(\,|s-t_{n,j}|\,) - \kappa_\delta(\,|s-t_{n,j}|\,)| \\
&\leq \sum_{|s-t_{n,j}|<\delta} w_{n,j} \big(\kappa_n(\,|s-t_{n,j}|\,) + \kappa(\delta)\big) \\
&\leq \sum_{|s-t_{n,j}|<\delta} w_{n,j} \kappa_n(\,|s-t_{n,j}|\,) + 2c\delta\kappa(\delta) \\
&\leq 6c \int_0^\delta \kappa(t)\,dt < \frac{\epsilon}{3}.
\end{aligned}$$

Hence for a given $\epsilon > 0$, there exists $n_1 := \max\{n_0, [1/\delta] + 1\}$ such that for $n > n_1$ and all $s \in [0,1]$, we have $|(T_n x)(s) - (Tx)(s)| < \epsilon$. This proves that $T_n \xrightarrow{\text{p}} T$.

Next, we prove that the set $\widetilde{E} := \bigcup_{n=n_0}^{\infty} \{T_n x : x \in X, \|x\|_\infty \leq 1\}$ is bounded and equicontinuous in X for some n_0.

Since $T_n \xrightarrow{\text{p}} T$, the sequence $(\|T_n\|)$ is bounded. As a consequence, the set $\bigcup_{n=1}^{\infty} \{T_n x : x \in X, \|x\|_\infty \leq 1\}$ is bounded. The equicontinuity is proved as follows:

Let $\epsilon > 0$. By Lemma 4.15(a), there exists $\delta \in]0,1]$ such that

$$\int_0^\delta \kappa(t)\,dt < \frac{\epsilon}{40c}.$$

By the uniform continuity of κ in $[\delta, 1]$, there exists $\eta > 0$ such that for all $s, t \in [\delta, 1]$, $|t-s| < \eta$ implies that $|\kappa(t) - \kappa(s)| < \dfrac{\epsilon}{2c}$. Let $x \in X$ be such that $\|x\|_\infty \leq 1$, $n > n_0 := [1/\delta] + 1$ and $s, t \in [0,1]$ satisfy $|t-s| < \eta$. We have

$$|(T_n x)(t) - (T_n x)(s)| \leq B_{1,n}(s,t) + B_{2,n}(s,t) + B_{3,n}(s,t) + B_{4,n}(s,t),$$

where

$$B_{1,n}(s,t) := \sum_{\substack{|t-t_{n,i}|<\delta \\ |s-t_{n,i}|<\delta}} w_{n,i} |\kappa_n(\,|t-t_{n,i}|\,) - \kappa_n(\,|s-t_{n,i}|\,)|$$

$$
\begin{aligned}
&\leq \sum_{|t-t_{n,i}|<\delta} w_{n,i}\kappa_n(|t-t_{n,i}|) + \sum_{|s-t_{n,i}|<\delta} w_{n,i}\kappa_n(|s-t_{n,i}|) \\
&\leq 8c\int_0^\delta \kappa(t)\,dt < \frac{\epsilon}{5}, \\
B_{2,n}(s,t) &:= \sum_{\substack{|t-t_{n,i}|\geq\delta \\ |s-t_{n,i}|\geq\delta}} w_{n,i}|\kappa_n(|t-t_{n,i}|) - \kappa_n(|s-t_{n,i}|)| \\
&= \sum_{\substack{|t-t_{n,i}|\geq\delta \\ |s-t_{n,i}|\geq\delta}} w_{n,i}|\kappa(|t-t_{n,i}|) - \kappa(|s-t_{n,i}|)| < \frac{\epsilon}{2}, \\
B_{3,n}(s,t) &:= \sum_{\substack{|t-t_{n,i}|<\delta \\ |s-t_{n,i}|\geq\delta}} w_{n,i}|\kappa_n(|t-t_{n,i}|) - \kappa_n(|s-t_{n,i}|)| \\
&\leq \sum_{|t-t_{n,i}|<\delta} w_{n,i}\kappa_n(|t-t_{n,i}|) + \kappa(\delta)\sum_{|t-t_{n,i}|<\delta} w_{n,i} \\
&< 4c\int_0^\delta \kappa(t)\,dt + 2c\delta\kappa(\delta) \leq 6c\int_0^\delta \kappa(t)\,dt < \frac{3\epsilon}{20}, \\
B_{4,n}(s,t) &:= \sum_{\substack{|t-t_{n,i}|\geq\delta \\ |s-t_{n,i}|<\delta}} w_{n,i}|\kappa_n(|t-t_{n,i}|) - \kappa(|s-t_{n,i}|)| \\
&< 6c\int_0^\delta \kappa(t)\,dt < \frac{3\epsilon}{20}.
\end{aligned}
$$

Hence $|(T_n x)(t) - (T_n x)(s)| < \epsilon$. This shows that the set $\widetilde{E}$ is relatively compact in X. Thus $T_n \overset{\text{cc}}{\to} T$. ∎

We remark that T_n and T_n^K are related by the formula

$$T_n^K = T_n + U_n,$$

U_n being defined by

$$(U_n x)(s) := x(s)[(Te)(s) - (T_n e)(s)] \quad \text{for } x \in X \text{ and } s \in [0,1],$$

where

$$e(s) := 1 \quad \text{for } s \in [0,1].$$

The following result was proved in [1].

Theorem 4.19
Under the hypothesis (H), $T_n^K \xrightarrow{\nu} T$, but $T_n^K \overset{\text{n}}{\not\to} T$ and $T_n^K \overset{\text{cc}}{\not\to} T$.

Proof
For $x \in X$, we have $\|U_n x\|_\infty \leq \|x\|_\infty \|T_n e - Te\|_\infty$, so that $\|U_n x\|_\infty \leq \|T_n e - Te\|_\infty$ which tends to 0 since $T_n \xrightarrow{\text{p}} T$. Thus $U_n \xrightarrow{\text{n}} O$. Since $T_n \xrightarrow{\text{cc}} T$ and T is compact, it follows from Lemma 2.2 that $T_n^K := T_n + U_n \xrightarrow{\nu} T$.

Also, $U_n \overset{\text{cc}}{\not\to} O$ since otherwise $U_n - T$ would be compact and hence U_n would be compact. This is not the case as we now show.

If $(T_n - T)e$ is a (nonzero) constant function, then U_n is a (nonzero) scalar multiple of the identity operator, which cannot be compact. Assume now that $(T_n - T)e$ is not a constant function. Then $\text{sp}(U_n)$ is the range of the function $(T_n - T)e$, as we have seen in Example 1.17. Since $(T_n - T)e$ is a nonconstant continuous function, its range is an uncountable set. But the spectrum of a compact operator is countable (Remark 1.34). Hence, by Lemma 2.2(d), $T_n^K \overset{\text{cc}}{\not\to} T$.

Finally, since T and T_n satisfy the hypotheses of Proposition 4.6, $T_n \overset{\text{n}}{\not\to} T$. Again by Lemma 2.2(d), $T_n^K \overset{\text{n}}{\not\to} T$. ∎

The preceding result is one of the reasons for considering ν-convergence rather than norm convergence or collectively compact convergence in Chapter 2.

4.3 A Posteriori Error Estimates

Suppose we wish to find a nonzero isolated spectral value λ of $T \in \text{BL}(X)$. If $T_n \xrightarrow{\nu} T$, then we have seen in Corollary 2.13 that for each large n, there is some $\lambda_n \in \text{sp}(T_n)$ such that $\lambda_n \to \lambda$. Suppose we have calculated such λ_n for some n. It is important to know how close this λ_n is to λ.

Theorem 2.17 shows that if λ is a nonzero simple eigenvalue of T and if ϵ is such that $0 < \epsilon < \text{dist}(\lambda, \text{sp}(T) \setminus \{\lambda\})$, then for all large enough n, there is a simple eigenvalue λ_n of T_n which satisfies

$$|\lambda_n - \lambda| \leq 2\epsilon \max_{|z-\lambda|=\epsilon} ||R(T,z)|| \frac{||(T_n - T)\varphi_n||}{||\varphi_n||},$$

where φ_n is an eigenvector of T_n corresponding to λ_n. Since we may be able to calculate φ_n, the factor $||(T_n - T)\varphi_n||/||\varphi_n||$ in the preceding estimate is computable. However, it is difficult to estimate the other factor $\max\limits_{|z-\lambda|=\epsilon} ||R(T,z)||$, in general.

In case X is a Hilbert space and T is a normal operator, then for every $z \in \text{re}(T)$, the operator $R(T,z)$ is also normal and hence

$$||R(T,z)|| = \rho(R(T,z)) = \frac{1}{\text{dist}(z, \text{sp}(T))}$$

by Proposition 1.41(a) and 1.10(c). Further, if we choose $\epsilon > 0$ such that $\epsilon \leq \text{dist}(\lambda, \text{sp}(T) \setminus \{\lambda\})/2$, then $\text{dist}(z, \text{sp}(T)) = \epsilon$ for every z satisfying $|z - \lambda| = \epsilon$, so that $||R(T,z)|| = 1/\epsilon$. Thus in this case,

$$|\lambda_n - \lambda| \leq 2\frac{||(T_n - T)\varphi_n||}{||\varphi_n||}$$

for all large n. However, it is difficult to say exactly how large the integer n need be for the preceding inequality to hold. Further, the effectiveness of this inequality depends on our being able to calculate an eigenvector φ_n of T_n corresponding to λ_n.

Let us consider the following question: Let $\widetilde{T} \in \text{BL}(X)$ be an approximation of T and suppose that a nonzero eigenvalue $\widetilde{\lambda}$ of $\widetilde{T}$ is found. How close is $\widetilde{\lambda}$ to *some* eigenvalue of T? The Krylov-Weinstein Inequality 1.41(c) says that there is some $\lambda \in \text{sp}(T)$ such that

$$|\widetilde{\lambda} - \lambda| \leq \inf\{||Tx - \widetilde{\lambda}x|| \, : \, ||x|| = 1\}.$$

Again, it is not easy to find a sharp upper bound for the right side of this inequality.

In this respect, we give a result of Brakhage [24] for integral operators which does not require any calculation of an eigenvector of $\widetilde{T}$ corresponding to $\widetilde{\lambda}$. One may refer to Section 3.16 of [22] for other error bounds for computed eigenvalues of integral operators. Since these bounds are available only after the computation of an approximate eigenvalue $\widetilde{\lambda}$, they are called a posteriori error estimates.

Lemma 4.20
Let $X = C^0([a,b])$ and T be a Fredholm integral operator with a continuous kernel $k(\cdot\,,\cdot)$. Consider a Nyström approximation of T given by

$$\widetilde{T}x(s) := \sum_{j=1}^{r} w_j k(s,t_j)x(t_j), \quad x \in X,\ s \in [a,b],$$

where $a \le t_1 < \cdots < t_r \le b$ and $w_j \ge 0$ for each $j = 1, 2, \ldots, r$.

Let $\alpha := \sum_{j=1}^{r} w_j$ and for $s,t \in [a,b]$, define

$$d_1(s,t) := \sum_{j=1}^{r} w_j k(s,t_j)k(t_j,t) - \int_a^b k(s,u)k(u,t)\,du,$$
$$d_2(s,t) := \sum_{j=1}^{r} w_j \overline{k(t_j,s)}k(t_j,t) - \int_a^b \overline{k(u,s)}k(u,t)\,du.$$

Assume that

$$|d_1(s,t_j)| \le \beta \quad \text{for all } s \in [a,b] \text{ and } j = 1,\ldots,r,$$

and that

$$|d_2(t_i,t_j)| \le \gamma \quad \text{for all } i,j = 1,\ldots,r.$$

Then for every $x \in X$, we have

$$\text{(i)} \int_a^b |(\widetilde{T} - T)\widetilde{T}x(s)|^2\,ds \le \beta^2\alpha(b-a) \sum_{j=1}^{r} w_j |x(t_j)|^2$$
$$\text{(ii)} \int_a^b |\widetilde{T}x(u)|^2\,du \ge \sum_{j=1}^{r} w_j (|\widetilde{T}x(t_j)|^2 - \gamma\alpha|x(t_j)|^2).$$

Proof
Let $x \in X$.
(i) For $s \in [a, b]$, we have

$$
\begin{aligned}
(\widetilde{T} - T)\widetilde{T}x(s) &= \sum_{i=1}^{r} w_i k(s, t_i)(\widetilde{T}x)(t_i) - \int_a^b k(s, u)(\widetilde{T}x)(u)\, du \\
&= \sum_{i=1}^{r} w_i k(s, t_i) \sum_{j=1}^{r} w_j k(t_i, t_j) x(t_j) - \int_a^b k(s, u) \sum_{j=1}^{r} w_j k(u, t_j) x(t_j)\, du \\
&= \sum_{j=1}^{r} \Big[\sum_{i=1}^{r} w_i k(s, t_i) k(t_i, t_j) - \int_a^b k(s, u) k(u, t_j)\, du \Big] w_j x(t_j) \\
&= \sum_{j=1}^{r} d_1(s, t_j) w_j x(t_j),
\end{aligned}
$$

so that

$$
|(\widetilde{T} - T)\widetilde{T}x(s)|^2 \le \Big(\sum_{j=1}^{r} w_j \, |d_1(s, t_j)|^2 \Big) \Big(\sum_{j=1}^{r} w_j \, |x(t_j)|^2 \Big).
$$

Since $w_j \ge 0$ for each $j = 1, \ldots, r$ and $\sum\limits_{j=1}^{r} w_j = \alpha$, we have

$$
\int_a^b |(\widetilde{T} - T)\widetilde{T}x(s)|^2 ds \le \beta^2 \alpha (b - a) \sum_{j=1}^{r} w_j |x(t_j)|^2.
$$

(ii) We note that

$$
\begin{aligned}
\int_a^b |\widetilde{T}x(u)|^2 du &= \int_a^b \Big| \sum_{i=1}^{r} w_i k(u, t_i) x(t_i) \Big|^2 du \\
&= \int_a^b \Big(\sum_{i=1}^{r} w_i \overline{k(u, t_i)}\, \overline{x(t_i)} \Big) \Big(\sum_{\ell=1}^{r} w_\ell k(u, t_\ell) x(t_\ell) \Big) du \\
&= \sum_{i=1}^{r} \sum_{\ell=1}^{r} w_i w_\ell \overline{x(t_i)} x(t_\ell) \int_a^b \overline{k(u, t_i)} k(u, t_\ell) du.
\end{aligned}
$$

Similarly, for each $j = 1, \ldots, r$, we have

$$
|\widetilde{T}x(t_j)|^2 = \Big| \sum_{i=1}^{r} w_i k(t_j, t_i) x(t_i) \Big|^2
$$

$$= \Big(\sum_{i=1}^{r} w_i \overline{k(t_j,t_i)}\,\overline{x(t_i)}\Big)\Big(\sum_{\ell=1}^{r} w_\ell k(t_j,t_\ell)x(t_\ell)\Big)$$

$$= \sum_{i=1}^{r}\sum_{\ell=1}^{r} w_i w_\ell \overline{x(t_i)} x(t_\ell) \overline{k(t_j,t_i)} k(t_j,t_\ell),$$

so that

$$\sum_{j=1}^{r} w_j |\widetilde{T}x(t_j)|^2 = \sum_{i=1}^{r}\sum_{\ell=1}^{r} w_i w_\ell \overline{x(t_i)} x(t_\ell) \sum_{j=1}^{r} w_j \overline{k(t_j,t_i)} k(t_j,t_\ell).$$

Hence

$$\sum_{j=1}^{r} w_j |\widetilde{T}x(t_j)|^2 - \int_a^b |\widetilde{T}x(u)|^2\,du = \sum_{i=1}^{r}\sum_{\ell=1}^{r} w_i w_\ell \overline{x(t_i)} x(t_\ell) d_2(t_i,t_\ell)$$

$$\leq \sum_{i=1}^{r}\sum_{\ell=1}^{r} w_i w_\ell |x(t_i)|\,|x(t_\ell)|\,|d_2(t_i,t_\ell)| \leq \gamma\Big(\sum_{i=1}^{r} w_i |x(t_i)|\Big)^2$$

$$\leq \gamma\Big(\sum_{i=1}^{r} w_i\Big)\Big(\sum_{i=1}^{r} w_i |x(t_i)|^2\Big) = \gamma\alpha \sum_{i=1}^{r} w_i |x(t_i)|^2,$$

so that

$$\int_a^b |\widetilde{T}x(u)|^2 du \geq \sum_{j=1}^{r} w_j \left(|\widetilde{T}x(t_j)|^2 - \gamma\alpha |x(t_j)|^2\right).$$

as desired. ∎

Proposition 4.21
Let $X, T, \widetilde{T}, \alpha, \beta, \gamma$ be as in Lemma 4.20. Assume further that

$$\int_a^b k(s,u)\overline{k(t,u)}du = \int_a^b \overline{k(u,s)} k(u,t) du$$

for all $s,t \in [a,b]$. Let $\widetilde{\lambda}$ be an eigenvalue of $\widetilde{T}$ such that

$$|\widetilde{\lambda}|^2 > \frac{1}{2}\Big(\gamma\alpha + \sqrt{\gamma^2\alpha^2 + 4\beta^2\alpha(b-a)}\Big).$$

Then there is a nonzero eigenvalue λ of T such that

$$|\widetilde{\lambda} - \lambda| \leq \widetilde{\delta} := \beta\left[\frac{\alpha(b-a)}{|\widetilde{\lambda}|^2 - \gamma\alpha}\right]^{1/2}.$$

Further, we have the following bounds for the relative error:

$$\frac{\widetilde{\delta}}{|\widetilde{\lambda}|+\widetilde{\delta}} \le \frac{|\widetilde{\lambda}-\lambda|}{|\lambda|} \le \frac{\widetilde{\delta}}{|\widetilde{\lambda}|-\widetilde{\delta}}.$$

Proof

Let $Y := L^2([a,b])$. For $x \in Y$ and $s \in [a,b]$, let

$$Ux(s) = \int_a^b k(s,t)x(t)dt.$$

Then $U \in \mathrm{BL}(Y)$, $\mathcal{R}(U) \subset X$ and $U|_{X,X} = T$. Hence a nonzero complex number λ is an eigenvalue of U if and only if λ is an eigenvalue of T and then $\mathcal{N}(U-\lambda I_Y) = \mathcal{N}(T-\lambda I)$, where I_Y is the identity operator on Y. Our assumption on the kernel $k(\cdot\,,\cdot)$ shows that $U^*U = UU^*$, that is, U is a normal operator on Y.

Let $\widetilde{\lambda}$ be an eigenvalue of $\widetilde{T}$ satisfying the stated inequality and let $\widetilde{\varphi}$ be a corresponding eigenvector. Then $\widetilde{\varphi} \in Y$ and by the Krylov-Weinstein Inequality (Proposition 1.41(c)), there exists $\lambda \in \mathrm{sp}(U)$ such that

$$|\widetilde{\lambda}-\lambda| \le \frac{\|U\widetilde{\varphi}-\widetilde{\lambda}\widetilde{\varphi}\|_2}{\|\widetilde{\varphi}\|_2}.$$

But by Lemma 4.20, we have

$$\begin{aligned}\frac{\|U\widetilde{\varphi}-\widetilde{\lambda}\widetilde{\varphi}\|_2^2}{\|\widetilde{\varphi}\|_2^2} &= \frac{\|(T-\widetilde{T})\widetilde{\varphi}\|_2^2}{\|\widetilde{\varphi}\|_2^2} = \frac{\|(T-\widetilde{T})\widetilde{T}\widetilde{\varphi}\|_2^2}{\|\widetilde{T}\widetilde{\varphi}\|_2^2}\\ &\le \frac{\beta^2\alpha(b-a)\sum\limits_{j=1}^{r} w_j|\widetilde{\varphi}(t_j)|^2}{\sum\limits_{j=1}^{r} w_j\left(|\widetilde{T}\widetilde{\varphi}(t_j)|^2-\gamma\alpha|\widetilde{\varphi}(t_j)|^2\right)} = \frac{\beta^2\alpha(b-a)}{|\widetilde{\lambda}|^2-\gamma\alpha},\end{aligned}$$

since $\widetilde{T}\widetilde{\varphi}(t_j) = \widetilde{\lambda}\widetilde{\varphi}(t_j)$ for $j = 1,\ldots,r$. Hence

$$|\widetilde{\lambda}-\lambda| \le \beta\left[\frac{\alpha(b-a)}{|\widetilde{\lambda}|^2-\gamma\alpha}\right]^{1/2}.$$

Note that $|\widetilde{\lambda}|^2-\gamma\alpha > 0$ because of the assumed lower bound for $|\widetilde{\lambda}|^2$. It remains to show that λ is an eigenvalue of T. First we claim that $\lambda \ne 0$. For if $\lambda = 0$, then

$$|\widetilde{\lambda}| \le \beta\Big[\frac{\alpha(b-a)}{|\widetilde{\lambda}|^2-\gamma\alpha}\Big]^{1/2}, \text{ that is, } |\widetilde{\lambda}|^4-\gamma\alpha|\widetilde{\lambda}|^2-\beta^2\alpha(b-a) \le 0.$$

But this is impossible since $|\widetilde{\lambda}|^2$ is greater than each of the roots of the preceding quadratic polynomial in $|\widetilde{\lambda}|^2$. Thus $\lambda \neq 0$. Being a nonzero spectral value of the compact operator U on Y, we see that λ is an eigenvalue of Y and hence of T.

Further, since $|\widetilde{\lambda}| > \widetilde{\delta}$ we have $0 < |\widetilde{\lambda}| - \widetilde{\delta} \leq |\lambda| \leq |\widetilde{\lambda}| + \widetilde{\delta}$. Consequently,

$$\frac{\widetilde{\delta}}{|\widetilde{\lambda}| + \widetilde{\delta}} \leq \frac{|\widetilde{\lambda} - \lambda|}{|\lambda|} \leq \frac{\widetilde{\delta}}{|\widetilde{\lambda}| - \widetilde{\delta}},$$

as desired. ∎

For $k(\cdot\,,\cdot) \in C^0([a,b]\times[a,b])$ and fixed $s, t \in [a,b]$, define

$$k_{s,t}(u) := k(s,u)k(u,t), \quad \breve{k}_{s,t}(u) := \overline{k(u,s)}k(u,t), \quad u \in [a,b].$$

Let $\widetilde{Q}$ denote the quadrature rule used in defining a Nyström approximation $\widetilde{T}$ of T, that is,

$$\widetilde{Q}(x) = \sum_{j=1}^{r} w_j x(t_j), \quad x \in C^0([a,b]).$$

Then the errors $d_1(s,t)$ and $d_2(s,t)$ defined in Lemma 4.20 can be written as follows:

$$d_1(s,t) = \widetilde{Q}(k_{s,t}) - \int_a^b k_{s,t}(u)du, \quad d_2(s,t) = \widetilde{Q}(\breve{k}_{s,t}) - \int_a^b \breve{k}_{s,t}(u)du.$$

This observation allows us to obtain an upper bound β for $|d_1(s,t)|$ and an upper bound γ for $|d_2(s,t)|$ in respect of several well-known quadrature rules, provided the kernel $k(\cdot\,,\cdot)$ is sufficiently smooth. (See Example 4.9 and Exercise 4.10.) For a positive integer p, assume that all partial derivatives of $k(\cdot\,,\cdot)$ up to the pth order exist and are continuous on $[a,b]\times[a,b]$. For $i, j = 0, \ldots, p$ and $i + j \leq p$, let

$$\alpha_{i,j} := \sup\left\{\left|\frac{\partial^{i+j}k}{\partial t^j \partial s^i}(s,t)\right| : s,t \in [a,b]\right\},$$

as in Lemma 4.11. Then for all $s, t \in [a,b]$, it follows that $k_{s,t}, \breve{k}_{s,t} \in C^p([a,b])$ and

$$\|k_{s,t}^{(p)}\|_\infty \leq \sum_{j=0}^{p} \binom{p}{j} \alpha_{p-j,0}\, \alpha_{0,j}, \quad \|\breve{k}_{s,t}^{(p)}\|_\infty \leq \sum_{j=0}^{p} \binom{p}{j} \alpha_{p-j,0}\, \alpha_{j,0}.$$

$\widetilde{Q}$	p	β	γ
Compound Rectangular Rule	1	$\frac{1}{2}\frac{\beta_1}{r}$	$\frac{1}{2}\frac{\gamma_1}{r}$
Compound Mid-Point Rule	2	$\frac{1}{24}\frac{\beta_2}{r^2}$	$\frac{1}{24}\frac{\gamma_2}{r^2}$
Compound Trapezoidal Rule	2	$\frac{1}{12}\frac{\beta_2}{(r-1)^2}$	$\frac{1}{12}\frac{\gamma_2}{(r-1)^2}$
Compound Simpson Rule	4	$\frac{1}{180}\frac{\beta_4}{(r-1)^4}$	$\frac{1}{180}\frac{\gamma_4}{(r-1)^4}$
Compound Gauss Two-Point Rule	4	$\frac{1}{270}\frac{\beta_4}{r^4}$	$\frac{1}{270}\frac{\gamma_4}{r^4}$
Gauss-Legendre Rule	$2r$	$\frac{(r!)^3}{(2r!)^4}\frac{\beta_{2r}}{(2r+1)}$	$\frac{(r!)^3}{(2r!)^4}\frac{\gamma_{2r}}{(2r+1)}$

Table 4.1

Let

$$\beta_p := \sup\left\{\left\|k_{s,t}^{(p)}\right\|_\infty : s,\, t \in [a,b]\right\}, \quad \gamma_p := \sup\left\{\left\|\breve{k}_{s,t}^{(p)}\right\|_\infty : s,\, t \in [a,b]\right\}.$$

If $a = 0$, $b = 1$ and $\widetilde{Q}$ denotes a quadrature rule with r nodes, $r \geq 2$, then the constants β and γ can be chosen as in Table 4.1. Note that in each of the cases mentioned above, $w_j \geq 0$ for each j and $\alpha = \sum_{j=1}^{r} w_j = 1$.

Example 4.22
Let $X = C^0([0,1])$ and for $x \in X$, consider

$$Tx(s) := \int_0^1 e^{st} x(t)dt, \quad s \in [0,1].$$

Since $k(s,t) = k(t,s) = e^{st} \in \mathbb{R}$ for all $s, t \in [0,1]$, the kernel $k(\cdot\,,\cdot)$ satisfies the requirements of Proposition 4.21. For fixed $s, t \in [0,1]$, we have

$$k_{s,t}(u) = e^{(s+t)u} = \breve{k}_{s,t}(u), \quad u \in [0,1],$$

and hence for any positive integer p,

$$k_{s,t}^{(p)}(u) = (s+t)^p e^{(s+t)u} = \breve{k}_{s,t}^{(p)}(u), \quad u \in [0,1].$$

Thus

$$\|k_{s,t}^{(p)}\|_\infty = \|\breve{k}_{s,t}^{(p)}\|_\infty \le 2^p e^2.$$

Let $\widetilde{T}$ be a Nyström approximation of T based on one of the quadrature rules $\widetilde{Q}$ with r nodes, $r \ge 2$, mentioned before. The constant $\beta = \gamma$ can be chosen as in Table 4.2.

Since $\alpha = \sum_{j=1}^{r} w_j = b - a = 1$, Proposition 4.21 shows that if $\widetilde{\lambda}$ is a nonzero eigenvalue of $\widetilde{T}$ such that $|\widetilde{\lambda}|^2 > (1+\sqrt{5})\beta/2$, then there is a nonzero eigenvalue λ of T such that

$$|\widetilde{\lambda} - \lambda| \le \frac{\beta}{\sqrt{|\widetilde{\lambda}|^2 - \beta}}.$$

In particular, let $\widetilde{Q}$ be the Gauss Two-Point rule with 100 nodes. Then the four largest eigenvalues of $\widetilde{T} := T_{100}^N$ are computed as follows:

$$\widetilde{\lambda}(1) := (0.13530301645)10^1 \qquad \widetilde{\lambda}(2) := 0.105983223$$
$$\widetilde{\lambda}(3) := (0.3560748)10^{-2} \qquad \widetilde{\lambda}(4) := (0.763795)10^{-4}.$$

(See Example 5.22.) In this case, we can let $\beta = \gamma := \dfrac{2^4 e^2}{270} 100^{-4} \le (0.44)10^{-8}$. For $j \in \{1,2,3\}$, we see that $|\widetilde{\lambda}(j)| > (1+\sqrt{5})\beta/2$. Hence there is a nonzero eigenvalue $\lambda(j)$ of T such that

$$|\widetilde{\lambda}(j) - \lambda(j)| \le \frac{\beta}{\sqrt{|\widetilde{\lambda}(j)|^2 - \beta}} \le \begin{cases} (0.4)10^{-8} & \text{if } j = 1, \\ (0.5)10^{-7} & \text{if } j = 2, \\ (0.2)10^{-5} & \text{if } j = 3. \end{cases}$$

$\widetilde{Q}$	$\beta = \gamma$
Compound Rectangular Rule	$\dfrac{e^2}{r}$
Compound Mid-Point Rule	$\dfrac{e^2}{6r^2}$
Compound Trapezoidal Rule	$\dfrac{e^2}{3(r-1)^2}$
Compound Simpson Rule	$\dfrac{4e^2}{45(r-1)^4}$
Compound Gauss Two-Point Rule	$\dfrac{8e^2}{135r^4}$
Gauss-Legendre Rule	$\dfrac{(r!)^3}{(2r!)^4}\dfrac{4^r e^2}{(2r+1)}$

Table 4.2

This shows that for $j \in \{1, 2, 3\}$, the integral part of $\lambda(j)$ is the same as the integral part of $\widetilde{\lambda}(j)$, and the first 8 decimal digits of $\lambda(1)$ are the same as those of $\widetilde{\lambda}(1)$, the first 7 decimal digits of $\lambda(2)$ are the same as those of $\widetilde{\lambda}(2)$, and the first 5 decimal digits of $\lambda(3)$ are the same as those of $\widetilde{\lambda}(3)$. Thus we obtain

$$\lambda(1) = 1.35303016\ldots \quad \lambda(2) = 0.1059832\ldots \quad \lambda(3) = 0.00356\ldots$$

On the other hand, since $|\widetilde{\lambda}(4)| \leq (1+\sqrt{5})\beta/2$, we are not in a position to say, on the basis of the approximation T_{100}^N of T, whether T has other nonzero eigenvalues. (See Example 4.17.)

Let $j \in \{1, 2, 3\}$. Consider the relative error $\dfrac{|\widetilde{\lambda}(j) - \lambda(j)|}{|\lambda(j)|}$ and define

$\delta_j := \dfrac{\beta}{\sqrt{|\widetilde{\lambda}(j)|^2 - \beta}}$, so that

$$\frac{\delta_j}{|\widetilde{\lambda}(j)| + \delta_j} \leq \frac{|\widetilde{\lambda}(j) - \lambda(j)|}{|\lambda(j)|} \leq \frac{\delta_j}{|\widetilde{\lambda}(j)| - \delta_j}.$$

Thus

$$(0.240)10^{-8} \leq \frac{|\widetilde{\lambda}(1) - \lambda(1)|}{|\lambda(1)|} \leq (0.241)10^{-8},$$

$$(0.392)10^{-6} \leq \frac{|\widetilde{\lambda}(2) - \lambda(2)|}{|\lambda(2)|} \leq (0.393)10^{-6},$$

$$(0.348)10^{-3} \leq \frac{|\widetilde{\lambda}(3) - \lambda(3)|}{|\lambda(3)|} \leq (0.349)10^{-3},$$

and we obtain a fairly good idea about the relative errors in approximating three eigenvalues of the operator T by the three largest eigenvalues of the operator T_{100}^N. ▮

4.4 Exercises

Unless otherwise stated, X denotes a Banach space over $\mathbb{C}$, $X \neq \{0\}$ and $T \in \mathrm{BL}(X)$. Prove the following assertions.

4.1 Let (π_n) be a sequence of projections in $\mathrm{BL}(X)$ such that $\pi_n y \to y$ for every $y \in \mathcal{R}(T)$. Then $T_n^P \xrightarrow{\text{p}} T$. If T is compact, then $T_n^P \xrightarrow{\text{n}} T$; and if in addition $(\|\pi_n\|)$ is bounded, $T_n^S \xrightarrow{\nu} T$, $T_n^G \xrightarrow{\nu} T$, $\|T_n^G - T_n^S\| = \|(T_n^G - T)\pi_n\| = \|(T_n^P - T)\pi_n\| \to 0$.

4.2 Let (π_n) be a sequence of projections in $\mathrm{BL}(X)$. Let Y be a subspace of X such that $\mathcal{R}(T) \subset Y$ and F be a linear map from Y to X such that $F \circ T \in \mathrm{BL}(X)$. If for some constant c and a sequence $\delta_n \to 0$, we have $\|\pi_n y - y\| \leq c\|F(y)\|\delta_n$ for all $y \in Y$, then $\|T_n^P - T\|$, $\|(T_n^S - T)T\|$, and $\|(T_n^G - T)T\|$ are all $O(\delta_n)$; and if in addition (π_n) is bounded, then $\|(T_n^S - T)T_n^S\|$ and $\|(T_n^G - T)T_n^G\|$ are also $O(\delta_n)$.

4.3 Let X be a Hilbert space over $\mathbb{C}$ and (π_n) be a sequence of orthogonal projections defined on X such that $\pi_n \xrightarrow{\text{p}} I$. If T is a compact operator on X, then $T_n^P \xrightarrow{\text{n}} T$, $T_n^S \xrightarrow{\text{n}} T$ and $T_n^G \xrightarrow{\text{n}} T$.

4.4 Let $X := L^\infty([a,b])$, $a = t_{n,0} < t_{n,1} < \cdots < t_{n,n} = b$, and $h_n := \max\{t_{n,j} - t_{n,j-1} \ : \ j = 1, \ldots, n\} \to 0$. For $x \in X$, let

$$(\pi_n x)(t) := \frac{1}{t_{n,j} - t_{n,j-1}} \int_{t_{n,j-1}}^{t_{n,j}} x(s)\, ds, \quad t \in [t_{n,j-1}, t_{n,j}[,\ j = 1, \ldots, n.$$

Then $\pi_n^2 = \pi_n$, $\|\pi_n\| = 1$ and if $x \in C^0([a,b])$, then $\|\pi_n x - x\|_\infty \to 0$. Let T be such that $Tx \in C^0([a,b])$ for every $x \in X$. Then $T_n^P \xrightarrow{\text{p}} T$, and if T is compact, $T_n^P \xrightarrow{\text{n}} T$, $T_n^S \xrightarrow{\nu} T$ and $T_n^G \xrightarrow{\nu} T$. (Hint: Exercise 4.1.)

4.5 Let X be the linear space of all complex-valued bounded functions on $[a,b]$ with the sup norm $\|\cdot\|_\infty$. Let $a = t_{n,0} < t_{n,1} < \cdots < t_{n,n} = b$, $h_n := \max\{t_{n,j} - t_{n,j-1} \ : \ j = 1, \ldots, n\} \to 0$ and $s_{n,j} \in [t_{n,j-1}, t_{n,j}[$, $j = 1, \ldots, n$. For $x \in X$, let

$$\begin{aligned} (\pi_n x)(t) &:= x(s_{n,j}), \quad t \in [t_{n,j-1}, t_{n,j}[,\ j = 1, \ldots, n, \\ (\pi_n x)(b) &:= x(b). \end{aligned}$$

Then $\pi_n^2 = \pi_n$, $\|\pi_n\| = 1$, and if $x \in C^0([a,b]$, $\|\pi_n x - x\|_\infty \to 0$.

Let $X := L^\infty([a,b])$ and T be such that $Tx \in C^0([a,b])$ for every $x \in X$.

(a) We have $T_n^P \xrightarrow{\text{p}} T$, and if T is compact, $T_n^P \xrightarrow{\text{n}} T$.

(b) Let $X_0 := C^0([a,b])$ and $T_0 = T_{|X_0,X_0}$. Then T_0, $T_0^S \in \text{BL}(X_0)$, $T_0^S \xrightarrow{\text{p}} T_0$ and if T is compact, $T_0^S \xrightarrow{\nu} T_0$. (Hint: Exercise 4.1.)

4.6 Let $k(\cdot,\cdot)$ be a complex-valued Lebesgue measurable function on $[a,b] \times [a,b]$ such that

(i) $\sup\left\{ \int_a^b |k(s,t)| dt : s \in [a,b] \right\} < \infty$ and

(ii) $\lim_{\delta \to 0} \int_a^b \Big| k(s+\delta, t) - k(s,t) \Big| dt = 0$ for every $s \in [a,b]$.

(For example, let $k(s,t) = \ln|s-t|$ or $k(s,t) = |s-t|^{-\alpha}$ for $s,t \in [-1,1]$, where ln denotes the natural logarithm and $0 < \alpha < 1$.)
Let $X = L^\infty([a,b])$ and for $x \in X$,

$$(Tx)(s) = \int_a^b k(s,t) x(t) dt, \quad s \in [a,b].$$

Then $Tx \in C^0([a,b])$ for every $x \in X$ and $T : X \to X$ is a compact linear operator. As such, the results of Exercises 4.4 and 4.5 apply in this case.

4.7 (Petrov-Galerkin) Let (π_n), $(\widetilde{\pi}_n)$ be sequences of projections in BL(X) such that $\pi_n \xrightarrow{\text{p}} I$, $\widetilde{\pi}_n \xrightarrow{\text{p}} I$. Let $T_n^{PG} := \widetilde{\pi}_n T \pi_n$. Then $T_n^{PG} \xrightarrow{\text{p}} T$. If T is a compact operator, then $T_n^{PG} \xrightarrow{\nu} T$; and if in addition $\pi_n^* \xrightarrow{\text{p}} I^*$, $\widetilde{\pi}_n^* \xrightarrow{\text{p}} I^*$, then $T_n^{PG} \xrightarrow{\text{n}} T$.

4.8 Let $X = C^0([a,b]), a = t_{n,1} < \cdots < t_{n,n} = b$ and $h_n = \max\{t_{n,j} - t_{n,j-1} : j = 1, \ldots, n\}$. If π_n denotes the piecewise linear interpolatory projection given in part (i) of Subsection 4.1.2, then

$$\|\pi_n x - x\|_\infty \leq \frac{1}{8}\|x''\|_\infty (h_n)^2 \quad \text{for all } x \in C^2([a,b]).$$

(Hint: If $t \in [t_{n,j-1}, t_{n,j}]$, then $\pi_n x(t) = x(t_{n,j-1}) + [x(t_{n,j}) - x(t_{n,j-1})] \times \dfrac{t - t_{n,j-1}}{t_{n,j} - t_{n,j-1}} = x(t) + \dfrac{x''(s)}{2}(t - t_{n,j-1})(t_{n,j} - t)$, where $s \in [t_{n,j-1}, t_{n,j}]$, by applying Rolle's Theorem twice.)

4.9 Let $X := C^0([a,b])$, n and $r(n)$ be positive integers, $a = t_{n,0} < t_{n,1} < \cdots < t_{n,r(n)} = b$ and $\widetilde{t}_{n,j} \in]t_{n,j-1}, t_{n,j}[$ for $j = 1, \ldots, r(n)$. Define the piecewise quadratic interpolatory projection $\pi_n : X \to X$ as follows. For $x \in X$, let $(\pi_n x)_{|[t_{n,j-1}, t_{n,j}]}$ be the unique quadratic polynomial on $[t_{n,j-1}, t_{n,j}]$ which agrees with x at $t_{n,j-1}, \widetilde{t}_{n,j}$ and $t_{n,j}$, $j = 1, \ldots, r(n)$. If $h_n := \max\{t_{n,j} - t_{n,j-1} : j = 1, \ldots, r(n)\} \to 0$ and there are constants γ and δ satisfying $0 < \delta \leq \gamma$ such that $\delta \leq \dfrac{t_{n,j} - \widetilde{t}_{n,j}}{\widetilde{t}_{n,j} - t_{n,j-1}} \leq \gamma$ for all n and $j = 1, \ldots, r(n)$, then $\pi_n \xrightarrow{\text{p}} I$. For integers j, k and ℓ satisfying $0 \leq j, k, \ell \leq r(n)$, let $p_{k,\ell}^{n,j}$ denote the quadratic polynomial which takes the value 1 at $t_{n,j}$ and the value 0 at $t_{n,k}$ and $\widetilde{t}_{n,\ell}$; and let $\widetilde{p}_{k,\ell}^{n,j}$ denote the quadratic polynomial which takes the value 1 at $\widetilde{t}_{n,j}$ and the value 0 at $t_{n,k}$ and $t_{n,\ell}$. Define $e_{n,0}, \ldots, e_{n,r(n)}$ and $\widetilde{e}_{n,1}, \ldots, \widetilde{e}_{n,r(n)}$ in $C^0([a,b])$ as follows: For $j = 0, \ldots, r(n)$, let

$$e_{n,j}(t) := \begin{cases} p_{j-1,j}^{n,j}(t) & \text{if } t \in [t_{n,j-1}, t_{n,j}], \\ p_{j+1,j+1}^{n,j}(t) & \text{if } t \in [t_{n,j}, t_{n,j+1}], \\ 0 & \text{otherwise,} \end{cases}$$

where $t_{n,-1} := a - 2$, $\widetilde{t}_{n,-1} := a - 1$, $\widetilde{t}_{n,r(n)+1} := b + 1$ and $t_{n,r(n)+1} := b + 2$, and for $j = 1, \ldots, r(n)$, let

$$\widetilde{e}_{n,j}(t) := \begin{cases} \widetilde{p}_{j-1,j}^{n,j}(t) & \text{if } t \in [t_{n,j-1}, t_{n,j}], \\ 0 & \text{otherwise.} \end{cases}$$

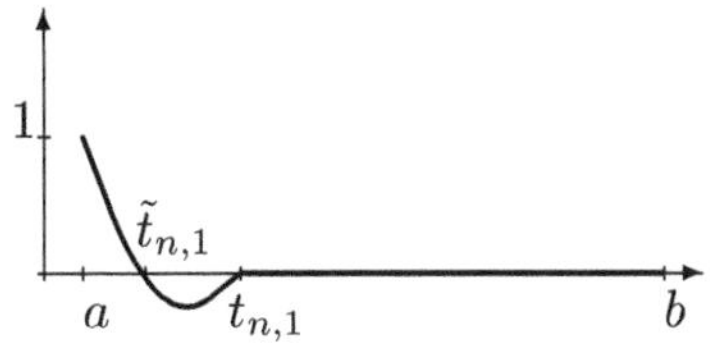

the function $e_{n,0}$

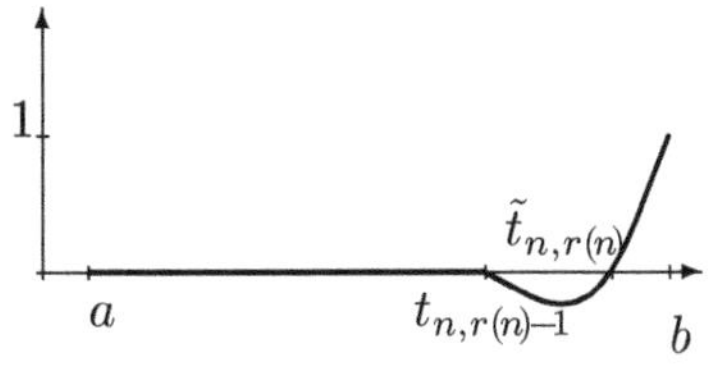

the function $e_{n,r(n)}$

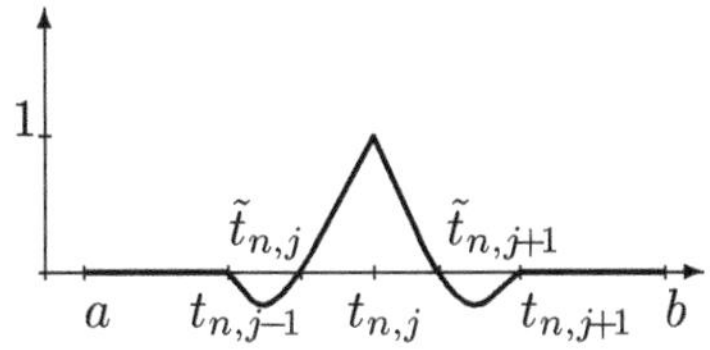

the function $e_{n,j}$, $0 < j < r(n)$

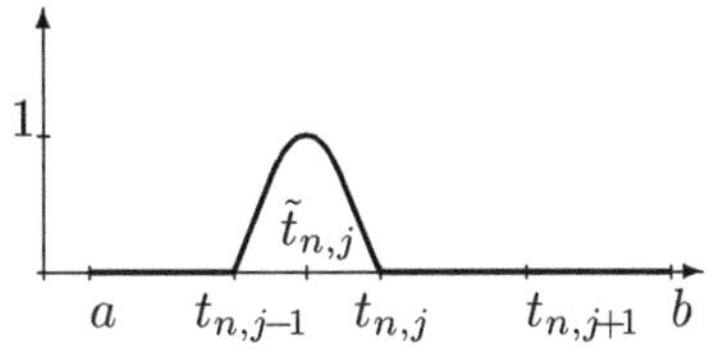

the function $\tilde{e}_{n,j}$, $1 \le j \le r(n)$

Then for $x \in C^0([a,b])$,

$$\pi_n x = \sum_{j=0}^{r(n)} x(t_{n,j}) e_{n,j} + \sum_{j=1}^{r(n)} x(\widetilde{t}_{n,j}) \widetilde{e}_{n,j}.$$

Also, we have

$$\|\pi_n x - x\|_\infty \le \frac{1}{6} \|x^{(3)}\|_\infty (h_n)^3 \quad \text{for all } x \in C^3([a,b]).$$

4.10 Let $a := 0$, $b := 1$, n an odd integer ≥ 3, $r(n) := (n-1)/2$, $t_{n,j} := 2j/(n-1)$ for $j = 0, \ldots, (n-1)/2$, and $\widetilde{t}_{n,j} := (2j-1)/(n-1)$ for $j = 1, \ldots, (n-1)/2$ in Exercise 4.9. Then the sequence (π_n) of piecewise quadratic interpolatory projections induces a sequence (Q_n) of convergent quadrature formulæ known as the Compound Simpson Rule. In fact, for $x \in C^0([a,b])$, we have

$$Q_n(x) = \frac{1}{3(n-1)} \left[x(0) + 4 \sum_{j=1}^{(n-1)/2} x\Big(\frac{2j-1}{n-1}\Big) + 2 \sum_{j=1}^{(n-3)/2} x\Big(\frac{2j}{n-1}\Big) + x(1) \right].$$

We also have

$$\left| Q_n(x) - \int_a^b x(t)dt \right| \le \frac{\|x^{(4)}\|_\infty}{180(n-1)^4} \text{ for } n = 1, 2, \ldots \text{ and } x \in C^4([a,b]).$$

4.11 For a positive integer n and nodes $a = t_{n,0} \le t_{n,1} < \cdots < t_{n,n} \le t_{n,n+1} = b$, let X_n denote the subspace of $X := C^0([a,b])$ consisting of all $x \in C^2([a,b])$ such that x is a polynomial of degree ≤ 3 on each of the subintervals $[t_{n,1}, t_{n,2}], \ldots, [t_{n,n-1}, t_{n,n}]$.

(a) Let $a = t_{n,1}$ and $b = t_{n,n}$. Then for each $j = 1, \ldots, n$, there is a unique $e_{n,j} \in X_n$ such that $e_{n,j}(t_{n,k}) = \delta_{j,k}$ for $k = 1, \ldots, n$ and $e''_{n,j}(t_{n,1}) = 0 = e''_{n,j}(t_{n,n})$. The projection π_n given by $\pi_n x = \sum_{j=1}^{n} x(t_{n,j}) e_{n,j}$, $x \in X$ is known as a natural cubic spline interpolatory projection. (Hint: $\dim X_n = n + 2$.)

(b) Let $a < t_{n,1}$ and $b > t_{n,n}$. Then for each $j = 0, 1, \ldots, n+1$, there is a unique $e_{n,j} \in X_n$ such that $e_{n,j}(t_{n,k}) = \delta_{n,k}$ for $k = 0, 1, \ldots, n+1$, and $e_{n,j}$ is a polynomial of degree ≤ 1 on each of the two subintervals $[a, t_{n,1}]$ and $[t_{n,n}, b]$. We have $e''_{n,j}(t_{n,1}) = 0 = e''_{n,j}(t_{n,n})$ for each $j = 0, \ldots, n+1$. (Hint: If $\widetilde{X}_n := \{x \in X_n : x$ is a polynomial of degree ≤ 1 on each of the two subintervals $[a, t_{n,1}]$ and $[t_{n,n}, b]\}$, then $\dim \widetilde{X}_n = 4(n-1) + 2 + 2 - 3n = n$.)

4.12 Let $X = C^0([a,b])$ and p be a positive integer. Suppose that $\mathcal{R}(T) \subset C^p([a,b])$ and $\|(Tx)^{(p)}\|_\infty \le \alpha_p \|x\|_\infty$ for some constant α_p and all $x \in X$. Let (π_n) be a sequence of interpolatory projections defined on X and $h_n \to 0$, where h_n is the mesh of the partition $a \le t_{n,1} < \cdots < t_{n,n} \le b$. If $\|\pi_n y - y\|_\infty \le c_p \|y^{(p)}\|_\infty (h_n)^p$ for some constant c_p and all $y \in C^p([a,b])$, then $\|T_n^P - T\|$, $\|(T_n^S - T)T\|$ and $\|(T_n^G - T)T\|$ are all $O((h_n)^p)$. If $\pi_n \xrightarrow{\text{p}} I$, then $\|(T_n^S - T)T_n^S\|$ and $\|(T_n^G - T)T_n^G\|$ are also $O((h_n)^p)$. In particular, if T is a Fredholm integral operator on X with a continuous kernel $k(\cdot,\cdot)$ whose partial derivatives up to the pth order with respect to the first variable exist and are continuous on $[a,b]\times[a,b]$, then the preceding results hold for $p = 1$ if π_n is a piecewise linear interpolatory projection; for $p = 2$ if π_n is a piecewise linear interpolatory projection with $t_{n,1} = a$, $t_{n,n} = b$; for $p = 3$ if π_n is a piecewise quadratic interpolatory projection as given in Exercise 4.9; and for $p = 4$ if π_n is a cubic spline interpolatory projection as given in part (ii) of Subsection 4.1.2. (Hint: Exercise 4.2 with $Y = C^p([a,b])$ and $Fx = x^{(p)}$, parts (i) and (ii) of Subsection 4.1.2, Exercise 4.8 and Exercise 4.9.)

4.13 Let $t_{n,1}, \ldots, t_{n,n} \in [-1, 1]$ be the roots of the nth Legendre polynomial and let π_n denote the Lagrange interpolatory projection with nodes at $t_{n,1}, \ldots, t_{n,n}$.

(a) (Vainikko) If $T : L^2([-1,1]) \to L^2([-1,1])$ is a compact operator, then $T_n^P \xrightarrow{\text{n}} T$.

(b) (Sloan-Burn) If $T : C^0([-1,1]) \to C^0([-1,1])$ is a Fredholm integral operator with a continuous kernel, then $T_n^S \xrightarrow{\text{cc}} T$ and hence $T_n^S \xrightarrow{\nu} T$. (Hint: Result of Erdös and Turan stated in the text.)

4.14 Let $a = t_{n,0} < t_{n,1} < \cdots < t_{n,n-1} < t_{n,n} = b$ and d be a positive integer. For $j = 1, \ldots, n$, let $\pi_{n,j}^{(d)}$ denote the orthogonal projection defined on $L^2([t_{n,j-1}, t_{n,j}[)$ whose range equals the set of all polynomials on $[t_{n,j-1}, t_{n,j}[$ of degree $\leq d$. Define $\pi_n^{(d)} : L^2([a,b]) \to L^2([a,b])$ as follows. For $x \in L^2([a,b])$,

$$(\pi_n^{(d)} x)(t) := \left(\pi_{n,j}^{(d)} x_{|[t_{n,j-1}, t_{n,j}[}\right)(t), \quad t \in [t_{n,j-1}, t_{n,j}[,$$

$$(\pi_n^{(d)} x)(b) := (\pi_n^{(d)} x)(t_{n,n-1}).$$

Then $\pi_n^{(d)}$ is an orthogonal projection defined on $L^2([a,b])$ whose range is contained in the set of all piecewise polynomials of degree $\leq d$ and $\pi_n^{(d)} \xrightarrow{\text{p}} I$. (Hint: If π_n denotes the orthogonal projection on piecewise constant functions as given in Subsection 4.1.3, then $\|\pi_n^{(d)} x - x\|_2 \leq \|\pi_n x - x\|_2$ for every $x \in L^2([a,b])$.)

4.15 Let $X := C^0([a,b])$ and T be a Fredholm integral operator with a continuous kernel $k(\cdot,\cdot)$. For each positive integer n, let $a \leq t_{n,1} < \cdots < t_{n,r(n)} \leq b$ and $e_{n,1}, \ldots, e_{n,r(n)} \in X$ such that $e_{n,j}(t_{n,k}) = \delta_{j,k}$, $j,k = 1, \ldots, r(n)$. Let π_n denote the corresponding interpolatory projection. For a fixed $s \in [a,b]$, define $k_s(t) := k(s,t)$, $t \in [a,b]$, and let

$$k_n(s,t) := (\pi_n k_s)(t) = \sum_{j=1}^{r(n)} k(s, t_{n,j}) e_{n,j}(t), \quad s,t \in [a,b].$$

For a fixed $t \in [a,b]$, define $k^t(s) := k(s,t)$, $s \in [a,b]$, and let

$$l_n(s,t) := (\pi_n k^t)(s) = \sum_{j=1}^{r(n)} k(t_{n,j}, t) e_{n,j}(s), \quad s,t \in [a,b].$$

The degenerate kernels $k_n(\cdot,\cdot)$ and $l_n(\cdot,\cdot)$ are obtained by interpolating the kernel $k(\cdot,\cdot)$ in the second variable and in the first variable, respectively.

(a) If $\|\pi_n(k_s) - k_s\|_\infty \to 0$ uniformly in $s \in [a, b]$, then $\|k_n(\cdot,\cdot) - k(\cdot,\cdot)\|_\infty \to 0$, so that $T_n^D \xrightarrow{\text{n}} T$. (cf. (i) of Subsection 4.2.1.)

(b) For the kernel $l_n(\cdot,\cdot)$, $T_n^D = T_n^P$, so that $T_n^D \xrightarrow{\text{n}} T$ if $\pi_n \xrightarrow{\text{p}} I$.

4.16 Let $X := L^2([a, b])$ and $e_1, e_2, \ldots$ form an orthonormal basis for X. For $n = 1, 2, \ldots$ let $\pi_n x = \sum_{j=1}^{n} \langle x\,,\, e_j \rangle e_j$. If T is a Fredholm integral operator on X with a square-integrable kernel, then T_n^P and T_n^S are degenerate kernel approximations of T. (cf. (iv) of Subsection 4.2.1.)

4.17 Let T denote the Fredholm integral operator on $C^0([0, 1])$ with kernel $k(s, t) = \exp(st)$, $0 \le s, t \le 1$, and T_{120}^N denote the Nyström approximation of T based on the Gauss Two-Point Rule with 120 nodes. The fourth largest eigenvalue of T_{120}^N is $\breve{\lambda}(4) = (.763796)10^{-4}$. (See Example 5.22.) Let $\breve{\beta} := \dfrac{2^4 e^2}{270}(120)^{-4}$. Then $\breve{\beta} \le (0.22)10^{-8}$ and $|\breve{\lambda}(4)|^2 > (1 + \sqrt{5})\breve{\beta}/2$. Hence there is a nonzero eigenvalue λ of T such that $|\breve{\lambda}(4) - \lambda| \le \dfrac{\breve{\beta}}{\sqrt{|\breve{\lambda}(4)|^2 - \breve{\beta}}} \le 0.58$. If $\lambda(1) = 1.35303016\ldots$ is the eigenvalue of T obtained in Example 4.22, then $\lambda \ne \lambda(1)$.

Chapter 5

Matrix Formulations

In Chapter 4 we saw how to approximate a bounded operator T on a complex Banach space X by a sequence (T_n) of bounded finite rank operators on X as well as how to find an approximate solution of the eigenvalue problem for T by solving the eigenvalue problem for T_n. In the present section we show that the eigenvalue problem for a bounded finite rank operator $\widetilde{T}$ can be solved by reducing it to a matrix eigenvalue problem in a canonical way. For a bounded finite rank operator $\widetilde{T}$, solutions of the operator equation $\widetilde{T}x - x = y$ (where $y \in X$ is given and $x \in X$ is to be found) or of the eigenvalue problem $\widetilde{T}\varphi = \lambda\varphi$ (where $0 \neq \varphi \in X$ and $0 \neq \lambda \in \mathbb{C}$ are to be found) have long been obtained with the help of matrix computations. This is usually done in a variety of ways, depending on the specific nature of the finite rank operator $\widetilde{T}$. A unified treatment for solutions of operator equations involving finite rank operators was given in [73]. It was extended to eigenvalue problems for finite rank operators in [34] and [53]. Our treatment here is along those lines. Although the operator T_n^K which appears in the singularity subtraction technique discussed in Subsection 4.2.3 is not of finite rank, the eigenvalue problem for it can still be reduced to matrix computations. We discuss a related question about finding a basis for a finite dimensional spectral subspace for T_n in such a way that, as n tends to infinity, each element of the basis is bounded and is bounded away from the span of the other elements. We call such bases uniformly well-conditioned. We also give matrix formulations for the iterative refinement schemes and for the acceleration procedure discussed in Chapter 3 when the approximate operator is of finite rank. We illustrate the implementation of these matrix formulations by giving some numerical examples.

5.1 Finite Rank Operators

A bounded finite rank operator is typically presented to us in the following form.

Proposition 5.1
A map $\widetilde{T} : X \to X$ is a bounded finite rank operator if and only if there are $\widetilde{x}_1, \ldots, \widetilde{x}_n$ in X and $\widetilde{f}_1, \ldots, \widetilde{f}_n$ in X^* such that

$$\widetilde{T}x = \sum_{j=1}^{n} \langle x \, , \, \widetilde{f}_j \rangle \widetilde{x}_j = \underline{\widetilde{x}} \, (x, \underline{\widetilde{f}}\,), \quad x \in X,$$

where $\underline{\widetilde{x}} := [\widetilde{x}_1, \ldots, \widetilde{x}_n] \in X^{1 \times n}$ *and* $\underline{\widetilde{f}} := [\widetilde{f}_1, \ldots, \widetilde{f}_n] \in (X^*)^{1 \times n}$.

Proof
It is clear that if $\widetilde{T}$ is defined as above, then it is a bounded operator since $\widetilde{f}_1, \ldots, \widetilde{f}_n$ are continuous conjugate-linear functionals on X, and the rank of $\widetilde{T}$ is finite since $\mathcal{R}(\widetilde{T}) \subset \text{span}\{\widetilde{x}_1, \ldots, \widetilde{x}_n\}$.

Conversely, let the rank of $\widetilde{T} \in \text{BL}(X)$ be finite. Then there is a finite set $\{\widetilde{x}_1, \ldots, \widetilde{x}_n\}$ in X such that $\mathcal{R}(\widetilde{T}) \subset \text{span}\{\widetilde{x}_1, \ldots, \widetilde{x}_n\}$. Renumbering $\widetilde{x}_1, \ldots, \widetilde{x}_n$, if necessary, we may assume that the set $\{\widetilde{x}_1, \ldots, \widetilde{x}_m\}$ is linearly independent and $\mathcal{R}(\widetilde{T}) \subset \text{span}\{\widetilde{x}_1, \ldots, \widetilde{x}_m\}$ for some $m \leq n$. Then there are unique complex numbers $c_1(x), \ldots, c_m(x)$ such that

$$\widetilde{T}x = c_1(x)\widetilde{x}_1 + \cdots + c_m(x)\widetilde{x}_m.$$

For $j = 1, \ldots, m$, define $\widetilde{f}_j : X \to \mathbb{C}$ by $\widetilde{f}_j(x) = \overline{c_j(x)}$, $x \in X$. It is easy to see that $\widetilde{f}_1, \ldots, \widetilde{f}_m$ are conjugate-linear functionals on X and

$$\widetilde{T}x = \sum_{j=1}^{m} \langle x \, , \, \widetilde{f}_j \rangle \widetilde{x}_j, \quad x \in X.$$

To see that each $\widetilde{f}_j$ is continuous, let

$$\delta_j := \text{dist}(\widetilde{x}_j, \text{span}\{\widetilde{x}_i : i = 1, \ldots, m, \, i \neq j\}), \quad j = 1, \ldots, m.$$

Since $\widetilde{x}_j$ does not belong to the closed subset $\text{span}\{\widetilde{x}_i : i = 1, \ldots, m, \, i \neq j\}$, we have $\delta_j > 0$. Now for all $x \in X$, we have

$$|\langle x \, , \, \widetilde{f}_j \rangle| \delta_j \leq \| \langle x \, , \, \widetilde{f}_1 \rangle \widetilde{x}_1 + \cdots + \langle x \, , \, \widetilde{f}_m \rangle \widetilde{x}_m \| = \|\widetilde{T}x\| \leq \|\widetilde{T}\| \, \|x\|.$$

Thus $|\widetilde{f}_j(x)| \le (\|\widetilde{T}\|/\delta_j)\|x\|$ for all $x \in X$, so that $\widetilde{f}_j$ is continuous for $j = 1, \ldots, m$. Let $\widetilde{f}_j := 0$ if $m < j \le n$. Then $\widetilde{f}_1, \ldots, \widetilde{f}_n \in X^*$ and

$$\widetilde{T}x = \sum_{j=1}^{n} \langle x\,,\, \widetilde{f}_j\rangle \widetilde{x}_j, \quad x \in X,$$

as desired. ∎

It may be noted that neither the elements $\widetilde{x}_1, \ldots, \widetilde{x}_n$ of X nor the elements $\widetilde{f}_1, \ldots, \widetilde{f}_n$ of X^* in the representation of a bounded finite rank operator $\widetilde{T}$ are required to be linearly independent and that the choice of these elements is not unique. One can easily see how each of the bounded finite rank operators T_n^P, T_n^S, T_n^G, T_n^D, T_n^N and T_n^F considered in Chapter 4 can be represented in this manner.

Let $\widetilde{T}$ be a bounded finite rank operator on X as given in the statement of Proposition 5.1. Then for each $x \in X$, $\widetilde{T}x$ is determined by $\underline{\widetilde{x}} = [\widetilde{x}_1, \ldots, \widetilde{x}_n] \in X^{1\times n}$ and $\underline{\widetilde{f}} = [\widetilde{f}_1, \ldots, \widetilde{f}_n] \in (X^*)^{1\times n}$. We show that $\widetilde{T}$ can be written as a composition of two operators, one of them being determined by $\underline{\widetilde{x}}$ and the other by $\underline{\widetilde{f}}$.

Recall that $\mathbb{C}^{n\times 1} := \{\mathsf{u} := [\mathsf{u}(1), \ldots, \mathsf{u}(n)]^\top : \mathsf{u}(j) \in \mathbb{C},\ j = 1, \ldots, n\}$ denotes the linear space of all $n\times 1$ matrices with complex entries.

Define $\widetilde{K} : X \to \mathbb{C}^{n\times 1}$ by

$$\widetilde{K}x := [\langle x\,,\, \widetilde{f}_1\rangle, \ldots, \langle x\,,\, \widetilde{f}_n\rangle]^\top = (\!(x, \underline{\widetilde{f}})\!), \quad x \in X,$$

and $\widetilde{L} : \mathbb{C}^{n\times 1} \to X$ by

$$\widetilde{L}\mathsf{u} := \sum_{j=1}^{n} \mathsf{u}(j)\widetilde{x}_j = \underline{\widetilde{x}}\,\mathsf{u}, \quad \mathsf{u} := [\mathsf{u}(1), \ldots, \mathsf{u}(n)]^\top \in \mathbb{C}^{n\times 1}.$$

Clearly, $\widetilde{K}$ and $\widetilde{L}$ are linear maps and

$$\widetilde{T} = \widetilde{L}\widetilde{K}.$$

We define an operator $\widetilde{A} : \mathbb{C}^{n\times 1} \to \mathbb{C}^{n\times 1}$ by

$$\widetilde{A} := \widetilde{K}\widetilde{L}.$$

For all $\mathsf{u} \in \mathbb{C}^{n\times 1}$, we have

$$\widetilde{A}\mathsf{u} = \widetilde{K}(\widetilde{L}\mathsf{u}) = (\!(\widetilde{L}\mathsf{u}, \underline{\widetilde{f}})\!) = (\!(\underline{\widetilde{x}}\,\mathsf{u}, \underline{\widetilde{f}})\!) = (\!(\underline{\widetilde{x}}, \underline{\widetilde{f}})\!)\mathsf{u}.$$

Hence the $n \times n$ Gram matrix $\widetilde{\mathsf{A}} := (\underline{\widetilde{x}}, \underline{\widetilde{f}})$ represents the operator $\widetilde{A}$ with respect to the standard basis for $\mathbb{C}^{n\times 1}$. Also, if the set of elements in $\underline{\widetilde{x}}$ are linearly independent in X and span a subspace X_n, then the matrix $\widetilde{\mathsf{A}}$ represents the operator $\widetilde{T}_{|X_n,X_n}$ with respect to the ordered basis $\underline{\widetilde{x}}$ of X_n.

We shall now prove a crucial result which will allow us to find bases for eigenspaces and for spectral subspaces of $\widetilde{T}$.

Lemma 5.2
Let p be a positive integer, $\mathsf{Z} \in \mathbb{C}^{p\times p}$ such that $0 \notin \mathrm{sp}(\mathsf{Z})$, $\underline{y} \in X^{1\times p}$, and $\underline{\mathsf{v}} := \underline{\widetilde{K}}\, \underline{y} \in \mathbb{C}^{n\times p}$. Then for $\underline{x} \in X^{1\times p}$ and $\underline{\mathsf{u}} \in \mathbb{C}^{n\times p}$, the following holds: $\underline{\widetilde{T}}\, \underline{x} = \underline{x}\,\mathsf{Z} + \underline{y}$ and $\underline{\widetilde{K}}\, \underline{x} = \underline{\mathsf{u}}$ if and only if $\underline{\widetilde{A}}\, \underline{\mathsf{u}} = \underline{\mathsf{u}}\,\mathsf{Z} + \underline{\mathsf{v}}$ and $\underline{\widetilde{L}}\, \underline{\mathsf{u}} = \underline{x}\,\mathsf{Z} + \underline{y}$.

In particular, let $\underline{y} := \underline{0}$ (so that $\underline{\mathsf{v}} = \underline{0}$ also); and let $\underline{x}$, $\underline{\mathsf{u}}$ satisfy the above conditions. Then the set of p elements in $\underline{x}$ is linearly independent in X if and only if the set of p vectors in $\underline{\mathsf{u}}$ is linearly independent in $\mathbb{C}^{n\times 1}$.

Proof
Assume that $\underline{\widetilde{T}}\, \underline{x} = \underline{x}\,\mathsf{Z} + \underline{y}$ and $\underline{\widetilde{K}}\, \underline{x} = \underline{\mathsf{u}}$. Then $\underline{\widetilde{L}}\, \underline{\mathsf{u}} = \underline{\widetilde{L}}\, \underline{\widetilde{K}}\, \underline{x} = \underline{\widetilde{T}}\, \underline{x} = \underline{x}\,\mathsf{Z} + \underline{y}$. Also, $\underline{\widetilde{A}}\, \underline{\mathsf{u}} = \underline{\widetilde{A}}\, \underline{\widetilde{K}}\, \underline{x} = \underline{\widetilde{K}}\, \underline{\widetilde{T}}\, \underline{x} = \underline{\widetilde{K}}(\underline{x}\,\mathsf{Z} + \underline{y}) = \underline{\mathsf{u}}\,\mathsf{Z} + \underline{\mathsf{v}}$.

Conversely, assume that $\underline{\widetilde{A}}\, \underline{\mathsf{u}} = \underline{\mathsf{u}}\,\mathsf{Z} + \underline{\mathsf{v}}$ and $\underline{\widetilde{L}}\, \underline{\mathsf{u}} = \underline{x}\,\mathsf{Z} + \underline{y}$. Then $\underline{\widetilde{K}}\, \underline{x}\,\mathsf{Z} = \underline{\widetilde{K}}(\underline{\widetilde{L}}\, \underline{\mathsf{u}} - \underline{y}) = \underline{\widetilde{A}}\, \underline{\mathsf{u}} - \underline{\widetilde{K}}\, \underline{y} = \underline{\widetilde{A}}\, \underline{\mathsf{u}} - \underline{\mathsf{v}} = \underline{\mathsf{u}}\,\mathsf{Z}$. Since Z is nonsingular, it follows that $\underline{\widetilde{K}}\, \underline{x} = \underline{\mathsf{u}}$. Also, $\underline{\widetilde{T}}\, \underline{x} = \underline{\widetilde{L}}\, \underline{\widetilde{K}}\, \underline{x} = \underline{\widetilde{L}}\, \underline{\mathsf{u}} = \underline{x}\,\mathsf{Z} + \underline{y}$.

Now consider $\underline{y} := \underline{0}$, so that $\underline{\mathsf{v}} = \underline{\widetilde{K}}\, \underline{y} = \underline{0}$, and $\underline{\widetilde{K}}\, \underline{x} = \underline{\mathsf{u}}$, $\underline{\widetilde{L}}\, \underline{\mathsf{u}} = \underline{x}\,\mathsf{Z}$.

Let the set of p vectors in $\underline{\mathsf{u}}$ be linearly independent in $\mathbb{C}^{n\times 1}$, and $\mathsf{c} \in \mathbb{C}^{p\times 1}$ be such that $\underline{x}\,\mathsf{c} = 0$. Then $\underline{\mathsf{u}}\,\mathsf{c} = (\underline{\widetilde{K}}\, \underline{x})\mathsf{c} = \widetilde{K}(\underline{x}\,\mathsf{c}) = \widetilde{K}(0) = 0$. The linear independence of the set of p vectors in $\underline{\mathsf{u}}$ implies that $\mathsf{c} = 0$, as desired.

Let the set of p elements in $\underline{x}$ be linearly independent in X, and $\mathsf{c} \in \mathbb{C}^{p\times 1}$ be such that $\underline{\mathsf{u}}\,\mathsf{c} = 0$. Then $\underline{x}(\mathsf{Z}\mathsf{c}) = (\underline{x}\,\mathsf{Z})\mathsf{c} = (\underline{\widetilde{L}}\, \underline{\mathsf{u}})\mathsf{c} = \widetilde{L}(\underline{\mathsf{u}}\,\mathsf{c}) = \widetilde{L}(0) = 0$. The linear independence of the set of p elements in $\underline{x}$ implies that $\mathsf{Z}\mathsf{c} = 0$ and since Z is nonsingular, $\mathsf{c} = 0$, as desired. ∎

Proposition 5.3

$$\mathrm{sp}(\widetilde{T}) \setminus \{0\} = \mathrm{sp}(\widetilde{A}) \setminus \{0\}.$$

In particular, $\mathrm{sp}(\widetilde{T})$ *is a finite set. Let* $\widetilde{\Lambda} \subset \mathrm{sp}(\widetilde{T}) \setminus \{0\}$, $\widetilde{P} := P(\widetilde{T}, \widetilde{\Lambda})$ *and* $\widetilde{\mathsf{P}} := P(\widetilde{A}, \widetilde{\Lambda})$*. Then*

$$\widetilde{K}\widetilde{P} = \widetilde{\mathsf{P}}\widetilde{K}.$$

Also, $\widetilde{K}$ *maps* $\mathcal{R}(\widetilde{P})$ *into* $\mathcal{R}(\widetilde{\mathsf{P}})$ *in a one-to-one manner, and hence* $\widetilde{\Lambda}$ *is a spectral set of finite type for* $\widetilde{T}$.

Proof

We show that $\mathrm{re}(\widetilde{T}) \setminus \{0\} = \mathrm{re}(\widetilde{A}) \setminus \{0\}$.

Let $0 \neq z \in \mathrm{re}(\widetilde{A})$. Consider $y \in X$ and define $\mathsf{v} := \widetilde{K}y$, $\mathsf{u} := R(\widetilde{A}, z)\mathsf{v}$ and $x := (\widetilde{L}\mathsf{u} - y)/z$. Then $\widetilde{A}\mathsf{u} = z\mathsf{u} + \mathsf{v}$ and $\widetilde{L}\mathsf{u} = zx + y$. Letting $p := 1$ in Lemma 5.2, we see that $\widetilde{T}x - zx = y$. Thus $\mathcal{R}(\widetilde{T} - zI) = X$. Next, let $x \in X$ be such that $(\widetilde{T} - zI)x = 0$ and define $\mathsf{u} := \widetilde{K}x$. Letting $p := 1$ and $y := 0$ in Lemma 5.2, we see that $\widetilde{A}\mathsf{u} = z\mathsf{u}$ and $\widetilde{L}\mathsf{u} = zx$. Since $z \in \mathrm{re}(\widetilde{A})$, we have $\mathsf{u} = \mathsf{0}$ and so $zx = \widetilde{L}(\mathsf{0}) = 0$. As $z \neq 0$, we obtain $x = 0$. Hence $\mathcal{N}(\widetilde{T} - zI) = \{0\}$. Thus $z \in \mathrm{re}(\widetilde{T})$.

Conversely, let $0 \neq z \in \mathrm{re}(\widetilde{T})$. To show that $z \in \mathrm{re}(\widetilde{A})$, it is enough to show that $\mathcal{N}(\widetilde{A} - zI) = \{\mathsf{0}\}$. Let $\mathsf{u} \in \mathbb{C}^{n \times 1}$ be such that $(\widetilde{A} - zI)\mathsf{u} = \mathsf{0}$ and define $x := \widetilde{L}\mathsf{u}/z$. Letting $p := 1$ and $y := 0$ in Lemma 5.2, we see that $\widetilde{T}x = zx$ and $\widetilde{K}x = \mathsf{u}$. Since $z \in \mathrm{re}(\widetilde{T})$, we have $x = 0$ and so $\mathsf{u} = \widetilde{K}(0) = \mathsf{0}$. Hence $\mathcal{N}(\widetilde{A} - zI) = \{\mathsf{0}\}$, as desired.

Since $\mathrm{sp}(\widetilde{A})$ has at most n elements, $\mathrm{sp}(\widetilde{T})$ has at most $n+1$ elements. Let $\widetilde{\Lambda} \subset \mathrm{sp}(\widetilde{T}) \setminus \{0\}$. Then $\widetilde{\Lambda}$ is a spectral set for $\widetilde{T}$ as well as for $\widetilde{A}$. Consider $\mathrm{C} \in \mathcal{C}(\widetilde{T}, \widetilde{\Lambda})$ such that $0 \in \mathrm{ext}(\mathrm{C})$. Then $\mathrm{C} \in \mathcal{C}(\widetilde{A}, \widetilde{\Lambda})$ as well.

Let $z \in \mathrm{re}(\widetilde{T}) \setminus \{0\}$. Since $\widetilde{K}\widetilde{T} = \widetilde{K}\widetilde{L}\widetilde{K} = \widetilde{A}\widetilde{K}$, we have

$$\widetilde{K}\, R(\widetilde{T}, z) = R(\widetilde{A}, z)\, \widetilde{K}.$$

Integrating the identity given above over C, we obtain

$$\begin{aligned}
\widetilde{K}\widetilde{P} &= \widetilde{K}\Big(- \frac{1}{2\pi \mathrm{i}} \int_{\mathrm{C}} R(\widetilde{T}, z)\, dz \Big) = -\frac{1}{2\pi \mathrm{i}} \int_{\mathrm{C}} \widetilde{K}\, R(\widetilde{T}, z)\, dz \\
&= -\frac{1}{2\pi \mathrm{i}} \int_{\mathrm{C}} R(\widetilde{A}, z)\, \widetilde{K}\, dz = \Big(- \frac{1}{2\pi \mathrm{i}} \int_{\mathrm{C}} R(\widetilde{A}, z)\, dz \Big)\, \widetilde{K} \\
&= \widetilde{\mathsf{P}}\widetilde{K}.
\end{aligned}$$

The equation $\widetilde{K}\widetilde{P} = \widetilde{\mathsf{P}}\widetilde{K}$ shows that $\widetilde{K}$ maps $\mathcal{R}(\widetilde{P})$ into $\mathcal{R}(\widetilde{\mathsf{P}})$. Now let $x \in \mathcal{R}(\widetilde{P})$ such that $\widetilde{K}x = 0$. Then $\widetilde{T}x = \widetilde{L}\widetilde{K}x = \widetilde{L}(0) = 0$. But since $\mathrm{sp}(\widetilde{T}_{|\mathcal{R}(\widetilde{P}),\mathcal{R}(\widetilde{P})}) = \widetilde{\Lambda}$, and $0 \notin \widetilde{\Lambda}$, the operator $\widetilde{T}_{|\mathcal{R}(\widetilde{P}),\mathcal{R}(\widetilde{P})}$ is invertible. As $x \in \mathcal{R}(\widetilde{P})$ and $\widetilde{T}x = 0$, we obtain $x = 0$. Hence the map $\widetilde{K}_{|\mathcal{R}(\widetilde{P})}$ is one-to-one, and $\operatorname{rank} \widetilde{P} \leq \operatorname{rank} \widetilde{\mathsf{P}} \leq n$. Thus $\widetilde{\Lambda}$ is a spectral set of finite type for $\widetilde{T}$. ∎

Theorem 5.4

(a) *Let $\widetilde{\lambda} \in \mathbb{C} \setminus \{0\}$. Then $\widetilde{\lambda}$ is an eigenvalue of the operator $\widetilde{T}$ if and only if $\widetilde{\lambda}$ is an eigenvalue of the matrix $\widetilde{\mathsf{A}}$.*

In this case, let $\widetilde{\underline{\mathsf{u}}} \in \mathbb{C}^{n\times g}$ form an ordered basis for the eigenspace of $\widetilde{\mathsf{A}}$ corresponding to $\widetilde{\lambda}$. Then $\widetilde{\underline{\varphi}} := \underline{\widetilde{L}}\, \widetilde{\underline{\mathsf{u}}}/\widetilde{\lambda}$ forms an ordered basis for the eigenspace of $\widetilde{T}$ corresponding to $\widetilde{\lambda}$ and satisfies $\underline{\widetilde{K}}\, \widetilde{\underline{\varphi}} = \widetilde{\underline{\mathsf{u}}}$.

In particular, the geometric multiplicity of $\widetilde{\lambda}$ as an eigenvalue of $\widetilde{T}$ is the same as its geometric multiplicity of $\widetilde{\lambda}$ as an eigenvalue of $\widetilde{\mathsf{A}}$.

(b) *Let $\widetilde{\Lambda} \subset \mathrm{sp}(\widetilde{\mathsf{A}}) \setminus \{0\}$ and consider an ordered basis $\widetilde{\underline{\mathsf{u}}} \in \mathbb{C}^{n\times m}$ for the spectral subspace $M(\widetilde{\mathsf{A}}, \widetilde{\Lambda})$. Then there is a nonsingular matrix $\widetilde{\Theta} \in \mathbb{C}^{m\times m}$ such that $\widetilde{\mathsf{A}}\widetilde{\underline{\mathsf{u}}} = \widetilde{\underline{\mathsf{u}}}\widetilde{\Theta}$. Further, $\widetilde{\underline{\varphi}} := (\underline{\widetilde{L}}\, \widetilde{\underline{\mathsf{u}}})\widetilde{\Theta}^{-1}$ forms an ordered basis for the spectral subspace $M(\widetilde{T}, \widetilde{\Lambda})$ and satisfies $\underline{\widetilde{K}}\, \widetilde{\underline{\varphi}} = \widetilde{\underline{\mathsf{u}}}$.*

In particular, if $\widetilde{\lambda} \in \mathrm{sp}(\widetilde{\mathsf{A}}) \setminus \{0\}$, then the algebraic multiplicity of $\widetilde{\lambda}$ as an eigenvalue of $\widetilde{T}$ is the same as the algebraic multiplicity of $\widetilde{\lambda}$ as an eigenvalue of $\widetilde{\mathsf{A}}$.

Proof

Recall that $\widetilde{\mathsf{A}}$ represents the operator $\widetilde{A}$ with respect to the standard basis for $\mathbb{C}^{n\times 1}$.

(a) Letting $p := 1$, $\mathsf{Z} := [\widetilde{\lambda}]$ and $y := 0$ in Lemma 5.2, we see that $\widetilde{\lambda}$ is an eigenvalue of $\widetilde{T}$ if and only if $\widetilde{\lambda}$ is an eigenvalue of $\widetilde{A}$. In this case, letting $p := g$ and $\mathsf{Z} := \widetilde{\lambda}\mathsf{I}_g$ in Lemma 5.2, we note that if $\widetilde{\underline{\varphi}} := \underline{\widetilde{L}}\, \widetilde{\underline{\mathsf{u}}}/\widetilde{\lambda}$, then $\underline{\widetilde{T}}\, \widetilde{\underline{\varphi}} = \widetilde{\lambda}\widetilde{\underline{\varphi}}$, $\underline{\widetilde{K}}\, \widetilde{\underline{\varphi}} = \widetilde{\underline{\mathsf{u}}}$ and the set $\{\widetilde{\varphi}_1, \ldots, \widetilde{\varphi}_g\}$ of elements in $\widetilde{\underline{\varphi}}$ is a linearly independent subset of the eigenspace of $\widetilde{T}$ corresponding to $\widetilde{\lambda}$. That $\{\widetilde{\varphi}_1, \ldots, \widetilde{\varphi}_g\}$ also spans this eigenspace can be seen as follows.

Let $\widetilde{\varphi}_{g+1} \notin \text{span}\{\widetilde{\varphi}_1, \ldots, \widetilde{\varphi}_g\}$ be such that $\widetilde{T}\widetilde{\varphi}_{g+1} = \widetilde{\lambda}\widetilde{\varphi}_{g+1}$. If $\underline{\widetilde{\psi}} := [\widetilde{\varphi}_1, \ldots, \widetilde{\varphi}_{g+1}] \in X^{1\times(g+1)}$, then $\underline{\widetilde{T}}\,\underline{\widetilde{\psi}} = \widetilde{\lambda}\underline{\widetilde{\psi}}$; and again by Lemma 5.2, $\{\widetilde{K}\widetilde{\varphi}_1, \ldots, \widetilde{K}\widetilde{\varphi}_{g+1}\}$ would be a linearly independent subset of the eigenspace of $\widetilde{A}$ corresponding to $\widetilde{\lambda}$, contrary to our assumption that this eigenspace is of dimension g.

(b) Since $\underline{\widetilde{\mathsf{u}}} \in \mathbb{C}^{n\times m}$ forms an ordered basis for $M(\widetilde{A}, \widetilde{\Lambda})$, $\underline{\widetilde{A}}\,\underline{\widetilde{\mathsf{u}}} = \underline{\widetilde{\mathsf{u}}}\widetilde{\Theta}$ for some $\widetilde{\Theta} \in \mathbb{C}^{m\times m}$ such that $\text{sp}(\widetilde{\Theta}) = \widetilde{\Lambda}$. As $0 \notin \widetilde{\Lambda}$, $\widetilde{\Theta}$ is invertible. Letting $p := m$, $\mathsf{Z} := \widetilde{\Theta}$ and $\underline{y} := \underline{0}$ in Lemma 5.2, we see that $\underline{\widetilde{T}}\,\underline{\widetilde{\varphi}} = \underline{\widetilde{\varphi}}\widetilde{\Theta}$, $\underline{\widetilde{K}}\,\underline{\widetilde{\varphi}} = \underline{\widetilde{\mathsf{u}}}$ and the set of m elements in $\underline{\widetilde{\varphi}}$ is linearly independent in X. Consider the closed subspace $\widetilde{Y}$ of X spanned by the elements in $\underline{\widetilde{\varphi}}$. Since the matrix $\widetilde{\Theta}$ represents the operator $\widetilde{T}_{|\widetilde{Y},\widetilde{Y}}$ with respect to $\underline{\widetilde{\varphi}}$, we have $\text{sp}(\widetilde{T}_{|\widetilde{Y},\widetilde{Y}}) = \text{sp}(\widetilde{\Theta}) = \widetilde{\Lambda}$. Hence $\widetilde{Y} \subset M(\widetilde{T}, \widetilde{\Lambda})$ by Proposition 1.28, so that $m \le \dim M(\widetilde{T}, \widetilde{\Lambda})$. But as we have seen in Proposition 5.3, $\widetilde{K}$ maps $M(\widetilde{T}, \widetilde{\Lambda})$ into $M(\widetilde{A}, \widetilde{\Lambda})$ in a one-to-one manner, so that $\dim M(\widetilde{T}, \widetilde{\Lambda}) \le \dim M(\widetilde{A}, \widetilde{\Lambda}) \le m$. Thus $\widetilde{Y} = M(\widetilde{T}, \widetilde{\Lambda})$.

In particular, the case $\widetilde{\Lambda} := \{\widetilde{\lambda}\}$ shows that if $\widetilde{\lambda} \in \text{sp}(\widetilde{A}) \setminus \{0\}$, then the algebraic multiplicity of $\widetilde{\lambda}$ as an eigenvalue of $\widetilde{T}$ is the same as the algebraic multiplicity of $\widetilde{\lambda}$ as an eigenvalue of $\widetilde{A}$. ∎

We conclude that the spectral subspace problem for a finite rank operator $\widetilde{T} \in \text{BL}(X)$ can be solved in the following manner. Obtain a representation of $\widetilde{T}$:

$$\widetilde{T}x := \underline{\widetilde{x}}\,\langle x, \underline{\widetilde{f}}\,\rangle, \quad x \in X,$$

where $\underline{\widetilde{x}} \in X^{1\times n}$ and $\underline{\widetilde{f}} \in (X^*)^{1\times n}$. Form the $n\times n$ Gram matrix $\widetilde{\mathsf{A}}$ given by

$$\widetilde{\mathsf{A}}(i,j) := \langle \widetilde{x}_j\,,\, \widetilde{f}_i \rangle, \quad 1 \le i, j \le n.$$

To find a spectral set $\widetilde{\Lambda}$ for $\widetilde{T}$ such that $0 \notin \widetilde{\Lambda}$ and to obtain a basis for the associated spectral subspace $M(\widetilde{T}, \widetilde{\Lambda})$, we may look for a spectral set $\widetilde{\Lambda}$ for the matrix $\widetilde{\mathsf{A}}$ such that $0 \notin \widetilde{\Lambda}$ and find a basis $\underline{\widetilde{\mathsf{u}}}$ for the spectral subspace $M(\widetilde{A}, \widetilde{\Lambda})$. Then $\widetilde{\mathsf{A}}\underline{\widetilde{\mathsf{u}}} = \underline{\widetilde{\mathsf{u}}}\widetilde{\Theta}$, where the nonsingular matrix $\widetilde{\Theta}$ is given by

$$\widetilde{\Theta} = (\underline{\widetilde{\mathsf{u}}}^*\underline{\widetilde{\mathsf{u}}})^{-1}\underline{\widetilde{\mathsf{u}}}^*\widetilde{\mathsf{A}}\underline{\widetilde{\mathsf{u}}}.$$

Let

$$\underline{\widetilde{\varphi}} := (\underline{\widetilde{L}}\,\underline{\widetilde{\mathsf{u}}})\widetilde{\Theta}^{-1} = \underline{\widetilde{x}}\,\underline{\widetilde{\mathsf{u}}}\widetilde{\Theta}^{-1}.$$

Then $\underline{\widetilde{\varphi}}$ forms an ordered basis for $M(\widetilde{T}, \widetilde{\Lambda})$. Moreover, the Gram product $(\underline{\widetilde{\varphi}}, \underline{\widetilde{f}})$ of $\underline{\widetilde{\varphi}}$ with $\underline{\widetilde{f}}$ is equal to $\underline{\widetilde{K}}\, \underline{\widetilde{\varphi}} = \widetilde{\mathsf{u}}$.

In the special case $\widetilde{\Lambda} = \{\widetilde{\lambda}\}$, where $\widetilde{\lambda}$ is a nonzero eigenvalue of $\widetilde{T}$ of geometric multiplicity g and we are interested only in finding a basis $\underline{\widetilde{\varphi}}$ for the corresponding eigenspace, we may find a basis $\widetilde{\mathsf{u}} \in \mathbb{C}^{n \times g}$ of the eigenspace of $\widetilde{\mathsf{A}}$ corresponding to $\widetilde{\lambda}$, so that $\widetilde{\Theta} = \widetilde{\lambda} \mathsf{I}_g$; and then let

$$\underline{\widetilde{\varphi}} := \frac{\underline{\widetilde{L}}\, \mathsf{u}}{\widetilde{\lambda}} = \frac{\underline{\widetilde{x}}\, \mathsf{u}}{\widetilde{\lambda}}.$$

Often $\widetilde{T}$ is a member of a sequence (T_n) of bounded finite rank operators on X given by

$$T_n x := \sum_{j=1}^{r(n)} \langle x\,,\, f_{n,j} \rangle x_{n,j} = \underline{x_n}\, (x, \underline{f_n}\,), \quad x \in X,$$

where $\underline{x_n} := [x_{n,1}, \ldots, x_{n,r(n)}] \in X^{1 \times r(n)}$ and $\underline{f_n} := [f_{n,1}, \ldots, f_{n,r(n)}] \in (X^*)^{1 \times r(n)}$. Let T_n be one of the finite rank approximations of a bounded operator T given in Chapter 4. We give the expressions for $x_{n,j}$ and $f_{n,i}$, $i, j = 1, \ldots, r(n)$, appearing in the representation of T_n and also the entries $\langle x_{n,j}\,,\, f_{n,i} \rangle$, $i, j = 1, \ldots, r(n)$ of the corresponding matrix A_n.

If π_n is a bounded projection of rank $r(n)$, then it is easy to see that

$$\pi_n x = \sum_{j=1}^{r(n)} \langle x\,,\, e_{n,j}^* \rangle e_{n,j} = \underline{e_n}\, (x, \underline{e_n}^*), \quad x \in X,$$

where $\underline{e_n} := [e_{n,1}, \ldots, e_{n,r(n)}] \in X^{1 \times r(n)}$ and $\underline{e_n}^* := [e_{n,1}^*, \ldots, e_{n,r(n)}^*] \in (X^*)^{1 \times r(n)}$ satisfy $(\underline{e_n}\,, \underline{e_n}^*) = \mathsf{I}_{r(n)}$.

Further, for $X := L^2([a,b])$ or $X := C^0([a,b])$ and $x \in X$, let

$$T_n^D x := \int_a^b k_n(s,t) x(t)\, dt, \quad s \in [a,b],$$

where

$$k_n(s,t) := \sum_{j=1}^{r(n)} x_{n,j}(s) y_{n,j}(t), \quad s, t \in [a,b],$$

for some $x_{n,j}$ and $y_{n,j} \in X$ for $j = 1, \ldots, r(n)$.

	$\widetilde{x}_j$	$\langle x\,, \widetilde{f}_i\rangle$	$\widetilde{\mathsf{A}}(i,j)$
T_n^P	$\widetilde{e}_j$	$\langle Tx\,, \widetilde{e}_i^*\rangle$	$\langle T\widetilde{e}_j\,, \widetilde{e}_i^*\rangle$
T_n^S	$T\widetilde{e}_j$	$\langle x\,, \widetilde{e}_i^*\rangle$	$\langle T\widetilde{e}_j\,, \widetilde{e}_i^*\rangle$
T_n^G	$\widetilde{e}_j$ $\sum_{\ell=1}^{\tilde{r}}\langle T\widetilde{e}_j\,, \widetilde{e}_\ell^*\rangle\widetilde{e}_\ell$	$\sum_{\ell=1}^{\tilde{r}}\langle x\,, \widetilde{e}_\ell^*\rangle\langle T\widetilde{e}_\ell\,, \widetilde{e}_i^*\rangle$ $\langle x\,, \widetilde{e}_i^*\rangle$	$\langle T\widetilde{e}_j\,, \widetilde{e}_i^*\rangle$
T_n^D	$\widetilde{x}_j$	$\int_a^b x(t)\widetilde{y}_i(t)\,dt$	$\int_a^b \widetilde{x}_j(t)\widetilde{y}_i(t)\,dt$
T_n^N	$\widetilde{w}_j k(\cdot, \widetilde{t}_j)$	$x(\widetilde{t}_i)$	$\widetilde{w}_j k(\widetilde{t}_i, \widetilde{t}_j)$
T_n^F	$\widetilde{e}_j$ $\widetilde{w}_j\sum_{\ell=1}^{\tilde{r}} k(\widetilde{t}_\ell, \widetilde{t}_j)\widetilde{e}_\ell$	$\sum_{\ell=1}^{\tilde{r}}\widetilde{w}_\ell k(\widetilde{t}_i, \widetilde{t}_\ell)x(\widetilde{t}_\ell)$ $x(\widetilde{t}_i)$	$\langle T\widetilde{e}_j\,, \widetilde{e}_i^*\rangle$

Table 5.1

Also, if T is a Fredholm integral operator on $C^0([a,b])$ with a continuous kernel $k(\cdot,\cdot)$, we consider the approximations T_n^N and T_n^F based on a quadrature formula

$$Q_n(x) := \sum_{j=1}^{r(n)} w_{n,j}x(t_{n,j}), \quad x \in C^0([a,b]).$$

In Table 5.1 we denote $x_{n,j}$, $f_{n,i}$, $e_{n,j}$, $e_{n,i}^*$, $r(n)$, $y_{n,i}$, $w_{n,j}$, $t_{n,j}$ and $\mathsf{A}_n(i,j)$ simply by $\widetilde{x}_j$, $\widetilde{f}_i$, $\widetilde{e}_j$, $\widetilde{e}_i^*$, $\tilde{r}$, $\widetilde{y}_i$, $\widetilde{w}_j$, $\widetilde{t}_j$ and $\widetilde{\mathsf{A}}(i,j)$, respectively.

Remark 5.5 Finite element approximation:
We make a few comments about the finite element approximation described in Subsection 4.1.4. In this case the operator $T := A^{-1}B$ may not be explicitly known and hence it may not be possible to represent the projection approximation $\breve{T}_n^P := \breve{\pi}_n T$ in the usual manner. However, as we have seen earlier, if $X_n := \mathcal{R}(\breve{\pi}_n)$, then finding $0 \neq \varphi_n \in X_n$ and $0 \neq \lambda_n \in \mathbb{C}$ such that

$$\breve{T}_n^P \varphi_n = \frac{1}{\lambda_n}\varphi_n$$

is equivalent to finding $0 \neq \varphi_n \in X_n$ and $0 \neq \lambda_n \in \mathbb{C}$ such that

$$a(\varphi_n, y) = \lambda_n b(\varphi_n, y) \quad \text{for all } y \in X_n.$$

Let $[x_{n,1}, \ldots, x_{n,r(n)}]$ form an ordered basis for X_n. Then it is enough to consider $y := x_{n,i}$, $i = 1, \ldots, r(n)$, in the preceding equation. Also, since

$$\varphi_n := \sum_{j=1}^{r(n)} c_{n,j} x_{n,j}$$

for some $c_{n,j} \in \mathbb{C}$, $j = 1, \ldots, r(n)$, our problem reduces to finding $c_{n,j} \in \mathbb{C}$, not all zero, and $0 \neq \lambda_n \in \mathbb{C}$ such that

$$a\Big(\sum_{j=1}^{r(n)} c_{n,j} x_{n,j}\,,\, x_{n,i}\Big) = \lambda_n b\Big(\sum_{j=1}^{r(n)} c_{n,j} x_{n,j}\,,\, x_{n,i}\Big), \quad i = 1, \ldots, r(n),$$

that is, to the generalized eigenvalue problem

$$\mathsf{A}_n \mathsf{u}_n = \lambda_n \mathsf{B}_n \mathsf{u}_n, \quad 0 \neq \mathsf{u}_n \in \mathbb{C}^{r(n)\times 1}, 0 \neq \lambda_n \in \mathbb{C},$$

where

$$\mathsf{A}_n(i,j) := a(x_{n,j}, x_{n,i}), \; \mathsf{B}_n(i,j) := b(x_{n,j}, x_{n,i}) \; \text{ and } \; \mathsf{u}_n(i) := c_{n,i}.$$

It can be easily proved that B_n is a Hermitian positive definite matrix. Also, if the sesquilinear form $a(\cdot,\cdot)$ is Hermitian, then A_n is a Hermitian matrix. Such a generalized eigenvalue problem can be reduced to an ordinary eigenvalue problem as explained after the proof of Theorem 6.3. ∎

5.1.1 Singularity Subtraction

Consider the Kantorovich-Krylov approximation T_n^K of a compact integral operator T on $X := C^0([0,1])$ with a weakly singular kernel, discussed in Subsection 4.2.3. As we have noted just before stating Theorem 4.19,

$$T_n^K = T_n + U_n, \quad n = 1, 2, \ldots$$

where

$$T_n x := \sum_{j=1}^{n} x(t_{n,j}) x_{n,j}, \quad x \in X,$$

for some fixed $x_{n,j} \in X$, $t_{n,j} \in [0,1]$, and

$$U_n x := x_{n,0} x, \quad x \in X,$$

for a fixed $x_{n,0} \in X$. We have proved in Proposition 4.18 that $T_n \stackrel{\text{cc}}{\to} T$, and in Theorem 4.19 that $U_n \stackrel{\text{n}}{\to} O$ and $T_n^K \stackrel{\nu}{\to} T$. Note that T_n is a finite rank operator, but T_n^K is not a finite rank operator. We shall now give a matrix formulation of the spectral subspace problem for the operator T_n^K along the lines of [8].

Let Λ be a spectral set for T such that $0 \notin \Lambda$. Since the operator T is compact, Λ is of finite type. (See Remark 1.34.) Let $\mathrm{C} \in \mathcal{C}(T, \Lambda)$ such that $0 \in \mathrm{ext}(\mathrm{C})$. If $\Lambda_n := \mathrm{sp}(T_n^K) \cap \mathrm{int}(\mathrm{C})$, then by Theorem 2.12, Λ_n is a spectral set of finite type for T_n^K for all large n. Also, $\Lambda_n \subset \mathrm{int}(\mathrm{C})$ and $\delta(\mathrm{C}) := \min\{|z| : z \in \mathrm{C}\} > 0$. We are therefore interested in finding a spectral set of finite type for T_n^K which is bounded away from zero. In particular, since $\|x_{n,0}\|_\infty = \|U_n\| \to 0$, we are not interested in spectral values of T_n^K which belong to $x_{n,0}([0,1])$, that is, the range of the function $x_{n,0}$. As we have seen in Example 1.17,

$$x_{n,0}([0,1]) = \mathrm{sp}(U_n).$$

In view of these considerations, let

$$\widetilde{T}^K := \widetilde{T} + \widetilde{U},$$

where $\widetilde{T}$ is a bounded finite rank operator on $X := C^0([0,1])$ given by

$$\widetilde{T} x := \sum_{j=1}^{n} x(\widetilde{t}_j) \widetilde{x}_j, \quad x \in X,$$

for some fixed $\widetilde{x}_j \in X$, $\widetilde{t}_j \in [0,1]$, and $\widetilde{U}$ is a bounded operator on X given by

$$\widetilde{U}x := \widetilde{x}_0 x, \quad x \in X,$$

for a fixed $\widetilde{x}_0 \in X$. We wish to find spectral sets of finite type for $\widetilde{T}^K$ which do not intersect

$$\widetilde{E} := \mathrm{sp}(\widetilde{U}) = \{\widetilde{x}_0(t) \, : \, t \in [0,1]\}.$$

Let

$$\widetilde{K}x := (x, \underline{\widetilde{f}}\,) = [x(\widetilde{t}_1), \ldots, x(\widetilde{t}_n)]^\top \in \mathbb{C}^{n\times 1}, \quad x \in X,$$
$$\widetilde{L}\mathsf{u} := \underline{\widetilde{x}}\,\mathsf{u} = \mathsf{u}(1)\widetilde{x}_1 + \cdots + \mathsf{u}(n)\widetilde{x}_n \in X, \quad \mathsf{u} \in \mathbb{C}^{n\times 1},$$

and

$$\widetilde{A} := \widetilde{K}\widetilde{L},$$

as before. Then the $n \times n$ matrix $\widetilde{\mathsf{A}}$ given by

$$\widetilde{\mathsf{A}}(i,j) := \widetilde{x}_j(\widetilde{t}_i), \quad 1 \le i, j \le n,$$

represents the operator $\widetilde{A} : \mathbb{C}^{n\times 1} \to \mathbb{C}^{n\times 1}$ with respect to the standard basis of $\mathbb{C}^{n\times 1}$. Define $\widetilde{D} : \mathbb{C}^{n\times 1} \to \mathbb{C}^{n\times 1}$ by

$$\widetilde{D}\mathsf{u} := [\widetilde{x}_0(\widetilde{t}_1)\mathsf{u}(1), \ldots, \widetilde{x}_0(\widetilde{t}_n)\mathsf{u}(n)]^\top, \quad \mathsf{u} \in \mathbb{C}^{n\times 1},$$

and $\widetilde{A}^K : \mathbb{C}^{n\times 1} \to \mathbb{C}^{n\times 1}$ by

$$\widetilde{A}^K := \widetilde{A} + \widetilde{D}.$$

It is easy to see that $\widetilde{K}\widetilde{U} = \widetilde{D}\widetilde{K}$. Note that

$$\mathrm{sp}(\widetilde{D}) = \{\widetilde{x}_0(\widetilde{t}_j) \, : \, j = 1, \ldots, n\} \subset \widetilde{E}.$$

The matrix $\widetilde{\mathsf{D}} := \mathrm{diag}\,[\widetilde{x}_0(\widetilde{t}_1), \ldots, \widetilde{x}_0(\widetilde{t}_n)] \in \mathbb{C}^{n\times n}$ represents the operator $\widetilde{D}$, and the matrix

$$\widetilde{\mathsf{A}}^K := \widetilde{\mathsf{A}} + \widetilde{\mathsf{D}}$$

represents the operator $\widetilde{A}^K$ with respect to the standard basis for $\mathbb{C}^{n\times 1}$.

In order to relate the spectral subspace problem for the operator $\widetilde{T}^K$ to the spectral subspace problem for the matrix $\widetilde{\mathsf{A}}^K$, we shall prove a generalization of Lemma 5.2.

Lemma 5.6

Let p be a positive integer, $\mathsf{Z} \in \mathbb{C}^{p\times p}$ such that $\mathrm{sp}(\mathsf{Z}) \cap \mathrm{sp}(\widetilde{D}) = \emptyset$,

$\underline{y} \in X^{1\times p}$ *and* $\underline{\mathsf{v}} := \widetilde{\underline{K}}\,\underline{y} \in \mathbb{C}^{n\times p}$. *Then for* $\underline{x} \in X^{1\times p}$ *and* $\underline{\mathsf{u}} \in \mathbb{C}^{n\times p}$, *the following holds:* $\widetilde{\underline{T}}^K\,\underline{x} = \underline{x}\,\mathsf{Z} + \underline{y}$ *and* $\widetilde{\underline{K}}\,\underline{x} = \underline{\mathsf{u}}$ *if and only if* $\widetilde{\underline{A}}^K\,\underline{\mathsf{u}} = \underline{\mathsf{u}}\,\mathsf{Z} + \underline{\mathsf{v}}$ *and* $\widetilde{\underline{L}}\,\underline{\mathsf{u}} = \underline{x}\,\mathsf{Z} - \widetilde{\underline{U}}\,\underline{x} + \underline{y}$.

In particular, let $\underline{y} := \underline{0}$ *(so that* $\underline{\mathsf{v}} = \underline{0}$ *also), and let* $\underline{x}$, $\underline{\mathsf{u}}$ *satisfy the above conditions. Assume that* $\mathrm{sp}(\mathsf{Z}) \cap \widetilde{E} = \emptyset$. *Then the set of* p *elements in* $\underline{x}$ *is linearly independent in* X *if and only if the set of* p *vectors in* $\underline{\mathsf{u}}$ *is linearly independent in* $\mathbb{C}^{n\times 1}$.

Proof

Assume that $\widetilde{\underline{T}}^K\,\underline{x} = \underline{x}\,\mathsf{Z} + \underline{y}$ and $\widetilde{\underline{K}}\,\underline{x} = \underline{\mathsf{u}}$. Then $\widetilde{\underline{L}}\,\underline{\mathsf{u}} = \widetilde{\underline{L}}\,\widetilde{\underline{K}}\,\underline{x} = \widetilde{\underline{T}}\,\underline{x} = \widetilde{\underline{T}}^K\,\underline{x} - \widetilde{\underline{U}}\,\underline{x} = \underline{x}\,\mathsf{Z} - \widetilde{\underline{U}}\,\underline{x} + \underline{y}$. Also, $\widetilde{\underline{A}}^K\,\underline{\mathsf{u}} = \widetilde{\underline{A}}^K\,\widetilde{\underline{K}}\,\underline{x} = \widetilde{\underline{K}}\,\widetilde{\underline{T}}^K\,\underline{x} = \widetilde{\underline{K}}(\underline{x}\,\mathsf{Z} + \underline{y}) = \underline{\mathsf{u}}\,\mathsf{Z} + \underline{\mathsf{v}}$.

Conversely, assume that $\widetilde{\underline{A}}^K\,\underline{\mathsf{u}} = \underline{\mathsf{u}}\,\mathsf{Z} + \underline{\mathsf{v}}$ and $\widetilde{\underline{L}}\,\underline{\mathsf{u}} = \underline{x}\,\mathsf{Z} - \widetilde{\underline{U}}\,\underline{x} + \underline{y}$. Then $\widetilde{\underline{K}}\,\underline{x}\,\mathsf{Z} - \widetilde{\underline{D}}(\widetilde{\underline{K}}\,\underline{\widetilde{x}}) = \widetilde{\underline{K}}(\underline{\widetilde{x}}\,\mathsf{Z} - \widetilde{\underline{U}}\,\underline{\widetilde{x}}) = \widetilde{\underline{K}}(\widetilde{\underline{L}}\,\underline{\mathsf{u}} - \underline{y}) = \widetilde{\underline{A}}\,\underline{\mathsf{u}} - \underline{\mathsf{v}} = \widetilde{\underline{A}}^K\,\underline{\mathsf{u}} - \widetilde{\underline{D}}\,\underline{\mathsf{u}} - \underline{\mathsf{v}} = \underline{\mathsf{u}}\,\mathsf{Z} - \widetilde{\underline{D}}\,\mathsf{u}$. Since $\mathrm{sp}(\mathsf{Z}) \cap \mathrm{sp}(\widetilde{D}) = \emptyset$, it follows from Proposition 1.50 that $\widetilde{\underline{K}}\,\underline{x} = \underline{\mathsf{u}}$. Also, $\widetilde{\underline{T}}^K\,\underline{x} = \widetilde{\underline{L}}\,\widetilde{\underline{K}}\,\underline{x} + \widetilde{\underline{U}}\,\underline{x} = \widetilde{\underline{L}}\,\underline{\mathsf{u}} + \widetilde{\underline{U}}\,\underline{x} = \underline{x}\,\mathsf{Z} + \underline{y}$.

Now let $\underline{y} := \underline{0}$, so that $\underline{\mathsf{v}} = \widetilde{\underline{K}}\,\underline{y} = \underline{0}$, $\widetilde{\underline{L}}\,\underline{\mathsf{u}} = \underline{x}\,\mathsf{Z} - \widetilde{\underline{U}}\,\underline{x}$.

Let the set of p vectors in $\underline{\mathsf{u}}$ be linearly independent in $\mathbb{C}^{n\times 1}$ and $\mathsf{c} \in \mathbb{C}^{p\times 1}$ be such that $\underline{x}\,\mathsf{c} = 0$. Then $\underline{\mathsf{u}}\,\mathsf{c} = (\widetilde{\underline{K}}\,\underline{x})\mathsf{c} = \widetilde{K}(\underline{x}\,\mathsf{c}) = \widetilde{K}(0) = 0$. The linear independence of the set of p vectors in $\underline{\mathsf{u}}$ implies that $\mathsf{c} = 0$, as desired.

Conversely, assume that the set of p elements in $\underline{x}$ is linearly independent in X, and $\mathsf{c} \in \mathbb{C}^{p\times 1}$ is such that $\underline{\mathsf{u}}\,\mathsf{c} = 0$. By mathematical induction, we show that

$$\widetilde{x}_0^j\,\underline{x}\,\mathsf{c} = \underline{x}\,\mathsf{Z}^j\mathsf{c} \quad \text{for } j = 0, 1, \ldots$$

Clearly, this equality holds for $j = 0$. Assume that it holds for some nonnegative integer j. Since $\widetilde{K}\widetilde{U} = \widetilde{D}\widetilde{K}$, we have

$$\begin{aligned}\underline{\mathsf{u}}\,\mathsf{Z}^j\mathsf{c} &= (\widetilde{\underline{K}}\,\underline{x})\mathsf{Z}^j\mathsf{c} = \widetilde{K}(\underline{x}\,\mathsf{Z}^j\mathsf{c}) = \widetilde{K}(\widetilde{x}_0^j\,\underline{x}\,\mathsf{c}) = \widetilde{D}^j\widetilde{K}(\underline{x}\,\mathsf{c})\\ &= (\widetilde{\underline{D}}^j(\widetilde{\underline{K}}\,\underline{x}))\mathsf{c} = (\widetilde{\underline{D}}^j\,\underline{\mathsf{u}})\mathsf{c} = \widetilde{D}^j(\underline{\mathsf{u}}\,\mathsf{c}) = \widetilde{D}^j(0) = 0.\end{aligned}$$

As $\underline{x}\,\mathsf{Z} = \widetilde{\underline{T}}^K\,\underline{x} = \widetilde{\underline{L}}\,\widetilde{\underline{K}}\,\underline{x} + \widetilde{\underline{U}}\,\underline{x} = \widetilde{\underline{L}}\,\underline{\mathsf{u}} + \widetilde{x}_0\,\underline{x}$, we see that

$$\begin{aligned}\underline{x}\,\mathsf{Z}^{j+1}\mathsf{c} &= (\widetilde{\underline{L}}\,\underline{\mathsf{u}})\mathsf{Z}^j\mathsf{c} + \widetilde{x}_0\,\underline{x}\,\mathsf{Z}^j\mathsf{c} = \widetilde{L}(\underline{\mathsf{u}}\,\mathsf{Z}^j\mathsf{c}) + \widetilde{x}_0^{j+1}\,\underline{x}\,\mathsf{c}\\ &= \widetilde{L}(0) + \widetilde{x}_0^{j+1}\,\underline{x}\,\mathsf{c} = \widetilde{x}_0^{j+1}\,\underline{x}\,\mathsf{c}.\end{aligned}$$

Thus the induction argument is complete. It now follows that

$$\mathsf{p}(\widetilde{x}_0)\,\underline{x}\,\mathsf{c} = \underline{x}\,\mathsf{p}(\mathsf{Z})\mathsf{c}$$

for every polynomial p in one variable. Let p be the characteristic polynomial of Z. Then $\mathsf{p}(\mathsf{Z}) = \mathsf{O}$ by the Cayley-Hamilton Theorem (Remark 1.33), but $\mathsf{p}(\widetilde{x}_0)(t) \neq 0$ for each $t \in [0,1]$ since $\mathrm{sp}(\mathsf{Z})$ consists of the roots of p and $\mathrm{sp}(\mathsf{Z}) \cap \widetilde{x}_0([0,1]) = \emptyset$. Thus $\mathsf{p}(\widetilde{x}_0)\,\underline{x}\,\mathsf{c} = 0$. Dividing by $\mathsf{p}(\widetilde{x}_0)$, we obtain $\underline{x}\,\mathsf{c} = 0$. The linear independence of the set of p elements in $\underline{x}$ implies that $\mathsf{c} = 0$, as desired. ▌

Proposition 5.7

$$\mathrm{sp}(\widetilde{T}^K) \setminus \widetilde{E} = \mathrm{sp}(\widetilde{A}^K) \setminus \widetilde{E}.$$

In particular, $\mathrm{sp}(\widetilde{T}^K) \setminus \widetilde{E}$ *is a finite set. Let* $\widetilde{\Lambda} \subset \mathrm{sp}(\widetilde{T}^K) \setminus \widetilde{E}$, $\mathrm{C} \in \mathcal{C}(\widetilde{T}^K, \widetilde{\Lambda})$ *such that* $\widetilde{E} \subset \mathrm{ext}(\mathrm{C})$ *(so that* $\mathrm{C} \in \mathcal{C}(\widetilde{A}^K, \widetilde{\Lambda})$ *as well);* $\widetilde{P} := P(\widetilde{T}^K, \widetilde{\Lambda})$, *and* $\widetilde{\mathsf{P}} := P(\widetilde{A}^K, \widetilde{\Lambda})$. *Then*

$$\widetilde{K}\widetilde{P} = \widetilde{\mathsf{P}}\widetilde{K} \quad \text{and} \quad \widetilde{P} = \Big(\frac{1}{2\pi \mathrm{i}}\int_{\mathrm{C}} (\widetilde{U} - zI)^{-1}\,\widetilde{L}\,R(\widetilde{A}^K, z)\,dz\Big)\,\widetilde{K}.$$

Also, $\widetilde{K}$ *maps* $\mathcal{R}(\widetilde{P})$ *into* $\mathcal{R}(\widetilde{\mathsf{P}})$ *in a one-to-one manner, and hence* $\widetilde{\Lambda}$ *is a spectral set of finite type for* $\widetilde{T}^K$.

Proof

We show that $\mathrm{re}(\widetilde{T}^K) \setminus \widetilde{E} = \mathrm{re}(\widetilde{A}^K) \setminus \widetilde{E}$.

Let $z \in \mathrm{re}(\widetilde{A}^K) \setminus \widetilde{E}$. Consider $y \in X$ and define $\mathsf{v} := \widetilde{K}y$, $\mathsf{u} := R(\widetilde{A}^K, z)\mathsf{v}$ and $x := (\widetilde{L}\mathsf{u} - y)/(z - \widetilde{x}_0)$. Then $\widetilde{A}^K\mathsf{u} = z\mathsf{u} + \mathsf{v}$ and $\widetilde{L}\mathsf{u} = zx - \widetilde{U}x + y$. Letting $p := 1$ in Lemma 5.6, we see that $\widetilde{T}^K x - zx = y$. Thus $\mathcal{R}(\widetilde{T}^K - zI) = X$. Next, let $x \in X$ be such that $(\widetilde{T}^K - zI)x = 0$ and define $\mathsf{u} := \widetilde{K}x$. Letting $p := 1$ and $y := 0$ in Lemma 5.6, we see that $\widetilde{A}^K\mathsf{u} = z\mathsf{u}$ and $\widetilde{L}\mathsf{u} = zx - \widetilde{U}x$. Since $z \in \mathrm{re}(\widetilde{A}^K)$, we have $\mathsf{u} = 0$ and so $(z - \widetilde{x}_0)x = zx - \widetilde{U}x = \widetilde{L}(0) = 0$. As $z \notin \widetilde{E}$, we obtain $x = 0$. Hence $\mathcal{N}(\widetilde{T}^K - zI) = \{0\}$. Thus $z \in \mathrm{re}(\widetilde{T}^K)$.

Conversely, let $z \in \mathrm{re}(\widetilde{T}^K)$. To show that $z \in \mathrm{re}(\widetilde{A}^K)$, it is enough to show that $\mathcal{N}(\widetilde{A}^K - zI) = \{0\}$. Let $\mathsf{u} \in \mathbb{C}^{n\times 1}$ be such that $(\widetilde{A}^K - zI)\mathsf{u} = 0$ and define $x := \widetilde{L}\mathsf{u}/(z - \widetilde{x}_0)$. Letting $p := 1$ and $y := 0$ in Lemma 5.6, we see that $\widetilde{T}^K x = zx$ and $\widetilde{K}x = \mathsf{u}$. Since $z \in \mathrm{re}(\widetilde{T}^K)$, we have $x = 0$ and so $\mathsf{u} = \widetilde{K}(0) = 0$. Hence $\mathcal{N}(\widetilde{A}^K - zI) = \{0\}$, as desired.

Since $\mathrm{sp}(\widetilde{A}^K)$ has at most n elements, so does $\mathrm{sp}(\widetilde{T}^K) \setminus \widetilde{E}$. Let $\widetilde{\Lambda} \subset \mathrm{sp}(\widetilde{T}^K) \setminus \widetilde{E}$. Then $\widetilde{\Lambda}$ is a spectral set for $\widetilde{T}^K$ as well as for $\widetilde{A}^K$. Letting $E := \widetilde{E}$ in Proposition 1.29, it follows that there is $\mathrm{C} \in \mathcal{C}(\widetilde{T}^K, \widetilde{\Lambda})$ such that $\widetilde{E} \subset \mathrm{ext}(\mathrm{C})$ and then $\mathrm{C} \in \mathcal{C}(\widetilde{A}^K, \widetilde{\Lambda})$.

Let $z \in \mathrm{re}(\widetilde{T}^K) \setminus \widetilde{E}$. Since $\widetilde{K}\widetilde{T}^K = \widetilde{K}(\widetilde{T} + \widetilde{U}) = (\widetilde{A} + \widetilde{D})\widetilde{K} = \widetilde{A}^K \widetilde{K}$, we have

$$\widetilde{K}\, R(\widetilde{T}^K, z) = R(\widetilde{A}^K, z)\, \widetilde{K}.$$

Also, the proof of $\mathcal{R}(\widetilde{T}^K - zI) = X$ given before shows that for every $y \in X$,

$$R(\widetilde{T}^K, z)y = \frac{\widetilde{L}\, R(\widetilde{A}^K, z)\, \widetilde{K}y - y}{z - \widetilde{x}_0} = (\widetilde{U} - zI)^{-1}[I - \widetilde{L}\, R(\widetilde{A}^K, z)\, \widetilde{K}]y.$$

Hence

$$R(\widetilde{T}^K, z) = (\widetilde{U} - zI)^{-1}[I - \widetilde{L}\, R(\widetilde{A}^K, z)\, \widetilde{K}].$$

Integrating the two identities given above over C, we obtain

$$\begin{aligned}
\widetilde{K}\widetilde{P} &= \widetilde{K}\Big(-\frac{1}{2\pi \mathrm{i}} \int_{\mathrm{C}} R(\widetilde{T}^K, z)\, dz \Big) = -\frac{1}{2\pi \mathrm{i}} \int_{\mathrm{C}} \widetilde{K}\, R(\widetilde{T}^K, z)\, dz \\
&= -\frac{1}{2\pi \mathrm{i}} \int_{\mathrm{C}} R(\widetilde{A}^K, z)\, \widetilde{K}\, dz = \Big(-\frac{1}{2\pi \mathrm{i}} \int_{\mathrm{C}} R(\widetilde{A}^K, z)\, dz \Big) \widetilde{K} \\
&= \widetilde{\mathrm{P}}\widetilde{K}.
\end{aligned}$$

Since $\mathrm{sp}(\widetilde{U}) = \widetilde{E} \subset \mathrm{ext}(\mathrm{C})$, we have $\displaystyle\int_{\mathrm{C}} (\widetilde{U} - zI)^{-1}\, dz = O$, so that

$$\begin{aligned}
\widetilde{P} &= -\frac{1}{2\pi \mathrm{i}} \int_{\mathrm{C}} R(\widetilde{T}^K, z)\, dz = -\frac{1}{2\pi \mathrm{i}} \int_{\mathrm{C}} (\widetilde{U} - zI)^{-1}[I - \widetilde{L}\, R(\widetilde{A}^K, z)\, \widetilde{K}]\, dz \\
&= \Big(\frac{1}{2\pi \mathrm{i}} \int_{\mathrm{C}} (\widetilde{U} - zI)^{-1} \widetilde{L}\, R(\widetilde{A}^K, z)\, dz \Big) \widetilde{K}.
\end{aligned}$$

The equation $\widetilde{K}\widetilde{P} = \widetilde{\mathrm{P}}\widetilde{K}$ shows that $\widetilde{K}$ maps $\mathcal{R}(\widetilde{P})$ into $\mathcal{R}(\widetilde{\mathrm{P}})$. Now let $x \in \mathcal{R}(\widetilde{P})$ such that $\widetilde{K}x = 0$. Then

$$x = \widetilde{P}x = \Big(\frac{1}{2\pi \mathrm{i}} \int_{\mathrm{C}} (\widetilde{U} - zI)^{-1} \widetilde{L}\, R(\widetilde{A}^K, z)\, dz \Big) \widetilde{K}x = 0.$$

Hence the map $\widetilde{K}_{|\mathcal{R}(\widetilde{P})}$ is one-to-one, and $\mathrm{rank}\, \widetilde{P} \le \mathrm{rank}\, \widetilde{\mathrm{P}} \le n$. Thus $\widetilde{\Lambda}$ is a spectral set of finite type for $\widetilde{T}^K$. ∎

Theorem 5.8

(a) *Let $\widetilde{\lambda} \in \mathbb{C} \setminus \widetilde{E}$. Then $\widetilde{\lambda}$ is an eigenvalue of the operator $\widetilde{T}^K$ if and only if $\widetilde{\lambda}$ is an eigenvalue of the matrix $\widetilde{\mathsf{A}}^K$.*

In this case, let $\widetilde{\underline{\mathsf{u}}} \in \mathbb{C}^{n\times g}$ form an ordered basis for the eigenspace of $\widetilde{\mathsf{A}}^K$ corresponding to $\widetilde{\lambda}$. Then $\underline{\widetilde{\varphi}} := \underline{\widetilde{L}}\, \widetilde{\underline{\mathsf{u}}}/(\widetilde{\lambda} - \widetilde{x}_0)$ forms an ordered basis for the eigenspace of $\widetilde{T}^K$ corresponding to $\widetilde{\lambda}$ and satisfies $\underline{\widetilde{K}}\, \underline{\widetilde{\varphi}} = \widetilde{\underline{\mathsf{u}}}$.

In particular, the geometric multiplicity of $\widetilde{\lambda}$ as an eigenvalue of $\widetilde{T}^K$ is the same as its geometric multiplicity of $\widetilde{\lambda}$ as an eigenvalue of $\widetilde{\mathsf{A}}^K$.

(b) *Let $\widetilde{\Lambda} \subset \mathrm{sp}(\widetilde{\mathsf{A}}^K) \setminus \widetilde{E}$ and consider an ordered basis $\widetilde{\underline{\mathsf{u}}} \in \mathbb{C}^{n\times m}$ of the spectral subspace $M(\widetilde{\mathsf{A}}^K, \widetilde{\Lambda})$. Then there is a matrix $\widetilde{\Theta} \in \mathbb{C}^{m\times m}$ such that $\widetilde{\mathsf{A}}^K \widetilde{\underline{\mathsf{u}}} = \widetilde{\underline{\mathsf{u}}}\widetilde{\Theta}$ and $\mathrm{sp}(\widetilde{\Theta}) \cap \widetilde{E} = \emptyset$. Further, $\underline{\widetilde{\varphi}} := (\underline{\widetilde{L}}\, \widetilde{\underline{\mathsf{u}}})(\widetilde{\Theta} - \widetilde{x}_0 \mathsf{I}_m)^{-1} \in X^{1\times m}$ forms an ordered basis for the spectral subspace $M(\widetilde{T}^K, \widetilde{\Lambda})$ and satisfies $\underline{\widetilde{K}}\, \underline{\widetilde{\varphi}} = \widetilde{\underline{\mathsf{u}}}$.*

In particular, if $\widetilde{\lambda} \in \mathrm{sp}(\widetilde{\mathsf{A}}^K) \setminus \widetilde{E}$, then the algebraic multiplicity of $\widetilde{\lambda}$ as an eigenvalue of $\widetilde{T}^K$ is the same as the algebraic multiplicity of $\widetilde{\lambda}$ as an eigenvalue of $\widetilde{\mathsf{A}}^K$.

Proof

Recall that $\widetilde{\mathsf{A}}^K$ represents the operator $\widetilde{A}^K$ with respect to the standard basis for $\mathbb{C}^{n\times 1}$.

(a) Letting $p := 1$, $\mathsf{Z} := [\widetilde{\lambda}]$ and $\underline{y} := \underline{0}$ in Lemma 5.6, we see that $\widetilde{\lambda}$ is an eigenvalue of $\widetilde{T}^K$ if and only if $\widetilde{\lambda}$ is an eigenvalue of $\widetilde{A}^K$. In this case, letting $p := g$ and $\mathsf{Z} := \widetilde{\lambda}\mathsf{I}_g$ in Lemma 5.6, we note that if $\underline{\widetilde{\varphi}} := \underline{\widetilde{L}}\, \widetilde{\underline{\mathsf{u}}}/(\widetilde{\lambda} - \widetilde{x}_0)$, then $\underline{\widetilde{T}}^K \underline{\widetilde{\varphi}} = \widetilde{\lambda}\underline{\widetilde{\varphi}}$, $\underline{\widetilde{K}}\, \underline{\widetilde{\varphi}} = \widetilde{\underline{\mathsf{u}}}$; and the set $\{\widetilde{\varphi}_1, \ldots, \widetilde{\varphi}_g\}$ of elements in $\underline{\widetilde{\varphi}}$ is a linearly independent subset of the eigenspace of $\widetilde{T}^K$ corresponding to $\widetilde{\lambda}$. That $\{\widetilde{\varphi}_1, \ldots, \widetilde{\varphi}_g\}$ also spans this eigenspace can be seen as follows. Let $\widetilde{\varphi}_{g+1} \notin \mathrm{span}\{\widetilde{\varphi}_1, \ldots, \widetilde{\varphi}_g\}$ be such that $\widetilde{T}^K\widetilde{\varphi}_{g+1} = \widetilde{\lambda}\widetilde{\varphi}_{g+1}$. If $\underline{\widetilde{\psi}} := [\widetilde{\varphi}_1, \ldots, \widetilde{\varphi}_{g+1}] \in X^{1\times(g+1)}$, then $\underline{\widetilde{T}}^K \underline{\widetilde{\psi}} = \widetilde{\lambda}\underline{\widetilde{\psi}}$; and again by Lemma 5.6, $\{\widetilde{K}\widetilde{\varphi}_1, \ldots, \widetilde{K}\widetilde{\varphi}_{g+1}\}$ would be a linearly independent subset of the eigenspace of $\widetilde{A}^K$ corresponding to $\widetilde{\lambda}$, contrary to our assumption that this eigenspace is of dimension g.

(b) Since $\widetilde{\underline{\mathsf{u}}} \in \mathbb{C}^{n\times m}$ forms an ordered basis for the spectral subspace $M(\widetilde{A}^K, \widetilde{\Lambda})$, there exists a matrix $\widetilde{\Theta} \in \mathbb{C}^{m\times m}$ such that $\underline{\widetilde{A}}^K \widetilde{\underline{\mathsf{u}}} = \widetilde{\underline{\mathsf{u}}}\widetilde{\Theta}$

and $\text{sp}(\widetilde{\Theta}) = \widetilde{\Lambda}$. As $\text{sp}(\widetilde{\Theta}) \cap \text{sp}(\widetilde{U}) = \widetilde{\Lambda} \cap \widetilde{E} = \emptyset$, Proposition 1.50 shows that there is a unique $\underline{\widetilde{\varphi}} \in X^{1\times m}$ such that $\underline{\widetilde{\varphi}}\,\widetilde{\Theta} - \widetilde{x}_0\,\underline{\widetilde{\varphi}} = \underline{\widetilde{L}}\,\widetilde{\underline{\mathsf{u}}}$. In fact, $\underline{\widetilde{\varphi}} = (\underline{\widetilde{L}}\,\widetilde{\underline{\mathsf{u}}})(\widetilde{\Theta} - \widetilde{x}_0\mathsf{I}_m)^{-1}$. Letting $p := m$, $\mathsf{Z} := \widetilde{\Theta}$ and $\underline{y} := \underline{0}$ in Lemma 5.6, we see that $\underline{\widetilde{T}^K\,\widetilde{\varphi}} = \underline{\widetilde{\varphi}}\,\widetilde{\Theta}$, $\underline{\widetilde{K}\,\widetilde{\varphi}} = \widetilde{\underline{\mathsf{u}}}$ and that the set of m elements in $\underline{\widetilde{\varphi}}$ is linearly independent in X. Consider the closed subspace $\widetilde{Y}$ of X spanned by the elements in $\underline{\widetilde{\varphi}}$. Since the matrix $\widetilde{\Theta}$ represents the operator $\widetilde{T}^K_{|\widetilde{Y},\widetilde{Y}}$ with respect to $\underline{\widetilde{\varphi}}$, we have $\text{sp}(\widetilde{T}^K_{|\widetilde{Y},\widetilde{Y}}) = \text{sp}(\widetilde{\Theta}) = \widetilde{\Lambda}$. Hence $\widetilde{Y} \subset M(\widetilde{T}^K, \widetilde{\Lambda})$ by Proposition 1.28, so that $m \le \dim M(\widetilde{T}^K, \widetilde{\Lambda})$. But as we have seen in Proposition 5.7, $\widetilde{K}$ maps $M(\widetilde{T}^K, \widetilde{\Lambda})$ into $M(\widetilde{A}^K, \widetilde{\Lambda})$ in a one-to-one manner, so that $\dim M(\widetilde{T}^K, \widetilde{\Lambda}) \le \dim M(\widetilde{A}^K, \widetilde{\Lambda}) \le m$. Thus $\widetilde{Y} = M(\widetilde{T}^K, \widetilde{\Lambda})$.

In particular, the case $\widetilde{\Lambda} := \{\widetilde{\lambda}\}$ shows that if $\widetilde{\lambda} \in \text{sp}(\widetilde{A}^K) \setminus \widetilde{E}$, then the algebraic multiplicity of $\widetilde{\lambda}$ as an eigenvalue of $\widetilde{T}^K$ is the same as the algebraic multiplicity of $\widetilde{\lambda}$ as an eigenvalue of $\widetilde{A}^K$. ∎

We conclude that to find a spectral set $\widetilde{\Lambda}$ for $\widetilde{T}^K$ such that $\widetilde{\Lambda} \cap \widetilde{x}_0([0,1]) = \emptyset$ and to obtain a basis for the associated spectral subspace $M(\widetilde{T}^K, \widetilde{\Lambda})$, we may look for a spectral set $\widetilde{\Lambda}$ for the matrix $\widetilde{\mathsf{A}}^K$ such that $\widetilde{\Lambda} \cap \widetilde{x}_0([0,1]) = \emptyset$ and find a basis $\widetilde{\underline{\mathsf{u}}} \in \mathbb{C}^{n\times m}$ for the spectral subspace $M(\widetilde{\mathsf{A}}^K, \widetilde{\Lambda})$. Then $\widetilde{\mathsf{A}}^K\,\widetilde{\underline{\mathsf{u}}} = \widetilde{\underline{\mathsf{u}}}\,\widetilde{\Theta}$, where the $m\times m$ matrix $\widetilde{\Theta}$ satisfies $\text{sp}(\widetilde{\Theta}) \cap \widetilde{x}_0([0,1]) = \emptyset$ and is given by

$$\widetilde{\Theta} = (\widetilde{\underline{\mathsf{u}}}^*\,\widetilde{\underline{\mathsf{u}}})^{-1}\,\widetilde{\underline{\mathsf{u}}}^*\widetilde{\mathsf{A}}^K\,\widetilde{\underline{\mathsf{u}}}.$$

Let

$$\underline{\widetilde{\varphi}} := (\underline{\widetilde{L}}\,\widetilde{\underline{\mathsf{u}}})(\widetilde{\Theta} - \widetilde{x}_0\mathsf{I}_m)^{-1} = \underline{\widetilde{x}}\,\widetilde{\underline{\mathsf{u}}}(\widetilde{\Theta} - \widetilde{x}_0\mathsf{I}_m)^{-1}.$$

Then $\underline{\widetilde{\varphi}}$ forms an ordered basis for $M(\widetilde{T}^K, \widetilde{\Lambda})$. Moreover, the Gram product $(\underline{\widetilde{\varphi}}\,,\,\underline{\widetilde{f}}\,)$ of $\underline{\widetilde{\varphi}}$ with $\underline{\widetilde{f}}$ is equal to $\underline{\widetilde{K}\,\widetilde{\varphi}} = \widetilde{\underline{\mathsf{u}}}$.

In the special case $\widetilde{\Lambda} = \{\widetilde{\lambda}\}$, where $\widetilde{\lambda}$ is a nonzero eigenvalue of $\widetilde{T}^K$ of geometric multiplicity g and we are interested only in finding a basis $\underline{\widetilde{\varphi}}$ for the corresponding eigenspace, we may find a basis $\widetilde{\underline{\mathsf{u}}} \in \mathbb{C}^{n\times g}$ of the eigenspace of $\widetilde{\mathsf{A}}^K$ corresponding to $\widetilde{\lambda}$, so that $\widetilde{\Theta} = \widetilde{\lambda}\mathsf{I}_g$; and then let

$$\underline{\widetilde{\varphi}} := \frac{\underline{\widetilde{L}}\,\underline{\mathsf{u}}}{\widetilde{\lambda} - \widetilde{x}_0} = \frac{\underline{\widetilde{x}}\,\underline{\mathsf{u}}}{\widetilde{\lambda} - \widetilde{x}_0}.$$

5.1.2 Uniformly Well-Conditioned Bases

Let $T \in \mathrm{BL}(X)$ and Λ be a spectral set for T such that $0 \notin \Lambda$ and the corresponding spectral subspace $M := M(T, \Lambda)$ is of dimension $m < \infty$. Let (T_n) be a sequence of bounded finite rank operators such that $T_n \xrightarrow{\nu} T$. In Proposition 2.9 and Theorem 2.18, we have seen that for each large n, the operator T_n has a spectral set Λ_n such that $0 \notin \Lambda_n$, the corresponding spectral subspace $M_n := M(T_n, \Lambda_n)$ is also of dimension m, Λ_n 'approximates' Λ, and M_n 'approximates' M.

For simplicity we assume that $\operatorname{rank} T \leq n$, that is, we let $r(n) = n$. Let

$$T_n x := \sum_{j=1}^{n} \langle x\,,\, f_{n,j} \rangle x_{n,j} = \underline{x_n}\, (x,\, \underline{f_n}\,), \quad x \in X,$$

where $\underline{x_n} := [x_{n,1}, \ldots, x_{n,n}] \in X^{1 \times n}$ and $\underline{f_n} := [f_{n,1}, \ldots, f_{n,n}] \in (X^*)^{1 \times n}$, and $\mathsf{A}_n := (\,\underline{x_n}\,,\, \underline{f_n}\,)$, that is,

$$\mathsf{A}_n(i,j) := \langle x_{n,j}\,,\, f_{n,i} \rangle, \quad 1 \leq i, j \leq n.$$

Then Λ_n is a spectral set for A_n, and the associated spectral subspace $M(\mathsf{A}_n, \Lambda_n)$ is of dimension m.

Let $\underline{\mathsf{u}_n} := [\mathsf{u}_{n,1}, \ldots, \mathsf{u}_{n,m}]$ form an ordered basis for $M(\mathsf{A}_n, \Lambda_n)$. Then $\mathsf{A}_n\, \underline{\mathsf{u}_n} = \underline{\mathsf{u}_n}\, \Theta_n$, where $\Theta_n \in \mathbb{C}^{m \times m}$ and $\mathrm{sp}(\Theta_n) = \Lambda_n$. As $0 \notin \mathrm{sp}(\Theta_n)$, the matrix Θ_n is nonsingular.

Consider $L_n : \mathbb{C}^{n \times 1} \to X$ defined by

$$L_n \mathsf{u} := \sum_{j=1}^{n} \mathsf{u}(j) x_{n,j} = \underline{x_n}\, \mathsf{u}, \quad \mathsf{u} := [\mathsf{u}(1), \ldots, \mathsf{u}(n)]^\top \in \mathbb{C}^{n \times 1}.$$

Then $(\underline{L_n}\, \underline{\mathsf{u}_n}) \Theta_n^{-1}$ forms an ordered basis for the spectral subspace $M_n := M(T_n, \Lambda_n)$.

In general, the rank of the operator T_n will increase as n increases. From the point of view of computations, the orthonormal bases in a Hilbert space are the best among well-conditioned bases. It is desirable to obtain bases with similar properties in the general framework of a Banach space. Specifically, one would like to construct an ordered basis $\underline{\varphi_n} := [\varphi_{n,1}, \ldots, \varphi_{n,m}]$ for M_n such that for some constants γ and δ satisfying $0 < \delta \leq \gamma$, and for all large n and $j = 1, \ldots, m$,

$$\|\varphi_{n,j}\| \leq \gamma, \quad \mathrm{dist}(\varphi_{n,j}, \mathrm{span}\{\varphi_{n,i} : i = 1, \ldots, m,\, i \neq j\}) \geq \delta.$$

This requirement was also stated in Proposition 2.21 for the convergence to 0 of the sequences $(\|(\underline{T_n} - \underline{T})\, \underline{\varphi_n}\,\|_\infty)$ and $(\|\, \underline{\varphi_n} - \underline{P}\, \underline{\varphi_n}\,\|_\infty)$.

Further, we have come across the same requirement in Theorems 3.7 and 3.8 for the convergence of the iterative refinement schemes for a cluster of eigenvalues. Also, see Theorem 3.15(a) for a similar requirement in the acceleration technique.

We shall show that this can be accomplished if we choose a norm $\|\cdot\|_{[n]}$ on $\mathbb{C}^{n\times 1}$ such that the norms of the operator $L_n : \mathbb{C}^{n\times 1} \to X$ given above and of the operator $K_n : X \to \mathbb{C}^{n\times 1}$ given by

$$K_n x := [\langle x\,,\, f_{n,1}\rangle, \ldots, \langle x\,,\, f_{n,n}\rangle]^\top = (x, \underline{f_n}), \quad x \in X,$$

are bounded; and if we can find an ordered basis $[\mathsf{u}_{n,1}, \ldots, \mathsf{u}_{n,m}]$ of $M(\mathsf{A}_n, \Lambda_n)$ such that for some constants c and d satisfying $0 < d \leq c$, and for all large n and $j = 1, \ldots, m$,

$$\|\mathsf{u}_{n,j}\|_{[n]} \leq c, \quad \mathrm{dist}(\mathsf{u}_{n,j}, \mathrm{span}\{\mathsf{u}_{n,i} \,:\, i = 1, \ldots, m,\, i \neq j\}) \geq d.$$

Most of the discussion in this section has appeared in [13].

Theorem 5.9

Suppose that there is a norm $\|\cdot\|_{[n]}$ on $\mathbb{C}^{n\times 1}$ such that $\|K_n\| \leq \alpha$ and $\|L_n\| \leq \beta$ for some constants α, β and all positive integers n.

Let $\underline{\mathsf{u}_n} := [\mathsf{u}_{n,1}, \ldots, \mathsf{u}_{n,m}]$ form an ordered basis for the spectral subspace $M(\mathsf{A}_n, \Lambda_n)$ such that for all large n and $j = 1, \ldots, m$,

$$\|\mathsf{u}_{n,j}\|_{[n]} \leq c, \quad \mathrm{dist}(\mathsf{u}_{n,j}, \mathrm{span}\{\mathsf{u}_{n,i} \,:\, i = 1, \ldots, m,\, i \neq j\}) \geq d$$

for some constants c and d such that $0 < d \leq c$. Define

$$\underline{\varphi_n} := \underline{L_n}\, \underline{\mathsf{u}_n}\, \Theta_n^{-1} = [\varphi_{n,1}, \ldots, \varphi_{n,m}],$$

where $\mathsf{A}_n\, \underline{\mathsf{u}_n} = \underline{\mathsf{u}_n}\, \Theta_n$. Then $\|\Theta_n^{-1}\|_1 \leq \eta$ for some constant η and all large n, and $\underline{\varphi_n}$ forms an ordered basis for the spectral subspace $M(T_n, \Lambda_n)$ such that for all large n and $j = 1, \ldots, m$,

$$\|\varphi_{n,j}\| \leq \gamma, \quad \mathrm{dist}(\varphi_{n,j}, \mathrm{span}\{\varphi_{n,i} \,:\, i = 1, \ldots, m,\, i \neq j\}) \geq \delta,$$

where $\gamma := \beta c\, \eta$ and $\delta := d/\alpha$.

Proof

It follows from Theorem 5.4(b) that for all large n, $\underline{\varphi_n}$ forms an ordered basis for $M(T_n, \Lambda_n)$.

Let $A_n := K_n L_n$ and note that $\|A_n\| = \|K_n L_n\| \le \|K_n\|\,\|L_n\| \le \alpha\beta$. The matrix A_n represents the operator A_n with respect to the standard basis for $\mathbb{C}^{n\times 1}$. First we find an upper bound for $\|\Theta_n\|_1$. Let $\theta_{n,i,j}$ denote the (i,j)th entry of the $m\times m$ matrix Θ_n. For each $j = 1,\ldots,m$, we have

$$A_n \mathsf{u}_{n,j} = \theta_{n,1,j}\mathsf{u}_{n,1} + \cdots + \theta_{n,m,j}\mathsf{u}_{n,m},$$

so that for all $i = 1,\ldots,m$, we obtain

$$d|\theta_{n,i,j}| \le \|A_n \mathsf{u}_{n,j}\| \le \|A_n\|\,\|\mathsf{u}_{n,j}\|_{[n]} \le \alpha\beta c.$$

Hence

$$\|\Theta_n\|_1 = \max_{j=1,\ldots,m} \sum_{i=1}^{m} |\theta_{n,i,j}| \le \frac{m\alpha\beta c}{d}.$$

Recall that we are given a spectral set Λ for T such that $0 \notin \Lambda$ and the dimension of the corresponding spectral subspace $M(T,\Lambda)$ of T is $m < \infty$. By Theorem 1.32 Λ consists of a finite number of nonzero isolated spectral values of T. Let C be a Cauchy contour which separates Λ from $\mathrm{sp}(T)\setminus\Lambda$ as well as from 0. In particular, $0 \in \mathrm{ext}(\mathrm{C})$. Since $T_n \xrightarrow{\nu} T$, Proposition 2.9 shows that $\mathrm{C} \subset \mathrm{re}(T_n)$ and $\Lambda_n = \mathrm{sp}(T_n) \cap \mathrm{int}(\mathrm{C})$ for all large n. As $\mathrm{sp}(\Theta_n) = \Lambda_n$, it follows that if $\epsilon := \mathrm{dist}(0, \mathrm{C})$, then

$$\epsilon_n := \min\{|\lambda_n| \,:\, \lambda_n \in \mathrm{sp}(\Theta_n)\} \ge \epsilon > 0.$$

Now by Lemma 2.20(b),

$$\|\Theta_n^{-1}\|_1 \le \frac{m^{m/2}}{\epsilon_n^m}\|\Theta_n\|_1^{m-1} \le \eta := \frac{m^{m/2}}{\epsilon^m}\left(\frac{m\alpha\beta c}{d}\right)^{m-1}.$$

Hence for all large n and $j = 1,\ldots,m$,

$$\begin{aligned}\|\varphi_{n,j}\| \le \|\underline{\varphi_n}\|_\infty &\le \|\underline{L_n}\ \underline{\mathsf{u}_n}\|_\infty \|\Theta_n^{-1}\|_1 \\ &\le \|L_n\| \max_{j=1,\ldots,m} \|\mathsf{u}_{n,j}\|_{[n]}\, \|\Theta_n^{-1}\|_1 \\ &\le \gamma := \beta c \eta.\end{aligned}$$

Since $\underline{K_n}\ \underline{\varphi_n} = \underline{\mathsf{u}_n}$ by Theorem 5.4(b), $K_n\varphi_{n,j} = \mathsf{u}_{n,j}$ for each $j = 1,\ldots,m$. Thus for all large n, all complex numbers $c_1,\ldots,c_m$ and each $j = 1,\ldots,m$,

$$\begin{aligned} d &\le \Big\|\mathsf{u}_{n,j} - \sum_{i=1,\, i\neq j}^{m} c_i \mathsf{u}_{n,i}\Big\|_{[n]} = \Big\|K_n\Big(\varphi_{n,j} - \sum_{i=1,\, i\neq j}^{m} c_i\varphi_{n,i}\Big)\Big\|_{[n]} \\ &\le \|K_n\|\,\Big\|\varphi_{n,j} - \sum_{i=1,\, i\neq j}^{m} c_i\varphi_{n,i}\Big\| \le \alpha\Big\|\varphi_{n,j} - \sum_{i=1,\, i\neq j}^{m} c_i\varphi_{n,i}\Big\|,\end{aligned}$$

so that $\mathrm{dist}(\varphi_{n,j}, \mathrm{span}\{\varphi_{n,i} : i = 1, \ldots, m,\ i \neq j\}) \geq \delta := d/\alpha$. ∎

Letting $m := 1$ in the preceding theorem, we obtain the following result.

Corollary 5.10
Suppose that $\|\cdot\|_{[n]}$ is a norm on $\mathbb{C}^{n\times 1}$ such that $\|K_n\| \leq \alpha$ and $\|L_n\| \leq \beta$ for some constants α, β and all positive integers n. Let λ be a nonzero simple eigenvalue of T, $T_n \xrightarrow{\nu} T$ and λ_n be the simple eigenvalue of T_n such that $\lambda_n \to \lambda$ and $|\lambda_n| \geq \epsilon > 0$. Let u_n be an eigenvector of A_n corresponding to λ_n such that $\|\mathsf{u}_n\|_{[n]} = 1$ and let $\varphi_n := L_n\mathsf{u}_n/\lambda_n$. Then φ_n is an eigenvector of T_n such that

$$\frac{1}{\alpha} \leq \|\varphi_n\| \leq \frac{\beta}{\epsilon}.$$

The utility of the above-mentioned considerations depends on the possibility of

1. finding for each large n a norm $\|\cdot\|_{[n]}$ on $\mathbb{C}^{n\times 1}$ such that the sequences of operator norms $(\|K_n\|)$ and $(\|L_n\|)$ are bounded, and

2. finding an ordered basis $\underline{\mathsf{u}_n}$ of the spectral subspace $M(\mathsf{A}_n, \Lambda_n)$ satisfying the conditions stated in the theorem.

We now discuss these two points. First we give several examples where a suitable choice of a norm on $\mathbb{C}^{n\times 1}$ makes the sequences $(\|K_n\|)$ and $(\|L_n\|)$ bounded.

Example 5.11 2-norm on $\mathbb{C}^{n\times 1}$:
Let $X := L^2([a,b])$ with

$$\|x\|_2 := \left(\int_a^b |x(t)|^2\, dt\right)^{1/2}, \quad x \in X,$$

and $T \in \mathrm{BL}(X)$. Consider an orthonormal basis (e_j) for X and let π_n denote the orthogonal projection

$$\pi_n x := \sum_{j=1}^{n} \langle x\,,\, e_j\rangle e_j, \quad x \in X.$$

	Upper bound for $\|K_n\|$	Upper bound for $\|L_n\|$
T_n^P	$\|T\|$	1
T_n^S	1	$\|T\|$
T_n^G	1	$\|T\|$

Table 5.2

Choose the norm $\|\cdot\|_{[n],2}$ on $\mathbb{C}^{n\times 1}$ given by

$$\|\mathsf{u}\|_{[n],2} := \Big(\sum_{j=1}^{n} |\mathsf{u}(j)|^2\Big)^{1/2}, \quad \mathsf{u} := [\mathsf{u}(1), \ldots, \mathsf{u}(n)]^\top \in \mathbb{C}^{n\times 1}.$$

Then we have Table 5.2.
Note that these upper bounds depend only on T. ∎

Example 5.12 ∞-norm on $\mathbb{C}^{n\times 1}$:
Let $X := C^0([a,b])$ with the sup norm and T be a Fredholm integral operator with a continuous kernel $k(\cdot,\cdot)$. Consider $a \le t_{n,1} < \cdots < t_{n,n} \le b$ and $e_{n,1}, \ldots, e_{n,n}$ in X such that $e_{n,j}(t_{n,k}) = \delta_{j,k}$, j, $k = 1, \ldots, n$. Let π_n denote the interpolatory projection and Q_n the quadrature formula given by

$$\pi_n x := \sum_{j=1}^{n} x(t_{n,j}) e_{n,j}, \quad \text{and} \quad Q_n(x) := \sum_{j=1}^{n} w_{n,j} x(t_{n,j}), \quad x \in X.$$

Choose the norm $\|\cdot\|_{[n],\infty}$ on $\mathbb{C}^{n\times 1}$ given by

$$\|\mathsf{u}\|_{[n],\infty} := \sup\{|\mathsf{u}(j)| \, : \, j = 1, \ldots, n\}, \quad \mathsf{u} := [\mathsf{u}(1), \ldots, \mathsf{u}(n)]^\top \in \mathbb{C}^{n\times 1}.$$

Then we have Table 5.3.
Note that if $\pi_n \xrightarrow{\text{p}} I$, then the sequence $(\|\pi_n\|)$ is bounded; and if the sequence (Q_n) of quadrature formulæ is convergent, then the sequence $\Big(\sum_{j=1}^{n} |w_{n,j}|\Big)$ is bounded. ∎

	Upper bound for $\|K_n\|$	Upper bound for $\|L_n\|$
T_n^P	$\|T\|$	$\|\pi_n\|$
T_n^S	1	$\|\pi_n\|(b-a)\|k(\cdot,\cdot)\|_\infty$
T_n^G	1	$\|\pi_n\|(b-a)\|k(\cdot,\cdot)\|_\infty$
T_n^N	1	$\|k(\cdot,\cdot)\|_\infty \sum_{j=1}^{n} \lvert w_{n,j}\rvert$
T_n^F	1	$\|\pi_n\|\ \|k(\cdot,\cdot)\|_\infty \sum_{j=1}^{n} \lvert w_{n,j}\rvert$

Table 5.3

The preceding two examples may explain why the norm $\|\cdot\|_{[n],2}$ on $\mathbb{C}^{n\times 1}$ is usually employed while discretizing the infinite dimensional space $L^2([a,b])$ and why the norm $\|\cdot\|_{[n],\infty}$ on $\mathbb{C}^{n\times 1}$ is usually employed while discretizing the infinite dimensional space $C^0([a,b])$. However, the following example suggests that this should not be followed as a rule.

Example 5.13 1-norm on $\mathbb{C}^{n\times 1}$:
Let $X := C^0([a,b])$, $a = t_{n,0} \le t_{n,1} < \cdots < t_{n,n} \le t_{n,n+1} = b$, and $e_{n,1},\ldots,e_{n,n}$ be the corresponding hat functions. For an integral operator T with a continuous kernel $k(\cdot,\cdot)$, consider the degenerate kernel

$$k_n(s,t) := \sum_{j=1}^{n} k(s,t_{n,j})e_{n,j}(t), \quad s,t\in[a,b],$$

obtained by interpolating $k(\cdot,\cdot)$ in the second variable. Then for the degenerate kernel approximation T_n^D of T, we have

$$K_n x = \Big[\int_a^b x(t)e_{n,1}(t)\,dt,\ldots,\int_a^b x(t)e_{n,n}(t)\,dt\Big]^\top, \quad x\in X,$$

$$L_n\mathsf{u} = \mathsf{u}(1)k(\cdot, t_{n,1}) + \cdots + \mathsf{u}(n)k(\cdot, t_{n,n}), \quad \mathsf{u} \in \mathbb{C}^{n\times 1}.$$

Let us choose the norm $\|\cdot\|_{[n],\infty}$ on $\mathbb{C}^{n\times 1}$. Then

$$\|K_n\| \le \max\left\{\int_a^b e_{n,j}(t)\,dt \;:\; j = 1,\ldots,n\right\} \le \frac{3h_n}{2},$$

where $h_n := \max\{t_{n,j} - t_{n,j-1} \;:\; j = 1,\ldots,n+1\}$. If $h_n \to 0$, then $\|K_n\| \to 0$ and, in particular, the sequence $(\|K_n\|)$ is bounded. However, since we only have

$$\|L_n\| \le \big\|\, |k(\cdot, t_{n,1})| + \cdots + |k(\cdot, t_{n,m})|\,\big\|_\infty,$$

the sequence $(\|L_n\|)$ may not be bounded. For example, if $a := 0$, $b := 1$, $k(s,t) := e^{st}$ for $0 \le s, t \le 1$ and $t_{n,j} := (j-1)/n$, $j = 1,\ldots,n$, then for $\mathsf{u} := [1,\ldots,1]^\top \in \mathbb{C}^{n\times 1}$, we have

$$\|L_n\mathsf{u}\|_\infty = \sum_{j=1}^{n} e^{(j-1)/n} = \frac{e-1}{e^{1/n}-1}$$

and hence $\|L_n\| \to \infty$.

On the other hand, if we choose the norm $\|\cdot\|_{[n],1}$ on $\mathbb{C}^{n\times 1}$ given by

$$\|\mathsf{u}\|_{[n],1} := |\mathsf{u}(1)| + \cdots + |\mathsf{u}(n)|, \quad \mathsf{u} := [\mathsf{u}(1),\ldots,\mathsf{u}(n)]^\top \in \mathbb{C}^{n\times 1},$$

then it is easy to see that

$$\|K_n\| \le b - a \quad \text{and} \quad \|L_n\| \le \|k(\cdot,\cdot)\|_\infty.$$

Thus in this case, to obtain uniformly well-conditioned bases for the approximate spectral subspaces, we choose the norm $\|\cdot\|_{[n],1}$ on $\mathbb{C}^{n\times 1}$ rather than the norm $\|\cdot\|_{[n],\infty}$ on $\mathbb{C}^{n\times 1}$ while discretizing the infinite dimensional space $C^0([a,b])$. ∎

Remark 5.14

Given a norm $\|\cdot\|_{[n]}$ on $\mathbb{C}^{n\times 1}$, we discuss the question of finding an ordered basis $\underline{\mathsf{u}_n} := [\mathsf{u}_{n,1},\ldots,\mathsf{u}_{n,m}]$ for the spectral subspace $M(\mathsf{A}_n, \Lambda_n)$ such that for all large n and $j = 1,\ldots,m$,

$$\|\mathsf{u}_{n,j}\|_{[n]} \le c, \quad \operatorname{dist}(\mathsf{u}_{n,j}, \operatorname{span}\{\mathsf{u}_{n,i} \;:\; i = 1,\ldots,m,\; i \ne j\} \ge d,$$

where c and d are constants satisfying $0 < d \le c$.

First we consider the most simple case: Let $\|\cdot\|_{[n]}$ be the 2-norm $\|\cdot\|_{[n],2}$ on $\mathbb{C}^{n\times 1}$. Given any ordered basis for $M(\mathsf{A}_n, \Lambda_n)$, the Gram-Schmidt Process described in Proposition 1.39 yields an ordered basis $\underline{\mathsf{u}_n} := [\mathsf{u}_{n,1}, \ldots, \mathsf{u}_{n,m}]$ of $M(\mathsf{A}_n, \Lambda_n)$ such that, for $j = 1, \ldots, m$,

$$\|\mathsf{u}_{n,j}\|_{[n],2} = 1 = \mathrm{dist}(\mathsf{u}_{n,j}, \mathrm{span}\{\mathsf{u}_{n,i} : i = 1, \ldots, m,\ i \neq j\}.$$

Thus in this case, we can take $c = d = 1$. ∎

In the general case, the following result can be used.

Proposition 5.15
Let n and m be positive integers, $m \leq n$, $\|\cdot\|_{[n]}$ be a norm on $\mathbb{C}^{n\times 1}$ and let $\underline{\mathsf{v}} := [\mathsf{v}_1, \ldots, \mathsf{v}_m]$ form an ordered basis for a subspace S of $\mathbb{C}^{n\times 1}$. Then an ordered basis $\underline{\mathsf{u}} := [\mathsf{u}_1 \ldots, \mathsf{u}_m]$ for S may be constructed such that for each $j = 1, \ldots, m$,

$$\mathrm{span}\{\mathsf{u}_1, \ldots, \mathsf{u}_j\} = \mathrm{span}\{\mathsf{v}_1, \ldots, \mathsf{v}_j\},$$
$$\|\mathsf{u}_j\|_{[n]} = 1 \quad \text{and} \quad \mathrm{dist}(\mathsf{u}_j, \mathrm{span}\{\mathsf{u}_i : i = 1, \ldots, m,\ i \neq j\}) \geq \frac{1}{2^{m-1}}.$$

Proof
Define $\mathsf{u}_1 := \mathsf{v}_1/\|\mathsf{v}_1\|_{[n]}$. If $j = 2, \ldots, m$ and we have found $\mathsf{u}_1, \ldots, \mathsf{u}_{j-1}$ such that $\mathrm{span}\{\mathsf{u}_1, \ldots, \mathsf{u}_{j-1}\} = \mathrm{span}\{\mathsf{v}_1, \ldots, \mathsf{v}_{j-1}\}$ and $\|\mathsf{u}_1\|_{[n]} = \cdots = \|\mathsf{u}_{j-1}\|_{[n]} = 1$, consider the matrix

$$\mathsf{C}_j := [\mathsf{u}_1, \ldots, \mathsf{u}_{j-1}] \in \mathbb{C}^{n\times(j-1)}$$

and find $\mathsf{w}_j \in \mathbb{C}^{(j-1)\times 1}$ such that

$$\|\mathsf{v}_j - \mathsf{C}_j\mathsf{w}_j\|_{[n]} = \min\left\{\|\mathsf{v}_j - \mathsf{C}_j\mathsf{w}\|_{[n]} : \mathsf{w} \in \mathbb{C}^{(j-1)\times 1}\right\}.$$

Since $\left\{\mathsf{C}_j\mathsf{w} : \mathsf{w} \in \mathbb{C}^{(j-1)\times 1}\right\} = \mathrm{span}\{\mathsf{u}_1, \ldots, \mathsf{u}_{j-1}\}$, we have

$$\|\mathsf{v}_j - \mathsf{C}_j\mathsf{w}_j\|_{[n]} = \mathrm{dist}(\mathsf{v}_j, \mathrm{span}\{\mathsf{u}_1, \ldots, \mathsf{u}_{j-1}\}).$$

Let

$$\mathsf{u}_j := \frac{\mathsf{v}_j - \mathsf{C}_j\mathsf{w}_j}{\|\mathsf{v}_j - \mathsf{C}_j\mathsf{w}_j\|_{[n]}}.$$

Then it follows that

$$\operatorname{span}\{\mathsf{u}_1,\ldots,\mathsf{u}_j\} = \operatorname{span}\{\mathsf{v}_1,\ldots,\mathsf{v}_j\},$$

$$\|\mathsf{u}_j\|_{[n]} = 1 \quad \text{and} \quad \operatorname{dist}(\mathsf{u}_j, \operatorname{span}\{\mathsf{u}_1 \ldots, \mathsf{u}_{j-1}\}) = 1.$$

We prove that for each $j = m, m-1, \ldots, 1$,

$$\operatorname{dist}(\mathsf{u}_j, \operatorname{span}\{\mathsf{u}_i : i = 1, \ldots, m,\ i \neq j\}) \geq \frac{1}{2^{m-1}}.$$

By assumption, we have

$$\operatorname{dist}(\mathsf{u}_m, \operatorname{span}\{\mathsf{u}_i : i = 1, \ldots, m-1\}) \geq 1.$$

Hence the result holds for $j = m$. Next, fix j such that $m - 1 \geq j \geq 1$. For given complex numbers c_i, $i = 1, \ldots, m$, $i \neq j$, let

$$\alpha_0 := \Big\|\mathsf{u}_j - \sum_{i\neq j} c_i \mathsf{u}_i\Big\|_{[n]} \quad \text{and} \quad \alpha_1 := \Big\|\mathsf{u}_j - \sum_{i \neq j,m} c_i \mathsf{u}_i\Big\|_{[n]}.$$

If $|c_m| \leq \alpha_1/2$, then

$$\alpha_0 \geq \alpha_1 - |c_m|\, \|\mathsf{u}_m\|_{[n]} = \alpha_1 - |c_m| \geq \frac{\alpha_1}{2};$$

and if $|c_m| > \alpha_1/2$, then

$$\alpha_0 = |c_m|\, \Big\|\mathsf{u}_m + \sum_{i\neq j,m} \frac{c_i}{c_m}\mathsf{u}_i - \frac{1}{c_m}\mathsf{u}_j\Big\|_{[n]} \geq |c_m| > \frac{\alpha_1}{2}.$$

Thus in any case $\alpha_0 \geq \alpha_1/2$. Similarly, if we let

$$\alpha_2 := \Big\|\mathsf{u}_j - \sum_{i\neq j,m,m-1} c_i \mathsf{u}_i\Big\|_{[n]},$$

we obtain $\alpha_1 \geq \alpha_2/2$ and hence $\alpha_0 \geq \alpha_2/2^2$. Repeating this argument $m - j$ times, we see that

$$\alpha_0 \geq \frac{1}{2^{m-j}}\|\mathsf{u}_j - c_1\mathsf{u}_1 - \cdots - c_{j-1}\mathsf{u}_{j-1}\|_{[n]} \geq \frac{1}{2^{m-j}} \geq \frac{1}{2^{m-1}}.$$

Hence for each fixed $j = m, m-1, \ldots, 1$, we have

$$\operatorname{dist}(\mathsf{u}_j, \operatorname{span}\{\mathsf{u}_i : i = 1, \ldots, m,\ i \neq j\}) \geq \frac{1}{2^{m-1}},$$

as desired. ∎

Remark 5.16
The proof of Proposition 5.15 involves a construction of $\mathsf{w}_j \in \mathbb{C}^{(j-1)\times 1}$ such that

$$\|\mathsf{v}_j - \mathsf{C}_j\mathsf{w}_j\|_{[n]} = \min\left\{\|\mathsf{v}_j - \mathsf{C}_j\mathsf{w}\|_{[n]} \ : \ \mathsf{w} \in \mathbb{C}^{(j-1)\times 1}\right\},$$

where $\mathsf{v}_j \in \mathbb{C}^{n\times 1}$ and $\mathsf{C}_j := [\mathsf{u}_1, \ldots, \mathsf{u}_{j-1}] \in \mathbb{C}^{n\times(j-1)}$ are given, $j = 2, \ldots, m$. This construction is in fact an implementation of the well-known Riesz Lemma (Lemma 5.3 of [55] or Lemma 2.5-4 of [48]) for the finite dimensional normed space $\mathbb{C}^{n\times 1}$ with the norm $\|\cdot\|_{[n]}$. It is equivalent to the computation of a minimum norm solution of the over-determined system

$$\mathsf{C}_j\mathsf{w} = \mathsf{v}_j,$$

of n equations in $j-1$ unknowns, where $j = 2, \ldots, m$ and $m \leq n$. The reader is referred to [72] for various algorithms which yield such a solution if the norm $\|\cdot\|_{[n]}$ is the p-norm $\|\cdot\|_{[n],p}$ and the scalars are real numbers. (See Chapter 1 if $p = \infty$, Chapter 4 if $1 < p < \infty$, and Chapter 6 if $p = 1$.)

If the norm $\|\cdot\|_{[n]}$ is the p-norm $\|\cdot\|_{[n],p}$ and the scalars are complex numbers (as in our case), we may utilize the minimum norm solution for the real scalars to obtain a result similar to Proposition 5.15. We proceed as follows.

Suppose we are given a matrix $\mathsf{C} \in \mathbb{C}^{n\times k}$ and $\mathsf{v} \in \mathbb{C}^{n\times 1}$. We write $\mathsf{C} := \mathsf{C}_1 + \mathrm{i}\,\mathsf{C}_2$ and $\mathsf{v} := \mathsf{v}_1 + \mathrm{i}\,\mathsf{v}_2$, where $\mathsf{C}_1, \mathsf{C}_2 \in \mathbb{R}^{n\times k}$ and $\mathsf{v}_1, \mathsf{v}_2 \in \mathbb{R}^{n\times 1}$. For any $\mathsf{w} \in \mathbb{C}^{k\times 1}$, write $\mathsf{w} := \mathsf{w}_1 + \mathrm{i}\,\mathsf{w}_2$ with $\mathsf{w}_1, \mathsf{w}_2 \in \mathbb{R}^{k\times 1}$. Then

$$\begin{aligned}\mathsf{v} - \mathsf{C}\mathsf{w} &= (\mathsf{v}_1 + \mathrm{i}\,\mathsf{v}_2) - (\mathsf{C}_1 + \mathrm{i}\,\mathsf{C}_2)(\mathsf{w}_1 + \mathrm{i}\,\mathsf{w}_2)\\ &= (\mathsf{v}_1 - \mathsf{C}_1\mathsf{w}_1 + \mathsf{C}_2\mathsf{w}_2) + \mathrm{i}\,(\mathsf{v}_2 - \mathsf{C}_2\mathsf{w}_1 - \mathsf{C}_1\mathsf{w}_2).\end{aligned}$$

Let $1 \leq p \leq \infty$. By one of the algorithms cited above, find $\widetilde{\mathsf{w}}_1, \widetilde{\mathsf{w}}_2 \in \mathbb{R}^{k\times 1}$ such that

$$\begin{aligned}&\left\|\begin{bmatrix}\mathsf{v}_1\\ \mathsf{v}_2\end{bmatrix} - \begin{bmatrix}\mathsf{C}_1 & -\mathsf{C}_2\\ \mathsf{C}_2 & \mathsf{C}_1\end{bmatrix}\begin{bmatrix}\widetilde{\mathsf{w}}_1\\ \widetilde{\mathsf{w}}_2\end{bmatrix}\right\|_{[2n],\,p}\\ &\qquad\qquad = \min_{\mathsf{w}_1,\mathsf{w}_2\in\mathbb{R}^{k\times 1}}\left\|\begin{bmatrix}\mathsf{v}_1\\ \mathsf{v}_2\end{bmatrix} - \begin{bmatrix}\mathsf{C}_1 & -\mathsf{C}_2\\ \mathsf{C}_2 & \mathsf{C}_1\end{bmatrix}\begin{bmatrix}\mathsf{w}_1\\ \mathsf{w}_2\end{bmatrix}\right\|_{[2n],\,p}.\end{aligned}$$

Note that for $z \in \mathbb{C}$, we have the following inequalities: If $1 \le p \le 2$, then

$$\frac{(|\Re z|^p + |\Im z|^p)^{1/p}}{2^{(2-p)/2p}} \le |z| \le (|\Re z|^p + |\Im z|^p)^{1/p}$$

and if $2 < p < \infty$, then

$$(|\Re z|^p + |\Im z|^p)^{1/p} \le |z| \le 2^{(p-2)/2p}(|\Re z|^p + |\Im z|^p)^{1/p},$$

where we make the convention that $(p-2)/2p := 1/2$ if $p = \infty$. Hence if we let $\widetilde{\mathsf{w}} := \widetilde{\mathsf{w}}_1 + \mathrm{i}\,\widetilde{\mathsf{w}}_2 \in \mathbb{C}^{k\times 1}$, it can be seen that

$$\|\mathsf{v} - \mathsf{C}\widetilde{\mathsf{w}}\|_{[n],p} \le 2^{|p-2|/2p} \min_{\mathsf{w}\in \mathbb{C}^{k\times 1}} \|\mathsf{v} - \mathsf{C}\mathsf{w}\|_{[n],p}$$

for any p, $1 \le p \le \infty$.

This procedure allows us to find $\underline{\mathsf{u}} := [\mathsf{u}_1, \ldots, \mathsf{u}_m]$ such that

$$\operatorname{span}\{\mathsf{u}_1, \ldots, \mathsf{u}_j\} = \operatorname{span}\{\mathsf{v}_1, \ldots, \mathsf{v}_j\} \text{ and } \|\mathsf{u}_j\|_{[n]} = 1 \text{ for } j = 1, \ldots, m,$$
$$\operatorname{dist}(\mathsf{u}_j, \operatorname{span}\{\mathsf{u}_1, \ldots, \mathsf{u}_{j-1}\}) \ge \frac{1}{2^{|p-2|/2p}} \quad \text{for } j = 2, \ldots, m.$$

The method given in the proof of Proposition 5.15 then yields a positive constant d (independent of n but possibly depending on m) such that

$$\operatorname{dist}(\mathsf{u}_j, \operatorname{span}\{\mathsf{u}_i : i = 1, \ldots, m,\ i \ne j\}) \ge d$$

for all $j = 1, \ldots, m$. In fact, we may take $d = \dfrac{d_0^m}{(1+d_0)^{m+1}}$, where $d_0 = \dfrac{1}{2^{|p-2|/2p}}$. (See Exercise 5.8.) ∎

To conclude, we observe that starting with any ordered basis $\underline{\mathsf{v}_n}$ of the spectral subspace $M(\mathsf{A}_n, \Lambda_n)$, we can construct an ordered basis $\underline{\mathsf{u}_n} := [\mathsf{u}_{n,1}, \ldots, \mathsf{u}_{n,m}]$ for it such that for each $j = 1, \ldots, m$,

$$\|\mathsf{u}_{n,j}\|_{[n],p} = 1 \quad \text{and} \quad \operatorname{dist}(\mathsf{u}_j, \operatorname{span}\{\mathsf{u}_i : i = 1, \ldots, m,\ i \ne j\}) \ge d,$$

for some constant d independent of n, where $\|\cdot\|_{[n],p}$ denotes the p-norm on $\mathbb{C}^{n\times 1}$, $1 \le p \le \infty$. These uniformly well-conditioned bases for $M(\mathsf{A}_n, \Lambda_n)$ allow us to find uniformly well-conditioned bases $\underline{\varphi_n}$ for $M(T_n, \Lambda_n)$ as shown in Theorem 5.9.

5.2 Iterative Refinement

In this section we consider the implementation of the iterative refinement schemes discussed in Section 3.1 when the approximation T_n of T is of finite rank. For convenience of notation, assume that rank $T_n \leq n$, where n is a fixed relatively small positive integer. Let T_n be given by

$$T_n x := \sum_{j=1}^{n} \langle x \,,\, f_{n,j} \rangle x_{n,j} = \underline{x_n} \, (x, \underline{f_n} \,), \quad x \in X,$$

where

$$\underline{x_n} := [x_{n,1}, \ldots, x_{n,n}] \in X^{1 \times n} \quad \text{and} \quad \underline{f_n} := [f_{n,1}, \ldots, f_{n,n}] \in (X^*)^{1 \times n},$$

as in Proposition 5.1. Define $K_n : X \to \mathbb{C}^{n \times 1}$ by

$$K_n x := [\langle x \,,\, f_{n,1} \rangle, \ldots, \langle x \,,\, f_{n,n} \rangle]^\top = (x, \underline{f_n} \,), \quad x \in X,$$

and $L_n : \mathbb{C}^{n \times 1} \to X$ by

$$L_n \mathsf{u} := \sum_{j=1}^{n} \mathsf{u}(j) x_{n,j} = \underline{x_n} \, \mathsf{u}, \quad \mathsf{u} = [\mathsf{u}(1), \ldots, \mathsf{u}(n)]^\top \in \mathbb{C}^{n \times 1}.$$

Then, as we have seen in Section 5.1, $T_n = L_n K_n$; and if we let $A_n := K_n L_n$, then the $n \times n$ matrix A_n defined by $\mathsf{A}_n(i,j) := \langle x_{n,j} \,,\, f_{n,i} \rangle$ represents the operator A_n with respect to the standard basis for $\mathbb{C}^{n \times 1}$.

Since for all $x \in X$ and $f \in X^*$,

$$\langle T_n x \,,\, f \rangle = \langle \underline{x_n} \, (x, \underline{f_n} \,) \,,\, f \rangle = (\, \underline{x_n} \,, f \,) \, (x, \underline{f_n} \,) = \langle x \,,\, \underline{f_n} \, (\, \underline{x_n} \,, f \,)^* \rangle,$$

we conclude that $T_n^* : X^* \to X^*$ is given by

$$T_n^* f := \underline{f_n} \, (\, \underline{x_n} \,, f \,)^* = \sum_{j=1}^{n} \overline{\langle x_{n,j} \,,\, f \rangle} f_{n,j}.$$

Similarly, $L_n^* : X^* \to \mathbb{C}^{n \times 1}$ is given by

$$L_n^* f = (\, \underline{x_n} \,, f \,)^*, \quad f \in X^*,$$

and $K_n^* : \mathbb{C}^{n \times 1} \to X^*$ is given by

$$K_n^* \mathsf{v} = \underline{f_n} \, \mathsf{v}, \quad \mathsf{v} \in \mathbb{C}^{n \times 1}.$$

Hence $T_n^* = K_n^* L_n^*$, $A_n^* = L_n^* K_n^*$, and the matrix A_n^* represents the operator A_n^* with respect to the standard basis for $\mathbb{C}^{n\times 1}$.

Let us first take up the case of a nonzero simple eigenvalue λ of T and see how to implement the Elementary Iteration considered in Subsection 3.1.2:

$$(E) \qquad \begin{cases} \varphi_n^{(0)} := \varphi_n, \text{ and for } k = 1, 2, \ldots \\ \lambda_n^{(k)} := \langle T\varphi_n^{(k-1)}\,,\, \varphi_n^* \rangle, \\ \varphi_n^{(k)} := \varphi_n^{(k-1)} + S_n(\lambda_n^{(k)}\varphi_n^{(k-1)} - T\varphi_n^{(k-1)}). \end{cases}$$

As we have mentioned while deriving the relation $(E)_3$ of Subsection 3.1.2,

$$\varphi_n^{(k)} = \varphi_n + S_n[(\lambda_n^{(k)} - \lambda_n)\varphi_n^{(k-1)} + (T_n - T)\varphi_n^{(k-1)}].$$

This way of writing $\varphi_n^{(k)}$ avoids the computation of the residual $\lambda_n^{(k)}\varphi_n^{(k-1)} - T\varphi_n^{(k-1)}$.

Firstly, we need to find an eigenvector φ_n of T_n corresponding to a nonzero simple eigenvalue λ_n. For this purpose, we may find an eigenvector u_n of the matrix A_n corresponding a nonzero simple eigenvalue λ_n and define

$$\varphi_n := \frac{1}{\lambda_n} L_n \mathsf{u}_n = \frac{1}{\lambda_n} \underline{x_n}\, \mathsf{u}_n.$$

Then $T_n\varphi_n = \lambda_n\varphi_n$ and $K_n\varphi_n = \mathsf{u}_n$ as proved in Theorem 5.4(a) with $g := 1$. In particular, $\varphi_n \neq 0$ as $\mathsf{u}_n \neq 0$.

Secondly, we need to find the eigenvector φ_n^* of T_n^* corresponding to the simple eigenvalue $\overline{\lambda}_n$ such that $\langle \varphi_n\,,\, \varphi_n^* \rangle = 1$. Now since λ_n is a simple eigenvalue of A_n, there is a unique eigenvector v_n of A_n^* corresponding to $\overline{\lambda}_n$ such that $\mathsf{v}_n^*\mathsf{u}_n = \langle \mathsf{u}_n\,,\, \mathsf{v}_n \rangle = 1$. (See Theorem 1.52.) If we define

$$\varphi_n^* := K_n^*\mathsf{v}_n = \mathsf{v}_n(1) f_{n,1} + \cdots + \mathsf{v}_n(n) f_{n,n},$$

then $T_n^*\varphi_n^* = K_n^* L_n^* K_n^* \mathsf{v}_n = K_n^* A_n^* \mathsf{v}_n = \overline{\lambda}_n K_n^* \mathsf{v}_n = \overline{\lambda}_n \varphi_n^*$, and

$$\begin{aligned} \lambda_n \langle \varphi_n\,,\, \varphi_n^* \rangle &= \langle L_n \mathsf{u}_n\,,\, K_n^* \mathsf{v}_n \rangle = \langle K_n L_n \mathsf{u}_n\,,\, \mathsf{v}_n \rangle \\ &= \langle A_n \mathsf{u}_n\,,\, \mathsf{v}_n \rangle = \langle \lambda_n \mathsf{u}_n\,,\, \mathsf{v}_n \rangle = \lambda_n \mathsf{v}_n^* \mathsf{u}_n = \lambda_n, \end{aligned}$$

so that $\langle \varphi_n\,,\, \varphi_n^* \rangle = 1$, as desired.

Thirdly, to find $\varphi_n^{(k)}$ for $k = 1, 2, \ldots$ we observe that

$$\begin{aligned}\langle \lambda_n^{(k)} \varphi_n^{(k-1)} - T\varphi_n^{(k-1)} \,,\, \varphi_n^* \rangle &= \lambda_n^{(k)} \langle \varphi_n^{(k-1)} \,,\, \varphi_n^* \rangle - \langle T\varphi_n^{(k-1)} \,,\, \varphi_n^* \rangle \\ &= \lambda_n^{(k)} - \lambda_n^{(k)} = 0,\end{aligned}$$

that is, $P_n(\lambda_n^{(k)} \varphi_n^{(k-1)} - T\varphi_n^{(k-1)}) = 0$. Hence we can find $\varphi_n^{(k)}$ if we are able to find $S_n y$ whenever $y \in X$ and $P_n y = 0$. Let $y \in X$ be such that $P_n y := \langle y \,,\, \varphi_n^* \rangle \varphi_n = 0$. Then

$$\mathsf{v}_n^* K_n y = \langle K_n y \,,\, \mathsf{v}_n \rangle = \langle y \,,\, K_n^* \mathsf{v}_n \rangle = \langle y \,,\, \varphi_n^* \rangle = 0,$$

that is, $K_n y$ belongs to the null space of the spectral projection associated with A_n and λ_n. Since the operator $A_n - \lambda_n I$ maps the above null space onto itself in a one-to-one manner, there is a unique $\mathsf{w}_n \in \mathbb{C}^{n \times 1}$ such that

$$\mathsf{A}_n \mathsf{w}_n - \lambda_n \mathsf{w}_n = K_n y = (\!(y, \underline{f_n})\!), \quad \mathsf{v}_n^* \mathsf{w}_n = 0.$$

If we define

$$x := \frac{1}{\lambda_n}(L_n \mathsf{w}_n - y) = \frac{1}{\lambda_n}(\underline{x_n}\, \mathsf{w}_n - y),$$

then

$$K_n x = \frac{1}{\lambda_n}(K_n L_n \mathsf{w}_n - K_n y) = \frac{1}{\lambda_n}(\mathsf{A}_n \mathsf{w}_n - (\!(y, \underline{f_n})\!)) = \mathsf{w}_n,$$

so that

$$T_n x - \lambda_n x = L_n K_n x - (L_n \mathsf{w}_n - y) = L_n \mathsf{w}_n - L_n \mathsf{w}_n + y = y.$$

Further, since

$$\langle x \,,\, \varphi_n^* \rangle = \langle x \,,\, K_n^* \mathsf{v}_n \rangle = \langle K_n x \,,\, \mathsf{v}_n \rangle = \langle \mathsf{w}_n \,,\, \mathsf{v}_n \rangle = \mathsf{v}_n^* \mathsf{w}_n = 0,$$

we see that $x \in \mathcal{N}(P_n)$, and hence $x = S_n y$, as desired. (See Remark 1.25.)

We summarize this discussion by noting that we need to perform the following matrix computations to implement the Elementary Iteration (E) for the case of a nonzero simple eigenvalue λ of T:

(i) Solve the eigenvalue problem $\mathsf{A}_n \mathsf{u}_n = \lambda_n \mathsf{u}_n$. (See Section 6.3.) Ensure that $\|\mathsf{u}_n\|_{[n]} = 1$, where $\|\cdot\|_{[n]}$ is a norm on $\mathbb{C}^{n \times 1}$ such that the sequences $(\|K_n\|)$ and $(\|L_n\|)$ are bounded. (See Corollary 5.10 and Examples 5.11, 5.12, 5.13.)

(ii) Compute the unique vector v_n such that

$$\mathsf{A}_n^*\mathsf{v}_n = \overline{\lambda}_n\mathsf{v}_n, \quad \mathsf{u}_n^*\mathsf{v}_n = 1.$$

To find the vector v_n, we may find an eigenvector $\widetilde{\mathsf{v}}_n$ of the matrix A_n^* corresponding to $\overline{\lambda}_n$ in the same manner as we found an eigenvector u_n of the matrix A_n corresponding to λ_n, and let $\mathsf{v}_n := \widetilde{\mathsf{v}}_n/\mathsf{u}_n^*\widetilde{\mathsf{v}}_n$. Alternatively, one may solve the system

$$(\mathsf{A}_n^* - \overline{\lambda}_n\mathsf{I}_n)\mathsf{v} = 0, \quad \mathsf{u}_n^*\mathsf{v} = 1$$

of $n+1$ linear equations in n unknowns to find its unique solution $\mathsf{v} = \mathsf{v}_n$.

(iii) Assuming that $\varphi_n^{(k-1)}$ is already computed, find $\varphi_n^{(k)}$ as follows:

Compute

$$\begin{aligned}
\lambda_n^{(k)} &:= \langle T\varphi_n^{(k-1)}, \varphi_n^*\rangle = \mathsf{v}_n^*\,\big(T\varphi_n^{(k-1)}, \underline{f_n}\,\big),\\
y &:= (\lambda_n^{(k)} - \lambda_n)\varphi_n^{(k-1)} + (T_n - T)\varphi_n^{(k-1)},\\
\mathsf{b}_k &:= \big(y, \underline{f_n}\,\big).
\end{aligned}$$

Solve the system

$$[\mathsf{A}_n(\mathsf{I}_n - \mathsf{u}_n\mathsf{v}_n^*) - \lambda_n\mathsf{I}_n]\mathsf{w} = \mathsf{b}_k$$

of n linear equations in n unknowns to find its unique solution $\mathsf{w} = \mathsf{w}_n$ or equivalently, solve the system

$$(\mathsf{A}_n - \lambda_n\mathsf{I}_n)\mathsf{w} = \mathsf{b}_k, \quad \mathsf{v}_n^*\mathsf{w} = 0$$

of $n+1$ linear equations in n unknowns to find its unique solution $\mathsf{w} = \mathsf{w}_n$.

Then let

$$\varphi_n^{(k)} := \varphi_n + \frac{1}{\lambda_n}(\underline{x_n}\,\mathsf{w}_n - y).$$

We remark that the coefficient matrix $\mathsf{A}_n(\mathsf{I}_n - \mathsf{u}_n\mathsf{v}_n^*)$ of the $n\times n$ linear system given above is nonsingular since $\lambda_n \neq 0$ and

$$\mathrm{sp}(\mathsf{A}_n(\mathsf{I}_n - \mathsf{u}_n\mathsf{v}_n^*)) = \mathrm{sp}((\mathsf{I}_n - \mathsf{u}_n\mathsf{v}_n^*)\mathsf{A}_n(\mathsf{I}_n - \mathsf{u}_n\mathsf{v}_n^*)) = \mathrm{sp}(\mathsf{A}_n)\cup\{0\}\setminus\{\lambda_n\}.$$

Moreover, this matrix does not change with the iterate number k.

A similar procedure for implementing the Elementary Iteration (E) was outlined in Section 4 of [34].

It is easily seen that for implementing the Double Iteration (D) for a nonzero simple eigenvalue λ of T, the only extra computations consist of evaluations of Tx for $x \in X$. In fact, each step of the Double Iteration consists of (i) an application of T and (ii) the Elementary Iteration described above.

Let us now turn to the general case of a spectral set Λ of finite type for T such that $0 \notin \Lambda$. Let us assume that the spectral projection P associated with T and Λ is of rank m, and $n \geq m$. The considerations for the implementation of the Elementary Iteration

$$(\underline{E}) \qquad \begin{cases} \underline{\varphi_n}^{(0)} := \underline{\varphi_n}\,, \text{ and for } k = 1, 2, \ldots \\ \Theta_n^{(k)} := (\underline{T}\,\underline{\varphi_n}^{(k-1)}, \underline{\varphi_n^*}) \\ \underline{\varphi_n}^{(k)} := \underline{\varphi_n}^{(k-1)} + \mathcal{S}_n(\underline{\varphi_n}^{(k-1)}\Theta_n^{(k)} - \underline{T}\,\underline{\varphi_n}^{(k-1)}) \end{cases}$$

are similar to those for the case of a simple eigenvalue of λ of T.

Firstly, we need to find a basis $\underline{\varphi_n} = [\varphi_{n,1}, \ldots, \varphi_{n,m}]$ of the spectral subspace associated with T_n and a spectral set Λ_n such that $0 \notin \Lambda_n$. For this purpose, we may find a basis $\underline{\mathsf{u}_n} = [\mathsf{u}_{n,1}, \ldots, \mathsf{u}_{n,m}]$ of the spectral subspace associated with A_n and a spectral set Λ_n such that $0 \notin \Lambda_n$. Since the subspace of $\mathbb{C}^{n\times 1}$ generated by $\underline{\mathsf{u}_n}$ is invariant under A_n, we see that $\mathsf{A}_n\,\underline{\mathsf{u}_n} = \underline{\mathsf{u}_n}\,\Theta_n$ for some $m\times m$ matrix Θ_n. Further, since $\mathrm{sp}(\Theta_n) = \Lambda_n$ and $0 \notin \Lambda_n$, we see that Θ_n is nonsingular. One may find the matrix Θ_n as follows. Since the subset $\{\mathsf{u}_{n,1} \ldots, \mathsf{u}_{n,m}\}$ of $\mathbb{C}^{n\times 1}$ is linearly independent, the $m\times m$ matrix $\underline{\mathsf{u}_n^*}\,\underline{\mathsf{u}_n}$ is nonsingular. As $\underline{\mathsf{u}_n^*}\,\mathsf{A}_n\,\underline{\mathsf{u}_n} = \underline{\mathsf{u}_n^*}\,\underline{\mathsf{u}_n}\,\Theta_n$, we obtain

$$\Theta_n = (\underline{\mathsf{u}_n^*}\,\underline{\mathsf{u}_n})^{-1}\,\underline{\mathsf{u}_n^*}\,\mathsf{A}_n\,\underline{\mathsf{u}_n}\,.$$

Define

$$\underline{\varphi_n} := \underline{L_n}\,\underline{\mathsf{u}_n}\,\Theta_n^{-1} = \underline{x_n}\,\underline{\mathsf{u}_n}\,\Theta_n^{-1}.$$

Then $\underline{T_n}\,\underline{\varphi_n} = \underline{\varphi_n}\,\Theta_n$, and $\underline{K_n}\,\underline{\varphi_n} = \underline{\mathsf{u}_n}$ as proved in Theorem 5.4(b). In particular, since $\underline{\mathsf{u}_n}$ forms a basis for an m-dimensional subspace of $\mathbb{C}^{n\times 1}$, $\underline{\varphi_n}$ also forms a basis for an m-dimensional subspace of X.

Secondly, we need to find a basis $\underline{\varphi_n^*}$ of $R(P_n^*)$ such that $(\underline{\varphi_n}, \underline{\varphi_n^*}) = \mathsf{I}_m$. Now since $\underline{\mathsf{u}_n}$ forms a basis for the m-dimensional spectral subspace associated with A_n and the spectral set Λ_n, there is a unique adjoint basis $\underline{\mathsf{v}_n}$ of the spectral subspace associated with A_n^* and $\overline{\Lambda}_n$. If we define

$$\underline{\varphi_n^*} := \underline{K_n^*}\,\underline{\mathsf{v}_n} = \underline{f_n}\,\underline{\mathsf{v}_n},$$

then $\underline{T_n^*}\,\underline{\varphi_n^*} = \underline{K_n^*}\,\underline{L_n^*}\,\underline{K_n^*}\,\underline{\mathsf{v}_n} = \underline{K_n^*}\,\underline{A_n^*}\,\underline{\mathsf{v}_n} = \underline{K_n}^*\,\underline{\mathsf{v}_n}\,\Theta_n^* = \underline{\varphi_n^*}\,\Theta_n^*$ and

$$\begin{aligned}(\underline{\varphi_n}\,,\underline{\varphi_n^*}) &= (\underline{L_n}\,\underline{\mathsf{u}_n}\,\Theta_n^{-1},\underline{K_n^*}\,\underline{\mathsf{v}_n}) = (\underline{K_n}\,\underline{L_n}\,\underline{\mathsf{u}_n}\,\Theta_n^{-1},\underline{\mathsf{v}_n})\\ &= (\underline{A_n}\,\underline{\mathsf{u}_n}\,\Theta_n^{-1},\underline{\mathsf{v}_n}) = (\underline{\mathsf{u}_n}\,,\underline{\mathsf{v}_n}) = \underline{\mathsf{v}_n^*}\,\underline{\mathsf{u}_n} = \mathsf{I}_m,\end{aligned}$$

as desired.

Thirdly, to find $\underline{\varphi_n}^{(k)}$ for $k = 1, 2, \ldots$ we observe that

$$\begin{aligned}(\underline{\varphi_n}^{(k-1)}\Theta_n^{(k)} - \underline{T}\,\underline{\varphi_n}^{(k-1)},\underline{\varphi_n^*}) &= (\underline{\varphi_n}^{(k-1)},\underline{\varphi_n^*})\Theta_n^{(k)} - (\underline{T}\,\underline{\varphi_n}^{(k-1)},\underline{\varphi_n^*})\\ &= \Theta_n^{(k)} - \Theta_n^{(k)} = \mathsf{O},\end{aligned}$$

that is, $\underline{P_n}\,(\underline{\varphi_n}^{(k-1)}\Theta_n^{(k)} - \underline{T}\,\underline{\varphi_n}^{(k-1)}) = \underline{0}$. (See Theorem 1.52.) Hence we can find $\underline{\varphi_n}^{(k)}$ if we are able to find $\mathcal{S}_n\,\underline{y}$ whenever $\underline{y} \in X^{1\times m}$ and $\underline{P_n}\,\underline{y} = \underline{0}$. Let $\underline{y} \in X^{1\times m}$ with $\underline{P_n}\,\underline{y} = \underline{\varphi_n}\,(\underline{y}\,,\underline{\varphi_n^*}) = \underline{0}$. Then

$$\underline{\mathsf{v}_n^*}\,\underline{K_n}\,\underline{y} = (\underline{K_n}\,\underline{y}\,,\underline{\mathsf{v}_n}) = (\underline{y}\,,\underline{K_n^*}\,\underline{\mathsf{v}_n}) = (\underline{y}\,,\underline{\varphi_n^*}) = \mathsf{O},$$

that is, $\underline{K_n}\,\underline{y}$ belongs to the null space of the operator $\underline{\mathsf{w}} \mapsto \underline{A_n}\,\underline{\mathsf{w}} - \underline{\mathsf{w}}\,\Theta_n$ from $\mathbb{C}^{n\times m}$ to $\mathbb{C}^{n\times m}$. Proposition 1.56 shows that this operator maps the null space onto itself in a one-to-one manner. Hence there is a unique $\underline{\mathsf{w}_n} \in \mathbb{C}^{n\times m}$ such that

$$\mathsf{A}_n\,\underline{\mathsf{w}_n} - \underline{\mathsf{w}_n}\,\Theta_n = (\underline{y}\,,\underline{f_n}), \quad \underline{\mathsf{v}_n^*}\,\underline{\mathsf{w}_n} = \mathsf{O}.$$

If we define

$$\underline{x} := (\underline{L_n}\,\underline{\mathsf{w}_n} - \underline{y})\Theta_n^{-1} = (\underline{x_n}\,\underline{\mathsf{w}_n} - \underline{y})\Theta_n^{-1},$$

then

$$\underline{K_n}\,\underline{x} = (\underline{K_n}\,\underline{L_n}\,\underline{\mathsf{w}_n} - \underline{K_n}\,\underline{y})\Theta_n^{-1} = (\underline{A_n}\,\underline{\mathsf{w}_n} - \underline{K_n}\,\underline{y})\Theta_n^{-1} = \underline{\mathsf{w}_n},$$

so that

$$\underline{T_n}\,\underline{x} - \underline{x}\,\Theta_n = \underline{L_n}\,\underline{K_n}\,\underline{x} - (\underline{L_n}\,\underline{\mathsf{w}_n} - \underline{y}) = \underline{y}.$$

Further, since

$$(\underline{x}\,,\underline{\varphi_n^*}) = (\underline{x}\,,\underline{K_n^*}\,\underline{\mathsf{v}_n}) = (\underline{K_n}\,\underline{x}\,,\underline{\mathsf{v}_n}) = (\underline{\mathsf{w}_n}\,,\underline{\mathsf{v}_n}) = \mathsf{O},$$

we see that $\underline{x} \in \mathcal{N}(\underline{P_n})$, and hence by Proposition 1.56, $\underline{x} = \mathcal{S}_n \underline{y}$, as desired. (See Remark 1.58.)

Thus to implement the Elementary Iteration $(\underline{E})$, we need to perform the following matrix computations:

(i) Solve the spectral subspace problem $\mathsf{A}_n \underline{\mathsf{u}_n} = \underline{\mathsf{u}_n}\, \Theta_n$. (See Section 6.3.) Let $\underline{\mathsf{u}_n} := [\mathsf{u}_{n,1}, \ldots, \mathsf{u}_{n,m}]$. Ensure that $\|\mathsf{u}_{n,j}\|_{[n]} \leq c$ and $\text{dist}(\mathsf{u}_{n,j}, \text{span}\{\mathsf{u}_{n,i} : i = 1, \ldots, m,\, i \neq j\}) \geq d$ for some constants c and d independent of n and j, in a norm $\|\cdot\|_{[n]}$ on $\mathbb{C}^{n\times 1}$ such that the sequences $(\|K_n\|)$ and $(\|L_n\|)$ are bounded. (See Theorem 5.9 and Examples 5.11, 5.12, 5.13.)

(ii) Compute the unique vectors in $\underline{\mathsf{v}_n}$ such that

$$\mathsf{A}_n^* \underline{\mathsf{v}_n} = \underline{\mathsf{v}_n}\, \Theta_n^*, \quad \underline{\mathsf{u}_n^*}\, \underline{\mathsf{v}_n} = \mathsf{I}_m.$$

To find the m vectors in $\underline{\mathsf{v}_n}$, we may find a basis $\underline{\widetilde{\mathsf{v}}_n}$ for the spectral subspace $M(A_n^*, \overline{\Lambda}_n)$ in the same manner as we found a basis $\underline{\mathsf{u}_n}$ for $M(A_n, \Lambda_n)$, and let $\underline{\mathsf{v}_n} := \underline{\widetilde{\mathsf{v}}_n}\,(\underline{\mathsf{u}_n^*}\,\underline{\widetilde{\mathsf{v}}_n})^{-1}$. Alternatively, we may solve the coupled Sylvester equation

$$\mathsf{A}_n^* \underline{\mathsf{v}} - \underline{\mathsf{v}}\, \Theta_n^* = \underline{\mathsf{0}}, \quad \underline{\mathsf{u}_n^*}\, \underline{\mathsf{v}} = \mathsf{I}_m,$$

to find its unique solution $\underline{\mathsf{v}} = \underline{\mathsf{v}_n}$.

(iii) Assuming that $\underline{\varphi_n}^{(k-1)}$ is already computed, find $\underline{\varphi_n}^{(k)}$ as follows: Compute

$$\begin{aligned}
\Theta_n^{(k)} &:= (\underline{T}\, \underline{\varphi_n}^{(k-1)}, \underline{\varphi_n^*}) = \underline{\mathsf{v}_n^*}\,(\underline{T}\, \underline{\varphi_n}^{(k-1)}, \underline{f_n}),\\
\underline{y} &:= \underline{\varphi_n}^{(k-1)}(\Theta_n^{(k)} - \Theta_n) + (\underline{T_n} - \underline{T})\, \underline{\varphi_n}^{(k-1)},\\
\underline{\mathsf{b}}_k &:= (\underline{y}, \underline{f_n}).
\end{aligned}$$

Solve the single Sylvester equation

$$\mathsf{A}_n(\mathsf{I}_n - \underline{\mathsf{u}_n}\, \underline{\mathsf{v}_n^*})\, \underline{\mathsf{w}} - \underline{\mathsf{w}}\, \Theta_n = \underline{\mathsf{b}}_k$$

to find its unique solution $\underline{\mathsf{w}} = \underline{\mathsf{w}_n}$ or equivalently, solve the coupled Sylvester equation

$$\mathsf{A}_n \underline{\mathsf{w}} - \underline{\mathsf{w}}\, \Theta_n = \underline{\mathsf{b}}_k, \quad \underline{\mathsf{v}_n^*}\, \underline{\mathsf{w}} = \mathsf{O},$$

to find its unique solution $\underline{\mathsf{w}} = \underline{\mathsf{w}_n}$.

Then let

$$\underline{\varphi_n}^{(k)} := \underline{\varphi_n} + (\underline{x_n}\, \underline{\mathsf{w}_n} - \underline{y})\Theta_n^{-1}.$$

A similar procedure for the implementation of the Elementary Iteration ($\underline{E}$) was outlined in Section 4 of [53].

As before, the Double Iteration ($\underline{D}$) poses no new problems.

5.3 Acceleration

Let $T \in \mathrm{BL}(X)$ and $\widetilde{T} \in \mathrm{BL}(X)$ be a finite rank operator given by

$$\widetilde{T}x := \sum_{j=1}^{n} \langle x\,,\, \widetilde{f}_j \rangle \widetilde{x}_j = \underline{\widetilde{x}}\, \big(x, \underline{\widetilde{f}}\,\big) \quad \text{for} \quad x \in X.$$

Fix an integer $q \geq 2$ and let

$$\boldsymbol{X} := X^{q \times 1} = \left\{ \boldsymbol{x} := \begin{bmatrix} x_1 \\ \vdots \\ x_q \end{bmatrix} : x_i \in X \text{ for } 1 \leq i \leq q \right\}.$$

The operators $\boldsymbol{T} : \boldsymbol{X} \to \boldsymbol{X}$ and $\widetilde{\boldsymbol{T}} : \boldsymbol{X} \to \boldsymbol{X}$ given by

$$\boldsymbol{T} \begin{bmatrix} x_1 \\ \vdots \\ \vdots \\ x_q \end{bmatrix} := \begin{bmatrix} Tx_1 \\ x_1 \\ \vdots \\ x_{q-1} \end{bmatrix} \quad \text{and} \quad \widetilde{\boldsymbol{T}} \begin{bmatrix} x_1 \\ \vdots \\ \vdots \\ x_q \end{bmatrix} := \begin{bmatrix} \sum_{k=0}^{q-1} (T - \widetilde{T})^k \widetilde{T} x_{k+1} \\ x_1 \\ \vdots \\ x_{q-1} \end{bmatrix}$$

for $\boldsymbol{x} \in \boldsymbol{X}$ were considered while discussing the accelerated spectral approximation in Section 3.2.

If X is infinite dimensional, then $\widetilde{\boldsymbol{T}}$ need not be of finite rank. In the present section we shall show that spectral computations for $\widetilde{\boldsymbol{T}}$ can still be reduced to spectral computations for an $nq \times nq$ matrix. This was first shown in [11]. Our treatment is substantially different.

Recall the operators $\widetilde{K} : X \to \mathbb{C}^{n\times 1}$ and $\widetilde{L} : \mathbb{C}^{n\times 1} \to X$ given by

$$\widetilde{K}x := \begin{bmatrix} \langle x\,,\,\widetilde{f}_1\rangle \\ \vdots \\ \langle x\,,\,\widetilde{f}_n\rangle \end{bmatrix} = (x, \underline{\widetilde{f}}) \quad \text{for } x \in X, \quad \text{and}$$

$$\widetilde{L}\mathsf{u} := \mathsf{u}(1)\widetilde{x}_1 + \cdots + \mathsf{u}(n)\widetilde{x}_n = \underline{\widetilde{x}}\,\mathsf{u} \quad \text{for } \mathsf{u} := \begin{bmatrix} \mathsf{u}(1) \\ \vdots \\ \mathsf{u}(n) \end{bmatrix} \in \mathbb{C}^{n\times 1}.$$

Then $\widetilde{T} = \widetilde{L}\widetilde{K}$. As in Section 5.1, we let $\widetilde{A} := \widetilde{K}\widetilde{L} : \mathbb{C}^{n\times 1} \to \mathbb{C}^{n\times 1}$. Then $\widetilde{K}\widetilde{T} = \widetilde{K}\widetilde{L}\widetilde{K} = \widetilde{A}\widetilde{K}$ and $\widetilde{L}\widetilde{A} = \widetilde{L}\widetilde{K}\widetilde{L} = \widetilde{T}\widetilde{L}$.

We identify $(\mathbb{C}^{n\times 1})^{q\times 1}$ with $\mathbb{C}^{nq\times 1}$ and define $\widetilde{\boldsymbol{A}} : \mathbb{C}^{nq\times 1} \to \mathbb{C}^{nq\times 1}$ by

$$\widetilde{\boldsymbol{A}} \begin{bmatrix} \mathsf{u}_1 \\ \vdots \\ \mathsf{u}_q \end{bmatrix} := \begin{bmatrix} \sum_{k=0}^{q-1} \widetilde{K}(T-\widetilde{T})^k \widetilde{L}\mathsf{u}_{k+1} \\ \mathsf{u}_1 \\ \vdots \\ \mathsf{u}_{q-1} \end{bmatrix} \quad \text{for } \begin{bmatrix} \mathsf{u}_1 \\ \vdots \\ \mathsf{u}_q \end{bmatrix} \in \mathbb{C}^{nq\times 1}.$$

The operator $\widetilde{\boldsymbol{A}}$ is represented by the following $nq\times nq$ matrix with respect to the standard basis for $\mathbb{C}^{nq\times 1}$:

$$\widetilde{\mathsf{A}} := \begin{bmatrix} \widetilde{\mathsf{A}}^{\langle 0\rangle} & \widetilde{\mathsf{A}}^{\langle 1\rangle} & \cdots & \cdots & \widetilde{\mathsf{A}}^{\langle q-1\rangle} \\ \mathsf{I}_n & \mathsf{O} & \cdots & \cdots & \mathsf{O} \\ \mathsf{O} & \mathsf{I}_n & \ddots & \ddots & \vdots \\ \vdots & \vdots & \ddots & \ddots & \vdots \\ \mathsf{O} & \mathsf{O} & \cdots & \mathsf{I}_n & \mathsf{O} \end{bmatrix},$$

where for $k = 0, \ldots, q-1$, $\widetilde{\mathsf{A}}^{\langle k\rangle}$ is the $n\times n$ matrix given by

$$\widetilde{\mathsf{A}}^{\langle k\rangle}(i,j) := \langle (T-\widetilde{T})^k \widetilde{x}_j\,,\,\widetilde{f}_i\rangle, \quad 1 \le i,j \le n,$$

and I_n is the $n\times n$ identity matrix.

Let $\widetilde{\mathbf{K}} : \boldsymbol{X} \to \mathbb{C}^{nq\times 1}$ and $\widetilde{\mathbf{L}} : \mathbb{C}^{nq\times 1} \to \boldsymbol{X}$ be defined by

$$\widetilde{\mathbf{K}}\begin{bmatrix} x_1 \\ \vdots \\ x_q \end{bmatrix} := \begin{bmatrix} \widetilde{K}x_1 \\ \vdots \\ \widetilde{K}x_q \end{bmatrix} \quad \text{and} \quad \widetilde{\mathbf{L}}\begin{bmatrix} \mathsf{u}_1 \\ \vdots \\ \mathsf{u}_q \end{bmatrix} := \begin{bmatrix} \widetilde{L}\mathsf{u}_1 \\ \vdots \\ \widetilde{L}\mathsf{u}_q \end{bmatrix}$$

for $\begin{bmatrix} x_1 \\ \vdots \\ x_q \end{bmatrix} \in \boldsymbol{X}$ and $\begin{bmatrix} \mathsf{u}_1 \\ \vdots \\ \mathsf{u}_q \end{bmatrix} \in \mathbb{C}^{nq\times 1}$. It is easy to see that $\widetilde{\boldsymbol{T}} \neq \widetilde{\mathbf{L}}\widetilde{\mathbf{K}}$ and $\widetilde{\boldsymbol{A}} \neq \widetilde{\mathbf{K}}\widetilde{\mathbf{L}}$. However, we have

$$\widetilde{\mathbf{K}}\widetilde{\boldsymbol{T}} = \widetilde{\boldsymbol{A}}\widetilde{\mathbf{K}}$$

since $\widetilde{L}\widetilde{K} = \widetilde{T}$. On the other hand, $\widetilde{\mathbf{L}}\widetilde{\boldsymbol{A}} \neq \widetilde{\boldsymbol{T}}\widetilde{\mathbf{L}}$ since $\sum_{k=0}^{q-1} \widetilde{T}(T - \widetilde{T})^k \neq \sum_{k=0}^{q-1} (T - \widetilde{T})^k \widetilde{T}$ in general.

Let p be a positive integer and consider the natural extensions $\underline{T}$, $\underline{\widetilde{T}} : X^{1\times p} \to X^{1\times p}$, $\underline{\widetilde{A}} : \mathbb{C}^{n\times p} \to \mathbb{C}^{n\times p}$, $\underline{\widetilde{K}} : X^{1\times p} \to \mathbb{C}^{n\times p}$ and $\underline{\widetilde{L}} : \mathbb{C}^{n\times p} \to X^{1\times p}$ of the operators $T, \widetilde{T} : X \to X$, $\widetilde{A} : \mathbb{C}^{n\times 1} \to \mathbb{C}^{n\times 1}$, $\widetilde{K} : X \to \mathbb{C}^{n\times 1}$ and $\widetilde{L} : \mathbb{C}^{n\times 1} \to X$. Let $\mathsf{Z} \in \mathbb{C}^{p\times p}$ be a nonsingular matrix. We define an operator $\widetilde{\mathcal{L}}(\mathsf{Z}) : \mathbb{C}^{n\times p} \to X^{1\times p}$ by

$$\widetilde{\mathcal{L}}(\mathsf{Z})\,\underline{\mathsf{u}} := \sum_{k=0}^{q-1} \left[(\underline{T} - \underline{\widetilde{T}})^k\, \underline{\widetilde{L}}\, \underline{\mathsf{u}}\right] \mathsf{Z}^{-k} \quad \text{for } \underline{\mathsf{u}} \in \mathbb{C}^{n\times p}.$$

Next, we let $\underline{\boldsymbol{X}} := \boldsymbol{X}^{1\times p}$, and consider the natural extensions $\underline{\boldsymbol{T}}$, $\underline{\widetilde{\boldsymbol{T}}} : \underline{\boldsymbol{X}} \to \underline{\boldsymbol{X}}$, $\underline{\widetilde{\boldsymbol{A}}} : \mathbb{C}^{nq\times p} \to \mathbb{C}^{nq\times p}$, $\underline{\widetilde{\mathbf{K}}} : \underline{\boldsymbol{X}} \to \mathbb{C}^{nq\times p}$ of the operators $\boldsymbol{T}, \widetilde{\boldsymbol{T}} : \boldsymbol{X} \to \boldsymbol{X}, \widetilde{\boldsymbol{A}} : \mathbb{C}^{nq\times 1} \to \mathbb{C}^{nq\times 1}$ and $\widetilde{\mathbf{K}} : \boldsymbol{X} \to \mathbb{C}^{nq\times 1}$.

For $\underline{\boldsymbol{x}} := [\boldsymbol{x}_1, \ldots, \boldsymbol{x}_p] = \begin{bmatrix} \underline{x}_1 \\ \vdots \\ \underline{x}_q \end{bmatrix} \in \underline{\boldsymbol{X}}$ and $\underline{\mathbf{u}} := [\mathbf{u}_1, \ldots, \mathbf{u}_p] = \begin{bmatrix} \underline{\mathsf{u}}_1 \\ \vdots \\ \underline{\mathsf{u}}_q \end{bmatrix} \in \mathbb{C}^{nq\times p}$, we have

$$\underline{\boldsymbol{T}}\,\underline{\boldsymbol{x}} := [\boldsymbol{T}\boldsymbol{x}_1, \ldots, \boldsymbol{T}\boldsymbol{x}_p] = \begin{bmatrix} \underline{T}\,\underline{x}_1 \\ \underline{x}_1 \\ \vdots \\ \underline{x}_{q-1} \end{bmatrix},$$

$$\underline{\widetilde{\boldsymbol{T}}}\,\underline{\boldsymbol{x}} := [\widetilde{\boldsymbol{T}}\boldsymbol{x}_1, \ldots, \widetilde{\boldsymbol{T}}\boldsymbol{x}_p] = \begin{bmatrix} \sum_{k=0}^{q-1} (\underline{T} - \underline{\widetilde{T}})^k \, \underline{\widetilde{T}} \, \underline{x}_{k+1} \\ \underline{x}_1 \\ \vdots \\ \underline{x}_{q-1} \end{bmatrix},$$

$$\underline{\widetilde{\boldsymbol{A}}}\,\underline{\mathbf{u}} := [\widetilde{\boldsymbol{A}}\mathbf{u}_1, \ldots, \widetilde{\boldsymbol{A}}\mathbf{u}_p] = \begin{bmatrix} \sum_{k=0}^{q-1} \underline{\widetilde{K}} \, (\underline{T} - \underline{\widetilde{T}})^k \, \underline{\widetilde{L}} \, \underline{\mathsf{u}}_{k+1} \\ \underline{\mathsf{u}}_1 \\ \vdots \\ \underline{\mathsf{u}}_{q-1} \end{bmatrix},$$

$$\underline{\widetilde{\mathbf{K}}}\,\underline{\boldsymbol{x}} := [\widetilde{\mathbf{K}}\boldsymbol{x}_1, \ldots, \widetilde{\mathbf{K}}\boldsymbol{x}_p] = \begin{bmatrix} \underline{\widetilde{K}} \, \underline{x}_1 \\ \vdots \\ \underline{\widetilde{K}} \, \underline{x}_q \end{bmatrix}.$$

Finally, consider the operator $\widetilde{\boldsymbol{\mathcal{L}}}(\mathsf{Z}) : \mathbb{C}^{nq \times p} \to \underline{\boldsymbol{X}}$ defined by

$$\widetilde{\boldsymbol{\mathcal{L}}}(\mathsf{Z})\,\underline{\mathbf{u}} := \begin{bmatrix} \widetilde{\mathcal{L}}(\mathsf{Z})\,\underline{\mathsf{u}}_1 \\ \vdots \\ \widetilde{\mathcal{L}}(\mathsf{Z})\,\underline{\mathsf{u}}_q \end{bmatrix} \quad \text{for } \underline{\mathbf{u}} = \begin{bmatrix} \underline{\mathsf{u}}_1 \\ \vdots \\ \underline{\mathsf{u}}_q \end{bmatrix} \in \mathbb{C}^{nq \times p}.$$

In order to obtain an analog of Lemma 5.2 for the operators $\underline{\widetilde{\boldsymbol{T}}}$ and $\underline{\widetilde{\boldsymbol{A}}}$, we first prove some preliminary results.

We define the operator $\mathcal{F}_1 : \underline{\boldsymbol{X}} \to X^{1\times p}$ by

$$\mathcal{F}_1\,\underline{\boldsymbol{y}} := \sum_{k=1}^{q-1} (\underline{T} - \underline{\widetilde{T}})^k \, \underline{\widetilde{T}} \, \underline{\zeta}_k \quad \text{for } \underline{\boldsymbol{y}} := \begin{bmatrix} \underline{y}_1 \\ \vdots \\ \underline{y}_q \end{bmatrix} \in \underline{\boldsymbol{X}},$$

where $\underline{\zeta}_k := \sum_{\ell=1}^{k} \underline{y}_{\ell+1} \mathsf{Z}^{-(k-\ell+1)}$ for $k = 1, \ldots, q-1$.

For $i = 1, \ldots, q-1$, let $\mathcal{F}_{i+1} : \underline{\boldsymbol{X}} \to X^{1\times p}$ be defined by

$$\mathcal{F}_{i+1}\,\underline{\boldsymbol{y}} := \Big[\underline{y}_i + \mathcal{F}_i\,\underline{\boldsymbol{y}} - \sum_{k=0}^{q-1} (\underline{T} - \underline{\widetilde{T}})^k \, \underline{\widetilde{T}} \, \underline{y}_{i+1} \mathsf{Z}^{-k}\Big] \mathsf{Z}^{-1} \quad \text{for } \underline{\boldsymbol{y}} \in \underline{\boldsymbol{X}},$$

and $\boldsymbol{\mathcal{F}} : \underline{\boldsymbol{X}} \to \underline{\boldsymbol{X}}$ be defined by

$$\boldsymbol{\mathcal{F}}\,\underline{\boldsymbol{y}} := \begin{bmatrix} \mathcal{F}_1\,\underline{\boldsymbol{y}} \\ \vdots \\ \mathcal{F}_q\,\underline{\boldsymbol{y}} \end{bmatrix} \quad \text{for } \underline{\boldsymbol{y}} \in \underline{\boldsymbol{X}}.$$

Lemma 5.17
Let p be a positive integer, $\mathsf{Z} \in \mathbb{C}^{p\times p}$ be such that $0 \notin \mathrm{sp}(\mathsf{Z})$, and $\boldsymbol{\mathcal{F}}$ be defined as above. Then for $\underline{\boldsymbol{x}} := \begin{bmatrix} \underline{x}_1 \\ \vdots \\ \underline{x}_q \end{bmatrix} \in \underline{\mathbf{X}}$, we have $\underline{\widetilde{\boldsymbol{T}}}\,\underline{\boldsymbol{x}} = \underline{\boldsymbol{x}}\,\mathsf{Z} + \underline{\boldsymbol{y}}$ if and only if $\underline{x}_i = \underline{x}_{i+1}\mathsf{Z} + \underline{y}_{i+1}$ for $i = 1,\ldots,q-1$ and $\sum_{k=0}^{q-1}(\underline{T} - \underline{\widetilde{T}})^k \underline{\widetilde{T}}\,\underline{x}_1\mathsf{Z}^{-k} = \underline{x}_1\mathsf{Z} + \underline{y}_1 + \mathcal{F}_1\underline{\boldsymbol{y}}$. In this case, we also have $\sum_{k=0}^{q-1}(\underline{T} - \underline{\widetilde{T}})^k \underline{\widetilde{T}}\,\underline{x}_i\mathsf{Z}^{-k} = \underline{x}_i\mathsf{Z} + \underline{y}_i + \mathcal{F}_i\underline{\boldsymbol{y}}$ for $i = 2,\ldots,q-1$, $\underline{\widetilde{\mathbf{A}}}\,\underline{\widetilde{\mathbf{K}}}\,\underline{\boldsymbol{x}} = \underline{\widetilde{\mathbf{K}}}\,\underline{\boldsymbol{x}}\,\mathsf{Z} + \underline{\widetilde{\mathbf{K}}}\,\underline{\boldsymbol{y}}$ and $\widetilde{\boldsymbol{\mathcal{L}}}(\mathsf{Z})\,\underline{\widetilde{\mathbf{K}}}\,\underline{\boldsymbol{x}} = \underline{\boldsymbol{x}}\,\mathsf{Z} + \underline{\boldsymbol{y}} + \boldsymbol{\mathcal{F}}\,\underline{\boldsymbol{y}}$.

Proof
The definition of the operator $\widetilde{\boldsymbol{T}}$ shows that $\underline{\widetilde{\boldsymbol{T}}}\,\underline{\boldsymbol{x}} = \underline{\boldsymbol{x}}\,\mathsf{Z} + \underline{\boldsymbol{y}}$ if and only if $\sum_{k=0}^{q-1}(\underline{T} - \underline{\widetilde{T}})^k \underline{\widetilde{T}}\,\underline{x}_{k+1} = \underline{x}_1\mathsf{Z} + \underline{y}_1$ and $\underline{x}_i = \underline{x}_{i+1}\mathsf{Z} + \underline{y}_{i+1}$ for $i = 1,\ldots,q-1$. In this case, we obtain for $k = 1,\ldots,q-1$, $\underline{x}_{k+1} = (\underline{x}_k - \underline{y}_{k+1})\mathsf{Z}^{-1} = [(\underline{x}_{k-1} - \underline{y}_k)\mathsf{Z}^{-1} - \underline{y}_{k+1}]\mathsf{Z}^{-1} = \cdots = \underline{x}_1\mathsf{Z}^{-k} - \sum_{\ell=2}^{k+1}\underline{y}_\ell\mathsf{Z}^{-(k+2-\ell)}$. Hence

$$\begin{aligned}\underline{x}_1\mathsf{Z} + \underline{y}_1 &= \underline{\widetilde{T}}\,\underline{x}_1 + \sum_{k=1}^{q-1}(\underline{T} - \underline{\widetilde{T}})^k \underline{\widetilde{T}}\Big[\underline{x}_1\mathsf{Z}^{-k} - \sum_{\ell=1}^{k}\underline{y}_{\ell+1}\mathsf{Z}^{-(k-\ell+1)}\Big] \\ &= \sum_{k=0}^{q-1}(\underline{T} - \underline{\widetilde{T}})^k \underline{\widetilde{T}}\,\underline{x}_1\mathsf{Z}^{-k} - \sum_{k=1}^{q-1}(\underline{T} - \underline{\widetilde{T}})\underline{\widetilde{T}}\,\underline{\zeta}_k \\ &= \sum_{k=0}^{q-1}(\underline{T} - \underline{\widetilde{T}})^k \underline{\widetilde{T}}\,\underline{x}_1\mathsf{Z}^{-k} - \mathcal{F}_1\underline{\boldsymbol{y}},\end{aligned}$$

as desired. The converse follows by working backwards.

Assume now that we have proved

$$\sum_{k=0}^{q-1}(\underline{T} - \underline{\widetilde{T}})^k \underline{\widetilde{T}}\,\underline{x}_i\mathsf{Z}^{-k} = \underline{x}_i\mathsf{Z} + \underline{y}_i + \mathcal{F}_i\underline{\boldsymbol{y}}$$

for some i such that $1 \le i \le q-1$. Putting $\underline{x}_i = \underline{x}_{i+1}\mathsf{Z} + \underline{y}_{i+1}$ in this equation, we obtain

$$\sum_{k=0}^{q-1}(\underline{T} - \underline{\widetilde{T}})^k \underline{\widetilde{T}}\, \underline{x}_{i+1}\mathsf{Z}^{-k+1} = \underline{x}_{i+1}\mathsf{Z}^2 + \underline{y}_{i+1}\mathsf{Z} + \underline{y}_i + \mathcal{F}_i \underline{\boldsymbol{y}} - \sum_{k=0}^{q-1}(\underline{T} - \underline{\widetilde{T}})^k \underline{\widetilde{T}}\, \underline{y}_{i+1}\mathsf{Z}^{-k},$$

so that

$$\begin{aligned}
&\sum_{k=0}^{q-1}(\underline{T} - \underline{\widetilde{T}})^k \underline{\widetilde{T}}\, \underline{x}_{i+1}\mathsf{Z}^{-k} \\
&= \underline{x}_{i+1}\mathsf{Z} + \underline{y}_{i+1} + \Big[\underline{y}_i + \mathcal{F}_i \underline{\boldsymbol{y}} - \sum_{k=0}^{q-1}(\underline{T} - \underline{\widetilde{T}})^k \underline{\widetilde{T}}\, \underline{y}_{i+1}\mathsf{Z}^{-k}\Big]\mathsf{Z}^{-1} \\
&= \underline{x}_{i+1}\mathsf{Z} + \underline{y}_{i+1} + \mathcal{F}_{i+1}\underline{\boldsymbol{y}},
\end{aligned}$$

as desired. Next,

$$\underline{\widetilde{\mathbf{A}}}\, \underline{\widetilde{\mathbf{K}}}\, \underline{\boldsymbol{x}} = \underline{\widetilde{\mathbf{K}}}\, \underline{\widetilde{\boldsymbol{T}}}\, \underline{\boldsymbol{x}} = \underline{\widetilde{\mathbf{K}}}\, \underline{\boldsymbol{x}}\, \mathsf{Z} + \underline{\widetilde{\mathbf{K}}}\, \underline{\boldsymbol{y}}$$

and

$$\begin{aligned}
\widetilde{\boldsymbol{\mathcal{L}}}(\mathsf{Z})\, \underline{\widetilde{\mathbf{K}}}\, \underline{\boldsymbol{x}} &= \widetilde{\boldsymbol{\mathcal{L}}}(\mathsf{Z}) \begin{bmatrix} \underline{\widetilde{K}}\, \underline{x}_1 \\ \vdots \\ \underline{\widetilde{K}}\, \underline{x}_q \end{bmatrix} = \begin{bmatrix} \sum_{k=0}^{q-1}\Big[(\underline{T} - \underline{\widetilde{T}})^k \underline{\widetilde{L}}\, \underline{\widetilde{K}}\, \underline{x}_1\Big]\mathsf{Z}^{-k} \\ \vdots \\ \sum_{k=0}^{q-1}\Big[(\underline{T} - \underline{\widetilde{T}})^k \underline{\widetilde{L}}\, \underline{\widetilde{K}}\, \underline{x}_q\Big]\mathsf{Z}^{-k} \end{bmatrix} \\
&= \begin{bmatrix} \underline{x}_1\mathsf{Z} + \underline{y}_1 + \mathcal{F}_1\underline{\boldsymbol{y}} \\ \vdots \\ \underline{x}_q\mathsf{Z} + \underline{y}_q + \mathcal{F}_q\underline{\boldsymbol{y}} \end{bmatrix} = \underline{\boldsymbol{x}}\,\mathsf{Z} + \boldsymbol{\mathcal{F}}\underline{\boldsymbol{y}}
\end{aligned}$$

since $\underline{\widetilde{L}}\, \underline{\widetilde{K}} = \underline{\widetilde{T}}$. ∎

We now carry out a similar analysis in $\mathbb{C}^{nq\times p}$. Define an operator $\mathcal{G}_1 : \mathbb{C}^{nq\times p} \to \mathbb{C}^{n\times p}$ by

$$\mathcal{G}_1 \underline{\mathbf{v}} := \sum_{k=1}^{q-1} \underline{\widetilde{K}}(\underline{T} - \underline{\widetilde{T}})^k \underline{\widetilde{L}}\, \underline{\mathsf{w}}_k \quad \text{for } \underline{\mathbf{v}} := \begin{bmatrix} \underline{\mathsf{v}}_1 \\ \vdots \\ \underline{\mathsf{v}}_q \end{bmatrix} \in \mathbb{C}^{nq\times p},$$

where $\underline{\mathsf{w}}_k := \sum_{\ell=1}^{k} \underline{\mathsf{v}}_{\ell+1}\mathsf{Z}^{-(k-\ell+1)}$ for $k = 1, \ldots, q-1$.

For $i = 1, \ldots, q-1$, let $\mathcal{G}_{i+1} : \mathbb{C}^{nq\times p} \to \mathbb{C}^{n\times p}$ be defined by

$$\mathcal{G}_{i+1}\,\underline{\mathsf{v}} := \Big[\underline{\mathsf{v}}_i + \mathcal{G}_i\,\underline{\mathsf{v}} - \sum_{k=0}^{q-1} \underline{\widetilde{K}}\,(\underline{T} - \underline{\widetilde{T}})^k\,\underline{\widetilde{L}}\,\underline{\mathsf{v}}_{i+1}\mathsf{Z}^{-k}\Big]\mathsf{Z}^{-1} \text{ for } \underline{\mathsf{v}} \in \mathbb{C}^{nq\times p},$$

and $\boldsymbol{\mathcal{G}} : \mathbb{C}^{nq\times p} \to \mathbb{C}^{nq\times p}$ be defined by

$$\boldsymbol{\mathcal{G}}\,\underline{\mathsf{v}} := \begin{bmatrix} \mathcal{G}_1\,\underline{\mathsf{v}} \\ \vdots \\ \mathcal{G}_q\,\underline{\mathsf{v}} \end{bmatrix} \qquad \text{for } \underline{\mathsf{v}} \in \mathbb{C}^{nq\times p}.$$

It can be easily seen that for every $\underline{\boldsymbol{y}} \in \underline{\boldsymbol{X}}$, $\underline{\widetilde{K}}\,\mathcal{F}_1\,\underline{\boldsymbol{y}} = \mathcal{G}_1\,\underline{\widetilde{\mathbf{K}}}\,\underline{\boldsymbol{y}}$, and recursively, $\underline{\widetilde{K}}\,\mathcal{F}_{i+1}\,\underline{\boldsymbol{y}} = \mathcal{G}_{i+1}\,\underline{\widetilde{\mathbf{K}}}\,\underline{\boldsymbol{y}}$ for $i = 1, \ldots, q-1$, so that

$$\underline{\widetilde{\mathbf{K}}}\,\boldsymbol{\mathcal{F}} = \boldsymbol{\mathcal{G}}\,\underline{\widetilde{\mathbf{K}}}.$$

Lemma 5.18

Let p be a positive integer, $\mathsf{Z} \in \mathbb{C}^{p\times p}$ be such that $0 \notin \operatorname{sp}(\mathsf{Z})$, and $\boldsymbol{\mathcal{G}}$ be defined as above. Then for $\underline{\mathsf{u}} := \begin{bmatrix} \underline{\mathsf{u}}_1 \\ \vdots \\ \underline{\mathsf{u}}_q \end{bmatrix} \in \mathbb{C}^{nq\times p}$, we have $\underline{\widetilde{\boldsymbol{A}}}\,\underline{\mathsf{u}} = \underline{\mathsf{u}}\,\mathsf{Z} + \underline{\mathsf{v}}$ if and only if $\underline{\mathsf{u}}_i = \underline{\mathsf{u}}_{i+1}\mathsf{Z} + \underline{\mathsf{v}}_{i+1}$ for $i = 1, \ldots, q-1$ and $\sum_{k=0}^{q-1} \underline{\widetilde{K}}\,(\underline{T} - \underline{\widetilde{T}})^k\,\underline{\widetilde{L}}\,\underline{\mathsf{u}}_1\mathsf{Z}^{-k} = \underline{\mathsf{u}}_1\mathsf{Z} + \underline{\mathsf{v}}_1 + \mathcal{G}_1\,\underline{\mathsf{v}}$. In this case, we also have $\sum_{k=0}^{q-1} \underline{\widetilde{K}}\,(\underline{T} - \underline{\widetilde{T}})^k\,\underline{\widetilde{L}}\,\underline{\mathsf{u}}_i\mathsf{Z}^{-k} = \underline{\mathsf{u}}_i\mathsf{Z} + \underline{\mathsf{v}}_i + \mathcal{G}_i\,\underline{\mathsf{v}}$ for $i = 2, \ldots, q-1$,

$$\underline{\widetilde{\mathbf{K}}}\,\widetilde{\boldsymbol{\mathcal{L}}}(\mathsf{Z})\,\underline{\mathsf{u}} = \underline{\mathsf{u}}\,\mathsf{Z} + \underline{\mathsf{v}} + \boldsymbol{\mathcal{G}}\,\underline{\mathsf{v}} \text{ and } \underline{\widetilde{\boldsymbol{T}}}\,\widetilde{\boldsymbol{\mathcal{L}}}(\mathsf{Z})\,\underline{\mathsf{u}} = \widetilde{\boldsymbol{\mathcal{L}}}(\mathsf{Z})(\underline{\mathsf{u}}\,\mathsf{Z} + \underline{\mathsf{v}}) + \begin{bmatrix} \underline{\zeta} \\ 0 \\ \vdots \\ 0 \end{bmatrix},$$

where $\underline{\zeta} := \sum_{k=1}^{q-1} \Big[\widetilde{\mathcal{L}}(\mathsf{Z})\,\underline{\widetilde{K}}\,(\underline{T} - \underline{\widetilde{T}})^k\,\underline{\widetilde{L}} - (\underline{T} - \underline{\widetilde{T}})^k\,\underline{\widetilde{T}}\,\widetilde{\mathcal{L}}(\mathsf{Z})\Big]\,\underline{\mathsf{w}}_k$.

Proof

The definition of the operator $\widetilde{\boldsymbol{A}}$ shows that $\underline{\widetilde{\boldsymbol{A}}}\,\underline{\mathsf{u}} = \underline{\mathsf{u}}\,\mathsf{Z} + \underline{\mathsf{v}}$ if and only if $\sum_{k=0}^{q-1} \underline{\widetilde{K}}(\underline{T} - \underline{\widetilde{T}})^k \underline{\widetilde{L}}\, \underline{\mathsf{u}}_{k+1} = \underline{\mathsf{u}}_1 \mathsf{Z} + \underline{\mathsf{v}}_1$ and $\underline{\mathsf{u}}_i = \underline{\mathsf{u}}_{i+1}\mathsf{Z} + \underline{\mathsf{v}}_{i+1}$ for $i = 1, \ldots, q-1$. The proof of the remaining part of the equivalence of the stated conditions is very similar to the proof of Lemma 5.17 if we note that for $k = 1, \ldots, q-1$,

$$\underline{\mathsf{u}}_{k+1} = \underline{\mathsf{u}}_1 \mathsf{Z}^{-1} - \sum_{\ell=1}^{k} \underline{\mathsf{v}}_{\ell+1} \mathsf{Z}^{-(k-\ell+1)} = \underline{\mathsf{u}}_1 \mathsf{Z}^{-1} - \underline{\mathsf{w}}_k.$$

Next,

$$\underline{\widetilde{\mathbf{K}}}\widetilde{\boldsymbol{\mathcal{L}}}(\mathsf{Z})\,\underline{\mathbf{u}} = \begin{bmatrix} \underline{\widetilde{K}}\,\widetilde{\mathcal{L}}(\mathsf{Z})\,\underline{\mathsf{u}}_1 \\ \vdots \\ \underline{\widetilde{K}}\,\widetilde{\mathcal{L}}(\mathsf{Z})\,\underline{\mathsf{u}}_q \end{bmatrix} = \begin{bmatrix} \sum_{k=0}^{q-1} \left[\underline{\widetilde{K}}(\underline{T} - \underline{\widetilde{T}})^k \underline{\widetilde{L}}\,\underline{\mathsf{u}}_1\right] \mathsf{Z}^{-k} \\ \vdots \\ \sum_{k=0}^{q-1} \left[\underline{\widetilde{K}}(\underline{T} - \underline{\widetilde{T}})^k \underline{\widetilde{L}}\,\underline{\mathsf{u}}_q\right] \mathsf{Z}^{-k} \end{bmatrix}$$

$$= \begin{bmatrix} \underline{\mathsf{u}}_1 \mathsf{Z} + \underline{\mathsf{v}}_1 + \mathcal{G}_1 \underline{\mathbf{v}} \\ \vdots \\ \underline{\mathsf{u}}_q \mathsf{Z} + \underline{\mathsf{v}}_q + \mathcal{G}_q \underline{\mathbf{v}} \end{bmatrix} = \underline{\mathbf{u}}\,\mathsf{Z} + \underline{\mathbf{v}} + \boldsymbol{\mathcal{G}}\,\underline{\mathbf{v}}.$$

Lastly, we have

$$\underline{\widetilde{\boldsymbol{T}}}\widetilde{\boldsymbol{\mathcal{L}}}(\mathsf{Z})\,\underline{\mathbf{u}} = \underline{\widetilde{\boldsymbol{T}}} \begin{bmatrix} \widetilde{\mathcal{L}}(\mathsf{Z})\,\underline{\mathsf{u}}_1 \\ \vdots \\ \widetilde{\mathcal{L}}(\mathsf{Z})\,\underline{\mathsf{u}}_q \end{bmatrix} = \begin{bmatrix} \sum_{k=0}^{q-1} (\underline{T} - \underline{\widetilde{T}})^k \underline{\widetilde{T}}\,\widetilde{\mathcal{L}}(\mathsf{Z})\,\underline{\mathsf{u}}_{k+1} \\ \widetilde{\mathcal{L}}(\mathsf{Z})\,\underline{\mathsf{u}}_1 \\ \vdots \\ \widetilde{\mathcal{L}}(\mathsf{Z})\,\underline{\mathsf{u}}_{q-1} \end{bmatrix}.$$

Now, since $\widetilde{T} = \widetilde{L}\widetilde{K}$,

$$\begin{aligned}
&\sum_{k=0}^{q-1} (\underline{T} - \underline{\widetilde{T}})^k \underline{\widetilde{T}}\,\widetilde{\mathcal{L}}(\mathsf{Z})\,\underline{\mathsf{u}}_{k+1} \\
&= \underline{\widetilde{T}}\,\widetilde{\mathcal{L}}(\mathsf{Z})\,\underline{\mathsf{u}}_1 + \sum_{k=1}^{q-1} (\underline{T} - \underline{\widetilde{T}})^k \underline{\widetilde{T}}\,\widetilde{\mathcal{L}}(\mathsf{Z})\,\underline{\mathsf{u}}_1 \mathsf{Z}^{-k} - \sum_{k=1}^{q-1} (\underline{T} - \underline{\widetilde{T}})^k \underline{\widetilde{T}}\,\widetilde{\mathcal{L}}(\mathsf{Z})\,\underline{\mathsf{w}}_k \\
&= \widetilde{\mathcal{L}}(\mathsf{Z})\,\underline{\widetilde{K}}\,\widetilde{\mathcal{L}}(\mathsf{Z})\,\underline{\mathsf{u}}_1 - \sum_{k=1}^{q-1} (\underline{T} - \underline{\widetilde{T}})^k \underline{\widetilde{T}}\,\widetilde{\mathcal{L}}(\mathsf{Z})\,\underline{\mathsf{w}}_k \\
&= \widetilde{\mathcal{L}}(\mathsf{Z})\Big[\underline{\mathsf{u}}_1 \mathsf{Z} + \underline{\mathsf{v}}_1 + \sum_{k=1}^{q-1} \underline{\widetilde{K}}(\underline{T} - \underline{\widetilde{T}})^k \underline{\widetilde{L}}\,\underline{\mathsf{w}}_k\Big] - \sum_{k=1}^{q-1} (\underline{T} - \underline{\widetilde{T}})^k \underline{\widetilde{T}}\,\widetilde{\mathcal{L}}(\mathsf{Z})\,\underline{\mathsf{w}}_k \\
&= \widetilde{\mathcal{L}}(\mathsf{Z})[\underline{\mathsf{u}}_1 \mathsf{Z} + \underline{\mathsf{v}}_1] + \underline{\zeta}.
\end{aligned}$$

Thus

$$\underline{\widetilde{\boldsymbol{T}}}\,\widetilde{\boldsymbol{\mathcal{L}}}(\mathsf{Z})\,\underline{\mathbf{u}} = \widetilde{\mathcal{L}}(\mathsf{Z})\begin{bmatrix} \underline{\mathsf{u}}_1\mathsf{Z} + \underline{\mathsf{v}}_1 \\ \underline{\mathsf{u}}_1 \\ \vdots \\ \underline{\mathsf{u}}_{q-1} \end{bmatrix} + \begin{bmatrix} \underline{\varsigma} \\ 0 \\ \vdots \\ 0 \end{bmatrix} = \widetilde{\boldsymbol{\mathcal{L}}}(\mathsf{Z})(\underline{\mathbf{u}}\,\mathsf{Z} + \underline{\mathbf{v}}) + \begin{bmatrix} \underline{\varsigma} \\ 0 \\ \vdots \\ 0 \end{bmatrix},$$

as desired. ∎

We are now in a position to prove an analog of Lemma 5.2 for the operators $\widetilde{\boldsymbol{T}}$ and $\widetilde{\boldsymbol{A}}$.

Lemma 5.19
Let p be a positive integer, $\mathsf{Z} \in \mathbb{C}^{p\times p}$ such that $0 \notin \mathrm{sp}(\mathsf{Z})$, $\underline{y} \in \underline{\boldsymbol{X}}$, and $\underline{\mathbf{v}} := \underline{\widetilde{\mathbf{K}}}\,\underline{y} \in \mathbb{C}^{nq\times p}$. Then for $\underline{x} \in \underline{\boldsymbol{X}}$ and $\underline{\mathbf{u}} \in \mathbb{C}^{nq\times p}$, the following holds: $\underline{\widetilde{\boldsymbol{T}}}\,\underline{x} = \underline{x}\,\mathsf{Z} + \underline{y}$ and $\underline{\widetilde{\mathbf{K}}}\,\underline{x} = \underline{\mathbf{u}}$ if and only if $\underline{\widetilde{\boldsymbol{A}}}\,\underline{\mathbf{u}} = \underline{\mathbf{u}}\,\mathsf{Z} + \underline{\mathbf{v}}$ and $\widetilde{\boldsymbol{\mathcal{L}}}(\mathsf{Z})\,\underline{\mathbf{u}} = \underline{x}\,\mathsf{Z} + \underline{y} + \boldsymbol{\mathcal{F}}\,\underline{y}$.

In particular, let $\underline{y} := \underline{\mathbf{0}}$ (so that $\underline{\mathbf{v}} = \underline{\mathbf{0}}$ also), and let $\underline{x}$, $\underline{\mathbf{u}}$ satisfy the above conditions. Then the set of p elements in $\underline{x}$ is linearly independent in $\boldsymbol{X}$ if and only if the set of p vectors in $\underline{\mathbf{u}}$ is linearly independent in $\mathbb{C}^{nq\times 1}$.

Proof
Assume that $\underline{\widetilde{\boldsymbol{T}}}\,\underline{x} = \underline{x}\,\mathsf{Z} + \underline{y}$ and $\underline{\widetilde{\mathbf{K}}}\,\underline{x} = \underline{\mathbf{u}}$. Then by Lemma 5.17, $\widetilde{\boldsymbol{\mathcal{L}}}(\mathsf{Z})\,\underline{\mathbf{u}} = \widetilde{\boldsymbol{\mathcal{L}}}(\mathsf{Z})\,\underline{\widetilde{\mathbf{K}}}\,\underline{x} = \underline{x}\,\mathsf{Z} + \underline{y} + \boldsymbol{\mathcal{F}}\,\underline{y}$ and $\underline{\widetilde{\boldsymbol{A}}}\,\underline{\mathbf{u}} = \underline{\widetilde{\boldsymbol{A}}}\,\underline{\widetilde{\mathbf{K}}}\,\underline{x} = \underline{\widetilde{\mathbf{K}}}\,\underline{\widetilde{\boldsymbol{T}}}\,\underline{x} = \underline{\widetilde{\mathbf{K}}}\,(\underline{x}\,\mathsf{Z} + \underline{y}) = \underline{\mathbf{u}}\,\mathsf{Z} + \underline{\mathbf{v}}$.

Conversely, assume that $\underline{\widetilde{\boldsymbol{A}}}\,\underline{\mathbf{u}} = \underline{\mathbf{u}}\,\mathsf{Z} + \underline{\mathbf{v}}$ and $\widetilde{\boldsymbol{\mathcal{L}}}(\mathsf{Z})\,\underline{\mathbf{u}} = \underline{x}\,\mathsf{Z} + \underline{y} + \boldsymbol{\mathcal{F}}\,\underline{y}$. Then by Lemma 5.18, $\underline{\widetilde{\mathbf{K}}}\,\underline{x}\,\mathsf{Z} = \underline{\widetilde{\mathbf{K}}}\,[\widetilde{\boldsymbol{\mathcal{L}}}(\mathsf{Z})\,\underline{\mathbf{u}} - \underline{y} - \boldsymbol{\mathcal{F}}\,\underline{y}] = \underline{\mathbf{u}}\,\mathsf{Z} + \underline{\mathbf{v}} + \boldsymbol{\mathcal{G}}\,\underline{\mathbf{v}} - \underline{\widetilde{\mathbf{K}}}\,\underline{y} - \underline{\widetilde{\mathbf{K}}}\,\boldsymbol{\mathcal{F}}\,\underline{y} = \underline{\mathbf{u}}\,\mathsf{Z}$, because $\underline{\widetilde{\mathbf{K}}}\,\underline{y} = \underline{\mathbf{v}}$ and $\underline{\widetilde{\mathbf{K}}}\,\boldsymbol{\mathcal{F}}\,\underline{y} = \boldsymbol{\mathcal{G}}\,\underline{\widetilde{\mathbf{K}}}\,\underline{y} = \boldsymbol{\mathcal{G}}\,\underline{\mathbf{v}}$ as we have mentioned just before Lemma 5.18. Since Z is nonsingular, it follows that $\underline{\widetilde{\mathbf{K}}}\,\underline{x} = \underline{\mathbf{u}}$. Again, by Lemma 5.18,

$$\begin{aligned}
\underline{\widetilde{\boldsymbol{T}}}\,\underline{x}\,\mathsf{Z} &= \underline{\widetilde{\boldsymbol{T}}}\,[\widetilde{\boldsymbol{\mathcal{L}}}(\mathsf{Z})\,\underline{\mathbf{u}} - \underline{y} - \boldsymbol{\mathcal{F}}\,\underline{y}] \\
&= \widetilde{\boldsymbol{\mathcal{L}}}(\mathsf{Z})(\underline{\mathbf{u}}\,\mathsf{Z} + \underline{\mathbf{v}}) + \begin{bmatrix} \underline{\varsigma} \\ 0 \\ \vdots \\ 0 \end{bmatrix} - \underline{\widetilde{\boldsymbol{T}}}\,(\underline{y} + \boldsymbol{\mathcal{F}}\,\underline{y})
\end{aligned}$$

$$= (\underline{\boldsymbol{x}}\,\mathsf{Z} + \underline{\boldsymbol{y}})\mathsf{Z} + (\boldsymbol{\mathcal{F}}\,\underline{\boldsymbol{y}})\mathsf{Z} + \widetilde{\boldsymbol{\mathcal{L}}}(\mathsf{Z})\,\underline{\mathsf{v}} + \begin{bmatrix} \underline{\zeta} \\ 0 \\ \vdots \\ 0 \end{bmatrix} - \widetilde{\underline{\boldsymbol{T}}}(\underline{\boldsymbol{y}} + \boldsymbol{\mathcal{F}}\,\underline{\boldsymbol{y}}),$$

where $\underline{\zeta} := \sum_{k=1}^{q-1} \left[\widetilde{\mathcal{L}}(\mathsf{Z})\,\widetilde{\underline{K}}\,(\underline{T} - \widetilde{\underline{T}})^k\,\widetilde{\underline{L}} - (\underline{T} - \widetilde{\underline{T}})^k\,\widetilde{\underline{T}}\,\widetilde{\mathcal{L}}(\mathsf{Z})\right]\underline{\mathsf{w}}_k$, and $\underline{\mathsf{w}}_k := \sum_{\ell=1}^{k} \underline{\mathsf{v}}_{\ell+1}\mathsf{Z}^{-(k-\ell+1)}$ for $k = 1, \ldots, q-1$.

To show that $\widetilde{\underline{\boldsymbol{T}}}\,\underline{\boldsymbol{x}} = \underline{\boldsymbol{x}}\,\mathsf{Z} + \underline{\boldsymbol{y}}$ it is enough to prove that

$$\widetilde{\boldsymbol{\mathcal{L}}}(\mathsf{Z})\,\underline{\mathsf{v}} + \begin{bmatrix} \underline{\zeta} \\ 0 \\ \vdots \\ 0 \end{bmatrix} = \widetilde{\underline{\boldsymbol{T}}}(\underline{\boldsymbol{y}} + \boldsymbol{\mathcal{F}}\,\underline{\boldsymbol{y}}) - \boldsymbol{\mathcal{F}}\,\underline{\boldsymbol{y}}\,\mathsf{Z}.$$

Now

$$\widetilde{\boldsymbol{\mathcal{L}}}(\mathsf{Z})\,\underline{\mathsf{v}} = \begin{bmatrix} \sum_{k=0}^{q-1}\left[(\underline{T} - \widetilde{\underline{T}})^k\,\widetilde{\underline{L}}\,\underline{\mathsf{v}}_1\right]\mathsf{Z}^{-k} \\ \vdots \\ \sum_{k=0}^{q-1}\left[(\underline{T} - \widetilde{\underline{T}})^k\,\widetilde{\underline{L}}\,\underline{\mathsf{v}}_q\right]\mathsf{Z}^{-k} \end{bmatrix} = \begin{bmatrix} \sum_{k=0}^{q-1}\left[(\underline{T} - \widetilde{\underline{T}})^k\,\widetilde{\underline{T}}\,\underline{y}_1\right]\mathsf{Z}^{-k} \\ \vdots \\ \sum_{k=0}^{q-1}\left[(\underline{T} - \widetilde{\underline{T}})^k\,\widetilde{\underline{T}}\,\underline{y}_q\right]\mathsf{Z}^{-k} \end{bmatrix}$$

since $\widetilde{\underline{L}}\,\underline{\mathsf{v}}_i = \widetilde{\underline{L}}\,\widetilde{\underline{K}}\,\underline{y}_i = \widetilde{\underline{T}}\,\underline{y}_i$ for $i = 1, \ldots, q-1$. Also,

$$\begin{aligned} \underline{\zeta} &= \sum_{k=1}^{q-1}\sum_{j=0}^{q-1}(\underline{T} - \widetilde{\underline{T}})^j\,\widetilde{\underline{L}}\,\widetilde{\underline{K}}\,(\underline{T} - \widetilde{\underline{T}})^k\,\widetilde{\underline{L}}\,\underline{\mathsf{w}}_k\mathsf{Z}^{-j} \\ &\quad -\sum_{k=1}^{q-1}(\underline{T} - \widetilde{\underline{T}})^k\,\widetilde{\underline{T}}\sum_{j=0}^{q-1}(\underline{T} - \widetilde{\underline{T}})^j\,\widetilde{\underline{L}}\,\underline{\mathsf{w}}_k\mathsf{Z}^{-j} \\ &= \sum_{k=1}^{q-1}\sum_{j=0}^{q-1}\Big[(\underline{T} - \widetilde{\underline{T}})^j\,\widetilde{\underline{T}}\,(\underline{T} - \widetilde{\underline{T}})^k \\ &\quad -(\underline{T} - \widetilde{\underline{T}})^k\,\widetilde{\underline{T}}\,(\underline{T} - \widetilde{\underline{T}})^j\Big]\,\widetilde{\underline{T}}\,\underline{\zeta}_k\mathsf{Z}^{-j} \end{aligned}$$

since $\widetilde{\underline{L}}\,\underline{\mathsf{w}}_k = \sum_{\ell=1}^{k}\widetilde{\underline{L}}\,\underline{\mathsf{v}}_{\ell+1}\mathsf{Z}^{-(k-\ell+1)} = \sum_{\ell=1}^{k}\widetilde{\underline{T}}\,\underline{y}_{\ell+1}\mathsf{Z}^{-(k-\ell+1)} = \widetilde{\underline{T}}\,\underline{\zeta}_k$, where $\underline{\zeta}_k = \sum_{\ell=1}^{k}\underline{y}_{\ell+1}\mathsf{Z}^{-(k-\ell+1)}$ for $k = 1, \ldots, q-1$.

On the other hand,

$$\underline{\widetilde{\boldsymbol{T}}}(\underline{\boldsymbol{y}} + \boldsymbol{\mathcal{F}}\underline{\boldsymbol{y}}) - \boldsymbol{\mathcal{F}}\underline{\boldsymbol{y}}\mathsf{Z} =$$

$$\begin{bmatrix} \sum_{k=0}^{q-1}(\underline{T} - \underline{\widetilde{T}})^k \underline{\widetilde{T}}\,\underline{y}_{k+1} \\ \underline{y}_1 \\ \vdots \\ \underline{y}_{q-1} \end{bmatrix} + \begin{bmatrix} \sum_{k=0}^{q-1}(\underline{T} - \underline{\widetilde{T}})^k \underline{\widetilde{T}}\mathcal{F}_{k+1}\underline{\boldsymbol{y}} \\ \mathcal{F}_1\underline{\boldsymbol{y}} \\ \vdots \\ \mathcal{F}_{q-1}\underline{\boldsymbol{y}} \end{bmatrix} - \begin{bmatrix} \mathcal{F}_1\underline{\boldsymbol{y}}\mathsf{Z} \\ \mathcal{F}_2\underline{\boldsymbol{y}}\mathsf{Z} \\ \vdots \\ \mathcal{F}_q\underline{\boldsymbol{y}}\mathsf{Z} \end{bmatrix}$$

$$= \begin{bmatrix} \sum_{k=0}^{q-1}(\underline{T} - \underline{\widetilde{T}})^k \underline{\widetilde{T}}(\underline{y}_{k+1} - \mathcal{F}_{k+1}\underline{\boldsymbol{y}}) - \mathcal{F}_1\underline{\boldsymbol{y}}\mathsf{Z} \\ \underline{y}_1 + \mathcal{F}_1\underline{\boldsymbol{y}} - \mathcal{F}_2\underline{\boldsymbol{y}}\mathsf{Z} \\ \vdots \\ \underline{y}_{q-1} + \mathcal{F}_{q-1}\underline{\boldsymbol{y}} - \mathcal{F}_q\underline{\boldsymbol{y}}\mathsf{Z} \end{bmatrix}.$$

Let $2 \le i \le q$. The ith component of $\underline{\widetilde{\boldsymbol{T}}}(\underline{\boldsymbol{y}} + \boldsymbol{\mathcal{F}}\underline{\boldsymbol{y}}) - \boldsymbol{\mathcal{F}}\underline{\boldsymbol{y}}\mathsf{Z}$ is

$$\begin{aligned} \underline{y}_{i-1} + \mathcal{F}_{i-1}\underline{\boldsymbol{y}} - \mathcal{F}_i\underline{\boldsymbol{y}}\mathsf{Z} &= \underline{y}_{i-1} + \mathcal{F}_{i-1}\underline{\boldsymbol{y}} - \Big[\underline{y}_{i-1} + \mathcal{F}_{i-1}\underline{\boldsymbol{y}} \\ &\quad - \sum_{k=0}^{q-1}(\underline{T} - \underline{\widetilde{T}})^k \underline{\widetilde{T}}\,\underline{y}_i\mathsf{Z}^{-k}\Big]\mathsf{Z}^{-1}\mathsf{Z} \\ &= \sum_{k=0}^{q-1}(\underline{T} - \underline{\widetilde{T}})^k \underline{\widetilde{T}}\,\underline{y}_i\mathsf{Z}^{-k}, \end{aligned}$$

which equals the ith component of $\widetilde{\boldsymbol{\mathcal{L}}}(\mathsf{Z})\underline{\mathbf{v}} + \begin{bmatrix} \underline{\zeta} \\ 0 \\ \vdots \\ 0 \end{bmatrix}$. As for the first components, we note that

$$\mathcal{F}_1\underline{\boldsymbol{y}} = \sum_{j=1}^{q-1}(\underline{T} - \underline{\widetilde{T}})^j \underline{\widetilde{T}}\,\underline{\zeta}_j$$

and the recursive definition of $\mathcal{F}_{k+1}\underline{\boldsymbol{y}}$ gives

$$\begin{aligned} \mathcal{F}_{k+1}\underline{\boldsymbol{y}} &= \sum_{\ell=1}^{k} \underline{y}_\ell\mathsf{Z}^{-(k-\ell+1)} + \sum_{j=1}^{q-1}(\underline{T} - \underline{\widetilde{T}})^j \underline{\widetilde{T}}\,\underline{\zeta}_j\mathsf{Z}^{-k} \\ &\quad - \sum_{j=1}^{q-1}(\underline{T} - \underline{\widetilde{T}})^j \underline{\widetilde{T}}\,\underline{\zeta}_k\mathsf{Z}^{-j} - \underline{\widetilde{T}}\,\underline{\zeta}_k \end{aligned}$$

for $k = 1, \ldots, q-1$. Employing this formula, it can be (painstakingly) checked that for each $i = 1, \ldots, q$, the first component of $\underline{\widetilde{\boldsymbol{T}}}(\,\underline{y} + \boldsymbol{\mathcal{F}}\,\underline{y}\,) - \boldsymbol{\mathcal{F}}\,\underline{y}\,\mathsf{Z}$ as well as the first component of $\widetilde{\boldsymbol{\mathcal{L}}}(\mathsf{Z})\,\underline{\mathsf{v}} + \begin{bmatrix} \underline{\varsigma} \\ 0 \\ \vdots \\ 0 \end{bmatrix}$ have the same terms involving $\underline{\widetilde{T}}\,\underline{y}_i$ for $i = 1, \ldots, q$. Thus $\underline{\widetilde{\boldsymbol{T}}}\,\underline{\boldsymbol{x}} = \underline{\boldsymbol{x}}\,\mathsf{Z} + \underline{y}$.

Finally, let $\underline{y} := \underline{0}$, so that $\underline{\mathsf{v}} = \underline{\widetilde{\mathbf{K}}}\,\underline{y} = \underline{0}$, and $\underline{\widetilde{\mathbf{K}}}\,\underline{\boldsymbol{x}} = \underline{\mathsf{u}}$, $\underline{\widetilde{\boldsymbol{A}}}\,\underline{\mathsf{u}} = \underline{\mathsf{u}}\,\mathsf{Z}$, $\widetilde{\boldsymbol{\mathcal{L}}}(\mathsf{Z})\,\underline{\mathsf{u}} = \underline{\boldsymbol{x}}\,\mathsf{Z}$.

Let the p vectors in $\underline{\mathsf{u}}$ be linearly independent in $\mathbb{C}^{nq\times 1}$, and $\mathsf{c} \in \mathbb{C}^{p\times 1}$ be such that $\underline{\boldsymbol{x}}\,\mathsf{c} = \mathbf{0}$. Then $\underline{\mathsf{u}}\,\mathsf{c} = (\,\underline{\widetilde{\mathbf{K}}}\,\underline{\boldsymbol{x}}\,)\mathsf{c} = \widetilde{\mathbf{K}}(\,\underline{\boldsymbol{x}}\,\mathsf{c}) = \widetilde{\mathbf{K}}(\mathbf{0}) = \mathbf{0}$. The linear independence of the p vectors in $\underline{\mathsf{u}}$ implies that $\mathsf{c} = 0$, as desired.

Let the p elements in $\underline{\boldsymbol{x}}$ be linearly independent in $\boldsymbol{X}$, and $\mathsf{c} \in \mathbb{C}^{p\times 1}$ be such that $\underline{\mathsf{u}}\,\mathsf{c} = \mathbf{0}$. By mathematical induction, $\underline{\mathsf{u}}\,\mathsf{Z}^j\mathsf{c} = \mathbf{0}$ for $j = 0, 1, \ldots$ since $\underline{\mathsf{u}}\,\mathsf{Z}^{j+1}\mathsf{c} = \underline{\mathsf{u}}\,\mathsf{Z}\mathsf{Z}^j\mathsf{c} = \underline{\widetilde{\boldsymbol{A}}}\,\underline{\mathsf{u}}\,\mathsf{Z}^j\mathsf{c} = \underline{\widetilde{\boldsymbol{A}}}\,(\mathbf{0}) = \mathbf{0}$. It follows that $\underline{\mathsf{u}}\,\mathrm{p}(\mathsf{Z})\mathsf{c} = \mathbf{0}$ for every polynomial p in one variable. As the matrix Z is nonsingular, there is a polynomial p such that $\mathsf{Z}^{-1} = \mathrm{p}(\mathsf{Z})$ by the Cayley-Hamilton Theorem (Remark 1.33). Hence $\underline{\mathsf{u}}\,\mathsf{Z}^{-1}\mathsf{c} = \mathbf{0}$, and in turn $\underline{\mathsf{u}}\,\mathsf{Z}^{-k}\mathsf{c} = \mathbf{0}$ for all $k = 1, 2, \ldots$ Now

$$(\,\underline{\boldsymbol{x}}\,\mathsf{Z})\mathsf{c} = (\widetilde{\boldsymbol{\mathcal{L}}}(\mathsf{Z})\,\underline{\mathsf{u}}\,)\mathsf{c} = \begin{bmatrix} (\widetilde{\mathcal{L}}(\mathsf{Z})\,\underline{\mathsf{u}}_1)\mathsf{c} \\ \vdots \\ (\widetilde{\mathcal{L}}(\mathsf{Z})\,\underline{\mathsf{u}}_q)\mathsf{c} \end{bmatrix} = \begin{bmatrix} 0 \\ \vdots \\ 0 \end{bmatrix}$$

since for $i = 1, \ldots, q$,

$$(\widetilde{\mathcal{L}}(\mathsf{Z})\,\underline{\mathsf{u}}_i)\mathsf{c} = \sum_{k=0}^{q-1}(T - \widetilde{T})^k(\,\underline{\widetilde{L}}\,(\,\underline{\mathsf{u}}_i\mathsf{Z}^{-k}))\mathsf{c} = \sum_{k=0}^{q-1}(T - \widetilde{T})^k\widetilde{L}(0) = 0.$$

The linear independence of the set of p elements in $\underline{\boldsymbol{x}}$ implies that $\mathsf{Z}\mathsf{c} = 0$, and since Z is nonsingular, $\mathsf{c} = 0$, as desired. ∎

Proposition 5.20

$$\mathrm{sp}(\widetilde{\boldsymbol{T}}) \setminus \{0\} = \mathrm{sp}(\widetilde{\boldsymbol{A}}) \setminus \{0\}.$$

In particular, $\mathrm{sp}(\widetilde{\boldsymbol{T}})$ *is a finite set. Let* $\widetilde{\Lambda}^{[q]} \subset \mathrm{sp}(\widetilde{\boldsymbol{T}}) \setminus \{0\}$, $\widetilde{\boldsymbol{P}} := P(\widetilde{\boldsymbol{T}}, \widetilde{\Lambda}^{[q]})$ *and* $\widetilde{\mathsf{P}} := P(\widetilde{\boldsymbol{A}}, \widetilde{\Lambda}^{[q]})$. *Then*

$$\widetilde{\mathbf{K}}\widetilde{\boldsymbol{P}} = \widetilde{\mathsf{P}}\widetilde{\mathbf{K}}.$$

Also, $\widetilde{\mathbf{K}}$ maps $\mathcal{R}(\widetilde{\boldsymbol{P}})$ into $\mathcal{R}(\widetilde{\mathrm{P}})$ in a one-to-one manner, and hence $\widetilde{\Lambda}^{[q]}$ is a spectral set of finite type for $\widetilde{\boldsymbol{T}}$.

Proof

We show that $\mathrm{re}(\widetilde{\boldsymbol{T}}) \setminus \{0\} = \mathrm{re}(\widetilde{\boldsymbol{A}}) \setminus \{0\}$.

Let $0 \neq z \in \mathrm{re}(\widetilde{\boldsymbol{A}})$. Consider $\boldsymbol{y} \in \boldsymbol{X}$ and define $\mathbf{v} := \widetilde{\mathbf{K}}\boldsymbol{y}$, $\mathbf{u} := R(\widetilde{\boldsymbol{A}}, z)\mathbf{v}$ and $\boldsymbol{x} := (\widetilde{\boldsymbol{\mathcal{L}}}([z])\mathbf{u} - \boldsymbol{y})/z$. Then $\widetilde{\boldsymbol{A}}\mathbf{u} = z\mathbf{u} + \mathbf{v}$ and $\widetilde{\boldsymbol{\mathcal{L}}}([z])\mathbf{u} = z\boldsymbol{x} + \boldsymbol{y}$. Letting $p := 1$ in Lemma 5.19, we see that $\widetilde{\boldsymbol{T}}\boldsymbol{x} - z\boldsymbol{x} = \boldsymbol{y}$. Thus $\mathcal{R}(\widetilde{\boldsymbol{T}} - z\mathbf{I}) = \boldsymbol{X}$. Next, let $\boldsymbol{x} \in \boldsymbol{X}$ be such that $(\widetilde{\boldsymbol{T}} - z\mathbf{I})\boldsymbol{x} = \mathbf{0}$ and define $\mathbf{u} := \widetilde{\mathbf{K}}\boldsymbol{x}$. Letting $p := 1$ and $\boldsymbol{y} := \mathbf{0}$ in Lemma 5.19, we see that $\widetilde{\boldsymbol{A}}\mathbf{u} = z\mathbf{u}$ and $\widetilde{\boldsymbol{\mathcal{L}}}([z])\mathbf{u} = z\boldsymbol{x}$. Since $z \in \mathrm{re}(\widetilde{\boldsymbol{A}})$, we have $\mathbf{u} = \mathbf{0}$ and so $z\boldsymbol{x} = \widetilde{\boldsymbol{\mathcal{L}}}([z])(\mathbf{0}) = \mathbf{0}$. As $z \neq 0$, we obtain $\boldsymbol{x} = \mathbf{0}$. Hence $\mathcal{N}(\widetilde{\boldsymbol{T}} - z\mathbf{I}) = \{\mathbf{0}\}$. Thus $z \in \mathrm{re}(\widetilde{\boldsymbol{T}})$.

Conversely, let $0 \neq z \in \mathrm{re}(\widetilde{\boldsymbol{T}})$. To show that $z \in \mathrm{re}(\widetilde{\boldsymbol{A}})$, it is enough to show that $\mathcal{N}(\widetilde{\boldsymbol{A}} - z\mathbf{I}) = \{\mathbf{0}\}$. Let $\mathbf{u} \in \mathbb{C}^{nq\times 1}$ be such that $(\widetilde{\boldsymbol{A}} - z\mathbf{I})\mathbf{u} = \mathbf{0}$ and define $\boldsymbol{x} := \widetilde{\boldsymbol{\mathcal{L}}}([z])\mathbf{u}/z$. Letting $p := 1$ and $\boldsymbol{y} := \mathbf{0}$ in Lemma 5.19, we see that $\widetilde{\boldsymbol{T}}\boldsymbol{x} = z\boldsymbol{x}$ and $\widetilde{\mathbf{K}}\boldsymbol{x} = \mathbf{u}$. Since $z \in \mathrm{re}(\widetilde{\boldsymbol{T}})$, we have $\boldsymbol{x} = \mathbf{0}$ and so $\mathbf{u} = \widetilde{\mathbf{K}}(\mathbf{0}) = \mathbf{0}$. Hence $\mathcal{N}(\widetilde{\boldsymbol{A}} - z\mathbf{I}) = \{\mathbf{0}\}$, as desired.

Since $\mathrm{sp}(\widetilde{\boldsymbol{A}})$ has at most nq elements, $\mathrm{sp}(\widetilde{\boldsymbol{T}})$ has at most $nq+1$ elements. Let $\widetilde{\Lambda}^{[q]} \subset \mathrm{sp}(\widetilde{\boldsymbol{T}}) \setminus \{0\}$. Then $\widetilde{\Lambda}^{[q]}$ is a spectral set for $\widetilde{\boldsymbol{T}}$ as well as for $\widetilde{\boldsymbol{A}}$. Consider $\mathrm{C} \in \mathcal{C}(\widetilde{\boldsymbol{T}}, \widetilde{\Lambda}^{[q]})$ such that $0 \in \mathrm{ext}(\mathrm{C})$. Then $\mathrm{C} \in \mathcal{C}(\widetilde{\boldsymbol{A}}, \widetilde{\Lambda}^{[q]})$ as well.

Let $z \in \mathrm{re}(\widetilde{\boldsymbol{T}}) \setminus \{0\}$. Since $\widetilde{\mathbf{K}}\widetilde{\boldsymbol{T}} = \widetilde{\boldsymbol{A}}\widetilde{\mathbf{K}}$, we have

$$\widetilde{\mathbf{K}}\, R(\widetilde{\boldsymbol{T}}, z) = R(\widetilde{\boldsymbol{A}}, z)\, \widetilde{\mathbf{K}}.$$

Integrating the identity given above over C, we obtain

$$\begin{aligned}
\widetilde{\mathbf{K}}\widetilde{\boldsymbol{P}} &= \widetilde{\mathbf{K}}\Big(-\frac{1}{2\pi \mathrm{i}}\int_{\mathrm{C}} R(\widetilde{\boldsymbol{T}}, z)\, dz\Big) = -\frac{1}{2\pi \mathrm{i}}\int_{\mathrm{C}} \widetilde{\mathbf{K}}\, R(\widetilde{\boldsymbol{T}}, z)\, dz \\
&= -\frac{1}{2\pi \mathrm{i}}\int_{\mathrm{C}} R(\widetilde{\boldsymbol{A}}, z)\, \widetilde{\mathbf{K}}\, dz = \Big(-\frac{1}{2\pi \mathrm{i}}\int_{\mathrm{C}} R(\widetilde{\boldsymbol{A}}, z)\, dz\Big)\widetilde{\mathbf{K}} \\
&= \widetilde{\mathrm{P}}\widetilde{\mathbf{K}}.
\end{aligned}$$

The equation $\widetilde{\mathbf{K}}\widetilde{\boldsymbol{P}} = \widetilde{\mathrm{P}}\widetilde{\mathbf{K}}$ shows that $\widetilde{\mathbf{K}}$ maps $\mathcal{R}(\widetilde{\boldsymbol{P}})$ into $\mathcal{R}(\widetilde{\mathrm{P}})$. Now let $\boldsymbol{x} := \begin{bmatrix} x_1 \\ \vdots \\ x_q \end{bmatrix} \in \widetilde{\boldsymbol{M}} := \mathcal{R}(\widetilde{\boldsymbol{P}})$ such that $\widetilde{\mathbf{K}}\boldsymbol{x} = \mathbf{0}$. We first claim that

$(\widetilde{\boldsymbol{T}})^q\boldsymbol{x} = \mathbf{0}$. Since $\widetilde{T}x_{k+1} = \widetilde{L}\widetilde{K}x_{k+1} = \widetilde{L}(0) = 0$ for each $k = 0, \ldots, q-1$,

$$\widetilde{\boldsymbol{T}}\boldsymbol{x} = \begin{bmatrix} \sum_{k=0}^{q-1}(T-\widetilde{T})^k\widetilde{T}x_{k+1} \\ x_1 \\ \vdots \\ x_{q-1} \end{bmatrix} = \begin{bmatrix} 0 \\ x_1 \\ \vdots \\ x_{q-1} \end{bmatrix}.$$

Again, we have

$$(\widetilde{\boldsymbol{T}})^2\boldsymbol{x} = \begin{bmatrix} \sum_{k=1}^{q-1}(T-\widetilde{T})^k\widetilde{T}x_k \\ 0 \\ x_1 \\ \vdots \\ x_{q-2} \end{bmatrix} = \begin{bmatrix} 0 \\ 0 \\ x_1 \\ \vdots \\ x_{q-2} \end{bmatrix}.$$

Proceeding in this fashion, we obtain $(\widetilde{\boldsymbol{T}})^q\boldsymbol{x} = \mathbf{0}$, as claimed.

Now $\left(\widetilde{\boldsymbol{T}}_{|\widetilde{\boldsymbol{M}},\widetilde{\boldsymbol{M}}}\right)^q \boldsymbol{x} = \mathbf{0}$, since $\boldsymbol{x} \in \widetilde{\boldsymbol{M}}$ and $\widetilde{\boldsymbol{T}}$ maps $\widetilde{\boldsymbol{M}}$ into itself. But since $\mathrm{sp}(\widetilde{\boldsymbol{T}}_{|\widetilde{\boldsymbol{M}},\widetilde{\boldsymbol{M}}}) = \widetilde{\Lambda}^{[q]}$ and $0 \notin \widetilde{\Lambda}^{[q]}$, the operator $\widetilde{\boldsymbol{T}}_{|\widetilde{\boldsymbol{M}},\widetilde{\boldsymbol{M}}}$ is invertible. Hence $\left(\widetilde{\boldsymbol{T}}_{|\widetilde{\boldsymbol{M}},\widetilde{\boldsymbol{M}}}\right)^q$ is invertible. As $\boldsymbol{x} \in \widetilde{\boldsymbol{M}}$ and $(\widetilde{\boldsymbol{T}})^q\boldsymbol{x} = \mathbf{0}$, we obtain $\boldsymbol{x} = \mathbf{0}$. Thus the operator $\widetilde{\mathbf{K}}_{|\widetilde{\boldsymbol{M}}}$ is one-to-one, and $\operatorname{rank}\widetilde{\boldsymbol{P}} \leq \operatorname{rank}\widetilde{\mathrm{P}} \leq nq$. Thus $\widetilde{\Lambda}^{[q]}$ is a spectral set of finite type for $\widetilde{\boldsymbol{T}}$. ∎

Theorem 5.21

(a) *Let $\widetilde{\lambda}^{[q]} \in \mathbb{C} \setminus \{0\}$. Then $\widetilde{\lambda}^{[q]}$ is an eigenvalue of the operator $\widetilde{\boldsymbol{T}}$ if and only if $\widetilde{\lambda}^{[q]}$ is an eigenvalue of the matrix $\widetilde{\mathsf{A}}$.*

In this case, let $\underline{\widetilde{\mathbf{u}}} \in \mathbb{C}^{nq \times g}$ form an ordered basis for the eigenspace of $\widetilde{\mathsf{A}}$ corresponding to $\widetilde{\lambda}^{[q]}$. Then $\underline{\widetilde{\boldsymbol{\varphi}}} := \widetilde{\boldsymbol{\mathcal{L}}}([\widetilde{\lambda}^{[q]}])\, \underline{\widetilde{\mathbf{u}}}\,/\widetilde{\lambda}^{[q]}$ forms an ordered basis for the eigenspace of $\widetilde{\boldsymbol{T}}$ corresponding to $\widetilde{\lambda}^{[q]}$ and satisfies $\underline{\widetilde{\mathbf{K}}}\, \underline{\widetilde{\boldsymbol{\varphi}}} = \underline{\widetilde{\mathbf{u}}}$.

In particular, the geometric multiplicity of $\widetilde{\lambda}^{[q]}$ as an eigenvalue of $\widetilde{\boldsymbol{T}}$ is the same as its geometric multiplicity of $\widetilde{\lambda}^{[q]}$ as an eigenvalue of $\widetilde{\mathsf{A}}$.

(b) *Let $\widetilde{\Lambda}^{[q]} \subset \mathrm{sp}(\widetilde{\mathsf{A}}) \setminus \{0\}$ and consider an ordered basis $\underline{\widetilde{\mathbf{u}}} \in \mathbb{C}^{nq\times m}$ for the spectral subspace $M(\widetilde{\mathsf{A}}, \widetilde{\Lambda}^{[q]})$. Then there is a nonsingular matrix $\widetilde{\Theta}^{[q]} \in \mathbb{C}^{m\times m}$ such that $\widetilde{\mathsf{A}}\,\underline{\widetilde{\mathbf{u}}} = \underline{\widetilde{\mathbf{u}}}\,\widetilde{\Theta}^{[q]}$. Further, $\underline{\widetilde{\boldsymbol{\varphi}}} := (\widetilde{\boldsymbol{\mathcal{L}}}(\widetilde{\Theta}^{[q]})\,\underline{\widetilde{\mathbf{u}}})(\widetilde{\Theta}^{[q]})^{-1}$ forms an ordered basis for the spectral subspace $M(\widetilde{\boldsymbol{T}}, \widetilde{\Lambda}^{[q]})$ and satisfies $\underline{\widetilde{\mathbf{K}}}\,\underline{\widetilde{\boldsymbol{\varphi}}} = \underline{\widetilde{\mathbf{u}}}$.*

In particular, if $\widetilde{\lambda}^{[q]} \in \mathrm{sp}(\widetilde{\mathsf{A}}) \setminus \{0\}$, then the algebraic multiplicity of $\widetilde{\lambda}^{[q]}$ as an eigenvalue of $\widetilde{\boldsymbol{T}}$ is the same as the algebraic multiplicity of $\widetilde{\lambda}^{[q]}$ as an eigenvalue of $\widetilde{\mathsf{A}}$.

Proof

Recall that $\widetilde{\mathsf{A}}$ represents the operator $\widetilde{\boldsymbol{A}}$ with respect to the standard basis for $\mathbb{C}^{nq\times 1}$.

(a) Letting $p := 1$, $\mathsf{Z} := [\widetilde{\lambda}^{[q]}]$ and $\boldsymbol{y} := \mathbf{0}$ in Lemma 5.19, we see that $\widetilde{\lambda}^{[q]}$ is an eigenvalue of $\widetilde{\boldsymbol{T}}$ if and only if $\widetilde{\lambda}^{[q]}$ is an eigenvalue of $\widetilde{\boldsymbol{A}}$. In this case, letting $p := g$ and $\mathsf{Z} := \widetilde{\lambda}^{[q]}\mathsf{I}_g$ in Lemma 5.19, we note that if $\underline{\widetilde{\boldsymbol{\varphi}}} := \widetilde{\boldsymbol{\mathcal{L}}}(\mathsf{Z})\,\underline{\widetilde{\mathbf{u}}}/\widetilde{\lambda}^{[q]}$, then $\underline{\widetilde{\boldsymbol{T}}}\,\underline{\widetilde{\boldsymbol{\varphi}}} = \widetilde{\lambda}^{[q]}\,\underline{\widetilde{\boldsymbol{\varphi}}}$, $\underline{\widetilde{\mathbf{K}}}\,\underline{\widetilde{\boldsymbol{\varphi}}} = \underline{\widetilde{\mathbf{u}}}$, and the set $\{\widetilde{\boldsymbol{\varphi}}_1, \ldots, \widetilde{\boldsymbol{\varphi}}_g\}$ of elements in $\underline{\widetilde{\boldsymbol{\varphi}}}$ is a linearly independent subset of the eigenspace of $\widetilde{\boldsymbol{T}}$ corresponding to $\widetilde{\lambda}^{[q]}$. That $\{\widetilde{\boldsymbol{\varphi}}_1, \ldots, \widetilde{\boldsymbol{\varphi}}_g\}$ also spans this eigenspace can be seen as follows. Let $\widetilde{\boldsymbol{\varphi}}_{g+1} \notin \mathrm{span}\{\widetilde{\boldsymbol{\varphi}}_1, \ldots, \widetilde{\boldsymbol{\varphi}}_g\}$ be such that $\widetilde{\boldsymbol{T}}\widetilde{\boldsymbol{\varphi}}_{g+1} = \widetilde{\lambda}^{[q]}\widetilde{\varphi}_{g+1}$. If $\underline{\widetilde{\boldsymbol{\psi}}} := [\widetilde{\boldsymbol{\varphi}}_1, \ldots, \widetilde{\boldsymbol{\varphi}}_{g+1}] \in \boldsymbol{X}^{q\times(g+1)}$, then $\underline{\widetilde{\boldsymbol{T}}}\,\underline{\widetilde{\boldsymbol{\psi}}} = \widetilde{\lambda}^{[q]}\,\underline{\widetilde{\boldsymbol{\psi}}}$, and again by Lemma 5.19, $\{\widetilde{\mathbf{K}}\widetilde{\boldsymbol{\varphi}}_1, \ldots, \widetilde{\mathbf{K}}\widetilde{\boldsymbol{\varphi}}_{g+1}\}$ would be a linearly independent subset of the eigenspace of $\widetilde{\boldsymbol{A}}$ corresponding to $\widetilde{\lambda}^{[q]}$, contrary to our assumption that this eigenspace is of dimension g.

(b) Since $\underline{\widetilde{\mathbf{u}}} \in \mathbb{C}^{nq\times m}$ forms an ordered basis for $M(\widetilde{\boldsymbol{A}}, \widetilde{\Lambda}^{[q]})$, $\underline{\widetilde{\boldsymbol{A}}}\,\underline{\widetilde{\mathbf{u}}} = \underline{\widetilde{\mathbf{u}}}\,\widetilde{\Theta}^{[q]}$ for some $\widetilde{\Theta}^{[q]} \in \mathbb{C}^{m\times m}$ such that $\mathrm{sp}(\widetilde{\Theta}^{[q]}) = \widetilde{\Lambda}^{[q]}$. As $0 \notin \widetilde{\Lambda}^{[q]}$, $\widetilde{\Theta}^{[q]}$ is invertible. Letting $p := m$, $\mathsf{Z} := \widetilde{\Theta}^{[q]}$ and $\underline{\boldsymbol{y}} := \underline{\mathbf{0}}$ in Lemma 5.19, we see that $\underline{\widetilde{\boldsymbol{T}}}\,\underline{\widetilde{\boldsymbol{\varphi}}} = \underline{\widetilde{\boldsymbol{\varphi}}}\,\widetilde{\Theta}^{[q]}$, $\underline{\widetilde{\mathbf{K}}}\,\underline{\widetilde{\boldsymbol{\varphi}}} = \underline{\widetilde{\mathbf{u}}}$, and the set of m elements in $\underline{\widetilde{\boldsymbol{\varphi}}}$ is linearly independent in $\boldsymbol{X}$. Consider the closed subspace $\widetilde{\boldsymbol{Y}}$ of $\boldsymbol{X}$ spanned by the elements in $\underline{\widetilde{\boldsymbol{\varphi}}}$. Since the matrix $\widetilde{\Theta}^{[q]}$ represents the operator $\widetilde{\boldsymbol{T}}_{|\widetilde{\boldsymbol{Y}},\widetilde{\boldsymbol{Y}}}$ with respect to $\underline{\widetilde{\boldsymbol{\varphi}}}$, we have $\mathrm{sp}(\widetilde{\boldsymbol{T}}_{|\widetilde{\boldsymbol{Y}},\widetilde{\boldsymbol{Y}}}) = \mathrm{sp}(\widetilde{\Theta}^{[q]}) = \widetilde{\Lambda}^{[q]}$. Hence $\widetilde{\boldsymbol{Y}} \subset M(\widetilde{\boldsymbol{T}}, \widetilde{\Lambda}^{[q]})$ by Proposition 1.28, so that $m \le \dim M(\widetilde{\boldsymbol{T}}, \widetilde{\Lambda}^{[q]})$. But as we have seen in Proposition 5.20, $\widetilde{\mathbf{K}}$ maps $M(\widetilde{\boldsymbol{T}}, \widetilde{\Lambda}^{[q]})$ into $M(\widetilde{\boldsymbol{A}}, \widetilde{\Lambda}^{[q]})$ in a one-to-one manner, so that $\dim M(\widetilde{\boldsymbol{T}}, \widetilde{\Lambda}^{[q]}) \le \dim M(\widetilde{\boldsymbol{A}}, \widetilde{\Lambda}^{[q]}) \le m$. Thus $\widetilde{\boldsymbol{Y}} = M(\widetilde{\boldsymbol{T}}, \widetilde{\Lambda}^{[q]})$.

In particular, the case $\widetilde{\Lambda}^{[q]} := \{\widetilde{\lambda}^{[q]}\}$ shows that if $\widetilde{\lambda}^{[q]} \in \mathrm{sp}(\widetilde{\boldsymbol{A}}) \setminus \{0\}$, then the algebraic multiplicity of $\widetilde{\lambda}^{[q]}$ as an eigenvalue of $\widetilde{\boldsymbol{T}}$ is the same as the algebraic multiplicity of $\widetilde{\lambda}^{[q]}$ as an eigenvalue of $\widetilde{\boldsymbol{A}}$. ∎

The result in part (b) of Theorem 5.21 is stated without proof in [16].

Let $T \in \mathrm{BL}(X)$ and Λ be a spectral set of finite type for T such that $0 \notin \Lambda$ and $m := \dim M(T, \Lambda)$. Consider a sequence (T_n) of finite rank operators on X such that $T_n \xrightarrow{\nu} T$. As we have seen in Theorem 3.12, for each large n, there is a spectral set $\Lambda_n^{[q]}$ of finite type for $\boldsymbol{T}_n^{[q]}$ which does not contain 0 and approximates Λ. Fix n and write $\widetilde{T} := T_n$, $\widetilde{\boldsymbol{T}} := \boldsymbol{T}_n^{[q]}$ and $\widetilde{\Lambda}^{[q]} := \Lambda_n^{[q]}$. Theorem 5.21 says that $\widetilde{\Lambda}^{[q]}$ can be located as a spectral set for the matrix $\widetilde{\mathsf{A}}$. Further, Theorem 3.15 shows that if $\widetilde{\boldsymbol{\varphi}}$ forms an ordered basis for the associated spectral subspace $M(\widetilde{\boldsymbol{T}}, \widetilde{\Lambda}^{[q]})$, and $\underline{\widetilde{\boldsymbol{T}}}\, \underline{\widetilde{\boldsymbol{\varphi}}} = \underline{\widetilde{\boldsymbol{\varphi}}}\, \widetilde{\Theta}^{[q]}$, where $\widetilde{\Theta}^{[q]}$ is a nonsingular $m \times m$ matrix, then

$$\underline{\widetilde{\boldsymbol{\varphi}}} = \begin{bmatrix} \underline{\widetilde{\varphi}^{[q]}} \\ \underline{\widetilde{\varphi}^{[q]}}\, (\widetilde{\Theta}^{[q]})^{-1} \\ \vdots \\ \underline{\widetilde{\varphi}^{[q]}}\, (\widetilde{\Theta}^{[q]})^{-q+1} \end{bmatrix} \in \boldsymbol{X},$$

and its first component $\underline{\widetilde{\varphi}^{[q]}}$ provides an approximation of a basis for the spectral subspace $M(T, \Lambda)$. Theorem 5.21 says that $\underline{\widetilde{\varphi}^{[q]}}$ can be found in the following manner. Find a basis $\underline{\widetilde{\mathsf{u}}}$ for the spectral subspace $M(\widetilde{\mathsf{A}}, \widetilde{\Lambda}^{[q]})$. Then $\widetilde{\mathsf{A}}\, \underline{\widetilde{\mathsf{u}}} = \underline{\widetilde{\mathsf{u}}}\, \widetilde{\Theta}^{[q]}$, where the nonsingular matrix $\widetilde{\Theta}^{[q]}$ is given by

$$\widetilde{\Theta}^{[q]} = (\underline{\widetilde{\mathsf{u}}}^* \underline{\widetilde{\mathsf{u}}})^{-1} \underline{\widetilde{\mathsf{u}}}^* \widetilde{\mathsf{A}}\, \underline{\widetilde{\mathsf{u}}}.$$

Let

$$\underline{\widetilde{\mathsf{u}}} = [\widetilde{\mathsf{u}}_1, \ldots, \widetilde{\mathsf{u}}_m] = \begin{bmatrix} \underline{\widetilde{\mathsf{u}}}_1 \\ \vdots \\ \underline{\widetilde{\mathsf{u}}}_q \end{bmatrix} \in \mathbb{C}^{nq \times m}, \quad \widetilde{\boldsymbol{\mathcal{L}}} := \widetilde{\boldsymbol{\mathcal{L}}}(\widetilde{\Theta}^{[q]}),$$

and

$$\underline{\widetilde{\boldsymbol{\varphi}}} := (\widetilde{\boldsymbol{\mathcal{L}}}\, \underline{\widetilde{\mathsf{u}}})(\widetilde{\Theta}^{[q]})^{-1} = \begin{bmatrix} \widetilde{\mathcal{L}}\, \underline{\widetilde{\mathsf{u}}}_1 \\ \vdots \\ \widetilde{\mathcal{L}}\, \underline{\widetilde{\mathsf{u}}}_q \end{bmatrix} (\widetilde{\Theta}^{[q]})^{-1}.$$

Hence the accelerated approximation of a basis for $M(T, \Lambda)$ is given by

$$\begin{aligned} \underline{\widetilde{\varphi}^{[q]}} &:= (\widetilde{\mathcal{L}}\, \underline{\mathsf{u}}_1)(\widetilde{\Theta}^{[q]})^{-1} \\ &= (\underline{\widetilde{L}}\, \underline{\mathsf{u}}_1)(\widetilde{\Theta}^{[q]})^{-1} + \sum_{k=1}^{q-1} (\underline{T} - \underline{\widetilde{T}})^k\, \underline{\widetilde{L}}\, \underline{\mathsf{u}}_1 (\widetilde{\Theta}^{[q]})^{-k-1} \end{aligned}$$

$$= \underline{\widetilde{x}}\, \underline{\mathsf{u}}_1 (\widetilde{\Theta}^{[q]})^{-1} + \sum_{k=1}^{q-1} (\underline{T} - \underline{\widetilde{T}})^k (\underline{\widetilde{x}}\, \underline{\mathsf{u}}_1)(\widetilde{\Theta}^{[q]})^{-k-1}.$$

Moreover, since $\underline{\widetilde{\mathbf{K}}}\, \underline{\widetilde{\boldsymbol{\varphi}}} = \underline{\widetilde{\mathbf{u}}}$, the Gram product $\big(\underline{\widetilde{\varphi}^{[q]}}\, , \underline{\widetilde{f}}\big)$ is equal to $\underline{\widetilde{K}}\, \underline{\widetilde{\varphi}^{[q]}} = \underline{\mathsf{u}}_1$.

In the special case $\widetilde{\Lambda}^{[q]} = \{\widetilde{\lambda}^{[q]}\}$, where $\widetilde{\lambda}^{[q]}$ is a nonzero eigenvalue of $\widetilde{\boldsymbol{T}}$ of geometric multiplicity g and we are interested only in finding the first components $\underline{\widetilde{\varphi}^{[q]}}$ of a basis for the corresponding eigenspace of $\widetilde{\boldsymbol{T}}$, we may find a basis $\underline{\widetilde{\mathbf{u}}} = \begin{bmatrix} \underline{\widetilde{\mathsf{u}}}_1 \\ \vdots \\ \underline{\widetilde{\mathsf{u}}}_q \end{bmatrix} \in \mathbb{C}^{nq \times g}$ of the eigenspace of $\widetilde{\mathsf{A}}$ corresponding to $\widetilde{\lambda}^{[q]}$, so that $\widetilde{\Theta}^{[q]} = \widetilde{\lambda}^{[q]} \mathsf{I}_g$; and then let

$$\underline{\widetilde{\varphi}^{[q]}} := \frac{\underline{\widetilde{\mathcal{L}}}\, \underline{\mathsf{u}}_1}{\widetilde{\lambda}^{[q]}} = \frac{\underline{\widetilde{L}}\, \underline{\mathsf{u}}_1}{\widetilde{\lambda}^{[q]}} + \sum_{k=1}^{q-1} \frac{(\underline{T} - \underline{\widetilde{T}})^k\, \underline{\widetilde{L}}\, \underline{\mathsf{u}}_1}{(\widetilde{\lambda}^{[q]})^k}$$

$$= \frac{\underline{\widetilde{x}}\, \underline{\mathsf{u}}_1}{\widetilde{\lambda}^{[q]}} + \sum_{k=1}^{q-1} \frac{(\underline{T} - \underline{\widetilde{T}})^k (\underline{\widetilde{x}}\, \underline{\mathsf{u}}_1)}{(\widetilde{\lambda}^{[q]})^k}.$$

Let $T \in \mathrm{BL}(X)$ and (T_n) be a sequence of bounded finite rank operators considered in Chapter 4. In Table 5.4, we give the entries

$$\widetilde{\mathsf{A}}^{\langle k \rangle}(i,j) := \langle (T - T_n)^k x_{n,j}\, ,\, f_{n,i} \rangle, \quad 1 \leq i, j \leq n,$$

of the $n \times n$ matrices $\widetilde{\mathsf{A}}^{\langle k \rangle}$, $k = 0, \ldots, q-1$, which are involved in the $nq \times nq$ matrix $\widetilde{\mathsf{A}}$. Note that $\widetilde{\mathsf{A}}^{\langle 0 \rangle} = \mathsf{A}_n := \big(\underline{x_n}\, , \underline{f_n}\big)$. We use the same notation as in Table 5.1. Further, for a kernel $k(\cdot,\cdot)$ in $C^0([a,b] \times [a,b])$ and a fixed $\widetilde{t}_j \in [a,b]$, we let

$$\widetilde{h}_j(s) := k(s, \widetilde{t}_j) \quad \text{for } s \in [a,b].$$

It is possible to obtain uniformly well-conditioned bases for the approximate spectral subspaces $M(\boldsymbol{T}_n, \Lambda_n^{[q]})$ by choosing uniformly well-conditioned bases for the approximate spectral subspaces $M(\boldsymbol{A}_n, \Lambda_n^{[q]})$. See Theorem 5.9, Exercise 5.15 and Theorem 4.1 of [16]. Such uniformly well-conditioned bases are needed for deriving the error estimate $\| \underline{\varphi_n^{[q]}} - \underline{P}\, \underline{\varphi_n^{[q]}} \|_\infty = O(\|(T - T_n)^q T_n\|)$ stated in Theorem 3.15(a).

	$\widetilde{\mathsf{A}}^{\langle k\rangle}(i,j)$
T_n^P	$\langle T(T-T_n^P)^k\widetilde{e}_j\,,\,\widetilde{e}_i^*\rangle$
T_n^S	$\langle (T-T_n^S)^k T\widetilde{e}_j\,,\,\widetilde{e}_i^*\rangle$
T_n^G	$\sum_{\ell=1}^{\tilde{r}}\langle (T-T_n^G)^k\widetilde{e}_j\,,\,\widetilde{e}_\ell^*\rangle\langle T\widetilde{e}_\ell\,,\,\widetilde{e}_i^*\rangle$ $\sum_{\ell=1}^{\tilde{r}}\langle T\widetilde{e}_j\,,\,\widetilde{e}_\ell^*\rangle\langle (T-T_n^G)^k\widetilde{e}_j\,,\,\widetilde{e}_i^*\rangle$
T_n^D	$\int_a^b [(T-T_n^D)^k\widetilde{x}_j](t)\widetilde{y}_i(t)\,dt$
T_n^N	$\widetilde{w}_j[(T-T_n^N)^k\widetilde{h}_j](\widetilde{t}_i)$
T_n^F	$\sum_{\ell=1}^{\tilde{r}}\widetilde{w}_\ell k(\widetilde{t}_i,\widetilde{t}_j)[(T-T_n^F)^k\widetilde{e}_j](\widetilde{t}_\ell)$ $\widetilde{w}_j\sum_{\ell=1}^{\tilde{r}}k(\widetilde{t}_\ell,\widetilde{t}_j)[(T-T_n^F)^k\widetilde{e}_\ell](\widetilde{t}_i)$

Table 5.4

5.4 Numerical Examples

The matrix formulations discussed in this chapter can be implemented on a computer by making use of standard routines. In this section we present some illustrative numerical examples. We have made use of MATLAB for this purpose.

The underlying complex Banach space is $X := C^0([0,1])$ with the norm $\|\cdot\|_\infty$. For $x \in X$, let

$$(Tx)(s) = \int_0^1 k(s,t)x(t)\,dt, \quad s \in [0,1],$$

where the kernel $k(\cdot,\cdot)$ is defined on $[0,1]\times[0,1]$. Except in Example 5.26, we replace Tx by $T_N x$ in our computations, where T_N is a member of a specified approximation (T_n) of T, and N is large. Also, we replace $\|x\|_\infty$ by the ∞-norm of the vector $[x(t_{N,1}),\ldots,x(t_{N,N})]^\top$ in $\mathbb{C}^{N\times 1}$, where $t_{N,1},\ldots,t_{N,N}$ are the specified nodes in $[0,1]$.

By $\lambda(j)$ we denote the jth eigenvalue of T, and by $\lambda_n(j)$ the jth eigenvalue of T_n in the descending order of modulus.

Example 5.22

Kernel	$k(s,t) := e^{st}, \quad s,t \in [0,1]$
Approximation	Nyström (Subsection 4.2.2)
Quadrature formula	Gauss Two Point Rule
Numerical parameters	$n = 10\,i,\ i \in [\![1,12]\!]$
Eigenvalues	$\lambda_n(j),\ j \in [\![1,4]\!]$

The kernel is infinitely differentiable with respect to each of the two variables s and t.

n	$\lambda_n(1)$	$\lambda_n(2)$	$\lambda_n(3)$	$\lambda_n(4)$
10	$(.135302849429)10^1$	$(.1059756286)10^0$	$(.35524057)10^{-2}$	$(.744126)10^{-4}$
20	$(.135303006009)10^1$	$(.1059827474)10^0$	$(.35602204)10^{-2}$	$(.762500)10^{-4}$
30	$(.135303014406)10^1$	$(.1059831298)10^0$	$(.35606444)10^{-2}$	$(.763538)10^{-4}$
40	$(.135303015820)10^1$	$(.1059831942)10^0$	$(.35607159)10^{-2}$	$(.763715)10^{-4}$
50	$(.135303016206)10^1$	$(.1059832118)10^0$	$(.35607355)10^{-2}$	$(.763763)10^{-4}$
60	$(.135303016345)10^1$	$(.1059832181)10^0$	$(.35607425)10^{-2}$	$(.763780)10^{-4}$
70	$(.135303016404)10^1$	$(.1059832209)10^0$	$(.35607455)10^{-2}$	$(.763788)10^{-4}$
80	$(.135303016433)10^1$	$(.1059832222)10^0$	$(.35607470)10^{-2}$	$(.763791)10^{-4}$
90	$(.135303016449)10^1$	$(.1059832229)10^0$	$(.35607478)10^{-2}$	$(.763793)10^{-4}$
100	$(.135303016457)10^1$	$(.1059832233)10^0$	$(.35607482)10^{-2}$	$(.763795)10^{-4}$
110	$(.135303016463)10^1$	$(.1059832235)10^0$	$(.35607485)10^{-2}$	$(.763795)10^{-4}$
120	$(.135303016466)10^1$	$(.1059832237)10^0$	$(.35607487)10^{-2}$	$(.763796)10^{-4}$

A posteriori error bounds and the computations of $\lambda_{100}(1)$, $\lambda_{100}(2)$, $\lambda_{100}(3)$ are used in Example 4.22 to conclude that $\lambda(1) = 1.35303016\ldots$ $\lambda(2) = 0.1059832\ldots$ and $\lambda(3) = 0.00356\ldots$ The computation of $\lambda_{120}(4)$ is referred to in Exercise 4.17. ∎

Example 5.23

Kernel	$k(s,t) := \ln\|s-t\|, \quad 0 \leq s \neq t \leq 1$
Approximation	Kantorovich-Krylov (Subsection 4.2.3)
Quadrature formula	Compound Trapezoidal Rule
Numerical parameters	Uniform grids with n nodes for some $n \in [\![11, N]\!]$, where $N = 501$
Eigenvalue	$\lambda_n(3)$, $\lambda_N(3)$

The kernel is weakly singular. As in 5.1.1, we obtain an eigenvector φ_n of T_n^K by using an eigenvector of the matrix A_n^K corresponding to its eigenvalue $\lambda_n(3)$. Define $\mathsf{v}_n \in \mathbb{C}^{N\times 1}$ by $\mathsf{v}_n(j) := \varphi_n\left(\dfrac{j-1}{N-1}\right)$ for j in $[\![1, N]\!]$, and $\mu_n := \mathsf{v}_n^* \mathsf{A}_N^K \mathsf{v}_n / \mathsf{v}_n^* \mathsf{v}_n$.

The relative residual in the ∞-norm is shown in the following table.

n	$\lambda_n(3)$	μ_n	$\dfrac{\lVert T_N^K \varphi_n - \mu_n \varphi_n \rVert_\infty}{\lVert \varphi_n \rVert_\infty}$
11	$-.428340$	$-.43018041$	$(.28)10^{-01}$
51	$-.429776$	$-.42996942$	$(.17)10^{-02}$
101	$-.429908$	$-.42996587$	$(.50)10^{-03}$
301	$-.429959$	$-.42996464$	$(.47)10^{-04}$
451	$-.429963$	$-.42996453$	$(.67)10^{-05}$
496	$-.429964$	$-.42996452$	$(.70)10^{-06}$

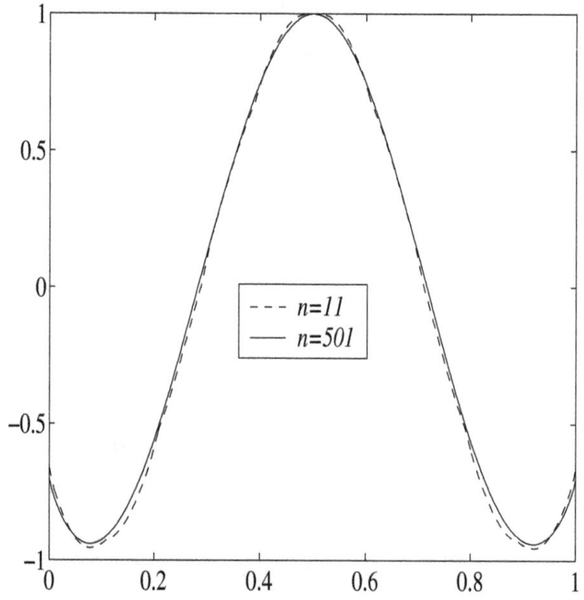

The figure above shows an eigenvector φ_{11} of T_{11}^K and an eigenvector φ_{501} of T_{501}^K normalized such that $\lVert \varphi_{11} \rVert_\infty = \lVert \varphi_{501} \rVert_\infty = 1$. ∎

Example 5.24

Kernel	$k(s,t) := \begin{cases} 3s/5 & \text{if } 0 \le s \le t \le 1 \\ t - 2s/5 & \text{if } 0 \le t \le s \le 1 \end{cases}$
Approximation	Fredholm (Subsection 4.2.2)
Quadrature formula	Compound Trapezoidal Rule
Numerical parameters	Uniform grids: $n = 11$ and $N = 1001$ nodes
Eigenvalues	$\lambda_n(5),\ \lambda_N(5)$

The kernel is the Green Function for the Sturm-Liouville Problem: $-\varphi'' = \mu\,\varphi$ on $[0,1]$, $\varphi(0) = 0$, $2\varphi'(0) + 3\varphi'(1) = 0$. It is a continuous function on $[0,1]\times[0,1]$, but it is not smooth.

All the eigenvalues of the corresponding Fredholm integral operator T are positive. Each one of them is a simple eigenvalue. In particular, the simple eigenvalue $\lambda(5)$ of T is approximated by a simple eigenvalue $\lambda_n(5)$ of T_n.

The results of iterative refinement with the Elementary Iteration (E) and with the Double Iteration (D) are shown in the following table. Computations are stopped when the relative residual in the ∞-norm is less than $(.5)10^{-9}$.

In this table k denotes the iterate number. We observe that the desired accuracy is achieved by the fourth iterate of the Double Iteration, while only the ninth iterate of the Elementary Iteration achieves this accuracy.

Since $T_n^F \xrightarrow{\nu} T$, the iterates of the Elementary Iteration are expected to converge in a semigeometric manner, while the iterates of the Double Iteration are expected to converge in a geometric manner.

k	(E) $\dfrac{\Vert T_N\varphi_n^{(k)} - \lambda_n^{(k)}\varphi_n^{(k)}\Vert_\infty}{\Vert\varphi_n^{(k)}\Vert_\infty}$	(D) $\dfrac{\Vert T_N\psi_n^{(k)} - \mu_n^{(k)}\psi_n^{(k)}\Vert_\infty}{\Vert\psi_n^{(k)}\Vert_\infty}$
0	$(.82)10^{-3}$	$(.82)10^{-3}$
1	$(.24)10^{-3}$	$(.17)10^{-4}$
2	$(.45)10^{-4}$	$(.27)10^{-6}$
3	$(.74)10^{-5}$	$(.44)10^{-8}$
4	$(.12)10^{-5}$	$(.75)10^{-10}$
5	$(.21)10^{-6}$	
6	$(.35)10^{-7}$	
7	$(.61)10^{-8}$	
8	$(.11)10^{-8}$	
9	$(.19)10^{-9}$	

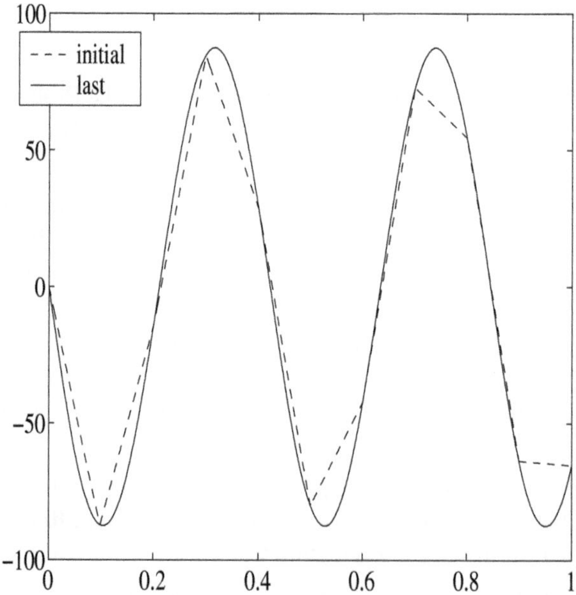

The figure above shows the initial approximation of an eigenvector corresponding to $\lambda_N(5)$ and the last iterate in the Elementary Iteration.

The initial approximation is a piecewise linear function on [0, 1] having ten linear pieces. ∎

Example 5.25

Kernel	$k(s,t) := \begin{cases} s/2 & \text{if } 0 \le s \le t \le 1 \\ t - s/2 & \text{if } 0 \le t \le s \le 1 \end{cases}$
Approximation	Fredholm (Subsection 4.2.2)
Quadrature formula	Compound Trapezoidal Rule
Numerical parameters	Uniform grids: $n = 21$ and $N = 2001$ nodes
Eigenvalues	Clusters $\{\lambda_n(5), \lambda_n(6)\}$, $\{\lambda_N(5), \lambda_N(6)\}$

The kernel is the Green Function for the Sturm-Liouville Problem: $-\varphi'' = \mu\,\varphi$ on $[0,1]$, $\varphi(0) = 0$, $\varphi'(0) + \varphi'(1) = 0$. It is a continuous function on $[0,1]\times[0,1]$, but it is not smooth.

The eigenvalues of the corresponding Fredholm integral operator T are $1/(2j-1)^2\pi^2$, $j = 1, 2, \ldots$ Each one of them is a double defective eigenvalue. In particular, the double eigenvalue $\lambda(3)$ of T is approximated by a cluster $\{\lambda_n(5), \lambda_n(6)\}$ of two simple eigenvalues of T_n.

The results of iterative refinement with the Elementary Iteration ($\underline{E}$) and with the Double Iteration ($\underline{D}$) are shown in the following table. Computations are stopped when the relative residual in the ∞-norm is less than $(.5)10^{-9}$.

In this table k denotes the iterate number. We observe that the desired accuracy is achieved by the third iterate of the Double Iteration, while only the seventh iterate of the Elementary Iteration achieves this accuracy.

Since $T_n^F \xrightarrow{\nu} T$, the iterates of the Elementary Iteration are expected to converge in a semigeometric manner, while the iterates of the Double Iteration are expected to converge in a geometric manner.

k	$(\underline{E})$ $\dfrac{\| \underline{T_N}\ \underline{\varphi_n}^{(k)} - \underline{\varphi_n}^{(k)}\Theta_n^{(k)} \|_\infty}{\| \underline{\varphi_n}^{(k)} \|_\infty}$	$(\underline{D})$ $\dfrac{\| \underline{T_N}\ \underline{\psi_n}^{(k)} - \underline{\psi_n}^{(k)}\Upsilon_n^{(k)} \|_\infty}{\| \underline{\psi_n}^{(k)} \|_\infty}$
0	$(.35)10^{-5}$	$(.35)10^{-5}$
1	$(.29)10^{-6}$	$(.82)10^{-7}$
2	$(.78)10^{-7}$	$(.19)10^{-8}$
3	$(.31)10^{-7}$	$(.29)10^{-9}$
4	$(.49)10^{-8}$	
5	$(.29)10^{-8}$	
6	$(.60)10^{-9}$	
7	$(.24)10^{-9}$	

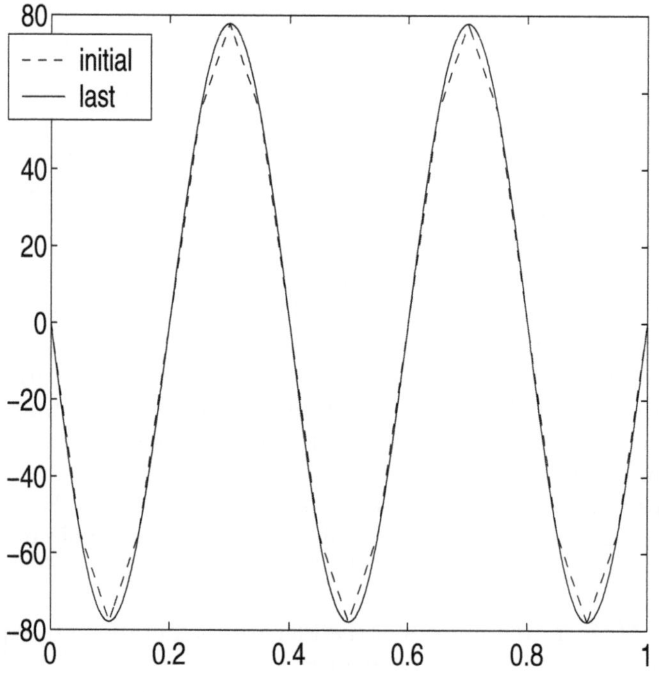

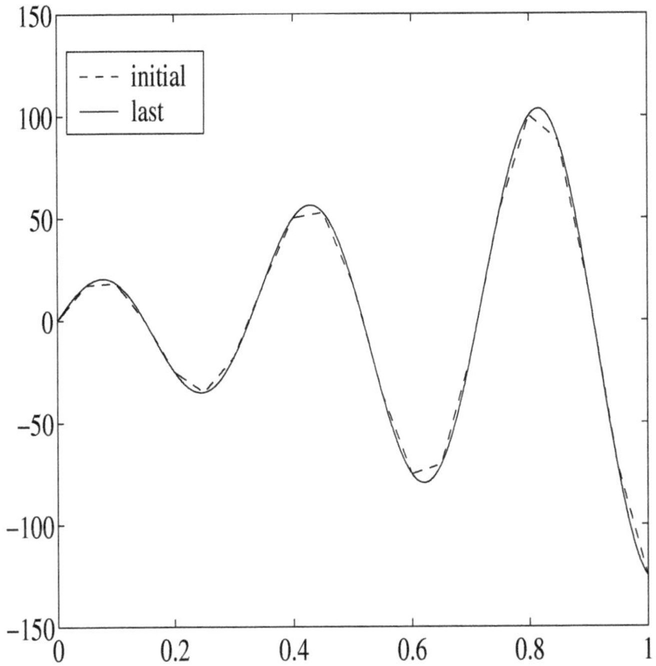

The spectral subspace associated with the operator T_N and the cluster $\{\lambda_N(5), \lambda_N(6)\}$ of its eigenvalues is two dimensional. Each of the preceding two figures show the initial approximation of an element of a basis for this subspace and the corresponding last iterate of the Double Iteration.

The initial approximations are piecewise linear functions on $[0, 1]$ having twenty linear pieces.

Further, since the 2×2 matrix Θ_n was chosen to be upper triangular, the initial approximation in the first of these figures is an eigenvector of the operator T_n corresponding to its eigenvalue $\lambda_n(5)$. The initial approximation in the second figure is a linear combination of two eigenvectors of the operator T_n, one corresponding to its eigenvalue $\lambda_n(5)$ and the other corresponding to its eigenvalue $\lambda_n(6)$. ∎

Example 5.26

Kernel	$k(s,t) := \begin{cases} s(1-t) & \text{if } 0 \le s \le t \le 1 \\ t(1-s) & \text{if } 0 \le t \le s \le 1 \end{cases}$
Approximation	Galerkin (Section 4.1)
Projection	Piecewise linear interpolatory
Numerical parameters	Uniform grids: $n = 10,\ 20,\ 30,\ 60$ nodes Acceleration order: $q = 1,\ 2,\ 3$
Eigenvalues	$\lambda_n(2),\ \lambda(2)$

The kernel is the Green Function for the Sturm-Liouville Problem: $-\varphi'' = \mu\,\varphi$ on $[0,1]$, $\varphi(0) = 0 = \varphi(1)$. The kernel is a continuous function on $[0,1]\times[0,1]$, but it is not smooth.

The eigenvalues of the corresponding Fredholm integral operator T are $1/j^2\pi^2$, $j = 1, 2, \ldots$ Each one of them is a simple eigenvalue.

In this example the computation of Tx for $x \in X$ is done in exact arithmetic.

Note that for the accelerated approximation of order q with n nodes, we solve a matrix eigenvalue problem of size nq.

The relative residuals in the ∞-norm for the usual approximation (that is, when $q = 1$) and for the accelerated approximation (that is, when $q > 1$) are shown in the following table.

q	n	$\dfrac{\|\|T\varphi_n^{[q]} - \lambda_n^{[q]}\varphi_n^{[q]}\|\|_\infty}{\|\|\varphi_n^{[q]}\|\|_\infty}$
1	60	$(.36)10^{-4}$
2	30	$(.99)10^{-4}$
3	20	$(.54)10^{-5}$

Note that if $q = 1$, then $\varphi_n^{[q]} = \varphi_n$. Also, as stated in Exercise 5.12, if $T_n = T_n^G$ and $q = 2$, then $\varphi_n^{[q]} = T\varphi_n$. This may explain why $\varphi_n^{[q]}$, when $q = 2$ and $n = 30$, does not compare well with $\varphi_n^{[q]}$, when $q = 1$ and $n = 60$, in this special case.

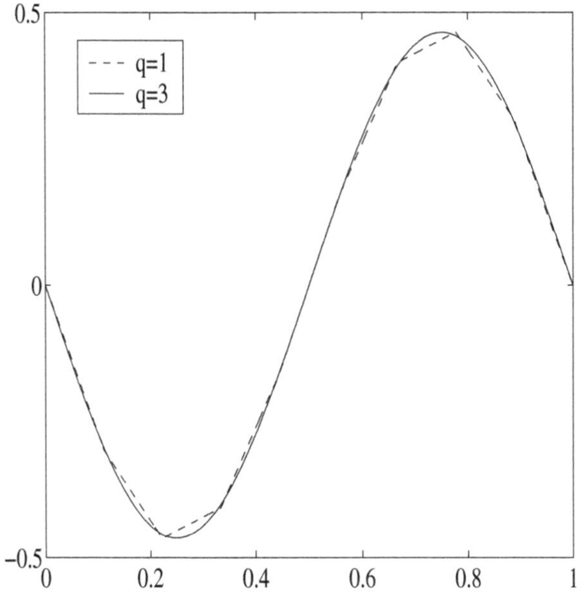

The figure above shows two approximate eigenvectors corresponding to $\lambda(2)$; one is computed with $n = 10$, $q = 1$ and the other with $n = 10$, $q = 3$. The first is a piecewise linear function on $[0, 1]$ having ten linear pieces. $\blacksquare$

5.5 Exercises

Unless otherwise stated, X denotes a Banach space over $\mathbb{C}$, $X \neq \{0\}$ and $T \in \mathrm{BL}(X)$. Prove the following assertions.

5.1 Let p be a positive integer, $\mathsf{Z} \in \mathbb{C}^{p\times p}$ such that $0 \notin \mathrm{sp}(\mathsf{Z})$, $\underline{\mathsf{v}} \in \mathbb{C}^{n\times p}$ and $\underline{y} := \underline{\widetilde{L}}\, \underline{\mathsf{v}} \in X^{1\times p}$. Then for $\underline{x} \in X^{1\times p}$ and $\underline{\mathsf{u}} \in \mathbb{C}^{n\times p}$, the following holds: $\underline{\widetilde{T}}\ \underline{x} = \underline{x}\,\mathsf{Z} + \underline{y}$ and $\underline{\widetilde{K}}\ \underline{x} = \underline{\mathsf{u}}\,\mathsf{Z} + \underline{\mathsf{v}}$ if and only if $\underline{\widetilde{A}}\ \underline{\mathsf{u}} = \underline{\mathsf{u}}\,\mathsf{Z} + \underline{\mathsf{v}}$ and $\underline{\widetilde{L}}\ \underline{\mathsf{u}} = \underline{x}$.

In particular, let $\underline{\mathsf{v}} := \underline{0}$ (so that $\underline{y} = \underline{0}$ also), and let $\underline{x}$, $\underline{\mathsf{u}}$ satisfy the above conditions. Then the set of p elements in $\underline{x}$ is linearly independent in X if and only if the set of p vectors in $\underline{\mathsf{u}}$ is linearly independent in $\mathbb{C}^{n\times 1}$. Also, if $z \in \mathrm{re}(\widetilde{T}) \setminus \{0\} = \mathrm{re}(\widetilde{A}) \setminus \{0\}$, then

$$R(\widetilde{T}, z) = \frac{\widetilde{L}\,R(\widetilde{A}, z)\widetilde{K} - I}{z} \quad \text{and} \quad R(\widetilde{A}, z) = \frac{\widetilde{K}\,R(\widetilde{T}, z)\widetilde{L} - I}{z}.$$

(Compare Lemma 5.2.)

5.2 (a) Let π_n be a projection in $\mathrm{BL}(X)$ given by $\pi_n x = \sum_{j=1}^{r(n)} \langle x\,,\, e_{n,j}^*\rangle e_{n,j}$, $x \in X$. If $\langle Te_{n,j}\,,\, e_{n,i}^*\rangle = \overline{\langle e_{n,i}\,,\, T^*e_{n,j}^*\rangle}$, $i, j = 1, \ldots, r(n)$, then every nonzero eigenvalue of each of the operators T_n^P, T_n^S and T_n^G is real, and its algebraic multiplicity equals its geometric multiplicity. In particular, this is the case if X is a Hilbert space, T is a self-adjoint operator and π_n is an orthogonal projection. (Hint: Theorem 5.4.)

(b) Let $X = C([a, b])$ and T be a Fredholm integral operator with a continuous kernel $k(\cdot, \cdot)$ such that $k(s, t) = \overline{k(t, s)}$ for all $s, t \in [a, b]$. Let Q_n be a quadrature rule given by $Q_n x = \sum_{j=1}^{r(n)} w_{n,j} x(t_{n,j})$ such that $w_{n,j} > 0$ for all $j = 1, \ldots, r(n)$. Then every nonzero eigenvalue of each of the operators T_n^N and T_n^F is real, and its algebraic multiplicity equals its geometric multiplicity. (Hint: The matrix $[w_{n,j} k(t_{n,i}, t_{n,j})]$ is similar to the matrix $[\sqrt{w_{n,i}}\, k(t_{n,i}, t_{n,j}) \sqrt{w_{n,j}}]$. Also, Remark 1.1 and Exercise 1.19.)

5.3 Let π_n be an interpolatory projection on $C^0([a, b])$ and Q_n be the quadrature formula induced by π_n. If T_n^N is the Nyström approximation of T based on Q_n, then $T_n^N = T_n^N \pi_n$. Let $T_n^F := \pi_n T_n^N$, λ_n be a nonzero simple eigenvalue of T_n^F and φ_n be a corresponding eigenvector. Define $\varphi_n^N := T_n^N \varphi_n / \lambda_n$ and $\varphi_n^{*N} = \varphi_n^*$. Then $T_n^N \varphi_n^N = \lambda_n \varphi_n^N$, $(T_n^N)^* \varphi_n^{*N} = \overline{\lambda}_n \varphi_n^{*N}$ and $\langle \varphi_n^N\,,\, \varphi_n^{*N}\rangle = 1$. Similar results hold for an ordered basis $\underline{\varphi_n}$ of $M(T_n^F, \Lambda_n)$, where Λ_n is a finite subset of $\mathrm{sp}(T_n^F)$ such that $0 \notin \Lambda_n$.

5.4 Let $\widetilde{T}$, $\widetilde{K}$, $\widetilde{L}$, $\widetilde{\mathsf{A}}$ and $\widetilde{\lambda}$ be as in Theorem 5.4(a), and $\widetilde{A} = \widetilde{K}\widetilde{L}$. For each positive integer j, let $Y_j := \mathcal{N}((\widetilde{T} - \widetilde{\lambda} I)^j)$ and $V_j := \mathcal{N}((\widetilde{A} - \widetilde{\lambda} I)^j)$. Then $\widetilde{K}$ maps Y_j into V_j in a one-to-one manner and $\widetilde{L}$ maps V_j into Y_j

in a one-to-one manner, so that $\dim Y_j = \dim V_j$. As a result, the ascent of $\widetilde{\lambda}$ as an eigenvalue of $\widetilde{T}$ is equal to its ascent as an eigenvalue of $\widetilde{\mathsf{A}}$.

5.5 Let $\widetilde{T}^K$, $\widetilde{U}$, $\widetilde{K}$, $\widetilde{L}$, $\widetilde{\mathsf{A}}^K$, $\widetilde{D}$ and $\widetilde{\lambda}$ be as in Theorem 5.8(a), and $\widetilde{A}^K = \widetilde{K}\widetilde{L} + \widetilde{D}$. For each positive integer j, let $Y_j^K := \mathcal{N}((\widetilde{T}^K - \widetilde{\lambda}I)^j)$ and $V_j^K := \mathcal{N}((\widetilde{A}^K - \widetilde{\lambda}I)^j)$. Then $\widetilde{K}$ maps Y_j^K into V_j^K in a one-to-one manner and $\widetilde{L}_j$ maps V_j^K into Y_j^K in a one-to-one manner, where

$$\widetilde{L}_j := \begin{cases} (\widetilde{\lambda}I - \widetilde{U})^{-1}\widetilde{L} & \text{if } j = 1, \\ (\widetilde{\lambda}I - \widetilde{U})^{-1}\left[\widetilde{L} - \widetilde{L}_{j-1}(\widetilde{A}^K - \widetilde{\lambda}I)\right] & \text{if } j \geq 2, \end{cases}$$

so that $\dim Y_j^K = \dim V_j^K$. As a result, the ascent of $\widetilde{\lambda}$ as an eigenvalue of $\widetilde{T}^K$ is equal to its ascent as an eigenvalue of $\widetilde{\mathsf{A}}^K$.

5.6 Let $\widetilde{T}^K$, $\widetilde{U}$, $\widetilde{A}^K$ and $\widetilde{E}$ be as in Theorem 5.8. Then $\mathrm{sp}(\widetilde{T}^K) = \mathrm{sp}(\widetilde{A}^K) \cup \widetilde{E}$. Also, if either (i) $\widetilde{\lambda} \in \mathrm{sp}(\widetilde{A}^K) \setminus \widetilde{E}$, or if (ii) $\widetilde{\lambda} \in \widetilde{E}$ and $\widetilde{\lambda}$ is an eigenvalue of $\widetilde{U}$, then $\widetilde{\lambda}$ is an eigenvalue of $\widetilde{T}^K$. (Hint: Proposition 5.7 and Exercise 1.9.)

5.7 Let K_n, $\underline{\mathsf{u}_n}$ and $\underline{\varphi_n}$ be as in Theorem 5.9. If $\|\cdot\|_{[n]}$ is a norm on $\mathbb{C}^{n\times 1}$ such that $\|K_n\| \to 0$, then for each $j = 1, \ldots, m$,

$$\mathrm{dist}(\varphi_{n,j}, \mathrm{span}\{\varphi_{n,i} : i = 1, \ldots, m, i \neq j\}) \to 0.$$

(Hint: $\varphi_{n,j} = K_n(\mathsf{u}_{n,j})$, $j = 1, \ldots, m$.)

5.8 Let $\|\cdot\|_{[n]}$ be a norm on $\mathbb{C}^{n\times 1}$ and $\{\mathsf{u}_1, \ldots, \mathsf{u}_m\}$ be a linearly independent subset of $\mathbb{C}^{n\times 1}$ such that $\|\mathsf{u}_j\|_{[n]} = 1$ for each $j = 1, \ldots, m$. If $\mathrm{dist}(\mathsf{u}_j, \mathrm{span}\{\mathsf{u}_i : i = 1, \ldots, j-1\}) \geq d_0 > 0$ for all $j = 2, \ldots, m$, then $\mathrm{dist}(\mathsf{u}_j, \mathrm{span}\{\mathsf{u}_i : i = 1, \ldots, m, i \neq j\}) \geq \dfrac{d_0^m}{(1+d_0)^{m+1}}$ for all $j = 1, \ldots, m$.

5.9 Let λ_n be a nonzero simple eigenvalue of T_n^G, φ_n be a corresponding eigenvector of T_n^G, and φ_n^* be the eigenvector of $(T_n^G)^*$ corresponding to $\overline{\lambda}_n$ such that $\langle \varphi_n, \varphi_n^* \rangle = 1$.

(a) Let $\varphi_n^P := \varphi_n$, $\varphi_n^{*P} := T^*\varphi_n^*/\overline{\lambda}_n$, $\varphi_n^S := T\varphi_n/\lambda_n$ and $\varphi_n^{*S} := \varphi_n^*$. Then $T_n^P\varphi_n^P = \lambda_n\varphi_n^P$, $\langle \varphi_n^P, \varphi_n^{*P} \rangle = 1$, $T_n^S\varphi_n^S = \lambda_n\varphi_n^S$, $\langle \varphi_n^S, \varphi_n^{*S} \rangle = 1$.

(b) For $k = 1, 2, \ldots$ let $\lambda_n^{(k)}(G)$, $\varphi_n^{(k)}(G)$; $\lambda_n^{(k)}(P)$, $\varphi_n^{(k)}(P)$ and $\lambda_n^{(k)}(S)$, $\varphi_n^{(k)}(S)$ be the kth iterates in the Elementary Iteration (E) for T_n^G, T_n^P

and T_n^S, respectively. Then $\lambda_n^{(1)}(G) = \lambda_n$, $\varphi_n^{(1)}(G) = \varphi_n^S$, $\lambda_n^{(2)}(G) = \lambda_n^1(P)$, and for $k = 1, 2, \ldots$ $\lambda_n^{(k)}(S) = \lambda_n^{(k)}(P)$, $\varphi_n^{(k)}(S) = T\varphi_n^{(k)}(P)/\lambda_n$. Also, if for $k = 1, 2, \ldots$ $\mu_n^{(k)}(P)$, $\psi_n^{(k)}(P)$ and $\mu_n^{(k)}(S)$, $\psi_n^{(k)}(S)$ denote the kth iterates in the Double Iteration (D) for T_n^P and T_n^S respectively, then $\mu_n^{(k)}(S) = \mu_n^{(k)}(P)$ and $\psi_n^{(k)}(S) = T\psi_n^{(k)}(P)/\lambda_n$. (Hint: $S_n^S T = TS_n^P$.)

5.10 Let Λ_n be a finite subset of $\mathrm{sp}(T_n^G)$ such that $0 \notin \Lambda_n$, $\underline{\varphi_n}$ be an ordered basis for the spectral subspace $M(T_n^G, \Lambda_n)$, $\underline{T_n^G}\ \underline{\varphi_n} = \underline{\varphi_n}\,\Theta_n$, and let $\underline{\varphi_n^*}$ be the adjoint ordered basis for $M((T_n^G)^*, \overline{\Lambda}_n)$.

(a) Let $\underline{\varphi_n}^P := \underline{\varphi_n}$, $\underline{\varphi_n^*}^P := \underline{T}^*\,\underline{\varphi_n^*}\,(\Theta_n^*)^{-1}$, $\underline{\varphi_n}^S = \underline{T}\,\underline{\varphi_n}\,\Theta_n^{-1}$ and $\underline{\varphi_n^*}^S := \underline{\varphi_n^*}$. Then $\underline{\varphi_n}^P$ (resp., $\underline{\varphi_n}^S$) forms an ordered basis for the spectral subspace $M(T_n^P, \Lambda_n)$ (resp., $M(T_n^S, \Lambda_n)$) and $\underline{T_n^P}\ \underline{\varphi_n}^P = \underline{\varphi_n}^P\Theta_n$, $(\underline{\varphi_n}^P, \underline{\varphi_n^*}^P) = \mathsf{I}$, $\underline{T_n^S}\ \underline{\varphi_n}^S = \underline{\varphi_n}^S\Theta_n$, $(\underline{\varphi_n}^S, \underline{\varphi_n^*}^S) = \mathsf{I}$.

(b) For $k = 1, \ldots$ let $\Theta_n^{(k)}(G)$, $\underline{\varphi_n}^{(k)}(G)$; $\Theta_n^{(k)}(P)$, $\underline{\varphi_n}^{(k)}(P)$ and $\Theta_n^{(k)}(S)$, $\underline{\varphi_n}^{(k)}(S)$ be the kth iterates in the Elementary Iteration $(\underline{E})$ for T_n^G, T_n^P and T_n^S respectively. Then $\Theta_n^{(1)}(G) = \Theta_n$, $\underline{\varphi_n}^{(1)}(G) = \underline{\varphi_n}^S$, $\Theta_n^{(2)}(G) = \Theta_n^{(1)}(P)$, and for $k = 1, 2, \ldots$ $\Theta_n^{(k)}(S)\Theta_n = \Theta_n\Theta_n^{(k)}(P)$, $\underline{\varphi_n}^{(k)}(S) = \underline{T}\,\underline{\varphi_n}^{(k)}(P)\Theta_n^{-1}$. Also, if for $k = 1, 2, \ldots$ $\Upsilon_n^{(k)}(P)$, $\underline{\psi_n}^{(k)}(P)$ and $\Upsilon_n^{(k)}(S)$, $\underline{\psi_n}^{(k)}(S)$ denote the kth iterates in the Double Iteration $(\underline{D})$, then $\Upsilon_n^{(k)}(S)\Theta_n = \Theta_n\Upsilon_n^{(k)}(P)$ and $\underline{\psi_n}^{(k)}(S) = \underline{T}\,\underline{\psi_n}^{(k)}(P)\Theta_n^{-1}$. (Hint: $\mathcal{S}_n^S\,\underline{T} = \underline{T}\,\mathcal{S}_n^P$.)

5.11 Let $T_n \in \mathrm{BL}(X)$, K_n, L_n, A_n, A_n, T_n^*, L_n^* and K_n^* be as in Section 5.2.

(a) Let λ_n be a nonzero simple eigenvalue of T_n, and let u_n, v_n be as in Section 5.2. For $y \in X$, we have

$$P_n y = \frac{\langle y\,,\, K_n^*\mathsf{v}_n\rangle}{\lambda_n} L_n\mathsf{u}_n \quad \text{and} \quad S_n y = \frac{1}{\lambda_n}(L_n\mathsf{w}_n - y + P_n y),$$

where w_n satisfies $\mathsf{A}_n\mathsf{w}_n - \lambda_n\mathsf{w}_n = K_n(y - P_n y)$ and $\mathsf{v}_n^*\mathsf{w}_n = 0$.

(b) Let Λ_n be a spectral set of finite type for T_n such that $0 \notin \Lambda_n$, and let $\underline{\mathsf{u}_n}$, $\underline{\mathsf{v}_n}$, Θ_n be as in Section 5.2. For $\underline{y} \in X^{1\times m}$, we have

$$\underline{P_n}\,\underline{y} = (\underline{y}\,,\,\underline{K_n^*}\,\underline{\mathsf{v}_n})\,\underline{L_n}\,\underline{\mathsf{u}_n}\,\Theta_n^{-1} \text{ and } \mathcal{S}_n\,\underline{y} = (\underline{L_n}\,\underline{\mathsf{w}_n} - \underline{y} + \underline{P_n}\,\underline{y})\Theta_n^{-1},$$

where $\underline{\mathsf{w}_n}$ satisfies $\mathsf{A}_n\,\underline{\mathsf{w}_n} - \underline{\mathsf{w}_n}\,\Theta_n = \underline{K_n}\,(\underline{y} - \underline{P_n}\,\underline{y})$ and $\underline{\mathsf{v}_n^*}\,\underline{\mathsf{w}_n} = \mathsf{O}$.

5.12 Let $T_n := T_n^G$, the Galerkin approximation of T, and $q = 2$.

(a) Let $\lambda_n \in \mathbb{C}$, $\lambda_n \neq 0$. Then λ_n is an eigenvalue of T_n if and only if λ_n itself is an eigenvalue of $\boldsymbol{T}_n$. Further, if φ_n is an eigenvector of

T_n corresponding to λ_n, then $\boldsymbol{\varphi}_n := \begin{bmatrix} T\varphi_n \\ T\varphi_n/\lambda_n \end{bmatrix}$ an eigenvector of $\boldsymbol{T}_n$ corresponding to λ_n. Also, if $\{\varphi_{n,1}, \ldots, \varphi_{n,g}\}$ is a linearly independent set of eigenvectors of T_n corresponding to λ_n, then $\{\boldsymbol{\varphi}_{n,1}, \ldots, \boldsymbol{\varphi}_{n,g}\}$ is a linearly independent set of eigenvectors of $\boldsymbol{T}_n$ corresponding to λ_n.

(b) Let $\Lambda \subset \mathbb{C}$ be a spectral set of finite type for T such that $0 \notin \Lambda$. Assume that $T_n^G \xrightarrow{\nu} T$. For large n, let Λ_n be the spectral set of finite type for T_n as given by Theorem 2.18. Then Λ_n itself is a spectral set of finite type for $\boldsymbol{T}_n$. Further, if $\underline{\varphi_n} \in X^{1\times m}$ forms an ordered basis for $M(T_n, \Lambda_n)$ and $\underline{T_n}\,\underline{\varphi_n} = \underline{\varphi_n}\,\Theta_n$, then $\underline{\boldsymbol{\varphi}_n} := \begin{bmatrix} \underline{T}\,\underline{\varphi_n} \\ \underline{T}\,\underline{\varphi_n}\,\Theta_n^{-1} \end{bmatrix}$ forms an ordered basis for $M(\boldsymbol{T}_n, \Lambda_n)$ and satisfies $\underline{\boldsymbol{T}_n}\,\underline{\boldsymbol{\varphi}_n} = \underline{\boldsymbol{\varphi}_n}\,\Theta_n$.

5.13 Let $\widetilde{T} \in \mathrm{BL}(X)$ and $q \geq 2$. Then the following conditions are equivalent: (i) $\mathbf{I} - \widetilde{\boldsymbol{T}}^{[q]}$ is invertible in $\mathrm{BL}(\boldsymbol{X}^{[q]})$; (ii) $I - \sum_{k=0}^{q-1} (T - \widetilde{T})^k \widetilde{T}$ is invertible in $\mathrm{BL}(X)$; (iii) $I - \sum_{k=0}^{q-1} \widetilde{K}(T - \widetilde{T})^k \widetilde{L}$ is invertible in $\mathrm{BL}(\mathbb{C}^{n\times 1})$; and (iv) $I - \widetilde{\boldsymbol{A}}$ is invertible in $\mathrm{BL}(\mathbb{C}^{nq\times 1})$.

As a result, $\mathrm{re}(\widetilde{\boldsymbol{T}}^{[q]}) \setminus \{0\} = \mathrm{re}(\widetilde{\boldsymbol{A}}) \setminus \{0\}$. (Hint: Let $w = 1/z$ and consider the operators $w\star \widetilde{\boldsymbol{T}}^{[q]}$, $w\star \widetilde{\boldsymbol{A}}$ obtained by replacing T, $\widetilde{T}$, $\widetilde{L}$ by wT, $w\widetilde{T}$, $w\widetilde{L}$, respectively.)

5.14 Let $\widetilde{\lambda}^{[q]}$, $\widetilde{\boldsymbol{T}}$, $\widetilde{\mathbf{K}}$, $\widetilde{\boldsymbol{\mathcal{L}}}([\widetilde{\lambda}^{[q]}])$, $\widetilde{\mathsf{A}}$ be as in Theorem 5.21(a), and $\widetilde{\boldsymbol{A}}$, $\boldsymbol{\mathcal{F}}$ be as in Lemma 5.19 with $p := 1$ and $\mathsf{Z} := [\widetilde{\lambda}^{[q]}]$. For each positive integer j, let $\boldsymbol{Y}_j := \mathcal{N}((\widetilde{\boldsymbol{T}} - \widetilde{\lambda}^{[q]}\mathbf{I})^j)$ and $\boldsymbol{V}_j := \mathcal{N}((\widetilde{\boldsymbol{A}} - \widetilde{\lambda}^{[q]}\mathbf{I})^j)$. Then $\widetilde{\mathbf{K}}$ maps $\boldsymbol{Y}_j$ into $\boldsymbol{V}_j$ in a one-to-one manner and $\widetilde{\boldsymbol{\mathcal{L}}}_j$ maps $\boldsymbol{V}_j$ into $\boldsymbol{Y}_j$ in a one-to-one manner, where

$$\widetilde{\boldsymbol{\mathcal{L}}}_j := \begin{cases} (\widetilde{\lambda}^{[q]})^{-1}\widetilde{\boldsymbol{\mathcal{L}}}([\widetilde{\lambda}^{[q]}]) & \text{if } j = 1, \\ (\widetilde{\lambda}^{[q]})^{-1}\left[\widetilde{\boldsymbol{\mathcal{L}}}([\widetilde{\lambda}^{[q]}]) - (\mathbf{I} - \boldsymbol{\mathcal{F}})\widetilde{\boldsymbol{\mathcal{L}}}_{j-1}(\widetilde{\boldsymbol{A}} - \widetilde{\lambda}^{[q]}\mathbf{I})\right] & \text{if } j \geq 2, \end{cases}$$

so that $\dim \boldsymbol{Y}_j = \dim \boldsymbol{V}_j$. As a result, the ascent of $\widetilde{\lambda}^{[q]}$ as an eigenvalue of $\widetilde{\boldsymbol{T}}$ is equal to its ascent as an eigenvalue of $\widetilde{\mathsf{A}}$.

5.15 **Uniformly well-conditioned bases for accelerated approximation:** Let $(T_n - T)^2 \xrightarrow{\nu} O$, where each T_n is a finite rank operator on X. Let $T_n = K_n L_n$ and $\|\cdot\|_{[n]}$ be a norm on $\mathbb{C}^{n\times 1}$ such that $(\|K_n\|)$ and $(\|L_n\|)$ are bounded sequences, as in Subsection 5.1.2. Consider a spectral set Λ of finite type for T such that $0 \notin \Lambda$. Let $q \geq 2$ be an integer and

$\underline{\mathbf{u}}_n = [\mathbf{u}_{n,1}, \ldots, \mathbf{u}_{n,m}]$ form an ordered basis for $M(\mathbf{A}_n, \Lambda_n^{[q]})$ such that for all large n and $j = 1, \ldots, m$,

$$\|\mathbf{u}_{n,j}\|_\infty \leq c, \quad \text{dist}(\mathbf{u}_{n,j}, \text{span}\{\mathbf{u}_{n,i} : i = 1, \ldots, m, \ i \neq j\}) \geq d$$

for some constants c and d satisfying $0 < d \leq c$.

Let $\mathbf{A}_n \underline{\mathbf{u}}_n = \underline{\mathbf{u}}_n \Theta_n^{[q]}$ for some nonsingular $\Theta_n^{[q]} \in \mathbb{C}^{m \times m}$, and define $\underline{\boldsymbol{\varphi}_n} := \boldsymbol{\mathcal{L}}_n(\Theta_n^{[q]}) \underline{\mathbf{u}}_n (\Theta_n^{[q]})^{-1} = [\boldsymbol{\varphi}_{n,1}, \ldots, \boldsymbol{\varphi}_{n,m}]$. Then $\underline{\boldsymbol{\varphi}_n}$ forms an ordered basis for $M(\boldsymbol{T}_n, \Lambda_n^{[q]})$ such that for all large n and $j = 1, \ldots, m$,

$$\|\boldsymbol{\varphi}_{n,j}\|_\infty \leq \gamma, \quad \text{dist}(\boldsymbol{\varphi}_{n,j}, \text{span}\{\boldsymbol{\varphi}_{n,i} : i = 1, \ldots, m, \ i \neq j\}) \geq \delta$$

for some constants γ and δ satisfying $0 < \delta \leq \gamma$. (Compare 5.9 and Theorem 3.15.)

Chapter 6

Matrix Computations

This chapter is devoted to some topics in Numerical Linear Algebra which are relevant to spectral computations and to the considerations of the accuracy of such computations. The main references for these topics are [74], [39] and [42].

The QR Method is the most widely used algorithm for computing eigenvalues and eigenvectors of large full matrices. We describe the Basic QR Method and prove its convergence under suitable hypotheses. A twofold discussion of the error analysis is given next. The forward error analysis involves finding a condition number, that is, a number which gives a bound for the relative error in the exact solution of a perturbed problem. The backward error analysis gives a measure of the stability of the computed solution of a problem by showing that the computed solution is in fact the exact solution of a perturbed problem. The condition number for a problem does not depend on the specific method that may be used to solve the problem, while the stability considerations depend on the specific algorithm used to solve that problem.

Condition numbers for the following problems are considered: solution of a linear system, computation of a multiple eigenvalue of a matrix, and solution of a Sylvester equation. Stability of the Basic QR Method is discussed. An algorithm, which involves the QR Method and Gaussian Elimination with partial pivoting, is suggested for solving a Sylvester Equation, and its stability is discussed. It may be recalled from Section 5.2 that we need to solve Sylvester equations for implementing the Elementary Iteration ($\underline{E}$) and the Double Iteration ($\underline{D}$) for the case of a cluster of eigenvalues.

The total relative error is less than or equal to the product of the condition number and the relative perturbation of the data. A stopping criterion for the Basic QR Method is given which ensures that the total relative error is less than a prescribed level.

6.1 QR factorization

Let X denote an n-dimensional Hilbert space with inner product $\langle \cdot, \cdot \rangle$ and let $\underline{e} := [e_1, \ldots, e_n]$ form an orthonormal basis for X. If $\underline{\varphi} := [\varphi_1, \ldots, \varphi_m]$ is an ordered set of elements in X, then each φ_i is a unique linear combination of the vectors in $\underline{e}$, that is, there exists a matrix $\mathsf{V} \in \mathbb{C}^{n \times m}$ such that

$$\underline{\varphi} = \underline{e}\,\mathsf{V}, \qquad \text{so that} \qquad \mathsf{V} = (\underline{\varphi}, \underline{e}).$$

Thus the jth column V consists of the coordinates of φ_j relative to the basis $\underline{e}$. If $\underline{\varphi}$ is an orthonormal set and $\mathsf{Q} \in \mathbb{C}^{n \times m}$ is such that $\underline{\varphi} = \underline{e}\,\mathsf{Q}$, then

$$\mathsf{I}_m = (\underline{\varphi}, \underline{\varphi}) = (\underline{e}\,\mathsf{Q}, \underline{e}\,\mathsf{Q}) = \mathsf{Q}^*(\underline{e}, \underline{e})\mathsf{Q} = \mathsf{Q}^*\mathsf{I}_n\mathsf{Q} = \mathsf{Q}^*\mathsf{Q},$$

that is, Q is a unitary matrix.

In this chapter, we let $X := \mathbb{C}^{n \times 1}$ with the canonical inner product

$$\langle \mathsf{x}, \mathsf{y} \rangle := \mathsf{y}^*\mathsf{x} \quad \text{for } \mathsf{x}, \mathsf{y} \in \mathbb{C}^{n \times 1}.$$

For $\mathsf{x} := [\mathsf{x}(1), \ldots, \mathsf{x}(n)]^\top \in \mathbb{C}^{n \times 1}$, let

$$\|\mathsf{x}\|_2 := \Big(\sum_{i=1}^{n} |\mathsf{x}(i)|^2\Big)^{1/2}$$

and for a matrix $\mathsf{A} \in \mathbb{C}^{m \times n}$, let

$$\|\mathsf{A}\|_2 := \sup\{\|\mathsf{A}\mathsf{x}\|_2 : \mathsf{x} \in \mathbb{C}^{n \times 1}, \|\mathsf{x}\|_2 \leq 1\}.$$

Unless otherwise specified, the orthonormal basis of reference will be assumed to be the **standard basis** $[\mathsf{e}_1, \ldots, \mathsf{e}_n]$, where

$$\mathsf{e}_i := [0, \ldots, 0, 1, 0, \ldots, 0]^\top \in \mathbb{C}^{n \times 1},$$

the ith entry being equal to 1 and the other entries equal to 0.

Lemma 6.1

Let n, m be integers such that $1 \leq m \leq n$. For every matrix $\mathsf{A} \in \mathbb{C}^{n \times m}$ such that $\operatorname{rank} \mathsf{A} = m$, *there are matrices $\mathsf{Q} \in \mathbb{C}^{n \times m}$ and $\mathsf{R} \in \mathbb{C}^{m \times m}$ such that*

(i) Q *is a unitary matrix,*

(ii) R *is a nonsingular upper triangular matrix, and*

(iii) $\mathsf{A} = \mathsf{QR}$.

If the entries of A *are real, those of* Q *and* R *can also be chosen real.*

Proof
This result is just a matrix version of the Gram-Schmidt Process presented in Proposition 1.39 and revisited in Example 1.45. Indeed, let $\mathsf{x}_j = \mathsf{A}(1,j)\mathsf{e}_1 + \cdots + \mathsf{A}(n,j)\mathsf{e}_n$ for $j \in [\![1,m]\!]$. The Gram-Schmidt process applied to this set will give rise to an ordered set of m orthonormal vectors $[\mathsf{q}_1,\ldots,\mathsf{q}_m]$ such that the jth vector in this set is a linear combination of the first j vectors of the former set. Let $\mathsf{x}_j := r_{1,j}\mathsf{q}_1 + \ldots + r_{j,j}\mathsf{q}_j$ for $j \in [\![1,m]\!]$, $\mathsf{Q} := [\mathsf{q}_1,\ldots,\mathsf{q}_m]$, $\mathsf{R}(i,j) := r_{i,j}$ for $i,j \in [\![1,m]\!]$ and $\mathsf{R}(i,j) := 0$ for $i \in [\![2,m]\!]$, $j \in [\![1,i-1]\!]$. Then $\mathsf{A} = \mathsf{QR}$, where Q is a unitary matrix since the set $\{\mathsf{q}_1,\ldots,\mathsf{q}_m\}$ is orthonormal, and the matrix R is upper triangular. If $\mathsf{A}(i,j) \in \mathbb{R}$ for all i,j, then clearly $\mathsf{Q}(i,j) \in \mathbb{R}$ and $\mathsf{R}(i,j) \in \mathbb{R}$ for all i,j. ∎

Theorem 6.2 QR Factorization:
Let n, m be integers such that $1 \le m \le n$. For every matrix $\mathsf{A} \in \mathbb{C}^{n\times m}$ such that $\operatorname{rank}\mathsf{A} = m$, *there is a unique pair of matrices* $(\mathrm{Q}(\mathsf{A}), \mathrm{R}(\mathsf{A}))$, *where* $\mathrm{Q}(\mathsf{A}) \in \mathbb{C}^{n\times m}$ *and* $\mathrm{R}(\mathsf{A}) \in \mathbb{C}^{m\times m}$ *such that*

(i) $\mathrm{Q}(\mathsf{A})$ *is unitary,*

(ii) $\mathrm{R}(\mathsf{A})$ *is upper triangular,*

(iii) $\mathsf{A} = \mathrm{Q}(\mathsf{A})\mathrm{R}(\mathsf{A})$, *and*

(iv) *all the diagonal entries of* $\mathrm{R}(\mathsf{A})$ *are positive.*

Proof
Let R and Q be matrices given by Lemma 6.1. Since R is nonsingular, $\mathsf{R}(i,i) \neq 0$ for $i \in [\![1,m]\!]$ and $d_i := \mathsf{R}(i,i)/|\mathsf{R}(i,i)|$ is well defined. Set $\mathsf{D} := \operatorname{diag}[d_1,\ldots,d_m]$, which is unitary. Clearly $\mathrm{Q}(\mathsf{A}) := \mathsf{QD}$ is unitary, $\mathrm{R}(\mathsf{A}) := \mathsf{D}^*\mathsf{R}$ is upper triangular with positive diagonal entries, and $\mathrm{Q}(\mathsf{A})\mathrm{R}(\mathsf{A}) = \mathsf{A}$.

Suppose now we have two pairs (Q,R) and $(\widetilde{\mathsf{Q}},\widetilde{\mathsf{R}})$ satisfying the same properties as $(\mathrm{Q}(\mathsf{A}),\mathrm{R}(\mathsf{A}))$. Since R and $\widetilde{\mathsf{R}}$ are nonsingular, $\widetilde{\mathsf{Q}}^*\mathsf{Q} =$

$\widetilde{R}R^{-1}$, which is a unitary upper triangular matrix with positive diagonal entries. Hence $\widetilde{Q}^*Q = I$. It follows that $\widetilde{Q} = Q$ and $R = \widetilde{R}$. ▮

Let $\mathcal{U} := \{U \in \mathbb{C}^{n\times m} : \operatorname{rank} U = m\}$. Clearly, $U \in \mathcal{U}$ if and only if $\det(U^*U) \neq 0$. Since $U \in \mathbb{C}^{n\times m} \mapsto \det(U^*U) \in \mathbb{R}$ is a continuous function, it follows that $\mathcal{U}$ is an open subset of $\mathbb{C}^{n\times m}$. The maps

$$Q : \mathcal{U} \to \mathbb{C}^{n\times m};\ A \mapsto Q(A) \quad \text{and} \quad R : \mathcal{U} \to \mathbb{C}^{m\times m};\ A \mapsto R(A)$$

are known as the unitary factor map and the upper triangular factor map, respectively. Any two matrices Q and R such that Q is unitary, R is upper triangular, and $QR = A$ will be called QR factors of A.

Theorem 6.3 Cholesky Factorization:
Let $B \in \mathbb{C}^{n\times n}$ *be a Hermitian positive definite matrix. Then there exists a unique upper triangular matrix* $R \in \mathbb{C}^{n\times n}$ *with positive diagonal entries such that* $B = R^*R$. *Moreover,* R *depends continuously on* B.

Proof
Let $B^{1/2}$ be the unique Hermitian positive definite square root of B (see Remark 1.3), and $Q := Q(B^{1/2})$, $R := R(B^{1/2})$. Then $B = R^*R$, where R is an upper triangular matrix with positive diagonal entries.

To prove the uniqueness of R, suppose that $\widetilde{R}$ is an upper triangular matrix with positive diagonal entries satisfying $B = \widetilde{R}^*\widetilde{R}$. Then $(\widetilde{R}R^{-1})^* = R\widetilde{R}^{-1}$, the matrix on the left being lower triangular and the one on the right upper triangular. This implies that $D := R\widetilde{R}^{-1}$ is a diagonal matrix. Moreover D has positive diagonal entries and $D = D^* = (\widetilde{R}R^{-1})^* = (R\widetilde{R}^{-1})^{-1} = D^{-1}$. Thus $D = I$. This shows that $\widetilde{R} = R$.

Finally, we prove that R depends continuously on B. Let $\widetilde{B}$ be a Hermitian positive matrix in $\mathbb{C}^{n\times n}$ and $\widetilde{B} = \widetilde{R}^*\widetilde{R}$, where $\widetilde{R}$ is upper triangular with positive diagonal entries. Define $H := \widetilde{R} - R$ and $K := \widetilde{B} - B$. Then H is an upper triangular matrix with real diagonal entries, K is a Hermitian matrix and

$$H^*H + R^*H + H^*R = K.$$

To show that $H \to O$ if $K \to O$, we proceed as follows. Suppose that $n = 1$, $H := [\eta]$, $R := [r]$ and $K := [\gamma]$, where $\eta \in \mathbb{R}$, $r > 0$, $r + \eta > 0$ and $\gamma \in \mathbb{R}$. Then

$$\eta^2 + 2\eta r - \gamma = 0,$$

and since $\eta + r > 0$, $\eta = -r + \sqrt{r^2 + \gamma}$ for all $\gamma \in \mathbb{R}$ such that $r^2 + \gamma > 0$. This shows that $\eta \to 0$ if $\gamma \to 0$. Suppose now that $n \geq 2$. For η, γ and r in $\mathbb{R}$, a, b and c in $\mathbb{C}^{(n-1)\times 1}$, H_0, R_0 and K_0 in $\mathbb{C}^{(n-1)\times(n-1)}$, define H, R and K as follows.

$$\mathsf{H} = \begin{bmatrix} \eta & \mathsf{a}^* \\ 0 & \mathsf{H}_0 \end{bmatrix}, \quad \mathsf{R} = \begin{bmatrix} r & \mathsf{b}^* \\ 0 & \mathsf{R}_0 \end{bmatrix}, \quad \mathsf{K} = \begin{bmatrix} \gamma & \mathsf{c}^* \\ \mathsf{c} & \mathsf{W}_0 \end{bmatrix}.$$

Then

$$\eta^2 + 2\eta r - \gamma = 0 \quad \text{and} \quad \eta\mathsf{a} + r\mathsf{a} + \eta\mathsf{b} = \mathsf{c},$$

where $\eta + r > 0$. Hence for $\gamma \in \mathbb{R}$ such that $r^2 + \gamma > 0$, $\eta = -r + \sqrt{r^2 + \gamma}$, as in the case $n = 1$. This shows that $\eta \to 0$ if $\gamma \to 0$. Also, $\mathsf{a} = \dfrac{\mathsf{c} - \eta\mathsf{b}}{\eta + r}$. Hence $\mathsf{a} \to 0$ if $\gamma \to 0$ and $\mathsf{c} \to 0$. Finally,

$$\mathsf{H}_0^*\mathsf{H}_0 + \mathsf{R}_0^*\mathsf{H}_0 + \mathsf{H}_0^*\mathsf{R}_0 = \mathsf{K}_0 := \mathsf{W}_0 - \mathsf{a}\mathsf{a}^* - \mathsf{b}\mathsf{a}^* - \mathsf{a}\mathsf{b}^*,$$

which allows us to complete the mathematical induction, since K_0 is Hermitian and $\mathsf{K}_0 \to \mathsf{O}$ if $\mathsf{K} \to \mathsf{O}$. ∎

The matrix R in the preceding theorem is called the Cholesky factor of B.

The Cholesky Factorization $\mathsf{B} = \mathsf{R}^*\mathsf{R}$ of B may be used to reduce a generalized eigenvalue problem $\mathsf{A}\mathsf{u} = \lambda\mathsf{B}\mathsf{u}$, $\mathsf{u} \neq 0$, where B is a Hermitian positive definite matrix, to an ordinary eigenvalue problem as follows. As R is a nonsingular matrix, the generalized eigenvalue problem can be rewritten as the ordinary eigenvalue problem $(\mathsf{R}^*)^{-1}\mathsf{A}\mathsf{R}^{-1}\mathsf{v} = \lambda\mathsf{v}$. Note that $\mathsf{v} := \mathsf{R}\mathsf{u} \neq 0$ if and only if $\mathsf{u} \neq 0$. If A is a Hermitian matrix, then $(\mathsf{R}^*)^{-1}\mathsf{A}\mathsf{R}^{-1}$ continues to be a Hermitian matrix.

Differential properties of the QR factors and the Cholesky factor are studied in [51].

6.1.1 Householder symmetries

A Householder symmetry is a matrix of the form

$$\mathsf{H} := \mathsf{I} - \beta\mathsf{v}\mathsf{v}^*,$$

where

$$\beta := \begin{cases} \dfrac{2}{\mathsf{v}^*\mathsf{v}} & \text{if } \mathsf{v} \neq 0 \\ 0 & \text{if } \mathsf{v} = 0. \end{cases}$$

A simple calculation shows that H is nonsingular and

$$\mathsf{H}^{-1} = \mathsf{H}^* = \mathsf{H},$$

that is, H is Hermitian and unitary.

The following result shows the utility of such a symmetry for calculating the QR factors of a full rank matrix.

Lemma 6.4
Let $\mathsf{x} := [\mathsf{x}(1), \ldots, \mathsf{x}(n)]^\top \in \mathbb{C}^{n\times 1}$ *and* $k \in [\![1, n-1]\!]$ *be such that there exists* $i \in [\![k+1, n]\!]$ *satisfying* $\mathsf{x}(i) \neq 0$. *Define*

$$\alpha := \sqrt{\sum_{j=k}^{n} |\mathsf{x}(j)|^2} \quad \textit{and} \quad s := \begin{cases} \alpha \dfrac{\mathsf{x}(k)}{|\mathsf{x}(k)|} & \textit{if } \mathsf{x}(k) \neq 0, \\ \alpha & \textit{if } \mathsf{x}(k) = 0. \end{cases}$$

Let $\mathsf{v} := [\mathsf{v}(1), \ldots, \mathsf{v}(n)]^\top \in \mathbb{C}^{n\times 1}$ *be defined by*

$$\mathsf{v}(j) := \begin{cases} 0 & \textit{for } j \in [\![1, k-1]\!], \\ \mathsf{x}(k) + s & \textit{for } j = k, \\ \mathsf{x}(j) & \textit{for } j \in [\![k+1, n]\!], \end{cases}$$

and

$$\beta := \frac{1}{\overline{s}\,\mathsf{v}(k)}.$$

Then $\mathsf{H} := \mathsf{I} - \beta \mathsf{v}\mathsf{v}^*$ *is a Householder symmetry and*

$$\mathsf{H}\mathsf{x} = \sum_{j=1}^{k-1} \mathsf{x}(j)\mathsf{e}_j - s\mathsf{e}_k.$$

Proof
From the definition of v and s, we get

$$\begin{aligned} \mathsf{v}^*\mathsf{v} &= \sum_{j=1}^{n} \overline{\mathsf{v}(j)}\mathsf{v}(j) = \overline{(\mathsf{x}(k) + s)}(\mathsf{x}(k) + s) + \sum_{j=k+1}^{n} |\mathsf{x}(j)|^2 \\ &= \overline{(\mathsf{x}(k) + s)}(\mathsf{x}(k) + s) + \overline{s}s - \overline{\mathsf{x}(k)}\mathsf{x}(k) = 2\overline{s}(s + \mathsf{x}(k)) \neq 0. \end{aligned}$$

Hence

$$\frac{2}{\mathsf{v}^*\mathsf{v}} = \frac{1}{\overline{s}(s + \mathsf{x}(k))} = \beta.$$

This shows that H is a Householder symmetry. On the other hand,

$$\begin{aligned} \mathsf{v}^*\mathsf{x} &= \overline{(\mathsf{x}(k)+s)}\mathsf{x}(k) + \sum_{j=k+1}^{n} \overline{\mathsf{x}(j)}\mathsf{x}(j) \\ &= \overline{(\mathsf{x}(k)+s)}\mathsf{x}(k) + |s|^2 - |\mathsf{x}(k)|^2 \\ &= \overline{(\mathsf{x}(k)+s)}s \;=\; (\mathsf{x}(k)+s)\overline{s}, \end{aligned}$$

and hence $\beta \mathsf{v}^*\mathsf{x} = 1$. Thus

$$\begin{aligned} \mathsf{H}\mathsf{x} &= \sum_{j=1}^{k-1}[\mathsf{x}(j) - 0]\mathsf{e}_j + [\mathsf{x}(k) - (\mathsf{x}(k)+s)]\mathsf{e}_k + \sum_{k+1}^{n}[\mathsf{x}(j) - \mathsf{x}(j)]\mathsf{e}_j \\ &= \sum_{j=1}^{k-1}\mathsf{x}(j)\mathsf{e}_j - s\mathsf{e}_k, \end{aligned}$$

and the proof is complete. ∎

Example 6.5 Computing QR factorizations:
Let

$$\mathsf{A} := \begin{bmatrix} 2 & 0 \\ 2 & -4 \\ \mathrm{i} & 9-\mathrm{i} \end{bmatrix}.$$

We first compute the QR factors using the Gram-Schmidt Process. Let

$$\mathsf{x}_1 := \begin{bmatrix} 2 \\ 2 \\ \mathrm{i} \end{bmatrix} \quad \text{and} \quad \mathsf{x}_2 := \begin{bmatrix} 0 \\ -4 \\ 9-\mathrm{i} \end{bmatrix}.$$

We compute

$$\|\mathsf{x}_1\|_2 = \sqrt{\mathsf{x}_1^*\mathsf{x}_1}, \; \mathsf{q}_1 = \frac{\mathsf{x}_1}{\|\mathsf{x}_1\|_2}, \; \widetilde{\mathsf{q}}_2 = \mathsf{x}_2 - \mathsf{x}_2^*\mathsf{q}_1\mathsf{q}_1 \;\text{ and }\; \mathsf{q}_2 = \frac{\widetilde{\mathsf{q}}_2}{\|\widetilde{\mathsf{q}}_2\|_2}.$$

Then

$$\mathsf{Q} := [\mathsf{q}_1 \; \mathsf{q}_2] = \begin{bmatrix} \frac{2}{3} & \frac{1+\mathrm{i}}{2\sqrt{5}} \\ \frac{2}{3} & \frac{\mathrm{i}-1}{2\sqrt{5}} \\ \frac{\mathrm{i}}{3} & \frac{2}{\sqrt{5}} \end{bmatrix}, \quad \mathsf{R} = \begin{bmatrix} \|\mathsf{x}_1\|_2 & \mathsf{x}_2^*\mathsf{q}_1 \\ 0 & \|\widetilde{\mathsf{q}}_2\|_2 \end{bmatrix} = \begin{bmatrix} 3 & -3(1+\mathrm{i}) \\ 0 & 4\sqrt{5} \end{bmatrix}.$$

We remark that the Gram-Schmidt process is a sequential algorithm in the sense that no orthonormalized vector can be produced before the preceding ones are computed. This procedure is not convenient for large full matrices since the last columns of Q may fail to be orthogonal to the initial ones due to roundoff errors, and hence Q will not be a unitary matrix.

We now compute the QR factors using complex rotations. The classical real 2×2 planar rotations used to annihilate the second coordinate of a given 2×1 column can be extended to the complex case as follows:

Let $\mathsf{y} := \begin{bmatrix} u \\ v \end{bmatrix} \in \mathbb{C}^{2\times 1}$ be such that $\|\mathsf{y}\|_2 \neq 0$, $\alpha := \dfrac{u}{\|\mathsf{y}\|_2}$ and $\beta := \dfrac{-v}{\|\mathsf{y}\|_2}$. Then the matrix $\mathsf{U} := \begin{bmatrix} \overline{\alpha} & -\overline{\beta} \\ \beta & \alpha \end{bmatrix}$ satisfies $\mathsf{U}\mathsf{y} = \begin{bmatrix} \|\mathsf{y}\|_2 \\ 0 \end{bmatrix}$.

Let $\mathsf{A} \in \mathbb{C}^{n\times m}$ be a given matrix. If the column index j is fixed, then we can annihilate the entries $\mathsf{A}(i,j)$ for $i \in [\![j+1, n]\!]$ by using these 'rotations'. It is easy to prove that if this process is carried out for a column index $j' > j$, then the entries $\mathsf{A}(i,j)$ for $i \in [\![j+1, n]\!]$ are not modified.

Consider the same matrix A as before. We obtain

$$\mathsf{U}_1 := \begin{bmatrix} \frac{1}{\sqrt{2}} & \frac{1}{\sqrt{2}} & 0 \\ \frac{-1}{\sqrt{2}} & \frac{1}{\sqrt{2}} & 0 \\ 0 & 0 & 1 \end{bmatrix} \quad \text{and} \quad \mathsf{U}_1\mathsf{A} := \begin{bmatrix} \frac{4}{\sqrt{2}} & \frac{-4}{\sqrt{2}} \\ 0 & \frac{-4}{\sqrt{2}} \\ \mathrm{i} & 9-\mathrm{i} \end{bmatrix}.$$

Now we set $\mathsf{A}_1 := \mathsf{U}_1\mathsf{A}$. Then

$$\mathsf{U}_2 := \begin{bmatrix} \frac{4}{3\sqrt{2}} & 0 & \frac{-\mathrm{i}}{3} \\ 0 & 1 & 0 \\ \frac{-\mathrm{i}}{\sqrt{2}} & 0 & \frac{4}{3\sqrt{2}} \end{bmatrix} \quad \text{and} \quad \mathsf{U}_2\mathsf{A}_1 := \begin{bmatrix} 3 & -3(1+\mathrm{i}) \\ 0 & \frac{-4}{\sqrt{2}} \\ 0 & \frac{12}{\sqrt{2}} \end{bmatrix}.$$

Finally, we set $\mathsf{A}_2 := \mathsf{U}_2\mathsf{A}_1$ and we have

$$\mathsf{U}_3 := \begin{bmatrix} 1 & 0 & 0 \\ 0 & \dfrac{-1}{\sqrt{10}} & \dfrac{3}{\sqrt{10}} \\ 0 & \dfrac{-3}{\sqrt{10}} & \dfrac{1}{\sqrt{10}} \end{bmatrix} \quad \text{and} \quad \mathsf{U}_3\mathsf{A}_2 := \begin{bmatrix} 3 & -3(1+\mathrm{i}) \\ 0 & 4\sqrt{5} \\ 0 & 0 \end{bmatrix}.$$

We recognize the upper triangular factor R as the submatrix

$$\mathsf{R} := \begin{bmatrix} 3 & -3(1+\mathrm{i}) \\ 0 & 4\sqrt{5} \end{bmatrix}$$

of $\mathsf{U}_3\mathsf{A}_2$ and note that the unitary factor Q consists of the first two columns of $(\mathsf{U}_3\mathsf{U}_2\mathsf{U}_1)^* = \mathsf{U}_1^*\mathsf{U}_2^*\mathsf{U}_3^*$. We remark that in general, each 'rotation' contributes with one zero entry in the construction of the upper triangular factor.

Next, let us construct a QR factorization of a matrix $\mathsf{A} \in \mathbb{C}^{n\times m}$ such that $\operatorname{rank}\mathsf{A} = m$, where $n \geq m \geq 2$, using Householder symmetries. Let $k := 1$ and $\mathsf{x} := \mathsf{A}(\cdot, k)$ in Lemma 6.4. If x satisfies the hypothesis of the Lemma, then define H_1 to be the Householder symmetry H given in the statement of the Lemma, and let $\mathsf{H}_1 := \mathsf{I}$ otherwise. Set $\mathsf{A}_1 := \mathsf{H}_1\mathsf{A}$. Then $\mathsf{A}_1(i,1) = 0$ for each $i \in [\![2, n]\!]$. If $n = m = 2$, then $\mathsf{R} := \mathsf{A}_1$ and $\mathsf{Q} := \mathsf{H}_1$ provide a QR factorization of A. Suppose that $m \geq 3$, and for some $\ell \in [\![1, m-2]\!]$ and each $j \in [\![1,\ell]\!]$, a matrix A_ℓ has been constructed such that $\mathsf{A}_\ell(i,j) = 0$ for each $i \in [\![j+1, n]\!]$. Let $k := \ell + 1$ and $\mathsf{x} := \mathsf{A}_\ell(\cdot, k)$ in Lemma 6.4. If x satisfies the hypothesis of the Lemma, then define $\mathsf{H}_{\ell+1}$ to be the Householder symmetry H given in the statement of the Lemma, and let $\mathsf{H}_{\ell+1} := \mathsf{I}$ otherwise. Set $\mathsf{A}_{\ell+1} := \mathsf{H}_{\ell+1}\mathsf{A}$. Then $\mathsf{A}_{\ell+1}(i,j) = 0$ for each $j \in [\![1, \ell+1]\!]$ and for each $i \in [\![j+1, n]\!]$. This means that

$$\mathsf{A}_{n-1} := \begin{bmatrix} \mathsf{R} \\ \mathsf{O} \end{bmatrix},$$

where $\mathsf{R} \in \mathbb{C}^{m\times m}$ is a nonsingular upper triangular matrix. If Q is defined to be the matrix formed by the first m columns of $\mathsf{H}_1 \cdots \mathsf{H}_{n-1}$, then $\mathsf{A} = \mathsf{QR}$.

Consider the same matrix A as before. Let the computations be carried out with 7 decimal figures. We obtain

$$\mathsf{H}_1 := \begin{bmatrix} -(.6666667)10^0 & -(.6666667)10^0 & (.3333333)10^0\mathrm{i} \\ -(.6666667)10^0 & (.7333333)10^0 & (.1333333)10^0\mathrm{i} \\ -(.3333333)10^0\mathrm{i} & -(.1333333)10^0\mathrm{i} & (.9333333)10^0 \end{bmatrix},$$

$$\mathsf{A}_1 := \begin{bmatrix} -(.3)10^1 & (.30)10^1 + (.30)10^1\mathrm{i} \\ 0 & (.28)10^1 + (.12)10^1\mathrm{i} \\ 0 & (.84)10^1 - (.40)10^0\mathrm{i} \end{bmatrix},$$

$$\mathsf{H}_2 := \begin{bmatrix} (.1)10^1 & 0 & 0 \\ 0 & -(.3405877)10^0 & -(.8455971)10^0 \\ 0 & -(.8455971)10^0 & (.3405877)10^0 \end{bmatrix} + \begin{bmatrix} 0 & 0 & 0 \\ 0 & 0 & -(.4110541)10^0\mathrm{i} \\ 0 & (.4110541)10^0\mathrm{i} & 0 \end{bmatrix},$$

$$\mathsf{A}_2 := \begin{bmatrix} -(.3)10^1 & (.3000000)10^1 + (.3000000)10^1\mathrm{i} \\ 0 & -(.8221083)10^1 - (.3523321)10^1\mathrm{i} \\ 0 & (a + b\mathrm{i}) \cdot 10^{-7} \end{bmatrix},$$

where $(.22)10^1 \leq \sqrt{|a|^2 + |b|^2} \leq (.25)10^1$. We recognize the upper triangular factor R as the submatrix

$$\mathsf{R} := \begin{bmatrix} -(.3)10^1 & (.3000000)10^1 + (.3000000)10^1\mathrm{i} \\ 0 & -(.8221083)10^1 - (.3523321)10^1\mathrm{i} \end{bmatrix}$$

of A_2. The presence of the 'spurious' value $\mathsf{A}_2(3,2) = (a + b\mathrm{i}) \cdot 10^{-7}$ which should be zero will be explained in Subsection 6.4.2. The unitary factor Q consists of the first two columns of $\mathsf{H}_1\mathsf{H}_2$. We remark that H_j contributes with $n - j$ zeros in constructing the upper triangular factor, and that the diagonal entry $\mathsf{R}(2,2)$ is not a real number. ∎

6.1.2 Hessenberg Matrices

A matrix $\mathsf{A} \in \mathbb{C}^{n \times n}$ is said to be an **upper Hessenberg matrix** if $\mathsf{A}(i,j) = 0$ when $i > j + 1$. A matrix $\mathsf{A} \in \mathbb{C}^{n \times n}$ is said to be a **lower Hessenberg matrix** if A^* is an upper Hessenberg matrix. We say that an upper Hessenberg matrix is **irreducible** if $\mathsf{A}(j+1,j) \neq 0$ for $j \in [\![1, n-1]\!]$.

Consider a polynomial $\mathrm{p}(t) := a_0 + \sum_{j=1}^{n-1} a_j t^j + t^n$. The irreducible upper Hessenberg matrix

$$\mathsf{A} := \left[\begin{array}{c|c} 0 & -a_0 \\ \hline & -a_1 \\ \mathsf{I}_{n-1} & \vdots \\ & -a_{n-1} \end{array}\right]$$

is called the **companion matrix** of the polynomial p. It can be proved that p is the characteristic polynomial of A. (See [43].)

Proposition 6.6
Every irreducible upper Hessenberg matrix is similar to the companion matrix of its characteristic polynomial.

Proof
Let A be an irreducible upper Hessenberg matrix. Then $\operatorname{rank}(\mathsf{A}) \geq n-1$. If λ is any eigenvalue of A, then $\dim \mathcal{N}(\mathsf{A} - \lambda \mathsf{I}) = 1$ since $\mathsf{A} - \lambda\mathsf{I}$ is also an irreducible upper Hessenberg matrix. Let $A : \mathbb{C}^{n\times 1} \to \mathbb{C}^{n\times 1}$ be the linear operator represented by A relative to the standard basis for $\mathbb{C}^{n\times 1}$. Set $\mathsf{u}_1 = \mathsf{e}_1$ and $\mathsf{u}_{i+1} = \mathsf{A}\mathsf{u}_i$ for $i \in [\![1, n-1]\!]$. Since u_i belongs to $\operatorname{span}\{\mathsf{e}_j : j \in [\![1,i]\!]\} \setminus \operatorname{span}\{\mathsf{e}_j : j \in [\![1,i-1]\!]\}$ for $i \in [\![1,n]\!]$, the ordered set $[\mathsf{u}_1, \ldots, \mathsf{u}_n]$ forms an ordered basis for $\mathbb{C}^{n\times 1}$. The matrix representing A relative to this basis is the companion matrix of the characteristic polynomial of A. ∎

Proposition 6.7
For $\mathsf{A} \in \mathbb{C}^{n\times n}$*, there are a unitary matrix* Q *and an upper Hessenberg matrix* A_0 *such that*

$$\mathsf{A}_0 = \mathsf{Q}^*\mathsf{A}\mathsf{Q},$$

so that every matrix $\mathsf{A} \in \mathbb{C}^{n\times n}$ *is unitarily similar to an upper Hessenberg matrix.*

Proof
Let $\mathsf{A} \in \mathbb{C}^{n\times n}$. We set

$$\mathsf{w}_2 := [0, \mathsf{A}(2,1), \ldots, \mathsf{A}(n,1)]^\top,\ \alpha_2 := \frac{\mathsf{A}(2,1)}{|\mathsf{A}(2,1)|}\|\mathsf{w}_2\|_2,\ \mathsf{v}_2 := \alpha_2 \mathsf{e}_2$$

and the Householder symmetry

$$\mathsf{H}_2 := \mathsf{I} - 2\frac{\mathsf{v}_2\mathsf{v}_2^*}{\mathsf{v}_2^*\mathsf{v}_2}.$$

It is easy to verify that $\mathsf{H}_2\mathsf{v}_2 = \mathsf{w}_2$ and $\mathsf{H}_2\mathsf{e}_1 = \mathsf{e}_1$. Hence

$$\mathsf{H}_2\mathsf{A}\mathsf{H}_2^* = \begin{bmatrix} \mathsf{w}_2 + \mathsf{A}(1,1)\mathsf{e}_1 & \begin{matrix} \widetilde{\mathsf{a}}^\top \\ \mathsf{A}_2 \end{matrix} \end{bmatrix},$$

where $\widetilde{\mathsf{a}} \in \mathbb{C}^{(n-1)\times 1}$ and $\mathsf{A}_2 \in \mathbb{C}^{(n-1)\times(n-1)}$. Now we suppose that for some $p \in [\![1, n-1]\!]$, there is a unitary matrix U_p such that $\mathsf{U}_p\mathsf{A}\mathsf{U}_p^*$ is partitioned as follows:

$$\mathsf{U}_p\mathsf{A}\mathsf{U}_p^* = \left[\begin{array}{c|c} \mathsf{B}^{(p)} & \widetilde{\mathsf{A}} \\ \hline \begin{matrix} 0 \dots 0 & \alpha_{p+1} \\ 0 \dots 0 & 0 \\ \vdots \ddots \vdots & \vdots \\ 0 \dots 0 & 0 \end{matrix} & \mathsf{A}^{(p+1)} \end{array}\right],$$

where $\mathsf{B}^{(p)}$ is an upper Hessenberg matrix in $\mathbb{C}^{p\times p}$ and $\widetilde{\mathsf{A}}$ and $\mathsf{A}^{(p+1)}$ respectively belong to $\mathbb{C}^{p\times(n-p)}$ and $\mathbb{C}^{(n-p)\times(n-p)}$. If $\mathsf{A}^{(p+1)}$ is not an upper Hessenberg matrix, then we set

$$\mathsf{w}_{p+2}^\top := [\underbrace{0,\dots,0}_{p+1 \text{ times}}, \mathsf{A}(p+2,p+1), \dots, \mathsf{A}(n,p+1)]^\top,$$

$$\alpha_{p+2} := \frac{\mathsf{A}(p+2,p+1)}{|\mathsf{A}(p+2,p+1)|}\|\mathsf{w}_{p+2}\|_2,$$

$$\mathsf{v}_{p+2} := \alpha_{p+2}\mathsf{e}_{p+2},$$

and

$$\mathsf{H}_{p+2} := \mathsf{I} - 2\frac{\mathsf{v}_{p+2}\mathsf{v}_{p+2}^*}{\mathsf{v}_{p+2}^*\mathsf{v}_{p+2}}.$$

It is clear that

$$\mathsf{H}_{p+2}\mathsf{U}_p\mathsf{A}(\mathsf{H}_{p+2}\mathsf{U}_p)^* = \left[\begin{array}{c|c} \mathsf{B}^{(p+1)} & \widetilde{\widetilde{\mathsf{A}}} \\ \hline \begin{matrix} 0 \dots 0 & \alpha_{p+2} \\ 0 \dots 0 & 0 \\ \vdots \ddots \vdots & \vdots \\ 0 \dots 0 & 0 \end{matrix} & \mathsf{A}^{(p+2)} \end{array}\right],$$

where

$$\mathsf{B}^{(p+1)} := \left[\begin{array}{ccc|c} & & & \widetilde{\mathsf{A}}(1,p+1) \\ & \mathsf{B}^{(p)} & & \vdots \\ & & & \widetilde{\mathsf{A}}(p,p+1) \\ \hline 0 & \dots\ 0 & \alpha_{p+1} & \mathsf{A}^{(p+1)}(1,1) \end{array}\right]$$

is an upper Hessenberg matrix.

By finite induction, there is a unitary matrix Q such that $\mathsf{A}_0 := \mathsf{Q}^*\mathsf{A}\mathsf{Q}$ is an upper Hessenberg matrix. ∎

Proposition 6.8

Let A *be a nonsingular irreducible upper Hessenberg complex matrix. Let* $\mathsf{A} = \mathsf{Q}\mathsf{R}$ *be a QR factorization of* A *and let* $\mathsf{B} = \mathsf{R}\mathsf{Q}$. *Then* Q *and* B *are nonsingular irreducible upper Hessenberg matrices.*

Proof

Since A and Q are nonsingular matrices, R is a nonsingular matrix and R^{-1} is an upper triangular matrix with nonzero diagonal entries. Let n be the order of A. Since $\mathsf{Q} = \mathsf{A}\mathsf{R}^{-1}$, for each $j \in [\![1, n-1]\!]$, the jth column of Q is a linear combination of the first j columns of A such that the scalar multiplying the jth column of A is nonzero. Hence Q is a nonsingular irreducible upper Hessenberg matrix. Also, for each $j \in [\![1, n-1]\!]$, the jth column of B is a linear combination of the first $j+1$ columns of R such that the scalar multiplying the $(j+1)$th column of R is nonzero. Hence B is an irreducible upper Hessenberg matrix. B is nonsingular because Q and R are. ∎

6.2 Convergence of a Sequence of Subspaces

6.2.1 Basic Definitions

Note that for a sequence (A_k) in $\mathbb{C}^{m\times n}$ and $\mathsf{A} \in \mathbb{C}^{m\times n}$,

$$\|\mathsf{A}_k - \mathsf{A}\|_2 \to 0 \quad \text{if and only if}$$

$$\mathsf{A}_k(i,j) \to \mathsf{A}(i,j) \text{ for every } i \in [\![1,m]\!] \text{ and every } j \in [\![1,n]\!].$$

Lemma 6.9
Let (R_k) *be a sequence of upper triangular matrices in* $\mathbb{C}^{m\times m}$ *with positive diagonal entries. If* $\|\mathsf{R}_k^*\mathsf{R}_k - \mathsf{I}\|_2 \to 0$, *then* $\|\mathsf{R}_k - \mathsf{I}\|_2 \to 0$.

Proof
Let $\mathsf{B}_k := \mathsf{R}_k^*\mathsf{R}_k$ and $\mathsf{B} := \mathsf{I} = \mathsf{I}^*\mathsf{I}$. The conclusion follows from the continuity of the Cholesky factor established in Theorem 6.3. ∎

In applications, we frequently need to approximate spectral subspaces of operators. Different definitions of the convergence of a sequence of subspaces have been proposed. (See for instance [60].) We shall adopt the following definition.

We denote by $\mathcal{M}$ the set of all linear subspaces of $\mathbb{C}^{n\times 1}$. Let (M_1, M_2) be a pair of elements of $\mathcal{M}$, and Π_1 (resp. Π_2) denote the matrix representing the orthogonal projection on M_1 (resp. M_2) relative to the standard basis for $\mathbb{C}^{n\times 1}$. We define

$$d(M_1, M_2) := \|\Pi_1 - \Pi_2\|_2.$$

Remark 6.10
It can be easily proved that d is a metric on $\mathcal{M}$, and that $d(M_1, M_2) = \text{gap}(M_1, M_2)$. (See Remark 2.16.) This statement is valid for closed subspaces M_1 and M_2 of any Hilbert space. (See page 198 of [47].) ∎

Since it is difficult to compute $d(M_1, M_2)$, we give some equivalent characterizations of the convergence of a sequence of subspaces in terms of ordered bases.

Let $M \in \mathcal{M}$. An ordered basis for M is represented by a matrix U whose columns contain the coordinates of the corresponding vectors relative to the standard basis for $\mathbb{C}^{n\times 1}$. If $m := \dim M$, then U belongs to $\mathbb{C}^{n\times m}$.

Proposition 6.11
Let $(M_k)_{k\geq 1}$ *be a sequence which converges to* M *in* $\mathcal{M}$. *Then*

$$\dim M_k = \dim M \quad \textit{for all large enough } k.$$

Proof
Let Π (resp. Π_k) denote the matrix representing the orthogonal projec-

tion onto M_k (resp. M) in the standard basis for $\mathbb{C}^{n\times 1}$. Since (M_k) converges to M, $d(M_k, M) := \|\Pi_k - \Pi\|_2 < 1$ for all large enough k. Hence $\rho(\Pi_k - \Pi) < 1$ for all large enough k, and by Lemma 2.11, $\dim M_k = \dim M$. ∎

Theorem 6.12
The following statements are equivalent:

(i) *The sequence of subspaces (M_k) converges to a subspace M.*

(ii) *Given any orthonormal basis for M, let Q be its matrix representation in the standard basis for $\mathbb{C}^{n\times 1}$. Then there is a sequence (Q_k) such that Q_k represents an orthonormal basis for M_k in the standard basis for $\mathbb{C}^{n\times 1}$, and*

$$\lim_{k\to\infty} \|\mathsf{Q}_k - \mathsf{Q}\|_2 = 0.$$

(iii) *Given any orthonormal basis for M, let Q be the matrix representing it in the standard basis for $\mathbb{C}^{n\times 1}$. Then there exists a matrix Y such that $[\mathsf{Q}, \mathsf{Y}]$ represents an ordered basis for $\mathbb{C}^{n\times 1}$ in its standard basis, and there exist matrices $\mathsf{X}_k \in \mathbb{C}^{n\times m}$, $\mathsf{C}_k \in \mathbb{C}^{m\times m}$, $\mathsf{D}_k \in \mathbb{C}^{(n-m)\times m}$ such that*

   (a) X_k *represents an ordered basis for M_k in the standard basis for $\mathbb{C}^{n\times 1}$ for all large enough k,*

   (b) C_k *is invertible for all large enough k,*

   (c) $\mathsf{X}_k = \mathsf{Q}\mathsf{C}_k + \mathsf{Y}\mathsf{D}_k$, *and*

   (d) $\|\mathsf{D}_k\mathsf{C}_k^{-1}\|_2 \to 0$.

(iv) *For each sequence (Q_k) in $\mathbb{C}^{n\times m}$ such that Q_k represents an orthonormal basis for M_k in the standard basis for $\mathbb{C}^{n\times 1}$, there is a sequence $(\mathsf{Q}_{[k]})$ in $\mathbb{C}^{n\times m}$ such that $\mathsf{Q}_{[k]}$ represents an orthonormal basis for M in the standard basis for $\mathbb{C}^{n\times 1}$, and*

$$\lim_{k\to\infty} \|\mathsf{Q}_k - \mathsf{Q}_{[k]}\|_2 = 0.$$

Proof

(i) implies (ii):

Suppose that $M_k \to M$. Let Q represent an orthonormal basis for M. Then $\mathsf{X}_k := \Pi_k\mathsf{Q} \to \Pi\mathsf{Q} = \mathsf{Q}$. Hence $\mathsf{X}_k^*\mathsf{X}_k \to \mathsf{Q}^*\mathsf{Q} = \mathsf{I}$. Let

$\mathsf{Q}_k := \mathrm{Q}(\mathsf{X}_k)$ and $\mathsf{R}_k := \mathrm{R}(\mathsf{X}_k)$ be the unitary and upper triangular factors of X_k, respectively. Then $\mathsf{R}_k^*\mathsf{R}_k \to \mathsf{I}$ and by Lemma 6.9, $\mathsf{R}_k \to \mathsf{I}$. Hence $\mathsf{R}_k^{-1} \to \mathsf{I}$ and

$$\|\mathsf{Q}_k - \mathsf{Q}\|_2 = \|\mathsf{X}_k\mathsf{R}_k^{-1} - \mathsf{Q}\|_2 \le \|\mathsf{X}_k - \mathsf{Q}\|_2\|\mathsf{R}_k^{-1}\|_2 + \|\mathsf{Q}\|_2\|\mathsf{R}_k^{-1} - \mathsf{I}\|_2 \to 0.$$

(ii) implies (iii):

Suppose that for each matrix Q representing an orthonormal basis for M, there exists a matrix Q_k representing an orthonormal basis for M_k such that $\|\mathsf{Q}_k - \mathsf{Q}\|_2 \to 0$. Let $\mathsf{Y} \in \mathbb{C}^{(n-m)\times n}$ represent an orthonormal basis for $M^\perp$. Then $[\mathsf{Q}, \mathsf{Y}]$ represents an orthonormal basis for $\mathbb{C}^{n\times 1}$, and there exist matrices $\mathsf{C}_k \in \mathbb{C}^{m\times m}$ and $\mathsf{D}_k \in \mathbb{C}^{(n-m)\times m}$ such that $\mathsf{Q}_k = \mathsf{Q}\mathsf{C}_k + \mathsf{Y}\mathsf{D}_k$. Since $\mathsf{Q}^*\mathsf{Y} = \mathsf{O}$ and $\mathsf{Y}^*\mathsf{Y} = \mathsf{I}$, then $\mathsf{Q}^*\mathsf{Q}_k = \mathsf{C}_k$ and $\mathsf{Y}^*\mathsf{Q}_k = \mathsf{D}_k$. Thus $\|\mathsf{C}_k - \mathsf{I}\|_2 \le \|\mathsf{Q}_k - \mathsf{Q}\|_2 \to 0$ and $\|\mathsf{D}_k\|_2 = \|\mathsf{Y}\mathsf{Q}_k\|_2 \to 0$. This shows that C_k is nonsingular for all large enough k and that $\|\mathsf{C}_k^{-1}\|_2 \to 1$. We conclude that $\|\mathsf{D}_k\mathsf{C}_k^{-1}\|_2 \to 0$.

(iii) implies (iv):

Let Q_k represent an orthonormal basis for M_k, Q represent an orthonormal basis for M, Y be such that $[\mathsf{Q}, \mathsf{Y}]$ represents a basis for $\mathbb{C}^{n\times 1}$, C_k and D_k be such that $\mathsf{X}_k := \mathsf{Q}\mathsf{C}_k + \mathsf{Y}\mathsf{D}_k$ represents a basis for M_k, and $\|\mathsf{D}_k\mathsf{C}_k^{-1}\|_2 \to 0$ as $k \to \infty$. There exists a nonsingular matrix B_k such that

$$\mathsf{Q}_k\mathsf{B}_k = \mathsf{X}_k\mathsf{C}_k^{-1} = \mathsf{Q} + \mathsf{Y}\mathsf{D}_k\mathsf{C}_k^{-1}$$

and hence $\|\mathsf{Q}_k\mathsf{B}_k - \mathsf{Q}\|_2 \to 0$ and $\|\mathsf{B}_k^*\mathsf{B}_k - \mathsf{I}\|_2 \to 0$ as $k \to \infty$. Let $\mathsf{B}_k = \mathrm{Q}(\mathsf{B}_k)\mathrm{R}(\mathsf{B}_k)$ be the QR factorization of B_k. Set $\mathsf{R}_k := \mathrm{R}(\mathsf{B}_k)$. Then by Lemma 6.9, $\|\mathsf{R}_k^*\mathsf{R}_k - \mathsf{I}\|_2 \to 0$ and $\|\mathsf{R}_k^{-1} - \mathsf{I}\|_2 \to 0$ as $k \to \infty$. From the identity

$$\mathsf{Q}_k - \mathsf{Q}\mathsf{B}_k^{-1} = (\mathsf{Q}_k\mathsf{B}_k - \mathsf{Q})\mathsf{B}_k^{-1} = (\mathsf{Q}_k\mathsf{B}_k - \mathsf{Q})\mathsf{R}_k^{-1}\mathrm{Q}(\mathsf{B}_k)^*,$$

we conclude that $\|\mathsf{Q}_k - \mathsf{Q}\mathsf{B}_k^{-1}\|_2 \to 0$ as $k \to \infty$. Now, let $\mathsf{R}_{[k]} := \mathrm{R}(\mathsf{Q}\mathsf{B}_k^{-1})$ and $\mathsf{Q}_{[k]} := \mathrm{Q}(\mathsf{Q}\mathsf{B}_k^{-1})$ be the QR factors of $\mathsf{Q}\mathsf{B}_k^{-1}$. We can prove, as before, that $\|\mathsf{R}_{[k]} - \mathsf{I}\|_2 \to 0$ and hence $\|\mathsf{Q}_k - \mathsf{Q}_{[k]}\|_2 \to 0$ as $k \to \infty$.

(iv) implies (i):

Suppose that for each matrix Q_k representing an orthonormal basis for M_k, there exists a matrix $\mathsf{Q}_{[k]}$ representing an orthonormal basis for M, such that $\|\mathsf{Q}_k - \mathsf{Q}_{[k]}\|_2 \to 0$. The matrices of the orthogonal projections onto M_k and M can be computed as $\Pi_k = \mathsf{Q}_k\mathsf{Q}_k^*$ and $\Pi = \mathsf{Q}_{[k]}\mathsf{Q}_{[k]}^*$, respectively. Hence

$$\|\Pi_k - \Pi\|_2 \le \|(\mathsf{Q}_k - \mathsf{Q}_{[k]})\mathsf{Q}_k^*\|_2 + \|\mathsf{Q}_{[k]}(\mathsf{Q}_k^* - \mathsf{Q}_{[k]}^*)\|_2 \le 2\|\mathsf{Q}_k - \mathsf{Q}_{[k]}\|_2 \to 0.$$

This proves that $M_k \to M$. ∎

Theorem 6.13
Let S be a proper subspace of M and S_k be a proper subspace of M_k. Suppose that $S_k \to S$ and $M_k \to M$ as $k \to \infty$. Let $\mathsf{U}_k := [\mathsf{X}_k, \mathsf{Y}_k]$ represent an orthonormal basis for M_k in the standard basis for $\mathbb{C}^{n\times 1}$, such that X_k represents an orthonormal basis for S_k in the standard basis for $\mathbb{C}^{n\times 1}$. Then there exists a matrix $\mathsf{U}_{[k]} := [\mathsf{X}_{[k]}, \mathsf{Y}_{[k]}]$ representing an orthonormal basis for M in the standard basis for $\mathbb{C}^{n\times 1}$, such that $\mathsf{X}_{[k]}$ represents an orthonormal basis for S in the standard basis for $\mathbb{C}^{n\times 1}$, and $\|\mathsf{U}_k - \mathsf{U}_{[k]}\|_2 \to 0$ as $k \to \infty$.

Proof
Let Π_S (resp. Π_M) be the matrix representing the orthogonal projection onto S (resp. M). Then $\|\Pi_S \mathsf{X}_k - \mathsf{X}_k\|_2 \to 0$ and $\|\Pi_M \mathsf{Y}_k - \mathsf{Y}_k\|_2 \to 0$. Hence, for large enough k, $\mathsf{Z}_{[k]} := [\Pi_S \mathsf{X}_k, \Pi_M \mathsf{Y}_k]$ represents a basis for M and $\Pi_S \mathsf{X}_k$ represents a basis for S. Let $\mathsf{U}_{[k]} := \mathrm{Q}(\mathsf{Z}_{[k]})$ and $\mathsf{R}_{[k]} := \mathrm{R}(\mathsf{Z}_{[k]})$ be the QR factors of $\mathsf{Z}_{[k]}$. Then $\mathsf{U}_{[k]} = [\mathsf{X}_{[k]}, \mathsf{Y}_{[k]}]$ represents an orthonormal basis for M such that $\mathsf{X}_{[k]}$ represents an orthonormal basis for S. Since $\|\mathsf{Z}_{[k]} - \mathsf{U}_k\|_2 \to 0$, we conclude that $\|\mathsf{R}_{[k]}^* \mathsf{R}_{[k]} - \mathsf{I}\|_2 \to 0$ and by Lemma 6.9, $\|\mathsf{R}_{[k]} - \mathsf{I}\|_2 \to 0$. Hence $\|\mathsf{U}_{[k]} - \mathsf{U}_k\|_2 \to 0$. ∎

6.2.2 Krylov Sequences

Let $A : \mathbb{C}^{n\times 1} \to \mathbb{C}^{n\times 1}$ be a linear map, $\mathsf{A} \in \mathbb{C}^{n\times n}$ be the matrix representing A in the standard basis for $\mathbb{C}^{n\times 1}$ and M_0 be a linear subspace of $\mathbb{C}^{n\times 1}$. The sequence of subspaces $(M_k)_{k\geq 1}$ defined by

$$M_k := A(M_{k-1})$$

is called the Krylov sequence corresponding to the pair (A, M_0).

We shall prove that under suitable hypotheses on A and M_0, the Krylov sequence converges to an invariant subspace of A.

Theorem 6.14
Assume that the eigenvalues λ_i, $i \in [\![1, n]\!]$, of A satisfy:

$$\text{(H1)} \qquad |\lambda_1| \geq \cdots \geq |\lambda_m| > |\lambda_{m+1}| \geq \cdots \geq |\lambda_n| > 0,$$

for some integer $m \le n-1$. *We set*

$$M := \bigoplus_{i=1}^{m} \mathcal{N}(A - \lambda_i I).$$

Let P *be the projection onto* M *along*

$$N := \bigoplus_{i=m+1}^{n} \mathcal{N}(A - \lambda_i I).$$

Let M_0 *be an* m*-dimensional subspace of* $\mathbb{C}^{n\times 1}$ *such that*

(H2) $$\dim(P(M_0)) = m.$$

Then the Krylov sequence corresponding to the pair (A, M_0) *converges to* M.

Proof

Let X_0 represent a basis for M_0, Q an orthonormal basis for M, and Y an orthonormal basis for N. Then $[\mathsf{Q}, \mathsf{Y}]$ forms a basis for $\mathbb{C}^{n\times 1}$ and hence there exist $\mathsf{C} \in \mathbb{C}^{m\times m}$ and $\mathsf{D} \in \mathbb{C}^{(n-m)\times m}$ such that

$$\mathsf{X}_0 = \mathsf{QC} + \mathsf{YD}.$$

Let P be the matrix representing P in the standard basis for $\mathbb{C}^{n\times 1}$. Hypothesis (H2) implies that C is nonsingular since $\mathsf{PY} = \mathsf{O}$ and PX_0 represents a basis for M. Since M and N are invariant subspaces under A, there exist $\Theta \in \mathbb{C}^{m\times m}$ and $\mathsf{Z} \in \mathbb{C}^{(n-m)\times(n-m)}$ such that

$$\mathsf{AQ} = \mathsf{Q}\Theta \quad \text{and} \quad \mathsf{AY} = \mathsf{YZ}.$$

Recursively, we get

$$\mathsf{A}^k\mathsf{X}_0 = \mathsf{Q}\Theta^k\mathsf{C} + \mathsf{YZ}^k\mathsf{D}.$$

Define $\mathsf{X}_k := \mathsf{A}^k\mathsf{X}_0$, $\mathsf{C}_k := \Theta^k\mathsf{C}$ and $\mathsf{D}_k := \mathsf{Z}^k\mathsf{D}$. Then conditions (a), (b) and (c) of the assertion (iii) of Theorem 6.12 are satisfied. We now prove that condition (d) is also satisfied. The matrix Θ represents the restriction of A to the spectral subspace associated with the subset of eigenvalues $\{\lambda_1, \ldots, \lambda_m\}$ in the orthonormal basis for this subspace represented by Q. Also, Z represents the restriction of A to the spectral subspace associated with the subset of eigenvalues $\{\lambda_{m+1}, \ldots, \lambda_n\}$ in the orthonormal basis for this subspace represented by Y. Hence

$$\rho(\Theta^{-1}) = \frac{1}{|\lambda_m|}, \quad \rho(\mathsf{Z}) = |\lambda_{m+1}| \quad \text{and} \quad \rho(\Theta^{-1})\rho(\mathsf{Z}) = \frac{|\lambda_{m+1}|}{|\lambda_m|} < 1.$$

On the other hand,

$$\|D_k C_k^{-1}\|_2 \leq \|Z^k\|_2 \|D\|_2 \|C^{-1}\|_2 \|\Theta^{-k}\|_2.$$

For $\epsilon := \dfrac{1 - \rho(\Theta^{-1})\rho(Z)}{2} > 0$, there exists an integer k_0 such that for all $k > k_0$, $\|\Theta^{-k}\|_2^{1/k} \|Z^k\|_2^{1/k} < \epsilon + \rho(\Theta^{-1})\rho(Z) = \dfrac{1 + \rho(\Theta^{-1})\rho(Z)}{2} < 1$, so that

$$\|\Theta^{-k}\|_2 \|Z^k\|_2 < \left(\frac{1 + \rho(\Theta^{-1})\rho(Z)}{2}\right)^k \to 0$$

as $k \to \infty$. Thus condition (d) is satisfied. Hence the Krylov sequence corresponding to the pair (A, M_0) converges to M. ▮

Remark 6.15
Let us regard the rate of convergence to 0 of $\|\Theta^{-k}\|_2 \|Z^k\|_2$ to be the rate of convergence of the Krylov sequence. The proof of Theorem 6.14 shows that this rate can be arbitrarily close to $\dfrac{|\lambda_{m+1}|}{|\lambda_m|}$ since ϵ can be chosen arbitrarily close to 0. This implies that if we apply the method to $A - \sigma I$ instead of A, then a suitable choice of $\sigma \in \mathbb{C}$ can substantially improve the rate of convergence. Such a parameter σ is called a shift. The optimal shift, if it exists, would be the solution of the following optimization problem:

$$\text{Find } \widehat{\sigma} \in \mathbb{C} \text{ such that } \quad \frac{|\lambda_{m+1} - \widehat{\sigma}|}{|\lambda_m - \widehat{\sigma}|} = \min\left\{\frac{|\lambda_{m+1} - \sigma|}{|\lambda_m - \sigma|} : \sigma \in \mathbb{C}\right\}.$$

However, since the eigenvalues $\{\lambda_1, \ldots, \lambda_n\}$, are not known, it is not possible to determine $\widehat{\sigma}$. ▮

Remark 6.16 The Power Method:
The most elementary application of the convergence of a Krylov sequence is the case $m = 1$ in Theorem 6.14.

Let $z_0 \in \mathbb{C}^{n \times 1}$ be such that $z_0 \neq 0$ and ϵ be a given positive number. The Power Method consists of the following algorithm:

```
Begin
  read(ε,MaxIter,z0);
  z := z0 / ||z0||_2
  k:=1;
  while Az ≠ 0, k<MaxIter and ||(I − zz*)Az||_2 > ε do
    begin
      z:= Az / ||Az||_2 ;
      k:=k+1
    end;
End.
```

The vector $(\mathsf{I} - \mathsf{z}\mathsf{z}^*)\mathsf{A}\mathsf{z}$ is the residual of z as an approximation to an eigenvector of A corresponding to an eigenvalue which is approximated by $\mathsf{z}^*\mathsf{A}\mathsf{z}$. Thus the stopping criterion of this algorithm is based on a maximal residual tolerance:

$$\|(\mathsf{I} - \mathsf{z}\mathsf{z}^*)\mathsf{A}\mathsf{z}\|_2 \leq \epsilon.$$

We remark that the sequence of vectors produced by the Power Method may not itself converge to a specific eigenvector of A. For example, if $\mathsf{A} := \begin{bmatrix} -1 & 0 \\ 0 & 0 \end{bmatrix}$ and $\mathsf{z}_0 := \begin{bmatrix} 1 \\ 0 \end{bmatrix}$, then the kth iterate is given by $\mathsf{z}_k = (-1)^k \mathsf{z}_0$ and is an eigenvector of A corresponding to the eigenvalue $\lambda = -1$.

A nonzero vector y is called a **left eigenvector** of A corresponding to an eigenvalue λ of A if $\mathsf{y}^*\mathsf{A} = \lambda \mathsf{y}^*$, that is, y is an eigenvector of A^* corresponding to the eigenvalue $\overline{\lambda}$ of A^*. Let us assume that the eigenvalues λ_i of A satisfy the hypothesis (H1) of Theorem 6.14 for $m = 1$:

$$|\lambda_1| > |\lambda_2| \geq |\lambda_3| \geq \ldots \geq |\lambda_n| > 0.$$

The hypothesis (H2) of Theorem 6.14 now reads as follows:

M_0 is spanned by a vector z_0 such that $\mathsf{y}^*\mathsf{z}_0 \neq 0$ for any left eigenvector y corresponding to λ_1.

If u_1 is a unit eigenvector of A corresponding to λ_1, then $\lambda_1 = \mathsf{u}_1^*\mathsf{A}\mathsf{u}_1$. According to Theorem 1.2, there is a unitary matrix $\mathsf{U} \in \mathbb{C}^{n \times n}$ such that

$$\mathsf{U}^*\mathsf{A}\mathsf{U} = \begin{bmatrix} \lambda_1 & \mathsf{v}^* \\ 0 & \mathsf{B} \end{bmatrix}$$

for some $\mathsf{v} \in \mathbb{C}^{(n-1)\times 1}$ and $\mathsf{B} \in \mathbb{C}^{(n-1)\times(n-1)}$.

A left eigenvector y of A^* corresponding to $\overline{\lambda}_1$ satisfies

$$\mathsf{U}^*\mathsf{y} = \begin{bmatrix} \alpha \\ \mathsf{w} \end{bmatrix}, \quad \text{where} \quad \begin{cases} \alpha \neq 0 \\ \text{and} \\ \mathsf{w} \text{ is such that } \alpha \mathsf{v} + \mathsf{B}^*\mathsf{w} = \overline{\lambda}_1 \mathsf{w}. \end{cases}$$

The eigenvalues of B are $\lambda_2, \ldots, \lambda_n$. Since $\dim \mathcal{N}(\mathsf{A} - \lambda_1 \mathsf{I}) = 1$, our hypothesis implies that $c := \mathsf{u}_1^* \mathsf{z}_0 \neq 0$. Hence the coordinates of z_0 relative to the basis U are $\begin{bmatrix} c \\ \mathsf{v} \end{bmatrix}$ and

$$\mathsf{U}^*\mathsf{A}^n\mathsf{U}\begin{bmatrix} c \\ \mathsf{v} \end{bmatrix} = \begin{bmatrix} \lambda_1^n c + \sum_{i=0}^{n-1} \lambda_1^i \mathsf{v}^* \mathsf{B}^{n-1-i}\mathsf{v} \\ \mathsf{B}^n \mathsf{v} \end{bmatrix}.$$

This procedure is known as the Inverse Iteration Algorithm when it is applied to the inverse of the shifted matrix $\mathsf{A} - \sigma\mathsf{I}$, where $\sigma \notin \mathrm{sp}(\mathsf{A})$ is an approximate eigenvalue of A. ∎

The rate of convergence of the Power Method can be arbitrarily close to $|\lambda_2|/|\lambda_1|$. It may be very poor, if $|\lambda_2|$ is close to $|\lambda_1|$.

Example 6.17 Actual rate of convergence of the Power Method:
Let

$$\mathsf{A} := \begin{bmatrix} 133.21 & 152.07 & -165.05 \\ -96.15 & -110.05 & 117.75 \\ 20.03 & 23.01 & -24.15 \end{bmatrix}.$$

This matrix is constructed in such a way that its eigenvalues are $\lambda := 1.01$ (which is a simple eigenvalue), and -1 (which is a double defective eigenvalue). The Power Method will give a sequence of one-dimensional subspaces converging to the eigenspace corresponding to λ and the expected rate of convergence is arbitrarily close to

$$r(\mathsf{A}) := \frac{1}{1.01} = 0.9900(9900)\ldots$$

The shifted matrix $\mathsf{A}' := \mathsf{A} + 1.01\mathsf{I}$ has the same spectral subspaces as A. If the Power Method is applied to A', then the expected rate of convergence is arbitrarily close to

$$r(\mathsf{A}') := \frac{0.01}{2.02} = 0.00495.$$

When we run the algorithm on a computer, starting with $z_0 := e_1$, we observe the following behavior:

Iteration	$\sin\theta$	$\sin\theta'$
1	$(.745)10^{-2}$	$(.518)10^{-2}$
2	$(.901)10^{-2}$	$(.159)10^{-3}$
3	$(.733)10^{-2}$	$(.121)10^{-5}$
4	$(.823)10^{-2}$	0
$\vdots$	$\vdots$	
1800	$(.997)10^{-6}$	

where θ (resp. θ') is the acute angle between the approximate eigenspace and the exact eigenspace when the method is applied to A (resp. A'). It can be shown that the sinus of this angle is equal to the gap between the subspaces. (See Remark 2.16 and [26].) Thus if we apply the Power Method to A, we need 1800 iterations to produce an approximate subspace with a gap less than 10^{-6}, while if we apply it to A', we only need 4 iterations to achieve this. ∎

6.3 QR Methods and Inverse Iteration

6.3.1 The Francis-Kublanovskaya QR Method

There are many algorithms to approximate the eigenvalues and eigenvectors of matrices. One of the most classical ones is the Basic QR Method:

```
Begin
  read(A);
  A_0:= an upper Hessenberg matrix unitarily similar to A;
  k := 1;
  while Not(Stopping Criterion) do
    begin
      A_{k-1} := Q_k R_k      % some QR factorization of A_{k-1}
      A_k := R_k Q_k
      k := k + 1
    end;
End.
```

Before presenting the stopping criterion for this method as well as some of the main variants of the algorithm, we shall give a convergence result for the basic version:

Theorem 6.18 Convergence of the Basic QR Method:
Let $\mathsf{A}_0 \in \mathbb{C}^{n\times n}$ *be a nonsingular irreducible upper Hessenberg matrix. Suppose that its eigenvalues, denoted by* λ_j, *satisfy*

$$|\lambda_1| > |\lambda_2| > \ldots > |\lambda_{n-1}| > |\lambda_n| > 0.$$

Let A_k, Q_k *and* R_k *be the matrices defined in the Basic QR Method. Define*

$$\mathsf{U}_k := \mathsf{Q}_1\mathsf{Q}_2\cdots\mathsf{Q}_k, \qquad \text{and} \qquad \mathsf{T}_k := \mathsf{R}_k\mathsf{R}_{k-1}\cdots\mathsf{R}_1.$$

For each $j \in [\![1,n]\!]$ *and each positive integer* k, *let* $M_k(j)$ *denote the subspace of* $\mathbb{C}^{n\times 1}$ *spanned by the first* j *columns of* A_0^k *and set*

$$M(j) := \bigoplus_{i=1}^{j} \mathcal{N}(\mathsf{A}_0 - \lambda_i\mathsf{I}).$$

Then the following results hold:

(a) *For each positive integer* k, Q_k *and* A_k *are irreducible upper Hessenberg matrices and* $\mathsf{A}_k = \mathsf{U}_k^*\mathsf{A}_0\mathsf{U}_k$;

(b) *for each positive integer* k, $\mathsf{U}_k\mathsf{T}_k = \mathsf{A}_0^k$ *and for each* $j \in [\![1,n]\!]$, *the first* j *columns of* U_k *form an orthonormal basis for* $M_k(j)$;

(c) *for each* $j \in [\![1,n]\!]$, *the sequence* $(M_k(j))_{k\geq 1}$ *converges to* $M(j)$;

(d) *for each positive integer* k, *there exists a unitary matrix* $\mathsf{U}_{[k]}$ *such that for each* $j \in [\![1,n]\!]$, *the first* j *columns of* $\mathsf{U}_{[k]}$ *span* $M(j)$, *and*

$$\lim_{k\to\infty} \|\mathsf{U}_{[k]} - \mathsf{U}_k\|_{\mathrm{F}} = 0;$$

(e) *for each positive integer* k, *the matrix*

$$\mathsf{A}_{[k]} := \mathsf{U}_{[k]}^*\mathsf{A}_0\mathsf{U}_{[k]}$$

is upper triangular, its diagonal coefficients are the eigenvalues of A_0 *arranged in the decreasing order of moduli and*

$$\lim_{k\to\infty} \|\mathsf{A}_{[k]} - \mathsf{A}_k\|_{\mathrm{F}} = 0;$$

(f) *the rate of convergence of the Basic QR Method is arbitrarily close to*

$$r := \max\left\{\frac{|\lambda_{i+1}|}{|\lambda_i|} : i \in [\![1, n-1]\!]\right\}.$$

Proof

(a) By Proposition 6.8, if A_k is a nonsingular irreducible upper Hessenberg matrix, then so are Q_k and A_k. Since for each positive integer k, $\mathsf{A}_k = \mathsf{Q}_k^* \mathsf{A}_{k-1} \mathsf{Q}_k$, we obtain $\mathsf{A}_k = \mathsf{U}_k^* \mathsf{A}_0 \mathsf{U}_k$.

(b) For each $k \geq 1$,

$$\mathsf{U}_k \mathsf{T}_k = \mathsf{Q}_1 \cdots \mathsf{Q}_{k-1} \mathsf{Q}_k \mathsf{R}_k \mathsf{R}_{k-1} \cdots \mathsf{R}_1.$$

Replacing the product $\mathsf{Q}_k \mathsf{R}_k$ by $\mathsf{R}_{k-1} \mathsf{Q}_{k-1}$, we obtain

$$\mathsf{U}_k \mathsf{T}_k = \mathsf{Q}_1 \cdots \mathsf{Q}_{k-2} \mathsf{Q}_{k-1} \mathsf{R}_{k-1} \mathsf{Q}_{k-1} \mathsf{R}_{k-1} \mathsf{R}_{k-2} \cdots \mathsf{R}_1$$

and now the product $\mathsf{Q}_{k-1} \mathsf{R}_{k-1}$ appears twice. Replacing it by $\mathsf{R}_{k-2} \mathsf{Q}_{k-2}$, we obtain

$$\mathsf{U}_k \mathsf{T}_k = \mathsf{Q}_1 \cdots \mathsf{Q}_{k-2} \mathsf{R}_{k-2} \mathsf{Q}_{k-2} \mathsf{R}_{k-2} \mathsf{Q}_{k-2} \mathsf{R}_{k-2} \cdots \mathsf{R}_1$$

and now the product $\mathsf{Q}_{k-2} \mathsf{R}_{k-2}$ appears three times.

Continuing in this way, the only product present in $\mathsf{U}_k \mathsf{T}_k$ will be $\mathsf{Q}_1 \mathsf{R}_1$ and it will appear k times. But $\mathsf{Q}_1 \mathsf{R}_1 = \mathsf{A}_0$. Thus $\mathsf{U}_k \mathsf{T}_k = \mathsf{A}_0^k$.

(c) Let $A_0 : \mathbb{C}^{n\times 1} \to \mathbb{C}^{n\times 1}$ be the linear operator represented by the matrix A_0 with respect to the standard basis for $\mathbb{C}^{n\times 1}$. Result (b) means that the Basic QR Method produces a set of n nested Krylov sequences corresponding to $(A_0, M_0(j))$, $j \in [\![1, n]\!]$, where

$$M_0(j) := \operatorname{span}[\mathsf{e}_1, \ldots, \mathsf{e}_j],$$

that is, for each $j \in [\![1, n]\!]$, $M_0(j)$ is the j dimensional subspace spanned by the first j vectors in the standard basis for $\mathbb{C}^{n\times 1}$.

The hypothesis on the eigenvalues of A_0 implies that hypothesis (H1) of Theorem 6.14 is satisfied for each $j \in [\![1, n-1]\!]$. We shall prove that hypothesis (H2) of Theorem 6.14 is satisfied for each $j \in [\![1, n]\!]$. For this purpose, we recall that for any linear space X, any subspace M of X, and any linear map $P : X \to X$, $\mathcal{N}(P_{|M}) = M \cap \mathcal{N}(P)$; also, if M is finite dimensional, then the restriction $P_{|M} : M \to X$ is injective if and only if $\operatorname{rank}(P_{|M}) = \dim M$. These facts imply that if $P(j)$ denotes the spectral projection onto $M(j)$, then hypothesis (H2) of Theorem 6.14 is equivalent to

$$\mathcal{N}(P(j)) \cap M_0(j) = \{0\} \quad \text{for each } j \in [\![1, n]\!].$$

Suppose this is not the case. Then for some j and some ℓ such that $1 \leq \ell \leq j$, one should have $\mathsf{e}_\ell \in N := \mathcal{N}(P(j)) = \mathcal{R}(I - P(j))$. Since N is invariant under A_0, the vectors $A_0\mathsf{e}_\ell, A_0^2\mathsf{e}_\ell, \ldots, A_0^{n-\ell}\mathsf{e}_\ell$ must belong to N. But since A_0 is bijective, all these vectors are linearly independent and hence $\dim N \geq n-\ell+1$. As $\dim N = n-j$, we obtain a contradiction $\ell \geq j+1$. The convergence of the sequence $(M_k(j))_{k\geq 1}$ to the subspace $M(j)$ for each $j \in [\![1, n]\!]$ follows from Theorem 6.14.

(d) We apply Theorem 6.13 recursively: First consider the pair of subspaces $M := M(n) = \mathbb{C}^{n\times 1}$ and $S := M(n-1)$ which are the limits of $M_k := M_k(n)$ and $S_k := M_k(n-1)$, respectively. Then let $M := M(n-1)$ and $S := M(n-2)$ which are the limits of $M_k := M_k(n-1)$ and $S_k := M_k(n-2)$, respectively. Finally, let $M := M(2)$ and $S := M(1)$ which are the limits of $M_k := M_k(2)$ and $S_k := M_k(1)$, respectively.

(e) For each $j \in [\![1, n]\!]$, the first j columns of $\mathsf{U}_{[k]}$ form an ordered basis for $M(j)$. Since $M(j)$ is invariant under A_0, we conclude that each of the first j columns of $\mathsf{A}_0\mathsf{U}_{[k]}$ is a linear combination of (only) the first j columns of $\mathsf{U}_{[k]}$, that is, there exists an upper triangular matrix $\mathsf{A}_{[k]}$ such that $\mathsf{A}_0\mathsf{U}_{[k]} = \mathsf{U}_{[k]}\mathsf{A}_{[k]}$. Since $\mathsf{U}_{[k]}$ is a unitary matrix, the upper triangular matrix $\mathsf{A}_{[k]}$ is unitarily similar to A_0 and has the eigenvalues of A_0 along its diagonal. The order in which the eigenvalues appear along the diagonal of $\mathsf{A}_{[k]}$ follows from the order of the columns of $\mathsf{U}_{[k]}$: for each $j \in [\![1, n]\!]$, the first j columns form an ordered basis for $M(j)$, the spectral subspace associated with the subset of eigenvalues $\{\lambda_1, \ldots, \lambda_j\}$ of A_0, where $|\lambda_1| > \cdots > |\lambda_j|$. Now, for $k \geq 2$, A_k is unitarily similar to A_0 since $\mathsf{A}_k = \mathsf{U}_k^*\mathsf{A}_0\mathsf{U}_k$. We thus have

$$\mathsf{A}_{[k]} - \mathsf{A}_k = (\mathsf{U}_{[k]} - \mathsf{U}_k)\mathsf{A}_0\mathsf{U}_{[k]} + \mathsf{U}_k\mathsf{A}_0(\mathsf{U}_{[k]} - \mathsf{U}_k).$$

Since $\|\mathsf{U}_{[k]}\|_{\mathrm{F}} = \|\mathsf{U}_k\|_{\mathrm{F}} = \sqrt{n}$ (because $\mathsf{U}_{[k]}$ and U_k are unitary matrices), we conclude that $\|\mathsf{A}_{[k]} - \mathsf{A}_k\|_{\mathrm{F}}$ tends to 0 as k tends to infinity.

(f) Since the n Krylov sequences on which the Basic QR Method is based are nested, Remark 6.15 gives the rate of convergence of the method. ∎

Note that all that can be said about the superdiagonal entries of A_k is that they remain bounded in k, since $\|\mathsf{A}_k\|_F = \|\mathsf{A}_0\|_F$ for all k. This means that the sequence (A_k) may not converge to a specific matrix.

For the same reasons which motivated shifting in the Power Method, we may consider shifting the Basic QR Method at each step.

The Single Shift QR Method reads as follows:

```
Begin
  read(A);
  A0:= an upper Hessenberg matrix similar to A
  k := 1;
  while Not(Stopping criterion) do
    begin
      Choice of the shift σk
      Ã(k-1) := A(k-1) − σk I
      Ã(k-1) := Q̃k R̃k      % some QR factorization of Ã(k-1)
      Ak := R̃k Q̃k + σk I
      k := k + 1
    end;
End.
```

Example 6.19 Usual single shifts for the QR Method:

Rayleigh's shift: $\sigma_k := \mathsf{A}_{k-1}(n,n)$,

Wilkinson's shift: $\sigma_k := \widetilde{\lambda}_{k-1}$,

where $\widetilde{\lambda}_{k-1}$ is the eigenvalue of

$$\mathsf{B}_{k-1} := \begin{bmatrix} \mathsf{A}_{k-1}(n-1,n-1) & \mathsf{A}_{k-1}(n-1,n) \\ \mathsf{A}_{k-1}(n,n-1) & \mathsf{A}_{k-1}(n,n) \end{bmatrix}$$

which is closest to $\mathsf{A}_{k-1}(n,n)$.

These shifts lead, under certain hypotheses, to a quadratic convergence of the Single Shift QR Method. (See [74], [63], [37] and [71].) However, the Single Shift QR Method is almost never used. ∎

To complete this subsection we present an implicit Double Shift QR Method.

Lemma 6.20

Let B *be an upper Hessenberg matrix and* p *a polynomial none of whose roots is an eigenvalue of* B*. Then there is a unitary matrix* Q *such that* $Q^*p(B)$ *is an upper triangular matrix and* Q^*BQ *is an upper Hessenberg matrix.*

Proof

Suppose $p(t) := a\prod_{i=1}^{d}(t - \tau_i)$. Define $B_1 := B$. Let Q_1R_1 be a QR factorization of $B_1 - \tau_1 I$ and let $B_2 := Q_1^* B_1 Q_1$. We obtain

$$p(B) = aQ_1Q_1^*Q_1\prod_{i=2}^{d}(B_1 - \tau_i I)Q_1^*(B_1\tau_1 I) = aQ_1\prod_{i=2}^{d}(B_2 - \tau_i I)R_1,$$

and $B_2 = Q_1^*[(B_1 - \tau_1 I) + \tau_1 I]Q_1 = R_1Q_1 + \tau_1 I$.

Suppose that we have constructed a unitary matrix Q_j and an upper triangular matrix R_j such that $B_j - \tau_j I = Q_jR_j$ for $j \in [\![1, \ell]\!]$. Let $B_{j+1} := R_jQ_j$ and $V_j := Q_1 \cdots Q_j$. Then

$$B_{j+1} = Q_j^*(Q_jR_j + \tau_j I)Q_j = Q_j^*B_jQ_j = V_jB_1V_j,$$

and

$$\begin{aligned} p(B) &= a\prod_{i=\ell+1}^{d}(B - \tau_i I)\prod_{k=0}^{\ell-1}V_{\ell-k}V_{\ell-k}^*(B - \tau_{\ell-k} I) \\ &= a\prod_{i=\ell+1}^{d}(B - \tau_i I)V_\ell\prod_{k=0}^{\ell-1}Q_{\ell-k}^*(B_{\ell-k} - \tau_{\ell-k} I) \\ &= a\prod_{i=\ell+1}^{d}(B - \tau_i I)V_\ell\prod_{k=0}^{\ell-1}R_{\ell-k}. \end{aligned}$$

Hence

$$\begin{aligned} p(B) &= a\prod_{i=\ell+2}^{d}(B - \tau_i I)V_\ell[V_\ell^*(B - \tau_{\ell+1} I)V_\ell]\prod_{k=0}^{\ell-1}R_{\ell-k} \\ &= a\prod_{i=\ell+2}^{d}(B - \tau_i I)V_\ell(B_{\ell+1} - \tau_{\ell+1} I)\prod_{k=0}^{\ell-1}R_{\ell-k}. \end{aligned}$$

The QR factorization $B_{\ell+1} - \tau_{\ell+1} I = Q_{\ell+1}R_{\ell+1}$ concludes the finite induction. Setting

$$Q := Q_1Q_2\cdots Q_d \quad \text{and} \quad R := R_dR_{d-1}\cdots R_1,$$

we obtain a QR factorization for $\mathsf{p}(\mathsf{B})$: $\mathsf{p}(\mathsf{B}) = \mathsf{QR}$. ∎

The QR factorization of $\mathsf{p}(\mathsf{B})$ exhibited in Lemma 6.20 is called an explicit multiple iteration of order d. It involves the QR factorization of d matrices and needs the knowledge of the roots of p. However, the following theorem shows that an implicit multiple iteration is possible, that is to say, QR factors of $\mathsf{p}(\mathsf{B})$ can be computed implicitly, which amounts to using multiple shifts implicitly.

Theorem 6.21 Implicit Multiple Shift QR Theorem:
Let A *be an* $n{\times}n$ *matrix. Let* Q *and* $\widetilde{\mathsf{Q}}$ *be unitary matrices such that* $\mathsf{A}_0 := \mathsf{Q}^*\mathsf{A}\mathsf{Q}$ *and* $\widetilde{\mathsf{A}}_0 := \widetilde{\mathsf{Q}}^*\mathsf{A}\widetilde{\mathsf{Q}}$ *are both irreducible upper Hessenberg matrices. If* $\mathsf{Q}\mathsf{e}_1$ *and* $\widetilde{\mathsf{Q}}\mathsf{e}_1$ *are linearly dependent, then there is a unitary diagonal matrix* D *such that* $\mathsf{Q} = \widetilde{\mathsf{Q}}\mathsf{D}$.

Proof
If $\mathsf{Q}\mathsf{e}_1$ and $\widetilde{\mathsf{Q}}\mathsf{e}_1$ are linearly dependent, there is $\theta_1 \in I\!R$ such that $\mathsf{Q}\mathsf{e}_1 = e^{\mathrm{i}\,\theta_1}\widetilde{\mathsf{Q}}\mathsf{e}_1$. Suppose there exists an integer p, $1 \le p < n$, and p real numbers θ_j, $j \in [\![1,p]\!]$, such that $\mathsf{Q}\mathsf{e}_j = e^{\mathrm{i}\,\theta_j}\widetilde{\mathsf{Q}}\mathsf{e}_j$. Since $\mathsf{A} = \mathsf{Q}\mathsf{A}_0\mathsf{Q}^* = \widetilde{\mathsf{Q}}\widetilde{\mathsf{A}}_0\widetilde{\mathsf{Q}}^*$, we have $\mathsf{A}\mathsf{Q}\mathsf{e}_p = \mathsf{Q}\mathsf{A}_0\mathsf{e}_p = e^{\mathrm{i}\,\theta_p}\mathsf{A}\widetilde{\mathsf{Q}}\mathsf{e}_p = \widetilde{\mathsf{Q}}\widetilde{\mathsf{A}}_0\mathsf{e}_p$ and therefore $\sum_{j=1}^{p+1} \mathsf{A}_0(j,p)\mathsf{Q}\mathsf{e}_j = e^{\mathrm{i}\,\theta_p}\sum_{j=1}^{p+1} \widetilde{\mathsf{A}}_0(j,p)\widetilde{\mathsf{Q}}\mathsf{e}_j$. According to our hypothesis, we get $\mathsf{A}_0(p+1,p)\mathsf{Q}\mathsf{e}_{p+1} = e^{\mathrm{i}\,\theta_p}\widetilde{\mathsf{A}}_0(p+1,p)\widetilde{\mathsf{Q}}\mathsf{e}_{p+1}$, and since $\mathsf{A}_0(p+1,p)$ and $\widetilde{\mathsf{A}}_0(p+1,p)$ are nonzero, there exists $\theta_{p+1} \in I\!R$ such that $\mathsf{Q}\mathsf{e}_{p+1} = e^{\mathrm{i}\,\theta_{p+1}}\widetilde{\mathsf{Q}}\mathsf{e}_{p+1}$. By finite induction, the proof is complete. ∎

One of the advantages of the Implicit Double Shift QR Method is that when A is real, one complex shift and its conjugate can be employed without using complex arithmetic computations.

6.3.2 Simultaneous Inverse Iteration Method

The approximate eigenvalues produced by the QR Method can be used as shifts for the Simultaneous Inverse Iteration Method which we now describe.

Let $\sigma \in C\!\!\!\!C$ be an approximate eigenvalue of A in the sense that there exists an eigenvalue λ of A such that

$$0 < |\lambda - \sigma| < \min\{|\mu - \sigma| \,:\, \mu \in \mathrm{sp}(\mathsf{A}),\ \mu \neq \lambda\}.$$

If the algebraic multiplicity of λ is m, σ may be defined as the arithmetic mean of the cluster of approximations of λ produced by the QR Method. Let Z_0 be a given matrix in $\mathbb{C}^{n\times m}$ of rank m and ϵ a given positive number. The Simultaneous Inverse Iteration Method consists of the following algorithm:

```
Begin
   read(ε,MaxIter,Z0);
   Z := Z0
   k:=1;
   while k<MaxIter and ||(I − ZZ*)AZ||_F > ε do
      begin
         Solve (A − σI)Y = Z;
         Z:=Orthonormalization of Y;
         k:=k+1
      end;
End.
```

The orthonormalization step may be accomplished by a QR factorization of Y. We emphasize that the iterations are stopped with a residual test.

Remark 6.22
The preceding algorithm generates the Krylov sequence corresponding to the matrix $(\mathsf{A}-\sigma\mathsf{I})^{-1}$ and the starting basis Z_0. Hence it converges if the hypothesis H2 of Theorem 6.14 is satisfied. In the case of a simple eigenvalue, this happens if we start the algorithm with $\mathsf{z}_0 := \mathsf{e}_n$.

In case of convergence, the rate of convergence can be arbitrarily close to

$$r(\sigma) := \max\left\{\frac{|\lambda-\sigma|}{|\mu-\sigma|} \; : \; \mu \in \mathrm{sp}(\mathsf{A})\,, \mu \neq \lambda\,, \frac{|\lambda-\sigma|}{|\mu-\sigma|} < 1\right\}.$$

Since σ is close to an eigenvalue of A, we may expect that the matrix $(\mathsf{A}-\sigma\mathsf{I})^{-1}$ be ill conditioned with respect to the linear system involved in the computation of Y satisfying $(\mathsf{A}-\sigma\mathsf{I})\mathsf{Y} = \mathsf{Z}$; but fortunately, at least when λ is a simple eigenvalue, the error involved in computing $\mathsf{Y} \in \mathbb{C}^{n\times 1}$ has its major component in the direction of the eigenspace associated with λ. (See [26].) However, when λ is a multiple defective eigenvalue, this error may affect the precision of the generalized approximate eigenvectors associated with λ. In order to overcome this lack of precision, some variants of the Newton Method have been proposed. (See Sections 2.10 and 5.9 of [26]). ∎

6.4 Error Analysis

6.4.1 Condition Numbers or Forward Error Analysis

Roughly speaking, the forward error analysis attempts to measure the sensitivity of the solution of an equation under small perturbations of the parameters appearing in the equation. These parameters are known as the data. If the relative error of the solution of the perturbed equation is bounded by some constant times the relative perturbation of the data, then that constant is called a condition number of the data relative to the equation.

Remark 6.23 Condition number for linear systems:
Let $\mathsf{A} \in \mathbb{C}^{n\times n}$ be a nonsingular matrix and $\mathsf{b} \in \mathbb{C}^{n\times 1}$. Consider the problem of computing the (unique) solution $\mathsf{x} = \mathsf{u} \in \mathbb{C}^{n\times 1}$ of the linear system $\mathsf{Ax} = \mathsf{b}$.

Let $F : \mathbb{C}^{n\times n} \times \mathbb{C}^{n\times 1} \times \mathbb{C}^{n\times 1} \to \mathbb{C}^{n\times 1}$ be the operator defined by

$$F(\mathsf{E},\mathsf{x},\mathsf{y}) := \mathsf{Ex} - \mathsf{y}, \quad (\mathsf{E},\mathsf{x},\mathsf{y}) \in \mathbb{C}^{n\times n} \times \mathbb{C}^{n\times 1} \times \mathbb{C}^{n\times 1}.$$

Then $F(\mathsf{A},\mathsf{u},\mathsf{b}) = 0$.

The partial Fréchet derivatives of F with respect to (E,y) and with respect to x at $(\mathsf{A},\mathsf{u},\mathsf{b})$ are the linear operators $\mathrm{D}_{(\mathsf{E},\mathsf{y})}F(\mathsf{A},\mathsf{u},\mathsf{b})$ and $\mathrm{D}_{\mathsf{x}}F(\mathsf{A},\mathsf{u},\mathsf{b})$ given by

$$\mathrm{D}_{(\mathsf{E},\mathsf{y})}F(\mathsf{A},\mathsf{u},\mathsf{b})(\mathsf{H},\mathsf{k}) := \mathsf{Hu} - \mathsf{k} \quad \text{for } (\mathsf{H},\mathsf{k}) \in \mathbb{C}^{n\times n} \times \mathbb{C}^{n\times 1},$$

and

$$\mathrm{D}_{\mathsf{x}}F(\mathsf{A},\mathsf{u},\mathsf{b})\mathsf{h} := \mathsf{Ah} \quad \text{for } \mathsf{h} \in \mathbb{C}^{n\times 1},$$

respectively. Since A is nonsingular, $\mathrm{D}_{\mathsf{x}}F(\mathsf{A},\mathsf{u},\mathsf{b})$ is bijective. The equation $F(\mathsf{E},\mathsf{x},\mathsf{y}) = 0$ implicitly defines a Fréchet differentiable function $(\mathsf{E},\mathsf{y}) \mapsto \mathsf{x}(\mathsf{E},\mathsf{y})$ for (E,y) in some open neighborhood $\mathcal{O}$ of (A,b), $\mathsf{x}(\mathsf{A},\mathsf{b}) = \mathsf{u} = \mathsf{A}^{-1}\mathsf{b}$ and

$$\begin{aligned}\mathrm{D}\mathsf{x}(\mathsf{A},\mathsf{b})(\mathsf{H},\mathsf{k}) &= -\mathrm{D}_{\mathsf{x}}F(\mathsf{A},\mathsf{A}^{-1}\mathsf{b},\mathsf{b})^{-1}(\mathrm{D}_{(\mathsf{E},\mathsf{y})}F(\mathsf{A},\mathsf{A}^{-1}\mathsf{b},\mathsf{b})(\mathsf{H},\mathsf{k}))\\ &= -\mathsf{A}^{-1}(\mathsf{HA}^{-1}\mathsf{b} - \mathsf{k}) = -\mathsf{A}^{-1}(\mathsf{Hu} - \mathsf{k}).\end{aligned}$$

Let $\|\cdot\|_2$ be the matrix subordinated norm in $\mathbb{C}^{n\times n}$. Since $\|\mathsf{b}\|_2 =$

$\|\mathsf{Au}\|_2 \le \|\mathsf{A}\|_2\|\mathsf{u}\|_2$ we have $\dfrac{\|\mathsf{b}\|_2}{\|\mathsf{u}\|_2} \le \|\mathsf{A}\|_2$. Hence if

$$\epsilon(\widetilde{\mathsf{A}},\widetilde{\mathsf{b}}) := \frac{\|\widetilde{\mathsf{A}}-\mathsf{A}\|_2}{\|\mathsf{A}\|_2} + \frac{\|\widetilde{\mathsf{b}}-\mathsf{b}\|_2}{\|\mathsf{b}\|_2}$$

is so small that $(\widetilde{\mathsf{A}},\widetilde{\mathsf{b}}) \in \mathcal{O}$, then there exists a unique vector $\widetilde{\mathsf{u}} := \mathsf{x}(\widetilde{\mathsf{A}},\widetilde{\mathsf{b}})$ such that $\widetilde{\mathsf{A}}\widetilde{\mathsf{u}} = \widetilde{\mathsf{b}}$ and

$$\frac{\|\widetilde{\mathsf{u}}-\mathsf{u}\|_2}{\|\mathsf{u}\|_2} \le \|\mathsf{A}\|_2\|\mathsf{A}^{-1}\|_2\,\epsilon(\widetilde{\mathsf{A}},\widetilde{\mathsf{b}}) + o(\epsilon(\widetilde{\mathsf{A}},\widetilde{\mathsf{b}})).$$

Thus

$$\kappa_2(\mathsf{A}) := \|\mathsf{A}\|_2\,\|\mathsf{A}^{-1}\|_2$$

is a condition number of A relative to the linear system $\mathsf{Ax} = \mathsf{b}$. ∎

Lemma 6.24
Let $A : \mathbb{C}^{n\times 1} \to \mathbb{C}^{n\times 1}$ be a bijective linear map, $\lambda \in \mathrm{sp}(A)$, $P := P(A,\{\lambda\})$, $M := \mathcal{R}(P)$, $m := \dim M$, $\mathsf{A} \in \mathbb{C}^{n\times n}$ be the matrix which represents A, and $\mathsf{P} \in \mathbb{C}^{n\times n}$ be the matrix which represents P with respect to the standard basis for $\mathbb{C}^{n\times 1}$. Consider a matrix $\mathsf{U} \in \mathbb{C}^{n\times m}$ representing an orthonormal basis for M and a matrix $\mathsf{Y} \in \mathbb{C}^{n\times m}$ such that $\mathsf{Y}^*\mathsf{U} = \mathsf{I}_m$. Then $\lambda \notin \mathrm{sp}((\mathsf{I}_n - \mathsf{UY}^*)\mathsf{A})$.

Proof
Let $\mathsf{V} \in \mathbb{C}^{n\times(n-m)}$ be such that $[\mathsf{U},\mathsf{V}] \in \mathbb{C}^{n\times n}$ is a nonsingular matrix, $\mathsf{Z} \in \mathbb{C}^{n\times(n-m)}$ be such that $[\mathsf{Y},\mathsf{Z}]^* = [\mathsf{U},\mathsf{V}]^{-1}$ and $\Theta := \mathsf{Y}^*\mathsf{AU}$. Since $\mathsf{Y}^*\mathsf{U} = \mathsf{I}_m$, the matrix UY^* represents a projection onto M in the standard basis for $\mathbb{C}^{n\times 1}$ and $\mathrm{sp}(\Theta) = \{\lambda\}$. Also, $\mathsf{Z}^*\mathsf{V} = \mathsf{I}_n$ and $\mathsf{I}_n - \mathsf{UY}^* = \mathsf{VZ}^*$. But

$$\begin{aligned}\mathsf{A} &= \left[\mathsf{UY}^* + (\mathsf{I}_n - \mathsf{UY}^*)\right]\mathsf{A}\left[\mathsf{UY}^* + (\mathsf{I}_n - \mathsf{UY}^*)\right]\\ &= [\mathsf{U},\mathsf{V}]\begin{bmatrix}\Theta & \mathsf{Y}^*\mathsf{AV}\\ 0 & \mathsf{Z}^*\mathsf{AV}\end{bmatrix}\begin{bmatrix}\mathsf{Y}^*\\ \mathsf{Z}^*\end{bmatrix} = [\mathsf{U},\mathsf{V}]\begin{bmatrix}\Theta & \mathsf{Y}^*\mathsf{AV}\\ 0 & \mathsf{Z}^*\mathsf{AV}\end{bmatrix}[\mathsf{U},\mathsf{V}]^{-1}.\end{aligned}$$

Hence $\lambda \notin \mathrm{sp}(\mathsf{Z}^*\mathsf{AV})$. To conclude that $\lambda \notin \mathrm{sp}((\mathsf{I}_n - \mathsf{UY}^*)\mathsf{A})$ we proceed as follows. Let $\mathsf{x} \in \mathbb{C}^{n\times 1}$ be such that $\mathsf{x} \ne 0$ and $(\mathsf{I}_n - \mathsf{UY}^*)\mathsf{A})\mathsf{x} = \lambda\mathsf{x}$. Then $\mathsf{x} = (\mathsf{I}_n - \mathsf{UY}^*)\mathsf{x}$ and hence $(\mathsf{I}_n - \mathsf{UY}^*)\mathsf{A}(\mathsf{I}_n - \mathsf{UY}^*)\mathsf{x} = \lambda\mathsf{x}$. Thus $\mathsf{x} = \mathsf{VZ}^*\mathsf{x} \ne 0$, and $\mathsf{VZ}^*\mathsf{AVZ}^*\mathsf{x} = \lambda\mathsf{x}$. Multiplying by Z^* on the left, we get $\mathsf{Z}^*\mathsf{AV}(\mathsf{Z}^*\mathsf{x}) = \lambda(\mathsf{Z}^*\mathsf{x})$. This implies that $\mathsf{Z}^*\mathsf{x} = 0$ because $\lambda \notin \mathrm{sp}(\mathsf{Z}^*\mathsf{AV})$. It follows that $\mathsf{x} = 0$ and hence $\lambda \notin \mathrm{sp}((\mathsf{I}_n - \mathsf{UY}^*)\mathsf{A})$. ∎

Consider now the equations $(\mathsf{I}_n - \mathsf{X}\mathsf{Y}^*)\mathsf{A}\mathsf{X} = \mathsf{O}$, $\mathsf{Y}^*\mathsf{X} = \mathsf{I}_m$.

Let $F : \mathbb{C}^{n\times n} \times \mathbb{C}^{n\times m} \to \mathbb{C}^{n\times n} \times \mathbb{C}^{n\times m}$ be the operator defined by

$$F(\mathsf{E},\mathsf{X}) := (\mathsf{I}_n - \mathsf{X}\mathsf{Y}^*)\mathsf{E}\mathsf{X} \quad \text{for } (\mathsf{E},\mathsf{X}) \in \mathbb{C}^{n\times n} \times \mathbb{C}^{n\times m}.$$

Then $F(\mathsf{A},\mathsf{U}) = (\mathsf{O},\mathsf{O})$ and the partial Fréchet derivatives of F at (A,U) are the linear operators $\mathrm{D}_{\mathsf{E}}F(\mathsf{A},\mathsf{U})$ and $\mathrm{D}_{\mathsf{X}}F(\mathsf{A},\mathsf{U})$ given by

$$\mathrm{D}_{\mathsf{E}}F(\mathsf{A},\mathsf{U})\mathsf{K} := (\mathsf{I}_n - \mathsf{U}\mathsf{Y}^*)\mathsf{K}\mathsf{U} \quad \text{for } \mathsf{K} \in \mathbb{C}^{n\times n},$$

and

$$\mathrm{D}_{\mathsf{X}}F(\mathsf{A},\mathsf{U})\mathsf{H} := (\mathsf{I}_n - \mathsf{U}\mathsf{Y}^*)\mathsf{A}\mathsf{H} - \mathsf{H}\Theta \quad \text{for } \mathsf{H} \in \mathbb{C}^{n\times m},$$

respectively. By Lemma 6.24, $\lambda \notin \mathrm{sp}((\mathsf{I}_n - \mathsf{U}\mathsf{Y}^*)\mathsf{A})$, so that $\mathrm{D}_{\mathsf{X}}F(\mathsf{A},\mathsf{U})$ is bijective. The equation $F(\mathsf{E},\mathsf{X}) = \mathsf{O}$ implicitly defines a Fréchet differentiable function $\mathsf{E} \mapsto \mathsf{X}(\mathsf{E})$ for E in some open neighborhood $\mathcal{O}_{\mathsf{A}}$ of A, $\mathsf{X}(\mathsf{A}) = \mathsf{U}$ and

$$\mathsf{D}\mathsf{X}(\mathsf{A})\mathsf{K} = -\mathrm{D}_{\mathsf{X}}F(\mathsf{A},\mathsf{U})^{-1}(\mathrm{D}_{\mathsf{E}}F(\mathsf{A},\mathsf{U})\mathsf{K}) = -\Sigma_{\mathsf{U},\mathsf{Y}}(\mathsf{K}\mathsf{U}),$$

by the Implicit Function Theorem.

Let $\|\cdot\|_2$ be the matrix subordinated norm in $\mathbb{C}^{n\times n}$ and $\mathbb{C}^{n\times m}$. It follows that if

$$\epsilon(\widetilde{\mathsf{A}}) := \frac{\|\widetilde{\mathsf{A}} - \mathsf{A}\|_2}{\|\mathsf{A}\|_2}$$

is so small that $\widetilde{\mathsf{A}} \in \mathcal{O}_{\mathsf{A}}$, then there exists a unique $\widetilde{\mathsf{U}} := \mathsf{X}(\widetilde{\mathsf{A}})$ such that

$$\|\widetilde{\mathsf{U}} - \mathsf{U}\|_2 \le \|\mathsf{A}\|_2 \|\Sigma_{\mathsf{U},\mathsf{Y}}\|\, \epsilon(\widetilde{\mathsf{A}}) + o(\epsilon(\widetilde{\mathsf{A}})),$$

where $\|\cdot\|$ is the operator norm in $\mathrm{BL}(\mathbb{C}^{n\times m})$ subordinated to the 2-norm in $\mathbb{C}^{n\times m}$. This suggests that

$$\mathrm{cond}_{\mathsf{U}}(\mathsf{A}) := \|\mathsf{A}\|_2\, \|\Sigma_{\mathsf{U},\mathsf{Y}}\|$$

is a condition number of A relative to the equations $(\mathsf{I}_n - \mathsf{X}\mathsf{Y}^*)\mathsf{A}\mathsf{X} = \mathsf{O}$, $\mathsf{Y}^*\mathsf{X} = \mathsf{I}_m$.

It can be shown that the norm of $\Sigma_{\mathsf{U},\mathsf{Y}}$ is minimized by the choice $\mathsf{Y} := \mathsf{U}$, that is, when the projection on M is orthogonal. (See [2].)

We consider the operator $G : \mathbb{C}^{n\times n} \to \mathbb{C}^{m\times m}$ defined by

$$G(\mathsf{E}) := \mathsf{Y}^*\mathsf{E}\mathsf{X}(\mathsf{E}) \quad \text{for } \mathsf{E} \in \mathbb{C}^{n\times n}.$$

Clearly, $G(\mathsf{A}) = \Theta$. The Fréchet derivative of G at A is given by

$$DG(\mathsf{A})\mathsf{K} := \mathsf{Y}^*\mathsf{K} + \mathsf{Y}^*\mathsf{A}\mathsf{D}\mathsf{X}(\mathsf{A})\mathsf{K} \quad \text{for } \mathsf{K} \in \mathbb{C}^{n\times n}.$$

Choose Y to be the matrix representing the unique ordered basis for $\mathcal{R}(P^*)$ such that $\mathsf{Y}^*\mathsf{U} = \mathsf{I}_m$. Then the matrix $\mathsf{P} := \mathsf{U}\mathsf{Y}^*$ represents the operator P with respect to the standard basis for $\mathbb{C}^{n\times 1}$. We have

$$\mathsf{U}\mathrm{D}G(\mathsf{A})\mathsf{K} = \mathsf{PKU} - \mathsf{PA}\Sigma_{\mathsf{U},\mathsf{Y}}(\mathsf{KU}) = \mathsf{PKU},$$

since P and A commute and $\mathsf{P}\Sigma_{\mathsf{U},\mathsf{Y}} = \mathsf{O}$. As a consequence,

$$\mathrm{D}G(\mathsf{A})\mathsf{K} = \mathsf{U}^*\mathsf{PKU}$$

and

$$\|\mathrm{D}G(\mathsf{A})\| \le \|\mathsf{P}\|_2.$$

Hence if $\epsilon(\widetilde{\mathsf{A}})$ is so small that $\widetilde{\mathsf{A}} \in \mathcal{O}_{\mathsf{A}}$, then letting $\widetilde{\Theta} := G(\widetilde{\mathsf{A}})$, we have

$$\|\widetilde{\Theta} - \Theta\|_2 \le \|\mathsf{P}\|_2\, \epsilon(\widetilde{\mathsf{A}}) + o(\epsilon(\widetilde{\mathsf{A}})).$$

Since $\mathrm{sp}(\Theta) = \{\lambda\}$, we have $\|\Theta\|_2^{-1} \le \dfrac{1}{|\lambda|}$, so that

$$\frac{\|\widetilde{\Theta} - \Theta\|_2}{\|\Theta\|_2} \le \kappa(\mathsf{A})\|\mathsf{P}\|_2\, \epsilon(\widetilde{\mathsf{A}}) + o(\epsilon(\widetilde{\mathsf{A}})).$$

Let $\widetilde{\lambda} := \dfrac{1}{m}\,\mathrm{tr}(\widetilde{\Theta})$. Then

$$|\widetilde{\lambda} - \lambda| = \frac{1}{m}\,|\,\mathrm{tr}(\widetilde{\Theta} - \Theta)| \le \rho(\widetilde{\Theta} - \Theta) \le \|\widetilde{\Theta} - \Theta\|_2$$

and

$$\frac{|\widetilde{\lambda} - \lambda|}{|\lambda|} \le \frac{\|\mathsf{A}\|_2}{|\lambda|}\|\mathsf{P}\|_2\, \epsilon(\widetilde{\mathsf{A}}) + o(\epsilon(\widetilde{\mathsf{A}})).$$

We remark that $\dfrac{\|\mathsf{A}\|_2}{|\lambda|} \in [1, \kappa_2(\mathsf{A})]$. Thus

$$\mathrm{cond}_\lambda(\mathsf{A}) := \kappa_2(\mathsf{A})\,\|\mathsf{P}\|_2$$

is a condition number of A relative to the computation of λ.

6.4.2 Stability or Backward Error Analysis

Roughly speaking, the **backward error** analysis consists of showing that the approximate results given by the computer are in fact the exact results corresponding to some perturbed data. It gives a measure of the stability of the computation.

Let $\mathbf{c}(\cdot)$ denote the computed value or result of the element in the brackets. For example, given real numbers x and y, $\mathbf{c}(x+y)$ denotes the computed result of their sum, given columns x and y in $\mathbb{C}^{n\times 1}$, $\mathbf{c}(\langle \mathsf{x}, \mathsf{y}\rangle)$ denotes the computed value of the inner product $\langle \mathsf{x}, \mathsf{y}\rangle$, $\mathbf{c}(\|\mathsf{y}\|_2)$ denotes the computed value of the norm $\|\mathsf{y}\|_2$, and $\mathbf{c}(\mathsf{Ax})$ denotes the computed result of the product Ax, where $\mathsf{A} \in \mathbb{C}^{n\times m}$.

The backward error analysis addresses the following problem:

Given normed linear spaces X and Y, a function $f : X \to Y$ and an element $x \in X$, find $h \in X$ such that

$$f(x+h) = \mathbf{c}(f(x)) \quad \text{and} \quad \|h\| \leq \|\Delta x\| \quad \text{for}$$
$$\text{each } \Delta x \in X \text{ such that } f(x+\Delta x) = \mathbf{c}(f(x)),$$

where $\|\cdot\|$ denotes the norm on the space X.

In most practical situations, we are not able to compute h and we restrict our attention to finding an element $\Delta x \in X$ such that $f(x+\Delta x) = \mathbf{c}(f(x))$ and then estimating its norm in terms of some 'reasonable' upper bound of the unit roundoff of the computer, which we now define.

Computers use a particular set of real numbers called floating point numbers, which we denote by $\mathbb{F}$.

From a mathematical point of view, the arithmetic unit of a computer is characterized by four positive integers: p the precision, β the arithmetic basis, L the underflow limit, U the overflow limit, and by the floating point function fl. Using the four positive integers mentioned above, let us define the following set:

$$\mathbb{F} := \Big\{ \pm (c_1\beta^{-1} + \ldots + c_p\beta^{-p})\,\beta^e \;:\; e \in [\![-L, U]\!],$$
$$c_i \in [\![0, \beta[\![\quad \text{for } i \in [\![1, p]\!], \quad c_1 \neq 0 \Big\} \cup \{0\}.$$

The set $\mathbb{F}$ is finite, symmetric with respect to 0, and the statistical distribution of its elements is far from being uniform.

Example 6.25 A set of floating point numbers:

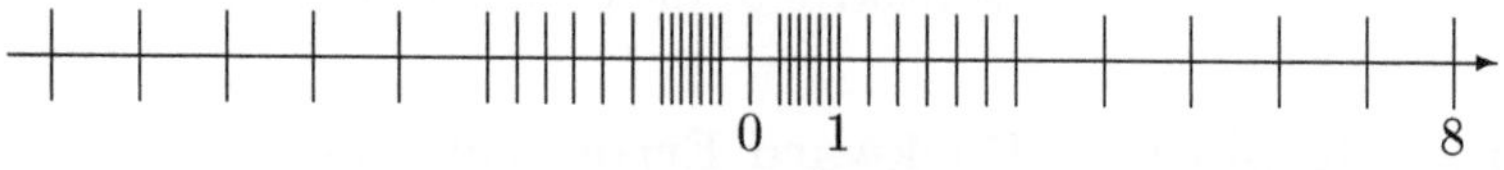

The above figure shows the set $\mathbb{F}$ corresponding to $p = 2$, $\beta = 3$, $L = 0$ and $U = 2$. ▮

Let
$$m := \beta^{-1-L} \quad \text{and} \quad M := \beta^U(1 - \beta^{-p}).$$

Then m is the smallest positive number in $\mathbb{F}$ and M is the largest one.

A floating point map

$$\text{fl} : [-M, -m] \cup [m, M] \cup \{0\} \to \mathbb{F}$$

may be of two different kinds: **chopping**, which is denoted it by fl_c, or **rounding**, which is denoted by fl_r. These maps are defined as follows.

If $m \leq x \leq M$, there exist $c_1, c_2, \ldots, c_p$ in $[\![0\,,\beta[\![$ and $e \in [\![-L\,,U]\!]$ such that $c_1 \neq 0$ and $x_- \leq x < x_+$, where

$$x_- := (c_1\beta^{-1} + \ldots + c_p\beta^{-p})\beta^e,$$
$$x_+ := (c_1\beta^{-1} + \ldots + c_p\beta^{-p})\beta^e + \beta^{e-p}.$$

Then

$$\text{fl}_c(x) := x_-,$$
$$\text{fl}_r(x) := \begin{cases} x_- \text{ if } x - x_- < x_+ - x, \\ x_+ \text{ if } x_+ - x \leq x - x_-. \end{cases}$$

These functions are extended to $[-M, -m]$ as odd functions and at 0 they are defined to be 0.

Computers based on fl_c are said to use the **chopping arithmetic** and those based on fl_r are said to use the **rounding arithmetic.**

We associate to the function fl, the **unit roundoff** of the computer:

$$\mathbf{u} := \inf\{v \in \mathbb{R} : v > 0 \text{ and } \text{fl}(1 + v) > 1\}.$$

Theorem 6.26
The unit roundoff in chopping arithmetic is

$$\mathbf{u}_c = \beta^{1-p}$$

and the unit roundoff in rounding arithmetic is

$$\mathbf{u}_r = \frac{1}{2}\beta^{1-p}.$$

Proof
The number following 1 in $\mathbb{F}$ is $(\beta^{-1}+\beta^{-p})\beta^1$. Hence if $0 < v < \beta^{1-p}$, then $\text{fl}_c(1+v) = 1$ and if $0 < v < \frac{1}{2}\beta^{1-p}$, then $\text{fl}_r(1+v) = 1$. Also,

$$\text{fl}_c(1+\beta^{1-p}) = \text{fl}_r(1+\frac{1}{2}\beta^{1-p}) = 1+\beta^{1-p} \in \mathbb{F}.$$

The proof is complete. ∎

In computers which optimize accuracy, $\mathbf{u}_c$ can be much smaller than β^{1-p} and $\mathbf{u}_r$ can be much smaller than $\frac{1}{2}\beta^{1-p}$.

Theorem 6.27
For each $x \in [m, M[$, there exists $\epsilon(x) \in [-\mathbf{u}, \mathbf{u}]$ such that

$$\text{fl}(x) = (1+\epsilon(x))x.$$

Proof
If $m \le x \le M$, then there exist $c_1, c_2, \ldots, c_p$ in $[\![0\,,\,\beta[\![$ and $e \in [\![-L\,,\,U]\!]$ such that $c_1 \neq 0$, $\text{fl}_c(x) = (c_1\beta^{-1}+\ldots+c_p\beta^{-p})\beta^e$ and

$$\text{fl}_c(x) \le x < (c_1\beta^{-1}+\ldots+c_p\beta^{-p})\beta^e + \beta^{e-p}.$$

Hence

$$-\beta^{e-p} < \text{fl}_c(x) - x \le 0.$$

If we define

$$\epsilon(x) := \frac{\text{fl}_c(x) - x}{x},$$

then

$$|\epsilon(x)| = \frac{\beta^{e-p}}{x} \le \frac{\beta^{e-p}}{\text{fl}_c(x)} \le \frac{\beta^{-p}}{\beta^{-1}} = \mathbf{u}_c.$$

In fact $\epsilon(x) \in [-\mathbf{u}_c, 0]$.

In the case of rounding arithmetic,

$$|\,\text{fl}_r(x) - x| \le \frac{1}{2}\beta^{e-p}.$$

If we define

$$\epsilon(x) := \frac{\text{fl}_r(x) - x}{x},$$

then

$$|\epsilon(x)| \le \frac{1}{2}\beta^{1-p} = \mathbf{u}_r.$$

The theorem is proved. ∎

Example 6.28 An example of a unit roundoff:
The IEEE standard 754 single- and double-precision values of the unit roundoff are

Precision	Size	$\mathbf{u}$
Single	32 bits	$(.596)10^{-7}$
Double	64 bits	$(.111)10^{-15}$

(See [44].) ∎

In order to estimate the roundoff error propagation, we need a mathematical model for the elementary arithmetic operations performed by the computer. This means that we need an hypothesis concerning $\mathrm{fl}(x \,\varphi\, y)$, when the operation φ is addition, subtraction, multiplication or division.

We shall assume that our ideal computer performs any of the four elementary operations φ in such a way that for each pair (x, y) satisfying $x \,\varphi\, y \in I\!\!F$, there is $\delta \in I\!\!R$ such that

$$\mathrm{fl}(x \,\varphi\, y) = (x \,\varphi\, y)(1+\delta) \quad \text{and} \quad |\delta| \le \mathbf{u}.$$

We assume in the same context that for each $x > 0$ such that $\sqrt{x} \in I\!\!F$, there is $\delta \in I\!\!R$ such that

$$\mathrm{fl}(\sqrt{x}) = \sqrt{x}(1+\delta) \quad \text{and} \quad |\delta| \le \mathbf{u}.$$

The function $\gamma_{\mathbf{u}}$, given by

$$\gamma_{\mathbf{u}}(k) := \frac{k\mathbf{u}}{1-k\mathbf{u}} \quad \text{for } k \in [0, 1/\mathbf{u}[,$$

has proved to be useful. It was introduced in [42].

For each integer k, the symbol θ_k will denote any complex number satisfying

$$|\theta_k| \le \gamma_{\mathbf{u}}(k).$$

In the following discussion, we attempt to give 'simple' bounds for the constants that appear in error estimates. More precisely, instead of producing an upper bound of the form $\alpha\gamma_{\mathbf{u}}(cj)$, where α is a given positive real number, c is a 'generic' constant (usually unspecified) and j a variable integer, we propose a 'simpler' upper bound $\gamma_{\mathbf{u}}(c_0 k)$ in which c_0 is an explicit integer and k a variable integer, possibly greater than j. In this sense, our bounds may not be 'optimal', but our conclusions remain valid and consistent with [42], which is one of the main references on this subject. (See also [7].)

Lemma 6.29
The function $\gamma_{\mathbf{u}}$ satisfies the following inequalities:

$$\begin{aligned} \gamma_{\mathbf{u}}(k+1) &\geq \gamma_{\mathbf{u}}(k) + \mathbf{u}, \\ \gamma_{\mathbf{u}}(k+j) &\geq \gamma_{\mathbf{u}}(k) + \gamma_{\mathbf{u}}(j) + \gamma_{\mathbf{u}}(k)\gamma_{\mathbf{u}}(j), \\ \gamma_{\mathbf{u}}(jk) &\geq j\gamma_{\mathbf{u}}(k), \end{aligned}$$

provided it is well defined.

Proof
First, if $k+1 < 1/\mathbf{u}$, then

$$\frac{k\mathbf{u}}{1-k\mathbf{u}} + \mathbf{u} = \frac{(k+1)\mathbf{u} - k\mathbf{u}^2}{1-k\mathbf{u}} \leq \frac{(k+1)\mathbf{u}}{1-(k+1)\mathbf{u}} = \gamma_{\mathbf{u}}(k+1).$$

Next, if $j+k < 1/\mathbf{u}$, then

$$\begin{aligned} \gamma_{\mathbf{u}}(j) + \gamma_{\mathbf{u}}(k) + \gamma_{\mathbf{u}}(j)\gamma_{\mathbf{u}}(k) &= \frac{j\mathbf{u}(1-k\mathbf{u}) + k\mathbf{u}(1-j\mathbf{u}) + jk\mathbf{u}^2}{1-j\mathbf{u}-k\mathbf{u}+jk\mathbf{u}^2} \\ &= \frac{j\mathbf{u}+k\mathbf{u}-jk\mathbf{u}^2}{1-j\mathbf{u}-k\mathbf{u}+jk\mathbf{u}^2} \leq \frac{j\mathbf{u}+k\mathbf{u}}{1-j\mathbf{u}-k\mathbf{u}} = \gamma_{\mathbf{u}}(j+k). \end{aligned}$$

Finally, if $jk < 1/\mathbf{u}$, then

$$\frac{jk\mathbf{u}}{1-k\mathbf{u}} \leq \frac{jk\mathbf{u}}{1-jk\mathbf{u}} = \gamma_{\mathbf{u}}(jk).$$

The proof is complete. ∎

Lemma 6.30
The following results hold:

$$\begin{aligned} (1+\theta_j)(1+\theta_k) &= 1 + \theta_{j+k}, \\ \frac{1+\theta_j}{1+\theta_k} &= 1 + \theta_{\max\{j,k\}+k}. \end{aligned}$$

Also, for each positive integer k satisfying $\gamma_{\mathbf{u}}(k) \le 3/4$, and each real $\theta_k \ge -1$, we have

$$|1 - \sqrt{1+\theta_k}| \le |\theta_k|.$$

Proof

First, for $j, k < 1/\mathbf{u}$, we have

$$(1+\theta_j)(1+\theta_k) - 1 = \theta_j + \theta_k + \theta_j\theta_k.$$

Next, let $j + k < 1/\mathbf{u}$. Then by Lemma 6.29,

$$|\theta_j + \theta_k + \theta_j\theta_k| \le \gamma_{\mathbf{u}}(j+k).$$

Also,

$$\left|\frac{1+\theta_j}{1+\theta_k} - 1\right| = \left|\frac{\theta_j - \theta_k}{1+\theta_k}\right| \le \frac{\left|\dfrac{j\mathbf{u}}{1-j\mathbf{u}} + \dfrac{k\mathbf{u}}{1-k\mathbf{u}}\right|}{\left|\dfrac{k\mathbf{u}}{1-k\mathbf{u}}\right|} = \left|\frac{(j+k)\mathbf{u} - 2jk\mathbf{u}^2}{(1-j\mathbf{u})(1-2k\mathbf{u})}\right|.$$

If $j > k$, then $(1-j\mathbf{u})(1-2k\mathbf{u}) \ge (1-(j+k)\mathbf{u})(1-k\mathbf{u})$ and $(j+k)\mathbf{u} - 2jk\mathbf{u}^2 \le (j+k)\mathbf{u}(1-k\mathbf{u})$, and hence

$$\left|\frac{1+\theta_j}{1+\theta_k} - 1\right| \le \frac{(j+k)\mathbf{u}}{1-(j+k)\mathbf{u}} = \gamma_{\mathbf{u}}(j+k).$$

If $j \le k$, then $(j+k)\mathbf{u} - 2jk\mathbf{u}^2 \le 2(1-j\mathbf{u})k\mathbf{u}$ and hence

$$\left|\frac{1+\theta_j}{1+\theta_k} - 1\right| \le \frac{2k\mathbf{u}}{1-2k\mathbf{u}} = \gamma_{\mathbf{u}}(2k).$$

Using the Mean Value Theorem for derivatives, there exists δ such that $|\delta| \le \theta_k$ and

$$|1 - \sqrt{1+\theta_k}| = \frac{|\theta_k|}{2\sqrt{1+\delta}}.$$

Thus

$$|1 - \sqrt{1+\theta_k}| \le \frac{|\theta_k|}{2\sqrt{1-3/4}} = |\theta_k|,$$

which completes the proof. ∎

The following results concern complex arithmetic and matrix operations. They allow us to estimate the propagation of roundoff errors in computations. They will be used in the backward error analysis of the Basic QR Method.

Proposition 6.31
Let x, y be complex numbers.
If $\mathbf{c}\,(x \pm y)$ exists, then there exists $\delta \in \mathbb{C}$ such that

$$\mathbf{c}\,(x \pm y) = (x \pm y)(1+\delta), \ |\delta| \le \mathbf{u}.$$

If $\mathbf{c}\,(x \times y)$ exists and $3\mathbf{u} < 1$, then there exists $\delta \in \mathbb{C}$ such that

$$\mathbf{c}\,(x{\times}y) = (x{\times}y)(1+\delta), \ |\delta| \le \gamma_{\mathbf{u}}(3).$$

If $\mathbf{c}\left(|y|^2\right)$ exists, then there exists $\delta \in \mathbb{R}$ such that

$$\mathbf{c}\left(|y|^2\right) = |y|^2(1+\delta), \ |\delta| \le \gamma_{\mathbf{u}}(2).$$

If $\mathbf{c}\,(x/y)$ exists, then there exists $\delta \in \mathbb{C}$ such that

$$\mathbf{c}\,(x/y) = (x/y)(1+\delta), \ |\delta| \le \gamma_{\mathbf{u}}(6).$$

Proof
Let us write a complex number x in its Cartesian form $x = x_{\Re} + \mathrm{i}\, x_{\Im}$, where $x_{\Re} \in \mathbb{R}$ is its real part and $x_{\Im} \in \mathbb{R}$ its imaginary part. Then

$$\begin{aligned}\mathbf{c}\,(x \pm y) &= \mathrm{fl}(x_{\Re} \pm y_{\Re}) + \mathrm{i}\ \mathrm{fl}(x_{\Im} \pm y_{\Im}) \\ &= (x_{\Re} \pm y_{\Re})(1+\delta_1) + \mathrm{i}\,(x_{\Im} \pm y_{\Im})(1+\delta_2)\end{aligned}$$

for some δ_1 and δ_2 such that $|\delta_1| \le \mathbf{u}$ and $|\delta_2| \le \mathbf{u}$. Suppose $x \pm y \ne 0$ and let $\delta := \dfrac{\mathbf{c}\,(x \pm y) - (x \pm y)}{x \pm y}$. Then

$$\begin{aligned}|\delta||x \pm y| = |\mathbf{c}\,(x \pm y) - (x \pm y)| &= |(x_{\Re} \pm y_{\Re})\delta_1 + \mathrm{i}\,(x_{\Im} \pm y_{\Im})\delta_2| \\ &= \sqrt{(x_{\Re} \pm y_{\Re})^2\delta_1^2 + (x_{\Im} \pm y_{\Im})^2\delta_2^2} \\ &\le \max\{|\delta_1|, |\delta_2|\}\,|x \pm y| \le \mathbf{u}|x \pm y|.\end{aligned}$$

Similarly, if $\mathbf{c}\,(x \times y)$ exists,

$$\begin{aligned}\mathbf{c}\,(x \times y) = &\ (x_{\Re}y_{\Re}(1+\delta_1) - x_{\Im}y_{\Im}(1+\delta_2))(1+\delta_3) \\ &+\mathrm{i}\,(x_{\Re}y_{\Im}(1+\delta_4) + x_{\Im}y_{\Re}(1+\delta_5))(1+\delta_6).\end{aligned}$$

Suppose $x \times y \neq 0$ and let $\delta := \dfrac{\mathbf{c}\,(x \times y) - (x \times y)}{x \times y}$. Then

$$\begin{aligned}|x \times y||\delta| &= |\mathbf{c}\,(x \times y) - (x \times y)| \\ &= |x_{\Re}y_{\Re}(\delta_1 + \delta_3 + \delta_1\delta_3) - x_{\Im}y_{\Im}(\delta_2 + \delta_3 + \delta_2\delta_3) \\ &\quad + \mathrm{i}\,(x_{\Re}y_{\Im}(\delta_4 + \delta_6 + \delta_4\delta_6) + x_{\Im}y_{\Re}(\delta_5 + \delta_6 + \delta_5\delta_6)|,\end{aligned}$$

for some δ_i such that $|\delta_i| \leq \mathbf{u}$, $i \in [\![1, 6]\!]$. Hence

$$\begin{aligned}&|x \times y|^2|\delta|^2 \leq \\ &(|x_{\Re}y_{\Re}|^2 + |x_{\Im}y_{\Im}|^2 + 4|x_{\Re}y_{\Re}x_{\Im}y_{\Im}| + |x_{\Re}y_{\Im}|^2 + |x_{\Im}y_{\Re}|^2)\mathbf{u}^2(2 + \mathbf{u})^2.\end{aligned}$$

But

$$2|x_{\Re}y_{\Re}x_{\Im}y_{\Im}| \leq |x_{\Re}y_{\Re}|^2 + |x_{\Im}y_{\Im}|^2, \quad 2|x_{\Re}y_{\Re}x_{\Im}y_{\Im}| \leq |x_{\Re}y_{\Im}|^2 + |x_{\Im}y_{\Re}|^2,$$

and

$$|x \times y|^2 = |x_{\Re}y_{\Re}|^2 + |x_{\Im}y_{\Im}|^2 + |x_{\Re}y_{\Im}|^2 + |x_{\Im}y_{\Re}|^2,$$

so that

$$|\delta|^2 \leq 2\mathbf{u}^2(2 + \mathbf{u})^2 \leq \gamma_{\mathbf{u}}(3)^2,$$

since $3\mathbf{u} < 1$ and the polynomial $\mathrm{p}(t) := 3t^2 + 5t - 2 + 3/\sqrt{2}$ has two negative roots. Also, if $\mathbf{c}\left(|y|^2\right)$ exists,

$$\mathbf{c}\left(|y|^2\right) = \mathbf{c}\,(y \times \overline{y}) = (y_{\Re}^2(1 + \theta_1) + y_{\Im}^2(1 + \theta_1))(1 + \theta_1) \leq |y|^2(1 + \theta_2).$$

Finally, if $\mathbf{c}\,(x/y)$ exists,

$$\mathbf{c}\,(x/y) = \mathbf{c}\left(\frac{x\overline{y}}{y\overline{y}}\right) = \frac{x\overline{y}(1 + \theta_3)}{y\overline{y}(1 + \theta_2)}(1 + \theta_1) = \frac{x}{y}(1 + \theta_6),$$

by Lemma 6.30. ∎

Example 6.32 Backward error analysis of elementary operations: The assertions

$$\begin{aligned}\mathbf{c}\,(x \pm y) &= (x + \Delta x) \pm (y + \Delta y), &&\text{where } |\Delta x| \leq \mathbf{u}|x|,\ |\Delta y| \leq \mathbf{u}|y|, \\ \mathbf{c}\,(x \times y) &= (x + \Delta x) \times y, &&\text{where } |\Delta x| \leq \gamma_{\mathbf{u}}(3)|x|, \\ \mathbf{c}\,(x/y) &= (x + \Delta x)/y, &&\text{where } |\Delta x| \leq \gamma_{\mathbf{u}}(6)|x|,\end{aligned}$$

follow from the proof of Proposition 6.31. ∎

Remark 6.33 Backward error analysis of Gaussian Elimination with partial pivoting:
Let A be a real matrix of order n and b be a real $n\times 1$ column matrix. Suppose that the Gaussian Elimination Algorithm with partial pivoting produces a computed solution $\mathbf{c}(\mathsf{x})$ of the linear system $\mathsf{Ax} = \mathsf{b}$. Then there exists a matrix $\Delta\mathsf{A}$ independent of b such that

$$(\mathsf{A} + \Delta\mathsf{A})\mathbf{c}(\mathsf{x}) = \mathsf{b} \quad \text{and} \quad ||\Delta\mathsf{A}||_\infty \leq 2n^2\gamma_{\mathbf{u}}(n)\rho_n||\mathsf{A}||_\infty,$$

where ρ_n is the growth factor. For proofs and details, see Sections 9.1 and 9.2 of [42]. ∎

In order to study the stability of the Basic QR Method, that is, the roundoff error propagation of this method when it runs on a computer, we need some results about the numerical computation of Householder symmetries.

Lemma 6.34
Let $\mathsf{x} := [\mathsf{x}(1), \ldots, \mathsf{x}(n)]^\top \in \mathbb{C}^{n\times 1}$ *and* $k \in [\![1, n-1]\!]$. *Define*

$$\mathsf{x}_k := \sum_{j=k}^{n} \mathsf{x}(j)\mathsf{e}_j.$$

Then

$$\mathbf{c}(||\mathsf{x}_k||_2) = ||\mathsf{x}_k||_2(1 + \theta_{3+n-k}),$$
$$\mathbf{c}(\langle \mathsf{y}, \mathsf{x}_k\rangle) = \langle y, \mathsf{x}_k\rangle(1 + \theta_{3+n-k}) \quad \textit{for each } \mathsf{y} \in \mathbb{C}^{n\times 1}.$$

Proof
Since

$$\mathbf{c}(||\mathsf{x}_k||_2) = \sqrt{\mathbf{c}(||\mathsf{x}_k||_2^2)}(1 + \theta_1),$$
$$\mathbf{c}(||\mathsf{x}_k||_2^2) = \left(|\mathsf{x}(k)|^2 + \mathbf{c}(||\mathsf{x}_{k+1}||_2^2)\right)(1 + \theta_1)$$

and by Proposition 6.31,

$$\mathbf{c}(|\mathsf{x}(k)|^2) = |\mathsf{x}(k)|^2(1 + \theta_2),$$

we get, by Lemma 6.30,

$$\begin{aligned}\mathbf{c}(||\mathsf{x}_k||_2^2) &= |\mathsf{x}(k)|^2(1 + \theta_3) + \mathbf{c}(||\mathsf{x}_{k+1}||_2^2)(1 + \theta_1)\\ &= \sum_{i=k}^{n-1} |\mathsf{x}(i)|^2(1 + \theta_{3+i-k}) + |\mathsf{x}(n)|^2(1 + \theta_{3+n-1-k}),\end{aligned}$$

that is

$$|\mathbf{c}\left(\|\mathsf{x}_k\|_2^2\right) - \|\mathsf{x}_k\|_2^2| \leq \|\mathsf{x}_k\|_2^2 \gamma_{\mathbf{u}}(2+n-k).$$

Thus $\mathbf{c}\left(\|\mathsf{x}_k\|_2\right) = \|\mathsf{x}_k\|_2(1+\theta_{3+n-k})$.

Similarly,

$$\begin{aligned}\mathbf{c}\left(\langle \mathsf{y}\,,\,\mathsf{x}_k\rangle\right) &= \left(\mathsf{y}(k)\overline{\mathsf{x}(k)}(1+\theta_3) + \mathbf{c}\left(\langle \mathsf{y}\,,\,\mathsf{x}_{k+1}\rangle\right)\right)(1+\theta_1)\\ &= \sum_{i=k}^{n-1} \mathsf{y}(i)\overline{\mathsf{x}(i)}(1+\theta_{4+i-k}) + \mathsf{y}(n)\overline{\mathsf{x}(n)}(1+\theta_{3+n-k}),\end{aligned}$$

so that

$$\mathbf{c}\left(\langle \mathsf{y}\,,\,\mathsf{x}_k\rangle\right) = \langle \mathsf{y}\,,\,\mathsf{x}_k\rangle(1+\theta_{3+n-k}),$$

as stated. ∎

Proposition 6.35
Let $\mathsf{x} := [\mathsf{x}(1),\ldots,\mathsf{x}(n)]^\top \in \mathbb{C}^{n\times 1}$ *be such that for some* $k \in [\![1, n-1]\!]$, $\mathsf{x}(k) \neq 0$. *Let* $\mathsf{H} := \mathsf{I} - \beta \mathsf{v}\mathsf{v}^*$ *be the Householder symmetry defined in Lemma 6.4. Then*

$$\begin{aligned}|\mathbf{c}\left(\beta\right) - \beta| &\leq \gamma_{\mathbf{u}}(2(n-k)+9)|\beta|,\\ |\mathbf{c}\left(\mathsf{v}\right) - \mathsf{v}| &\leq \gamma_{\mathbf{u}}(n-k+10)|\mathsf{v}|,\end{aligned}$$

where $|\mathsf{v}| := [\,|\mathsf{v}(1)|,\ldots,|\mathsf{v}(n)|\,]^\top$.

Proof
First note that β is real:

$$\beta = \frac{1}{(|\mathsf{x}(k)|+\alpha)\alpha}.$$

Next, by Lemma 6.34 and Lemma 6.30,

$$\mathbf{c}\left(\alpha\right) = \alpha(1+\theta_{3+n-k})$$

and hence

$$\begin{aligned}\mathbf{c}\left(\beta\right) &= \frac{1+\theta_1}{\mathbf{c}\left(\alpha\right)\left(\mathbf{c}\left(|\mathsf{x}(k)|\right) + \mathbf{c}\left(\alpha\right)\right)(1+\theta_2)}\\ &= \frac{1+\theta_1}{\alpha(1+\theta_{3+n-k})\left(|\mathsf{x}(k)|(1+\theta_3) + \alpha(1+\theta_{3+n-k})\right)(1+\theta_2)}\\ &= \frac{1+\theta_1}{\alpha\left(|\mathsf{x}(k)|+\alpha\right)(1+\theta_{3+n-k})^2(1+\theta_2)} = \beta(1+\theta_{2(n-k)+9}).\end{aligned}$$

Similarly,

$$\mathbf{c}\,(\mathsf{v}(k)) = (\mathsf{x}(k) + \mathbf{c}\,(s))(1+\theta_1),$$

and

$$\begin{aligned}\mathbf{c}\,(s) &= \mathbf{c}\left(\frac{\mathsf{x}(k)}{|\mathsf{x}(k)|}\right)\mathbf{c}\,(\alpha)(1+\theta_1)\\ &= \alpha\,\frac{\mathsf{x}(k)}{|\mathsf{x}(k)|}(1+\theta_4)(1+\theta_{3+n-k})(1+\theta_3)\\ &= s(1+\theta_{n-k+10}),\end{aligned}$$

since

$$\mathbf{c}\left(\frac{\mathsf{x}(k)}{|\mathsf{x}(k)|}\right) = \frac{\mathsf{x}(k)}{\mathbf{c}\,(|\mathsf{x}(k)|)}(1+\theta_1) = \frac{\mathsf{x}(k)(1+\theta_1)}{|\mathsf{x}(k)|(1+\theta_3)} = \frac{\mathsf{x}(k)}{|\mathsf{x}(k)|}(1+\theta_4).$$

The result now follows. ▮

Proposition 6.36
Let $\mathsf{y} \in \mathbb{C}^{n\times 1}$ and H be the Householder symmetry defined in Lemma 6.4. Then there exists $\Delta\mathsf{y} \in \mathbb{C}^{n\times 1}$ such that

$$\mathbf{c}\,(\mathsf{Hy}) = \mathsf{H}(\mathsf{y}+\Delta\mathsf{y}) \quad \textit{and} \quad \|\Delta\mathsf{y}\|_2 \le \gamma_{\mathbf{u}}(5(n-k)+37)\,\|\mathsf{y}\|_2.$$

Proof
We have

$$\begin{aligned}\mathbf{c}\,(\mathsf{Hy}) &= (\mathsf{y} - \mathbf{c}\,(\beta\mathsf{vv}^*\mathsf{y}))(1+\theta_1)\\ &= (\mathsf{y} - \mathbf{c}\,(\beta)\mathbf{c}\,(\mathsf{v})(1+\theta_1)\mathbf{c}\,(\mathsf{v}^*\mathsf{y})(1+\theta_3))\,(1+\theta_1)\\ &= (\mathsf{y} - \beta\mathsf{vv}^*\mathsf{y}(1+\theta_{2(n-k)+9})(1+\theta_{n-k+10})\\ &\quad (1+\theta_{n-k+3})(1+\theta_4))(1+\theta_1)\\ &= \mathsf{Hy}(1+\theta_{5(n-k)+37})\\ &= \mathsf{H}(\mathsf{y}+\Delta\mathsf{y}),\end{aligned}$$

where

$$\|\mathsf{H}\Delta\mathsf{y}\|_2 \le \gamma_{\mathbf{u}}(5(n-k)+37).$$

But H is a unitary matrix, so that $\|\mathsf{H}\Delta\mathsf{y}\|_2 = \|\Delta\mathsf{y}\|_2$. ▮

Proposition 6.37
Let $\mathsf{A} \in \mathbb{C}^{n\times n}$ and H be the Householder symmetry defined in Lemma

6.4. Then there exist $\mathsf{E} \in \mathbb{C}^{n\times n}$ *and* $\triangle\mathsf{A} \in \mathbb{C}^{n\times n}$ *such that*

$$\begin{aligned} \mathbf{c}\,(\mathsf{HA}) &= \mathsf{H}(\mathsf{A}+\mathsf{E}), & \|\mathsf{E}\|_{\mathrm{F}} &\le \gamma_{\mathbf{u}}(5(n-k)+37)\|\mathsf{A}\|_{\mathrm{F}}, \\ \mathbf{c}\,(\mathsf{HAH}) &= \mathsf{H}(\mathsf{A}+\triangle\mathsf{A})\mathsf{H}, & \|\triangle\mathsf{A}\|_{\mathrm{F}} &\le \gamma_{\mathbf{u}}(10(n-k)+74)\|\mathsf{A}\|_{\mathrm{F}}. \end{aligned}$$

Proof
For each $j \in [\![1,n]\!]$, define $\mathsf{y}_j := \mathsf{A}\mathsf{e}_j$. By Proposition 6.36,

$$\mathbf{c}\,(\mathsf{H}\mathsf{y}_j) = \mathsf{H}(\mathsf{y}_j + \triangle\mathsf{y}_j), \quad \|\triangle\mathsf{y}_j\|_2 \le \gamma_{\mathbf{u}}(5(n-k)+37)\,\|\mathsf{y}\|_2.$$

Hence

$$\begin{aligned} \mathbf{c}\,(\mathsf{HA}) &= [\mathbf{c}\,(\mathsf{H}\mathsf{y}_1),\ldots,\mathbf{c}\,(\mathsf{H}\mathsf{y}_n)] \;=\; \mathsf{H}[\mathsf{y}_1+\triangle\mathsf{y}_1,\ldots,\mathsf{y}_n+\triangle\mathsf{y}_n] \\ &= \mathsf{H}([\mathsf{y}_1,\ldots,\mathsf{y}_n]+[\triangle\mathsf{y}_1,\ldots,\triangle\mathsf{y}_n]). \end{aligned}$$

If we define

$$\mathsf{E} := [\triangle\mathsf{y}_1,\ldots,\triangle\mathsf{y}_n],$$

then

$$\|\mathsf{E}\|_{\mathrm{F}} = \sqrt{\sum_{j=1}^{n}\|\triangle\mathsf{y}_j\|_2^2} \le \gamma_{\mathbf{u}}(5(n-k)+37)\|\mathsf{A}\|_{\mathrm{F}}.$$

Also,

$$\mathbf{c}\,(\mathsf{AH}) = (\mathsf{A}+\mathsf{F})\mathsf{H} \quad \text{with} \quad \|\mathsf{F}\|_{\mathrm{F}} \le \gamma_{\mathbf{u}}(5(n-k)+37)\|\mathsf{A}\|_{\mathrm{F}},$$

and

$$\mathbf{c}\,(\mathsf{HAH}) = \mathsf{H}(\mathbf{c}\,(\mathsf{AH})+\mathsf{G}) \quad \text{with} \quad \|\mathsf{G}\|_{\mathrm{F}} \le \gamma_{\mathbf{u}}(5(n-k)+37)\|\mathbf{c}\,(\mathsf{AH})\|_{\mathrm{F}}.$$

Hence

$$\mathbf{c}\,(\mathsf{HAH}) = \mathsf{H}((\mathsf{A}+\mathsf{F})\mathsf{H}+\mathsf{G}\mathsf{H}^2) = \mathsf{H}(\mathsf{A}+\mathsf{F}+\mathsf{G}\mathsf{H})\mathsf{H},$$

since $\mathsf{H}^2 = \mathsf{I}$. If we define

$$\triangle\mathsf{A} := \mathsf{F}+\mathsf{G}\mathsf{H},$$

then

$$\|\triangle\mathsf{A}\|_{\mathrm{F}} \le \|\mathsf{F}\|_{\mathrm{F}} + \|\mathsf{G}\mathsf{H}\|_{\mathrm{F}} = \|\mathsf{F}\|_{\mathrm{F}} + \|\mathsf{G}\|_{\mathrm{F}},$$

since $\|\mathsf{G}\mathsf{H}\|_{\mathrm{F}} \le \|\mathsf{G}\|_{\mathrm{F}}\|\mathsf{H}\|_2$ and $\|\mathsf{H}\|_2 = 1$. For the same reason, by Lemma 6.29, $\|\mathsf{G}\|_{\mathrm{F}} \le \gamma_{\mathbf{u}}(5(n-k)+37)\|\mathsf{A}+\mathsf{F}\|_{\mathrm{F}} \le \gamma_{\mathbf{u}}(10(n-k)+74)\|\mathsf{A}\|_{\mathrm{F}}$. ∎

Proposition 6.38
Let $\mathsf{A} \in \mathbb{C}^{n\times n}$ *and* $\mathsf{A}_0 := \mathsf{U}^*\mathsf{A}\mathsf{U}$ *be the upper Hessenberg matrix obtained in Proposition 6.7 with the help of Householder symmetries. Then there exist* $\triangle\mathsf{A} \in \mathbb{C}^{n\times n}$ *and a unitary matrix* $\mathsf{V} \in \mathbb{C}^{n\times n}$ *such that* $\mathbf{c}(\mathsf{A}_0)$ *is an upper Hessenberg matrix satisfying*

$$\mathbf{c}(\mathsf{A}_0) = \mathsf{V}^*(\mathsf{A} + \triangle\mathsf{A})\mathsf{V}, \quad \|\triangle\mathsf{A}\|_{\mathrm{F}} \le \gamma_{\mathbf{u}}(13n^2)\|\mathsf{A}\|_{\mathrm{F}}.$$

Proof
Let H_k denote the Householder symmetry which at step k annihilates the entries of column k having a row index greater than $k+1$. Let us set

$$\mathsf{A}_1 := \mathbf{c}(\mathsf{H}_1\mathsf{A}\mathsf{H}_1) \quad \text{and} \quad \mathsf{A}_2 := \mathbf{c}(\mathsf{H}_2\mathsf{A}_1\mathsf{H}_2).$$

By Proposition 6.37,

$$\begin{aligned} \mathsf{A}_1 &= \mathsf{H}_1(\mathsf{A} + \triangle\mathsf{A}_1)\mathsf{H}_1, & \|\triangle\mathsf{A}_1\|_{\mathrm{F}} &\le \gamma_{\mathbf{u}}(10(n-1)+74)\|\mathsf{A}\|_{\mathrm{F}}, \\ \mathsf{A}_2 &= \mathsf{H}_2(\mathsf{A}_1 + \mathsf{E})\mathsf{H}_2, & \|\mathsf{E}\|_{\mathrm{F}} &\le \gamma_{\mathbf{u}}(10(n-1)+74)\|\mathsf{A}_1\|_{\mathrm{F}}, \end{aligned}$$

so that

$$\mathsf{A}_2 = \mathsf{H}_2\mathsf{H}_1(\mathsf{A} + \triangle\mathsf{A}_1 + \mathsf{H}_1\mathsf{E}\mathsf{H}_1)\mathsf{H}_1\mathsf{H}_2,$$

and since H_1 is a unitary matrix,

$$\triangle\mathsf{A}_2 := \triangle\mathsf{A}_1 + \mathsf{H}_1\mathsf{E}\mathsf{H}_1$$

satisfies

$$\begin{aligned} \|\triangle\mathsf{A}_2\|_{\mathrm{F}} &\le \|\triangle\mathsf{A}_1\|_{\mathrm{F}} + \|\mathsf{E}\|_{\mathrm{F}} \\ &\le \gamma_{\mathbf{u}}(10(n-1)+74)\|\mathsf{A}\|_{\mathrm{F}} + \gamma_{\mathbf{u}}(10(n-2)+74)\|\mathsf{A}_1\|_{\mathrm{F}}. \end{aligned}$$

But

$$\|\triangle\mathsf{A}_1\|_{\mathrm{F}} \le \|\mathsf{A}\|_{\mathrm{F}} + \|\triangle\mathsf{A}_1\|_{\mathrm{F}} \le (1 + \gamma_{\mathbf{u}}(10(n-1)+74))\|\mathsf{A}\|_{\mathrm{F}}.$$

Hence

$$\|\triangle\mathsf{A}_2\|_{\mathrm{F}} \le \gamma_{\mathbf{u}}(10[(n-1)+(n-2)]+128).$$

Repeating this process, we obtain

$$\mathsf{A}_{n-2} = \mathsf{H}_{n-2}\mathsf{H}_{n-3}\cdots\mathsf{H}_1(\mathsf{A} + \triangle\mathsf{A})\mathsf{H}_1\cdots\mathsf{H}_{n-3}\mathsf{H}_{n-2},$$

where

$$\|\triangle\mathsf{A}\|_{\mathrm{F}} \le \gamma_{\mathbf{u}}\Big(74(n-2) + 10\sum_{k=1}^{n-2}(n-k)\Big)\|\mathsf{A}\|_{\mathrm{F}} \le \gamma_{\mathbf{u}}(13n^2)\|\mathsf{A}\|_{\mathrm{F}},$$

since

$$74(n-2)+10\sum_{k=1}^{n-2}(n-k) = 5n^2+69n-158 \leq pn^2$$

for each positive integer n, if $p \geq 13$. ▮

Theorem 6.39
Let $\mathsf{A} \in \mathbb{C}^{n\times n}$ and $\mathsf{A}_0 := \mathsf{U}^*\mathsf{A}\mathsf{U}$ be the upper Hessenberg matrix obtained in Proposition 6.7 with the help of Householder symmetries. Let $k \geq 1$ be an integer and A_k the matrix defined by k iterations of the Basic QR algorithm. Then there exist $\triangle\mathsf{A} \in \mathbb{C}^{n\times n}$ and a unitary matrix $\mathsf{V} \in \mathbb{C}^{n\times n}$ such that $\mathbf{c}\,(\mathsf{A}_k)$ is an upper Hessenberg matrix satisfying

$$\mathbf{c}\,(\mathsf{A}_k) = \mathsf{V}^*(\mathsf{A}+\triangle\mathsf{A})\mathsf{V}, \quad \|\triangle\mathsf{A}\|_{\mathrm{F}} \leq \gamma_{\mathbf{u}}(22(k+1)n^2)\|\mathsf{A}\|_{\mathrm{F}}.$$

Proof
It suffices to analyze the backward error of one iteration of the Basic QR algorithm. Indeed,

$$\mathbf{c}\,(\mathsf{A}_k) = \mathbf{c}\left(\mathsf{Q}_k^*\mathbf{c}\,(\mathsf{A}_{k-1})\mathsf{Q}_m\right),$$

where Q_k is a product of $k-1$ Householder symmetries and A_0 is an upper Hessenberg matrix unitarily similar to A through Householder symmetries. Applying Proposition 6.38,

$$\mathbf{c}\,(\mathsf{A}_k) = \mathsf{V}^*(\mathbf{c}\,(\mathsf{A}_{k-1})+\triangle\mathsf{A}_{k-1})\mathsf{V}$$

for some unitary matrix V and for some matrix $\triangle\mathsf{A}_{k-1}$ satisfying

$$\|\triangle\mathsf{A}_{k-1}\|_{\mathrm{F}} \leq \gamma_{\mathbf{u}}\Big(74(n-1)+10\sum_{j=1}^{n-1}(n+1-k)\Big)\|\mathbf{c}\,(\mathsf{A}_{k-1})\|_{\mathrm{F}}.$$

Hence

$$\|\triangle\mathsf{A}_{k-1}\|_{\mathrm{F}} \leq \gamma_{\mathbf{u}}(22(k+1)n^2)\|\mathsf{A}\|_{\mathrm{F}}$$

since, as before,

$$74(n-2)+10\sum_{k=1}^{n-2}(n+1-k) = 5n^2+69n-74 \leq pn^2$$

for each positive integer n, if $p \geq 22$. ▮

6.4.3 Relative Error in Spectral Computations and Stopping Criteria for the Basic QR Method

Forward and backward error analyses taken together allow us to bound the relative error of a computed eigenvalue or a computed basis for a spectral subspace. In fact, if the algorithm is backward stable, the computed results correspond to exact computations on some perturbed data, and then

(Relative Error) ≤ (Condition Number) × (Relative Perturbation).

The backward error analysis provides a bound for the relative perturbation of the data, and the forward error analysis provides a bound for the condition number. However, the latter is usually expressed in terms of unknown quantities. That is why the preceding theoretical upper bound on the relative error is not useful in practice as a stopping criterion for a computer run.

This remark justifies our suggestion of a residual test to stop the Simultaneous Iteration Method in subsection 6.3.2.

Let us consider stopping criteria for the Basic QR Method. We first take up a backward error point of view. When the method is applied to an upper irreducible Hessenberg matrix, Theorem 6.18 suggests that the diagonal entries of an iterate A_k are 'good' approximate eigenvalues of A if the subdiagonal entries of A_k are 'small enough'; and that if a subdiagonal entry of A_k becomes 'almost' zero, we could decide to neglect it and then view A_k as a block triangular matrix and perform the next iterations in each diagonal block separately. This could be done using two different processors, for instance. The decoupling of the problem into two smaller problems is called deflation. These considerations are confirmed by the backward error standpoint as we see now.

With the notations of Theorem 6.39, let T_k and L_k be the matrices defined by

$$\mathsf{T}_k(i,j) := \begin{cases} \mathbf{c}\,(\mathsf{A}_k)(i,j) & \text{if } i \le j, \\ 0 & \text{otherwise,} \end{cases}$$

$$\mathsf{L}_k(i,j) := \begin{cases} \mathbf{c}\,(\mathsf{A}_k)(i,j) & \text{if } i = j+1, \\ 0 & \text{otherwise.} \end{cases}$$

Since $\mathbf{c}\,(\mathsf{A}_k)$ is an upper Hessenberg matrix, we have

$$\mathbf{c}\,(\mathsf{A}_k) = \mathsf{T}_k + \mathsf{L}_k.$$

Define

$$\widetilde{\mathsf{A}} := \mathsf{V}\mathsf{T}_k\mathsf{V}^*.$$

Then Theorem 6.39 gives

$$\widetilde{\mathsf{A}} = \mathsf{A} + \triangle\mathsf{A} - \mathsf{V}\mathsf{L}_k\mathsf{V}^*,$$

where

$$\|\triangle\mathsf{A}\|_{\mathrm{F}} \le \alpha_{n,k}\mathbf{u}\|\mathsf{A}\|_{\mathrm{F}},$$

for some constant $\alpha_{n,k}$ of 'moderate' size. Recall that $\|\mathsf{V}\mathsf{L}_k\mathsf{V}^*\|_{\mathrm{F}} = \|\mathsf{L}_k\|_{\mathrm{F}}$. We conclude that to guarantee stability of computations, we should stop the algorithm at an iteration k such that

$$\sqrt{\sum_{j=1}^{n-1} |\mathsf{L}(j+1,j)|^2} \le \beta_{n,k}\mathbf{u}\|\mathsf{A}\|_{\mathrm{F}}$$

for some constant $\beta_{n,k}$ of the order of $\alpha_{n,k}$. Then

$$\frac{\|\widetilde{\mathsf{A}} - \mathsf{A}\|_{\mathrm{F}}}{\|\mathsf{A}\|_{\mathrm{F}}} \le (\alpha_{n,k} + \beta_{n,k})\mathbf{u}.$$

This approach gives the type of stopping criteria used by several standard software packages. For instance, in EISPACK, if

$$|\mathsf{L}_k(j+1,j)| \le \beta\mathbf{u}(|\mathsf{T}_k(j,j)| + |\mathsf{T}_k(j+1,j+1)|)$$

for some 'small' constant β, then $\mathsf{L}_k(j+1,j)$ is redefined to be zero, and a deflation into two blocks takes place.

Since the relative error of computed eigenvalues is bounded by the product of both the condition number and the relative perturbation of the data, stopping criteria arising from a backward error analysis will be meaningful only for a matrix which is well conditioned relative to the computation of its eigenvalues.

This observation is confirmed by the following example:

Let ϵ be a "small" positive number, μ a "large" positive number, and let $\widetilde{\mu}_1$ and $\widetilde{\mu}_2$ be real numbers. Suppose that

$$\mathsf{A}_k := \begin{bmatrix} \widetilde{\mu}_1 & \mu \\ \epsilon & \widetilde{\mu}_2 \end{bmatrix}.$$

This matrix has eigenvalues

$$\mu_1 := \frac{\widetilde{\mu}_1 + \widetilde{\mu}_2 + \delta}{2}, \quad \mu_2 := \frac{\widetilde{\mu}_1 + \widetilde{\mu}_2 - \delta}{2},$$

where

$$\delta := \sqrt{(\widetilde{\mu}_1 - \widetilde{\mu}_2)^2 + 4\epsilon\mu}.$$

If ϵ is 'small enough', we would like to conclude that the diagonal entries of A_k, that is, $\mathsf{A}_k(1,1) = \widetilde{\mu}_1$ and $\mathsf{A}_k(2,2) = \widetilde{\mu}_2$, are 'good' approximate eigenvalues of A. However,

(i) if $\widetilde{\mu}_1 = \widetilde{\mu}_2 = \widetilde{\mu}$, then $\mu_1 = \widetilde{\mu} + \delta_1$, $\mu_2 = \widetilde{\mu} - \delta_1$, with $\delta_1 = \sqrt{\epsilon\widetilde{\mu}}$, and

(ii) if $\widetilde{\mu}_1 \neq \widetilde{\mu}_2$, then

$$\mu_1 = \frac{\widetilde{\mu}_1 + \widetilde{\mu}_2 + |\widetilde{\mu}_1 - \widetilde{\mu}_2|\delta_2}{2}, \quad \mu_2 = \frac{\widetilde{\mu}_1 + \widetilde{\mu}_2 - |\widetilde{\mu}_1 - \widetilde{\mu}_2|\delta_2}{2},$$

where

$$\delta_2 := \sqrt{1 + \frac{4\epsilon\mu}{(\widetilde{\mu}_1 - \widetilde{\mu}_2)^2}}.$$

We observe that the diagonal entries of A_k are 'good' approximate eigenvalues if and only if $\sqrt{\epsilon\widetilde{\mu}}$ is small enough in case (i), and if and only if $\sqrt{\epsilon\widetilde{\mu}}/|\widetilde{\mu}_1 - \widetilde{\mu}_2|$ is small enough in case (ii). For this reason, many implementations of the QR Method begin by 'scaling' A, that is, by dividing it by its own 1-norm, for example. Since $\rho(\mathsf{A}) \leq \|\mathsf{A}\|_1$, the scaled matrix will have all its eigenvalues inside the unit circle. But then the scaling operation may make convergence slow.

In fact, theoretical analysis suggests that the stopping criterion for the Basic QR Method from a relative error standpoint must take into account three aspects of the kth iterate A_k:

1. the size of the subdiagonal entries,
2. the size of the superdiagonal entries, and
3. the distance between two consecutive diagonal entries.

The next theorem suggests a similar conclusion for an $n \times n$ upper Hessenberg matrix with distinct diagonal entries.

Theorem 6.40

Let $\mathsf{B} \in \mathbb{C}^{n\times n}$ be an upper Hessenberg matrix such that all its diagonal entries are different. Let the matrix-valued function $\mathsf{B} : \mathbb{R} \to \mathbb{C}^{n\times n}$ be defined by

$$\mathsf{B}(t)(i,j) := \begin{cases} t\,\mathsf{B}(i,j) & \text{if } i-j=1, \\ \mathsf{B}(i,j) & \text{otherwise.} \end{cases}$$

If $\epsilon > 0$ is a small enough positive number and

$$\sum_{j=1}^{n-1} |\mathsf{B}(j+1,j)| < \epsilon,$$

then there exist n differentiable functions

$$t \in [-1,1] \mapsto \lambda_j(t) \in \mathbb{C}, \quad j \in [\![1,n]\!],$$

such that, for each $j \in [\![1,n]\!]$, $\lambda_j(t)$ is an eigenvalue of $\mathsf{B}(t)$ and

$$\lambda_j(0) = \mathsf{B}(j,j) \quad \textit{for } j \in [\![1,n]\!],$$

$$\lambda_j'(0) = \frac{\mathsf{B}(2,1)\mathsf{B}(1,2)}{\mathsf{B}(1,1) - \mathsf{B}(2,2)} \quad \textit{if } j = 1,$$

$$\lambda_j'(0) = \frac{\mathsf{B}(n,n-1)\mathsf{B}(n-1,n)}{\mathsf{B}(n-1,n-1) - \mathsf{B}(n,n)} \quad \textit{if } j = n, \quad \textit{and}$$

$$\lambda_j'(0) = \frac{\mathsf{B}(j,j-1)\mathsf{B}(j-1,j)}{\mathsf{B}(j-1,j-1) - \mathsf{B}(j,j)} + \frac{\mathsf{B}(j+1,j)\mathsf{B}(j,j+1)}{\mathsf{B}(j,j) - \mathsf{B}(j+1,j+1)} \quad \textit{if } 1<j<n.$$

Proof

The eigenvalues of $\mathsf{B}(0)$ are its diagonal entries which are all distinct. When a polynomial of degree n has n distinct roots, these are differentiable functions of the polynomial coefficients. Hence the eigenvalues of a matrix are continuous functions of its entries. Thus for ϵ small enough, $\mathsf{B}(1)$ has distinct eigenvalues.

Let us define the matrix-valued function $\mathsf{A} : \mathbb{R} \times \mathbb{C} \to \mathbb{C}^{n\times n}$ by

$$\mathsf{A}(t,x)(i,j) := \begin{cases} \mathsf{B}(i,j) - x & \text{if } i = j, \\ t\,\mathsf{B}(i,j) & \text{if } i - j = 1, \\ \mathsf{B}(i,j) & \text{otherwise.} \end{cases}$$

Then $\mathsf{A}(1,0) = \mathsf{B}$ and $\mathsf{A}(0,0)$ is the upper triangular part of B. A is an infinitely differentiable function and

$$\frac{\partial \mathsf{A}}{\partial t}(t,x) = \mathsf{B} - \mathsf{A}(0,0), \qquad \frac{\partial \mathsf{A}}{\partial x}(t,x) = -\mathsf{I}.$$

Since $\det : \mathbb{C}^{n\times 1} \times \ldots \times \mathbb{C}^{n\times 1} \to \mathbb{C}$ is an infinitely differentiable function, the chain rule gives

$$\frac{\partial \det \circ \mathsf{A}}{\partial x}(t,x) = \mathrm{D}(\det(\mathsf{A}(t,x)))\frac{\partial \mathsf{A}}{\partial x}(t,x)$$

$$= \sum_{j=1}^{n} \det \begin{bmatrix} \mathsf{B}(1,1)-x \cdots & & \cdots & \mathsf{B}(1,n) \\ t\mathsf{B}(2,1) & & & \mathsf{B}(2,n) \\ 0 & \vdots & \vdots & \vdots \\ \vdots & \vdots \ \ \mathsf{e}_j & \vdots & \mathsf{B}(j,n) \\ & \vdots & \vdots & \vdots \\ \vdots & & & \vdots \\ 0 & \cdots & \cdots & \mathsf{B}(n,n)-x \end{bmatrix}$$

$$= \sum_{j=1}^{n} \left[\prod_{\substack{\ell=1 \\ \ell \neq j}} (\mathsf{B}(\ell,\ell)-x) + t\,\Delta_j(t,x) \right]$$

$$= \sum_{j=1}^{n} \prod_{\substack{\ell=1 \\ \ell \neq j}} (\mathsf{B}(\ell,\ell)-x) + t\,\Delta(t,x),$$

where

$$|\Delta_j(t,x)| \leq \epsilon(2\|\mathsf{B}\|_\star)^{n-1}, \quad |\Delta(t,x)| \leq n\epsilon(2\|\mathsf{B}\|_\star)^{n-1}$$

and

$$\|\mathsf{B}\|_\star := \max\{|\mathsf{B}(i,j)| \, : \, i,j \in [\![1,n]\!]\}.$$

Let $j \in [\![1,n]\!]$. Since $\mathsf{B}(j,j) \neq \mathsf{B}(k,k)$ for all $k \neq j$,

$$\frac{\partial \det \circ \mathsf{A}}{\partial x}(0,\mathsf{B}(j,j)) = \prod_{\substack{\ell=1 \\ \ell \neq j}} (\mathsf{B}(\ell,\ell)-\mathsf{B}(j,j)) \neq 0,$$

and there exists a differentiable function $\lambda_j \, : \,]-1,1[\to \mathbb{C}$ such that

$$\det(\mathsf{A}(t,\lambda_j(t))) = 0 \ \text{ for } |t| < \eta, \quad \lambda_j(0) = \mathsf{B}(j,j)$$

and

$$\lambda_j'(0) = -\frac{\dfrac{\partial \det \circ \mathsf{A}}{\partial t}(0,\lambda_j(0))}{\dfrac{\partial \det \circ \mathsf{A}}{\partial x}(0,\lambda_j(0))}.$$

On the other hand,

$$\frac{\partial \det \circ \mathsf{A}}{\partial t}(t,x) = \mathrm{D}(\det(\mathsf{A}(t,x)))\frac{\partial \mathsf{A}}{\partial t}(t,x)$$

$$= \sum_{j=1}^{n} \det \begin{bmatrix} \mathsf{B}(1,1)-x \cdots & 0 & \cdots & \mathsf{B}(1,n) \\ t\mathsf{B}(2,1) \;\vdots & 0 & \vdots & \mathsf{B}(2,n) \\ 0 & & & \vdots \\ \vdots \quad \vdots & 0 & \vdots & \\ \vdots & \mathsf{B}(j{+}1,j) & \vdots & \mathsf{B}(j{+}1,n) \\ \vdots & 0 & & \vdots \\ \vdots & \vdots & & \vdots \\ 0 \quad \cdots & 0 & \cdots & \mathsf{B}(n,n)-x \end{bmatrix}$$

$$= \sum_{j=1}^{n} (-1)^{2j+1} \mathsf{B}(j+1,j) D_j,$$

where

$$D_j := \det \begin{bmatrix} \mathsf{B}(1,1)-x & \cdots & \mathsf{B}(1,j{-}1) & \mathsf{B}(1,j{+}1) & \cdots & \mathsf{B}(1,n) \\ 0 & \vdots & \vdots & \vdots & \vdots & \vdots \\ \vdots & & \mathsf{B}(j{-}1,j{-}1)-x & \mathsf{B}(j{-}1,j{+}1) & & \\ \vdots & \vdots & t\,\mathsf{B}(j,j{-}1) & \mathsf{B}(j,j{+}1) & \cdots & \mathsf{B}(j,n) \\ & \vdots & 0 & t\,\mathsf{B}(j{+}2,j{+}1) & \cdots & \mathsf{B}(j{+}2,n) \\ \vdots & & 0 & 0 & & \mathsf{B}(j{+}3,n) \\ \vdots & & \vdots & \vdots & \vdots & \vdots \\ 0 & \cdots & 0 & 0 & \cdots & \mathsf{B}(n,n)-x \end{bmatrix}$$

$$= -\sum_{j=1}^{n} \mathsf{B}(j+1,j)[A_j\, B_j - t\mathsf{B}(j,j-1)\widetilde{\Delta}_j - t\mathsf{B}(j+2,j+1)\breve{\Delta}_j],$$

A_j and B_j are given by

$$A_j := \det \begin{bmatrix} \mathsf{B}(1,1)-x & \cdots & \mathsf{B}(1,j{-}1) \\ t\,\mathsf{B}(2,1) & \vdots & \mathsf{B}(2,j{-}1) \\ & \ddots & \\ & t\,\mathsf{B}(j{-}1,j{-}2) & \mathsf{B}(j{-}1,j{-}1)-x \end{bmatrix},$$

$$B_j := \det \begin{bmatrix} \mathsf{B}(j+2,j+2)-x & \cdots & \mathsf{B}(j+2,n) \\ t\,\mathsf{B}(j+3,j+2) & \vdots & \mathsf{B}(j+3,n) \\ & \ddots & \\ & t\,\mathsf{B}(n,n-1) & \mathsf{B}(n,n)-x \end{bmatrix},$$

and $\widetilde{\Delta}_j$ and $\breve{\Delta}_j$ satisfy

$$\max\{|\widetilde{\Delta}_j|, |\breve{\Delta}_j|\} \le (2\|\mathsf{B}\|_*)^{n-2}.$$

Hence

$$\lambda_j'(0) = \frac{\mathsf{B}(2,1)\mathsf{B}(1,2)}{\mathsf{B}(1,1)-\mathsf{B}(2,2)} \quad \text{if } j=1,$$

$$\lambda_j'(0) = \frac{\mathsf{B}(n,n-1)\mathsf{B}(n-1,n)}{\mathsf{B}(n-1,n-1)-\mathsf{B}(n,n)} \quad \text{if } j=n, \quad \text{and}$$

$$\lambda_j'(0) = \frac{\mathsf{B}(j,j-1)\mathsf{B}(j-1,j)}{\mathsf{B}(j-1,j-1)-\mathsf{B}(j,j)} + \frac{\mathsf{B}(j+1,j)\mathsf{B}(j,j+1)}{\mathsf{B}(j,j)-\mathsf{B}(j+1,j+1)} \quad \text{if } 1<j<n.$$

This completes the proof. ∎

In Theorem 6.40, let B denote the iterate A_k of the Basic QR algorithm and $\epsilon > 0$ satisfy

$$\sum_{j=1}^{n-1} |\mathsf{A}_k(j+1,j)| < \epsilon.$$

Then, for ϵ small enough,

$$|\lambda_j(1) - \mathsf{A}_k(j,j)| \le |\lambda_j'(0)| + o(\epsilon), \qquad j \in [\![1,n]\!].$$

This leads to the following stopping criterion for the Basic QR Method which takes into account the total relative error:

Let $\frac{1}{2}10^{-p}$ be an upper bound of the permitted relative error in the approximate eigenvalues. Assume that, by introducing shifts if necessary,

$$|\lambda_j'(1)| < |\mathsf{A}_k(j,j)| \quad \text{for } j \in [\![1,n]\!],$$

where $\lambda_j'(1)$ is given by Theorem 6.40 with $\mathsf{B} := \mathsf{A}_k$. Then, the algorithm should be stopped at an iteration k such that

$$\frac{|\lambda_j'(1)|}{|\mathsf{A}_k(j,j)| - |\lambda_j'(1)|} < \frac{1}{2}10^{-p} \quad \text{for } j \in [\![1,n]\!], \quad \text{and}$$

$$\sum_{j=1}^{n-1} |\mathsf{A}_k(j+1,j)| < \frac{1}{2}10^{-p}.$$

This condition implies that the total relative error of each approximate eigenvalue $\mathsf{A}_k(j,j)$ is bounded as follows:

$$\frac{|\lambda_j(1) - \mathsf{A}_k(j,j)|}{|\lambda_j(1)|} < \frac{1}{2}10^{-p} + o(10^{-p}), \quad j \in [\![1,n]\!].$$

Similar criteria have been implemented in [7] and [10].

6.4.4 Relative Error in Solving a Sylvester Equation

Let m and n be positive integers such that $m \le n$. Consider $\mathsf{B} \in \mathbb{C}^{n\times n}$ and $\mathsf{Z} \in \mathbb{C}^{m\times m}$ such that $\mathrm{sp}(\mathsf{B}) \cap \mathrm{sp}(\mathsf{Z}) = \emptyset$. Let $\underline{\mathsf{y}} \in \mathbb{C}^{n\times m}$ be given. By Proposition 1.50, the Sylvester equation

$$\mathsf{B}\underline{\mathsf{x}} - \underline{\mathsf{x}}\mathsf{Z} = \underline{\mathsf{y}}$$

has a unique solution in $\mathbb{C}^{n\times m}$. In this section we give the forward error analysis as well as the backward error analysis for the computation of this solution.

The forward error analysis of this problem can be carried out as follows. Consider the Frobenius norm $\|\cdot\|_F$ on $\mathbb{C}^{n\times m}$ and let $\|\cdot\|$ denote the subordinated operator norm on $\mathrm{BL}(\mathbb{C}^{n\times m})$. Define a Sylvester operator $\mathcal{S} : \mathbb{C}^{n\times m} \to \mathbb{C}^{n\times m}$ by

$$\mathcal{S}(\underline{\mathsf{x}}) := \mathsf{B}\underline{\mathsf{x}} - \underline{\mathsf{x}}\mathsf{Z} \quad \text{for } \underline{\mathsf{x}} \in \mathbb{C}^{n\times m}.$$

Then, along the lines of Remark 6.23,

$$\kappa(\mathcal{S}) := \|\mathcal{S}\| \, \|\mathcal{S}^{-1}\|$$

is a condition number of $\mathcal{S} \in \mathrm{BL}(\mathbb{C}^{n\times m})$ relative to the equation $\mathcal{S}(\underline{\mathsf{x}}) = \underline{\mathsf{y}}$. In [39], $\|\mathcal{S}^{-1}\|$ is called the **separation** between the matrices B and Z. We have shown in Proposition 1.50 that $\mathrm{sp}(\mathcal{S}) = \{\mu - \lambda : \mu \in \mathrm{sp}(\mathsf{B}),\ \lambda \in \mathrm{sp}(\mathsf{Z})\}$. Thus

$$\rho(\mathcal{S}) = \max\{|\mu - \lambda| : \mu \in \mathrm{sp}(\mathsf{B}),\ \lambda \in \mathrm{sp}(\mathsf{Z})\} \quad \text{and}$$
$$\rho(\mathcal{S}^{-1}) = \frac{1}{\min\{|\mu - \lambda| : \mu \in \mathrm{sp}(\mathsf{B}),\ \lambda \in \mathrm{sp}(\mathsf{Z})\}}.$$

Hence

$$\kappa(\mathcal{S}) \geq \rho(\mathcal{S})\rho(\mathcal{S}^{-1}) \geq \frac{\max\{|\mu - \lambda| \,:\, \mu \in \operatorname{sp}(\mathsf{B}),\ \lambda \in \operatorname{sp}(\mathsf{Z})\}}{\min\{|\mu - \lambda| \,:\, \mu \in \operatorname{sp}(\mathsf{B}),\ \lambda \in \operatorname{sp}(\mathsf{Z})\}}.$$

This inequality shows that if B has at least two distinct eigenvalues and if Z changes in such a way that an eigenvalue of Z comes close to an eigenvalue of B, then $\kappa(\mathcal{S})$ tends to ∞. In other words, for $\mathcal{S}$ to be well conditioned it is necessary that $\operatorname{sp}(\mathsf{B})$ and $\operatorname{sp}(\mathsf{Z})$ be well separated.

The proof of Proposition 1.50 suggests the following method for solving the Sylvester equation $\mathcal{S}(\underline{\mathsf{x}}) = \underline{\mathsf{y}}$.

Compute a Schur form of Z, that is, find a unitary matrix $\mathsf{V} \in \mathbb{C}^{m\times m}$ such that $\mathsf{T} := \mathsf{V}^*\mathsf{Z}\mathsf{V}$ is an upper triangular matrix. Let $\underline{\mathsf{u}} := \underline{\mathsf{x}}\mathsf{V}$ and $\underline{\mathsf{v}} := \underline{\mathsf{y}}\mathsf{V}$. Then $\mathcal{S}(\underline{\mathsf{x}}) = \underline{\mathsf{y}}$ if and only if $\mathsf{B}\,\underline{\mathsf{u}} - \underline{\mathsf{u}}\,\mathsf{T} = \underline{\mathsf{v}}$. Let

$$\begin{aligned}
\mathsf{N} &:= \mathsf{T} - \operatorname{diag}\left[\mathsf{T}(1,1), \ldots, \mathsf{T}(m,m)\right] \\
\widehat{\lambda} &:= \frac{1}{m}\sum_{i=1}^{m} \mathsf{T}(i,i) \quad \text{and} \\
\widehat{\mathsf{T}} &:= \widehat{\lambda}\mathsf{I}_m + \mathsf{N},
\end{aligned}$$

that is, $\widehat{\mathsf{T}}$ is obtained from T by replacing each of the diagonal entries of T by their arithmetic mean. Suppose that there exists a positive constant γ of 'moderate' size such that the following 'cluster condition' is satisfied:

$$(C) \qquad \|\widehat{\mathsf{T}} - \mathsf{T}\|_{\mathrm{F}} = \Big(\sum_{i=1}^{m} |\mathsf{T}(i,i) - \widehat{\lambda}|^2\Big)^{1/2} \leq \gamma \mathbf{u}.$$

Now, instead of solving the equation $\mathsf{B}\,\underline{\mathsf{u}} - \underline{\mathsf{u}}\,\mathsf{T} = \underline{\mathsf{v}}$, we solve the perturbed equation $\mathsf{B}\,\underline{\mathsf{u}} - \underline{\mathsf{u}}\,\widehat{\mathsf{T}} = \mathbf{c}\,(\underline{\mathsf{v}})$.

Let $\underline{\mathsf{u}} := [\mathsf{u}_1, \ldots, \mathsf{u}_m]$ and $\mathbf{c}\,(\underline{\mathsf{v}}) := [\mathbf{c}\,(\mathsf{v}_1), \ldots, \mathbf{c}\,(\mathsf{v}_m)]$. Then $\mathsf{B}\,\underline{\mathsf{u}} - \underline{\mathsf{u}}\,\widehat{\mathsf{T}} = \mathbf{c}\,(\underline{\mathsf{v}})$ if and only if

$$\begin{aligned}
(\mathsf{B} - \widehat{\lambda}\mathsf{I}_n)\mathsf{u}_1 &= \mathbf{c}\,(\mathsf{v}_1), \\
(\mathsf{B} - \widehat{\lambda}\mathsf{I}_n)\mathsf{u}_j &= \mathbf{c}\,(\mathsf{v}_j) + \sum_{i=1}^{j-1} \mathsf{T}(i,j)\mathsf{u}_i \quad \text{for } j = 2, \ldots, m.
\end{aligned}$$

Suppose that the upper triangular matrix $\mathsf{T} = \mathsf{V}^*\mathsf{Z}\mathsf{V}$ is obtained as follows. First the matrix Z is transformed to a unitarily similar upper Hessenberg matrix Z_0 by employing Householder symmetries (Proposition 6.7). Then ℓ iterations of the QR Method are performed on the matrix Z_0 so as to satisfy the stopping criterion from the backward error standpoint presented in Subsection 6.4.3; the QR factors in each such iteration are also computed by employing the Householder symmetries. Further, suppose that the linear system satisfied by u_j, $j \in [\![1, m]\!]$, is solved by Gaussian Elimination with partial pivoting.

The backward error analysis of the above algorithm for solving a Sylvester equation can be developed as follows. As mentioned in Subsection 6.4.3, the matrix T satisfies

$$\frac{\|\mathsf{VTV}^* - \mathsf{Z}\|_{\mathrm{F}}}{\|\mathsf{Z}\|_{\mathrm{F}}} \le (\alpha_{m,\ell} + \beta_{m,\ell})\mathbf{u}$$

for constants $\alpha_{m,\ell}$ and $\beta_{m,\ell}$ of 'moderate' size. Let $\Delta\mathsf{Z} := \mathsf{V}\widehat{\mathsf{T}}\mathsf{V}^* - \mathsf{Z}$. Then, under the cluster condition (C),

$$\|\Delta\mathsf{Z}\|_{\mathrm{F}} \le (\alpha_{m,\ell} + \beta_{m,\ell} + \gamma)\mathbf{u}\|\mathsf{Z}\|_{\mathrm{F}}.$$

As stated in Remark 6.33, for each $j \in [\![1, m]\!]$, the solution $\mathbf{c}(\mathsf{u}_j) \in \mathbb{C}^{n\times 1}$, computed by Gaussian Elimination with partial pivoting, is such that there exists a matrix $\Delta\mathsf{B} \in \mathbb{C}^{n\times n}$ which satisfies

$$(\mathsf{B} + \Delta\mathsf{B} - \widehat{\lambda}\mathsf{I}_n)\mathbf{c}(\mathsf{u}_1) = \mathbf{c}(\mathsf{v}_1),$$

$$(\mathsf{B} + \Delta\mathsf{B} - \widehat{\lambda}\mathsf{I}_n)\mathbf{c}(\mathsf{u}_j) = \mathbf{c}(\mathsf{v}_j) + \sum_{i=1}^{j-1}\mathsf{T}(i,j)\mathbf{c}(\mathsf{u}_i) \quad \text{for } j = 2, \ldots, m,$$

and

$$\|\Delta\mathsf{B}\|_{\infty} \le 2n^2\gamma_{\mathbf{u}}(n)\rho_n\|\mathsf{B} - \widehat{\lambda}\mathsf{I}_n\|_{\infty}.$$

Hence

$$(\mathsf{B} + \Delta\mathsf{B})\mathbf{c}(\underline{\mathsf{u}}) - \mathbf{c}(\underline{\mathsf{u}})\widehat{\mathsf{T}} = \mathbf{c}(\underline{\mathsf{v}}),$$

and $\mathbf{c}(\underline{\mathsf{u}})\mathsf{V}^* \in \mathbb{C}^{n\times m}$ satisfies

$$(\mathsf{B} + \Delta\mathsf{B})\mathbf{c}(\underline{\mathsf{u}})\mathsf{V}^* - \mathbf{c}(\underline{\mathsf{u}})\mathsf{V}^*(\mathsf{Z} + \Delta\mathsf{Z}) = \mathbf{c}(\underline{\mathsf{v}})\mathsf{V}^*.$$

The computed solution of the original Sylvester equation is then

$$\mathbf{c}(\underline{\mathsf{x}}) := \mathbf{c}\left(\mathbf{c}(\underline{\mathsf{u}})\mathsf{V}^*\right).$$

Since the matrices V and V^* are constructed by employing Householder symmetries, the proof of Proposition 6.37 shows that there are $\delta\underline{\mathsf{y}}$ and $\underline{\mathsf{w}} \in \mathbb{C}^{n\times m}$ such that

$$\begin{aligned} \mathbf{c}(\underline{\mathsf{v}}) &= (\underline{\mathsf{y}} + \delta\underline{\mathsf{y}})\mathsf{V} &\quad \text{and} \quad& \|\delta\underline{\mathsf{y}}\|_{\mathrm{F}} \le c_1\mathbf{u}\|\underline{\mathsf{y}}\|_{\mathrm{F}}, \\ \mathbf{c}\left(\mathbf{c}(\underline{\mathsf{u}})\mathsf{V}^*\right) &= \left(\mathbf{c}(\underline{\mathsf{u}}) + \underline{\mathsf{w}}\right)\mathsf{V}^* &\quad \text{and} \quad& \|\underline{\mathsf{w}}\|_{\mathrm{F}} \le c_2\mathbf{u}\|\mathbf{c}(\underline{\mathsf{u}})\|_{\mathrm{F}}, \end{aligned}$$

for some constants c_1 and c_2 of 'moderate' size, depending only on n and m. Hence

$$(\mathsf{B} + \triangle\mathsf{B})\mathbf{c}(\underline{\mathsf{x}}) - \mathbf{c}(\underline{\mathsf{x}})(\mathsf{Z} + \triangle\mathsf{Z}) = \underline{\mathsf{y}} + \triangle\underline{\mathsf{y}},$$

where

$$\triangle\underline{\mathsf{y}} = \mathcal{S}(\underline{\mathsf{w}}\mathsf{V}^*) + \delta\underline{\mathsf{y}} + O((\triangle\mathsf{B})\underline{\mathsf{w}}) + O(\underline{\mathsf{w}}(\triangle\mathsf{Z})).$$

Since

$$\|\mathcal{S}(\underline{\mathsf{w}}\mathsf{V}^*)\|_{\mathrm{F}} \leq \|\mathcal{S}\|\,\|\underline{\mathsf{w}}\|_{\mathrm{F}} \quad \text{and} \quad \|\mathbf{c}(\underline{\mathsf{u}})\|_{\mathrm{F}} \leq \|\mathcal{S}^{-1}\|\,\|\underline{\mathsf{y}}\|_{\mathrm{F}} + O(\mathbf{u}^2),$$

it follows that

$$\|\triangle\underline{\mathsf{y}}\|_{\mathrm{F}} \leq (c_1 + c_2\kappa(\mathcal{S}))\mathbf{u}\|\underline{\mathsf{y}}\|_{\mathrm{F}} + O(\mathbf{u}^2).$$

Thus under the cluster condition stated above, the suggested algorithm for solving a Sylvester equation is backward stable, provided $\kappa(\mathcal{S})$ is of 'moderate' size. The condition number $\kappa(\mathcal{S})$ plays a role in the backward error analysis because we have performed a change of variable in the original equation.

Other numerical methods for solving Sylvester equations are presented in [26].

References

[1] Ahues M. (1987): A class of strongly stable operator approximations, *J. Austral. Math. Soc.* Ser. B, **28**, 435-442.

[2] Ahues M. (1989): Spectral condition numbers for defective eigenelements of linear operators in Hilbert spaces, *Numer. Funct. Anal. and Optimiz.* **10** (9 & 10), 843-861.

[3] Ahues M., Aranchiba S. and Telias M. (1990): Rayleigh-Schrödinger series for defective spectral elements of compact operators in Banach spaces First Part: Theoretical Aspects, *Numer. Funct. Anal. Optimiz.*, **11**, 839-850.

[4] Ahues M. and Hocine F. (1994): A note on spectral approximation of linear operations, *Appl. Math. Lett.*, **7**, 63-66.

[5] Ahues M. and Largillier A. (1990): Rayleigh-Schrödinger series for defective spectral elements of compact operators in Banach spaces Second Part: Numerical comparison with some inexact Newton methods, *Numer. Funct. Anal. Optimiz.*, **11**, 851-872.

[6] Ahues M. and Largillier A. (1995): A variant of the fixed tangent method for spectral computations on integral operators, *Numer. Funct. Anal. and Optimiz.* **16** (1 & 2), 1-17.

[7] Ahues M., Largillier A. and Tisseur F. (1996): Stability and precision of the QR eigenvalue computations, Householder Meeting, Pontresina, Switzerland.

[8] Ahues M. and Limaye B.V. (2000): Computation of spectral subspaces for weakly singular integral operators, to appear.

[9] Ahues M. and Telias M. (1986): Refinement methods of Newton type for approximate eigenelements of integral operators, *SIAM J. Numer. Anal.*, **23**, 144-159.

[10] Ahues M. and Tisseur F. (1997): A new deflation criterion for the QR algorithm, UT CS-97-353, (file: lawn122.ps), LAPACK.

[11] Alam, R. (1995): Accelerated spectral approximation, Ph.D. Thesis, Indian Institute of Technology Bombay, India.

[12] Alam R. (1998): On spectral approximation of linear operators, *J. Math. Anal. Appl.*, **226**, 229-244.

[13] Alam R., Kulkarni R.P. and Limaye B.V. (1996): Boundedness of adjoint bases of approximate spectral subspaces and of associated block reduced resolvents, *Numer. Funct. Anal. Optimiz.*, **17**, 473-501.

[14] Alam R., Kulkarni R.P. and Limaye, B.V. (1998): Accelerated spectral approximation, *Math. Comp.*, **67**, 1401-1422.

[15] Alam R., Kulkarni R.P. and Limaye, B.V. (2000): Accelerated spectral refinement. Part I: Simple eigenvalue, *J. Austral. Math. Soc.* Ser. B, **41**, 487-507.

[16] Alam R., Kulkarni R.P. and Limaye, B.V. (2000): Accelerated spectral refinement. Part II: Cluster of eigenvalues, *ANZIAM J.*, **42**, 224-243.

[17] Anselone P.M. (1971): *Collectively Compact Operator Approximation Theory and Applications to Integral Equations*, Prentice-Hall, Englewood Cliffs, N.J.

[18] Anselone P.M. (1981): Singularity subtraction in the numerical solution of integral equations, *J. Austral. Math. Soc.* Ser. B, **22**, 408-418.

[19] Anselone P.M. and Ansorge R. (1979): Compactness principle in nonlinear operator approximation theory, *Numer. Funct. Anal. Optimiz.*, **1**, 589-618.

[20] Anselone P.M. and Treuden M.L. (1996): Spectral analysis of asymptotically compact operator sequences, *Numer. Funct. Anal. Optimiz.*, **17**, 679-690.

[21] Atkinson K.E. (1967): The numerical solution of eigenvalue problem for compact integral operators, *Trans. Amer. Math. Soc.*, **129**, 458-465.

[22] Baker C.T.H. (1977): *The Numerical Treatment of Integral Equations*, Oxford Univ. Press, London.

[23] Bouldin R. (1990): Operator approximations with stable eigenvalues, *J. Austral. Math. Soc.*, Ser. A, **49**, 250-257.

[24] Brakhage H. (1961): Zur Feherabschätzung für die numerische Eigenwertbestimmung bei Integral-gleichungen, *Numer. Math.*, **3**, 174-179.

[25] Chatelin F. (1983): *Spectral Approximation of Linear Operators*, Academic Press, New York.

[26] Chatelin F. (1993): *Eigenvalues of Matrices*, (English edition) John Wiley and Sons, Chichester.

[27] Conte S.D. and de Boor C. (1980): *Elementary Numerical Analysis: An Algorithmic Approach*, Third Edition, McGraw Hill, New York.

[28] Cryer C.W. (1982): *Numerical Functional Analysis*, Oxford Univ. Press, London.

[29] de Boor C. (1978): *A Practical Guide to Splines*, Springer-Verlag, New York.

[30] de Boor C. and Swartz B. (1980): Collocation approximation to eigenvalues of an ordinary differential equation: The principle of the thing, *Math. Comp.*, **35**, 679-694.

[31] Dellwo D.R. and Friedman M.B. (1984): Accelerated spectral analysis of compact operators, *SIAM J. Numer. Anal.*, **21**, 1115-1131.

[32] Dellwo D.R. (1989): Accelerated spectral refinement with applications to integral equations, *SIAM J. Numer. Anal.*, **26**, 1184-1193.

[33] Deshpande L.N. and Limaye B.V. (1989): A fixed point technique to refine a simple approximate eigenvalue and a corresponding eigenvector, *Numer. Funct. Anal. Optimiz.*, **10**, 909-921.

[34] Deshpande L.N. and Limaye B.V. (1990): On the stability of singular finite-rank methods, *SIAM J. Numer. Anal.*, **27**, 792-803.

[35] Devore R.A. and Lorentz G.G. (1991): *Constructive Approximation*, Springer-Verlag, Berlin.

[36] Diestel J. (1984): *Sequences and Series in Banach Spaces*, Springer-Verlag, New York.

[37] Erxiong J. and Zhenyue Z.(1985): A new shift of the QL algorithm for irreducible symmetric tridiagonal matrices, *Linear Algebra Appl.*, **65**, 261-272.

[38] Gohberg I., Goldberg S. and Kaashoek M.A. (1990): *Classes of Linear Operators*, Vol. 1, Birkhäuser Verlag, Basel.

[39] Golub G.H. and Van Loan C. (1996): *Matrix Computations*, Third Edition, The Johns Hopkins University Press, Baltimore.

[40] Grigorieff R.D. (1975): Diskrete Approximation von Eigenwert problemen I: Qualitative Konvergenz, *Numer. Math.*, **24**, 355-374.

[41] Henrici P. (1962): Bounds for iterates, inverses, spectral variation and fields of values of nonnormal matrices, *Numer. Math.* **4**, 24-40.

[42] Higham N. (1996): *Accuracy and Stability of Numerical Algorithms*, SIAM, Philadelphia.

[43] Horn R. and Johnson C.R. (1985): *Matrix Analysis*, Cambridge Univ. Press, Cambridge.

[44] IEEE standard for binary floating point arithmetic, ANSI/IEEE Standard 754: (1985), Institute of Electrical and Electronics Engineers, New York. Reprinted in SIGPLAN Notices, **22**, (2), 9-25.

[45] Kantorovich L.V. and Krylov V.I. (1958): *Approximate Methods of Higher Analysis*, Interscience, New York.

[46] Kato T. (1960): Estimation of iterated matrices with application to the von Neumann condition, *Numer. Math.*, **2**, 22-29.

[47] Kato T. (1976): *Perturbation Theory for Linear Operators*, Second Edition, Springer-Verlag, Berlin.

[48] Kreyszig E. (1989): *Introductory Functional Analysis with Applications*, John Wiley and Sons, New York.

[49] Kulkarni R.P. and Limaye B.V. (1989): On the error estimates for the Rayleigh-Schrödinger series and the Kato-Rellich perturbation series, *J. Austral. Math. Soc.* Ser. A, **46**, 456-468.

[50] Kulkarni R.P. and Limaye B.V. (1990): Solution of a Schrödinger equation by iterative refinement, *J. Austral. Math. Soc.* Ser. B, **32**, 115-132.

[51] Largillier A. (1996): Bounds for relative errors of complex matrix factorizations, *Appl. Math. Lett.*, **6**, 79-84.

[52] Largillier A. and Levet M. (1993): A note on a collectively compact approximation for weakly singular integral operators, *Appl. Math. Lett.*, **4**, 87-90.

[53] Largillier A. and Limaye B.V. (1996): Finite-rank methods and their stability for coupled systems of operator equations, *SIAM J. Numer. Anal.*, **33**, 707-728.

[54] Limaye B.V. (1987): *Spectral Perturbation and Approximation with Numerical Experiments*, Proceedings of the Centre for Mathematical Analysis, Vol. 13, Austral. Nat. Univ., Canberra.

[55] Limaye B.V. (1996): *Functional Analysis*, Second Edition, New Age International, New Delhi.

[56] Limaye B.V. and Nair M.T. (1986): Rayleigh-Schrödinger procedure for iterative refinement of computed eigenelements under strong approximation, in the Proceedings of the International Conference on 'Methods of Functional Analysis in Approximation Theory', (C.A. Micchelli, D.V. Pai, B.V. Limaye, eds.), International Series of Numerical Mathematics, Vol. 76, 371-388, Birkhäuser Verlag, Basel.

[57] Limaye B.V. and Nair M.T. (1989): On the accuracy of the Rayleigh-Schrödinger approximations, *J. Math. Anal. Appl.*, **139**, 413-431.

[58] Limaye B.V. and Nair M.T. (1990): Eigenelements of perturbed operators, *J. Austral. Math. Soc.* Ser. A, **49**, 138-148.

[59] Limaye B.V. and Nair M.T. (1995): On the multiplicities and the ascent of an eigenvalue of a linear map, *Math. Student*, **64**, 162-166.

[60] Michael E. (1951): Topologies on spaces of subsets, *Trans. Amer. Math. Soc.*, **71**, 152-182.

[61] Nair M.T. (1992): On strongly stable approximations, *J. Austral. Math. Soc.* Ser. A, **52**, 251-260.

[62] Osborn J.E. (1975): Spectral approximation for compact operators, *Math. Comp.*, **29**, 712-725.

[63] Parlett B.N. (1980): *The Symmetric Eigenvalue Problem*, Prentice-Hall, Englewood Cliffs, N.J.

[64] Redont P. (1979): Application de la théorie de la perturbation des opérateurs linéaires à l'obtention de bornes d'erreur sur les éléments propres et à leur calcul, Thèse de docteur-ingénieur, Université de Grenoble I, France.

[65] Schumaker L.L. (1981): *Spline Functions: Basic Theory*, John Wiley and Sons, New York.

[66] Sloan, I.H. (1976): Convergence of degenerate kernel methods, *J. Austral. Math. Soc.*, Ser. B, **19**, 422-431.

[67] Sloan I.H. (1976): Iterated Galerkin method for eigenvalue problems, *SIAM J. Numer. Anal.*, **13**, 753-760.

[68] Vainikko G.M. (1969): The compact approximation principle in the theory of approximation methods (in Russian), *Ž. Vyčisl. Mat. i Mat. Fiz.*, **9**, 739-761.

[69] Vainikko G.M. (1977): Über die Konvergenz und Divergenz von Näherungs-methoden bei Eigenwert Problemen, *Math. Nachr.*, **78**, 145-164.

[70] Wang J.Y. (1976): On the numerical computation of eigenvalues and eigenfunctions of compact integral operators using spline functions, *J. Inst. Maths Applics*, **18**, 177-188.

[71] Watkins D.S. (1991): *Fundamentals of Matrix Computations*, John Wiley and Sons, New York.

[72] Watson G.A. (1980): *Approximation Theory and Numerical Methods*, John Wiley and Sons, Chichester.

[73] Whitley R. (1986): The stability of finite rank methods with applications to integral equations, *SIAM J. Numer. Anal.*, **23**, 118-134.

[74] Wilkinson J.H. (1965): *The Algebraic Eigenvalue Problem*, Oxford Univ. Press, London.

Index

Milton Keynes UK
Ingram Content Group UK Ltd.
UKHW020319111024
449327UK00040B/1413

9 780367 455354